Genome and Epigenome Editing for Stress-Tolerant Crops

Genome and Epigenome Editing for Stress-Tolerant Crops

Edited by

Jen-Tsung Chen
National University of Kaohsiung
Kaohsiung, Taiwan

WILEY

This edition first published 2025

Registered Offices
John Wiley & Sons, Inc., 111 River Street, Hoboken, NJ 07030, USA
John Wiley & Sons Ltd, New Era House, 8 Oldlands Way, Bognor Regis, West Sussex, PO22 9NQ, UK

For details of our global editorial offices, customer services, and more information about Wiley products visit us at www.wiley.com.

The manufacturer's authorized representative according to the EU General Product Safety Regulation is Wiley-VCH GmbH, Boschstr. 12, 69469 Weinheim, Germany, e-mail: Product_Safety@wiley.com.

Wiley also publishes its books in a variety of electronic formats and by print-on-demand. Some content that appears in standard print versions of this book may not be available in other formats.

Library of Congress Cataloging-in-Publication Data
Names: Chen, Jen-Tsung, editor.
Title: Genome and epigenome editing for stress-tolerant crops / edited by Jen-Tsung Chen.
Description: First edition. | Hoboken, NJ : Wiley, 2025. | Includes bibliographical references and index.
Identifiers: LCCN 2024058673 (print) | LCCN 2024058674 (ebook) | ISBN 9781394280018 (hardback) | ISBN 9781394280032 (adobe pdf) | ISBN 9781394280025 (epub)
Subjects: LCSH: Plant genetics. | Plants–Effect of stress on. | Epigenetics.
Classification: LCC QK981 .G484 2025 (print) | LCC QK981 (ebook) | DDC 581.3/5–dc23/eng/20250205
LC record available at https://lccn.loc.gov/2024058673
LC ebook record available at https://lccn.loc.gov/2024058674

Cover Design: Wiley
Cover Image: © Til Vogt/Shutterstock

Set in 9.5/12.5pt STIXTwoText by Straive, Pondicherry, India

Contents

List of Contributors

Mariam Akhtar
Agricultural Biotechnology Division
National Institute for Biotechnology and Genetic Engineering (NIBGE)
Constituent College of Pakistan Institute of Engineering and Applied Sciences (PIEAS)
Faisalabad, Pakistan

Imran Amin
Agricultural Biotechnology Division
National Institute for Biotechnology and Genetic Engineering (NIBGE)
Constitute College of Pakistan Institute of Engineering and Applied Sciences (PIEAS)
Faisalabad, Pakistan

Afreen Anis
Faculty of Agriculture and Allied Sciences
C.V. Raman Global University
Bhubaneswar, India

Muhammad Jawad Akbar Awan
Agricultural Biotechnology Division
National Institute for Biotechnology and Genetic Engineering (NIBGE)
Constituent College of Pakistan Institute of Engineering and Applied Sciences (PIEAS)
Faisalabad, Pakistan

Asad Azeem
Department of Plant Breeding and Genetics
University of Layyah
Layyah, Pakistan

Sana Basharat
College of Tropical Agriculture and Forestry
Hainan University
Haikou, Hainan, China

Key Laboratory of Tropical Horticultural Crop Quality Regulation
College of Horticulture
Hainan University
Haikou, Hainan, China

Hainan Yazhou Bay Seed Laboratory
Sanya Nanfan Research Institute of Hainan University
Sanya, China

Fang Zhiyuan Academician Team Innovation Center of Hainan Province
Sanya, China

Andriy Bilichak
Agriculture and Agri-Food Canada
Morden Research and Development Centre
Morden, Manitoba, Canada

Muhammad Ismail Buzdar
Agricultural Biotechnology Division
National Institute for Biotechnology and Genetic Engineering (NIBGE)
Constituent College of Pakistan Institute of Engineering and Applied Sciences (PIEAS)
Faisalabad, Pakistan

Himanshushekhar Chaurasia
Mechanical Processing Division
ICAR-Central Institute for Research on Cotton Technology
Mumbai, Maharashtra, India

Privilege Chikove
Faculty of Agriculture and Allied Sciences
C.V. Raman Global University
Bhubaneswar, India

Ananya Choudhury
Faculty of Agriculture and Allied Science
C. V. Raman Global University
Bhubaneswar, India

Parinita Das
Division of Agricultural Bioinformatics
ICAR-Indian Agricultural Statistics Research Institute
New Delhi, Delhi, India

Discipline of Bioinformatics
The Graduate School
ICAR-Indian Agricultural Research Institute
New Delhi, Delhi, India

Soumyashree Dash
Faculty of Agriculture and Allied Sciences
C.V. Raman Global University
Bhubaneswar, India

Swatismita Deo
Faculty of Agriculture and Allied Science
C. V. Raman Global University
Bhubaneswar, India

Dhaarani V.
Faculty of Agriculture and Allied Science
C. V. Raman Global University
Bhubaneswar, India

Tushar K. Dutta
Division of Nematology
ICAR-Indian Agricultural Research Institute
New Delhi, Delhi, India

Subham Ghosh
Division of Agricultural Bioinformatics
ICAR-Indian Agricultural Statistics Research Institute
New Delhi, Delhi, India

Discipline of Bioinformatics
The Graduate School
ICAR-Indian Agricultural Research Institute
New Delhi, Delhi, India

Ifrah Imran
Agricultural Biotechnology Division
National Institute for Biotechnology and Genetic Engineering (NIBGE)
Constitute College of Pakistan Institute of Engineering and Applied Sciences (PIEAS)
Faisalabad, Pakistan

Anam Ishtiaq
Agricultural Biotechnology Division
National Institute for Biotechnology and Genetic Engineering (NIBGE)
Constitute College of Pakistan Institute of Engineering and Applied Sciences (PIEAS)
Faisalabad, Pakistan

Abul Kalam Mohammad Aminul Islam
Department of Genetics and Plant Breeding
Faculty of Agriculture
Bangabandhu Sheikh Mujibur Rahman Agricultural University
Gazipur, Bangladesh

Faiz Ahmad Joyia
Centre of Agricultural Biochemistry and Biotechnology (CABB)
University of Agriculture
Faisalabad, Pakistan

Angeline Jurry
Faculty of Agriculture and Allied Sciences
C.V. Raman Global University
Bhubaneswar, India

Soumita Karmakar
Faculty of Agriculture and Allied Sciences
C.V. Raman Global University
Bhubaneswar, India

Emanpreet Kaur
Agriculture and Agri-Food Canada
Morden Research and Development Centre
Morden, Manitoba, Canada

Kadiyala Kavya
Department of Genetics and Plant Breeding
College of Post Graduate Studies in
Agricultural Sciences
Central Agricultural University (Imphal)
Umiam, Meghalaya, India

Ajay Kumar Keot
Biological Sciences and Technology Division
CSIR-North East Institute of Science and
Technology (NEIST)
Jorhat, Assam, India

Academy of Scientific and Innovative
Research (AcSIR)
Ghaziabad, India

Jahan Khan
Agricultural Biotechnology Division
National Institute for Biotechnology and
Genetic Engineering (NIBGE)
Constituent College of Pakistan Institute of
Engineering and Applied Sciences (PIEAS)
Faisalabad, Pakistan

Muhammad Sarwar Khan
Centre of Agricultural Biochemistry and
Biotechnology (CABB)
University of Agriculture
Faisalabad, Pakistan

Marium Khatun
Department of Genetics and Plant Breeding
Faculty of Agriculture
Bangabandhu Sheikh Mujibur Rahman
Agricultural University
Gazipur, Bangladesh

Abhishek Kumar
Faculty of Agriculture
Maharishi Markandeshwar University
Mullana, Ambala, Haryana, India

Niraj Kumar
Biological Sciences and Technology
Division, CSIR-North East Institute of Science
and Technology (NEIST)
Jorhat, Assam, India

Academy of Scientific and Innovative
Research (AcSIR)
Ghaziabad, India

Sharia Yesmin Labonno
Department of Genetics and Plant Breeding
Faculty of Agriculture
Bangabandhu Sheikh Mujibur Rahman
Agricultural University
Gazipur, Bangladesh

Louie Cris Lopos
Agriculture and Agri-Food Canada
Morden Research and Development Centre
Morden, Manitoba, Canada

Muhammad Arslan Mahmood
Agricultural Biotechnology Division
National Institute for Biotechnology and
Genetic Engineering (NIBGE)
Constitute College of Pakistan Institute of
Engineering and Applied Sciences (PIEAS)
Faisalabad, Pakistan

Alagu Manickavelu
Department of Genomic Science
Central University of Kerala
Kasaragod, Kerala, India

Riwandahun Marwein
Biological Sciences and Technology Division
CSIR-North East Institute of Science and
Technology (NEIST)
Jorhat, Assam, India

Academy of Scientific and Innovative Research
(AcSIR)
Ghaziabad, India

Srutirekha Mishra
Faculty of Agriculture and Allied Science
C. V. Raman Global University
Bhubaneswar, India

Sneha Murmu
Division of Agricultural Bioinformatics
ICAR-Indian Agricultural Statistics Research Institute
New Delhi, Delhi, India

Ghulam Mustafa
Centre of Agricultural Biochemistry and Biotechnology (CABB)
University of Agriculture
Faisalabad, Pakistan

Rubab Zahra Naqvi
Agricultural Biotechnology Division
National Institute for Biotechnology and Genetic Engineering (NIBGE)
Constitute College of Pakistan Institute of Engineering and Applied Sciences (PIEAS)
Faisalabad, Pakistan

Bhupati Nayak
Faculty of Agriculture and Allied Science
C. V. Raman Global University
Bhubaneswar, India

Fernanda Freitas de Oliveira
Genetics Department
Federal University of Parana
Curitiba-PR, Brazil

Jardel de Oliveira
Department of Agronomy - Postgraduate Program in Agronomy - Plant Production
São Paulo Western University
Presidente Prudente-SP, Brazil

Pritilagna Panigrahi
Faculty of Agriculture and Allied Science
C. V. Raman Global University
Bhubaneswar, India

Aqsa Parvaiz
Department of Biochemistry and Biotechnology
The Women University Multan
Multan, Pakistan

Laxmipriya Pati
Faculty of Agriculture and Allied Science
C. V. Raman Global University
Bhubaneswar, India

Liu Pingwu
College of Tropical Agriculture and Forestry
Hainan University
Haikou, Hainan, China

Key Laboratory of Tropical Horticultural Crop Quality Regulation
College of Horticulture
Hainan University
Haikou, Hainan, China

Hainan Yazhou Bay Seed Laboratory
Sanya Nanfan Research Institute of Hainan University
Sanya, China

Fang Zhiyuan Academician Team Innovation Center of Hainan Province
Sanya, China

Pooja
Department of Plant Pathology
Chaudhary Charan Singh (CCS) Haryana Agricultural University
Hisar, Haryana, India

Janapareddy Rajesh
Department of Seed Science and Technology
Hemvati Nandan Bahuguna Garhwal University
Srinagar, Uttarakhand, India

Noru Rajasekhar Reddy
Department of Genetics and Plant Breeding
College of Agriculture
Kerala Agricultural University
Thiruvananthapuram, Kerala, India

Maria Rehman
Centre of Agricultural Biochemistry and Biotechnology (CABB)
University of Agriculture
Faisalabad, Pakistan

Department of Biotechnology
University of Sargodha
Sargodha, Pakistan

Rabia Rehman
Agricultural Biotechnology Division
National Institute for Biotechnology and Genetic Engineering (NIBGE)
Constitute College of Pakistan Institute of Engineering and Applied Sciences (PIEAS)
Faisalabad, Pakistan

Anna Rera
Faculty of Agriculture and Allied Sciences
C.V. Raman Global University
Bhubaneswar, India

Lellapalli Rithesh
Department of Plant Pathology
Kerala Agricultural University
Thiruvananthapuram, Kerala, India

Aryadeep Roychoudhury
Discipline of Life Sciences
School of Sciences
Indira Gandhi National Open University
New Delhi, Delhi, India

Jyoti Prakash Sahoo
Faculty of Agriculture and Allied Sciences
C.V. Raman Global University
Bhubaneswar, India

Subhashree Priyadarshinee Sahoo
Faculty of Agriculture and Allied Science
C. V. Raman Global University
Bhubaneswar, India

Arfa Saifullah
Centre of Agricultural Biochemistry and Biotechnology (CABB)
University of Agriculture
Faisalabad, Pakistan

Muhammad Waseem Sajjad
Agricultural Biotechnology Division
National Institute for Biotechnology and Genetic Engineering (NIBGE)
Constitute College of Pakistan Institute of Engineering and Applied Sciences (PIEAS)
Faisalabad, Pakistan

Santanu Samanta
Department of Biotechnology
St. Xavier's College (Autonomous)
Kolkata, West Bengal, India

Abhik Sarkar
Division of Agricultural Bioinformatics
ICAR-Indian Agricultural Statistics Research Institute
New Delhi, Delhi, India

Discipline of Bioinformatics
The Graduate School
ICAR-Indian Agricultural Research Institute
New Delhi, Delhi, India

Sumi Sarkar
Department of Genetics and Plant Breeding
Faculty of Agriculture
Bangabandhu Sheikh Mujibur Rahman Agricultural University
Gazipur, Bangladesh

Manash Pratim Sarmah
Department of Zoology
University of Science & Technology
Baridua, Meghalaya, India

Tiago Benedito dos Santos
Department of Agronomy - Program in Agronomy - Plant Production
São Paulo Western University
Presidente Prudente-SP, Brazil

Iffat Shaheen
Department of Agriculture
Bahauddin Zakariya University
Multan, Pakistan

Nimra Shahid
Department of Botany
University of Narowal
Narowal, Punjab, Pakistan

Muhammad Shareef
Department of Botany
University of Narowal
Narowal, Punjab, Pakistan

Sehrish Shehzadi
Centre of Agricultural Biochemistry and Biotechnology (CABB)
University of Agriculture
Faisalabad, Pakistan

Nyashadzashe Shengezi
Faculty of Agriculture and Allied Sciences
C.V. Raman Global University
Bhubaneswar, India

Saqib Siddique
Agricultural Biotechnology Division
National Institute for Biotechnology and Genetic Engineering (NIBGE)
Constitute College of Pakistan Institute of Engineering and Applied Sciences (PIEAS)
Faisalabad, Pakistan

Hidam Bishworjit Singh
Department of Biotechnology
Gauhati University
Guwahati, Assam, India

Dhanawantari L. Singha
Rabindranath Tagore University
Hojai, Assam, India

Dipro Sinha
Division of Agricultural Bioinformatics
ICAR-Indian Agricultural Statistics Research Institute
New Delhi, Delhi, India

Subhalaxmi Swain
Faculty of Agriculture and Allied Science
C. V. Raman Global University
Bhubaneswar, India

Swarnalata Tripathy
Faculty of Agriculture and Allied Science
C. V. Raman Global University
Bhubaneswar, India

Swapan Kumar Tripathy
College of Agriculture
Odisha University of Agriculture and Technology
Bhawanipatna, India

Muhammad Waseem
College of Tropical Agriculture and Forestry
Hainan University
Haikou, Hainan, China

Key Laboratory of Tropical Horticultural Crop Quality Regulation
College of Horticulture
Hainan University
Haikou, Hainan, China

Hainan Yazhou Bay Seed Laboratory
Sanya Nanfan Research Institute of Hainan University
Sanya, China

Fang Zhiyuan Academician Team Innovation Center of Hainan Province
Sanya, China

Diliane Harumi Yaguinuma
Department of Agronomy - Postgraduate Program in Agronomy - Plant Production
São Paulo Western University
Presidente Prudente-SP, Brazil

Rida Zahid
Agricultural Biotechnology Division
National Institute for Biotechnology and Genetic Engineering (NIBGE)
Constitute College of Pakistan Institute of Engineering and Applied Sciences (PIEAS)
Faisalabad, Pakistan

Preface

In recent decades, the world has faced serious crises mainly caused by the continuously expanding human population, the resulting environmental pollution, and the shortage of food and natural resources. Undoubtedly, these conditions can be worse under the scenario of climate change globally, such as a rising temperature that can lead to shortages of water resources and, consequently, increase the risk of agricultural and ecological droughts. To ensure sufficient agricultural nutrition and food production, scientists are expected to solve the problems by developing new crop breeding methods, particularly focusing on identifying the stress-tolerant traits and uncovering underlying in-depth machinery. Importantly, these approaches can advance agricultural biotechnology and crop breeding focusing on acquiring future crops with climate-resilient capacities, which might greatly ensure nutrition and food security under the challenge of upcoming global climate change. Therefore, nowadays, the demand to organize crop breeding programs for developing climate-resilient crops is inevitably increasing.

To support such demand, modern plant molecular biotechnology has been upgraded fast to achieve high-throughput, high-resolution, smart, and precision manners such as multiple omics-based functional genomics, CRISPR/Cas9-mediated genome editing, RNA technology, and so on. In the post-genomics era, the approaches of multiple omics continue to mine huge amounts of genetic resources and accumulate interesting key genes, and definitely need efficient ways of genetic engineering to benefit crop breeding and agricultural production. In the past decade, an emerging genetic engineering technology, that is, genome editing technology, has the power to release the potential of plant genomes through the precise delivery of tolerant genes for crop improvement. The system of genome editing has upgraded fast, and now, CRISPR/Cas9 has become the dominant technology. Thus, the integration and coordination of multiple omics and CRISPR/Cas9 for organizing omics-CRISPR breeding strategies inevitably become the next wave of important tasks in plant science. In addition, in recent years, RNA technology/epigenetic regulation has been introduced into diverse fields of plant science and has been proven to its great potential in mitigating plant stress responses as well as being a fine-tuning regulator in some ways to advance crop improvement and breeding. The role of noncoding RNAs (ncRNAs) in the management of CRISPR-edited crop production has become an important topic. In the future, achieving climate-resilient agriculture needs the coordination of some crucial technologies involving multiple omics, CRISPR/Cas9, and functional RNA technology/epigenetics.

In this book, the ways for vertical integration and horizontal coordination of the crucial technologies were comprehensively discussed to advance crop breeding programs toward stress-resilient agriculture.

This book presents the coordinated CRISPR-noncoding RNAs (ncRNAs) strategies for combating diverse stressors and complicated or multiple stress conditions. It covers both abiotic and biotic stress, including stressors of salinity, temperature, drought, heavy metals, pests, pathogens, and so on, and proposes strategies to develop stress-tolerant crops with high-yield and high-quality traits through the integration or coordination of the mainstream technologies, that is, multiple omics, CRISPR/Cas, and ncRNA-based epigenetics.

This book is an ideal reference to integrate the emerging field of multiple omics, CRISPR/Cas, and ncRNA-based epigenetics by sharing crucial aspects of methods, applications, and future directions. It opens doors for students and researchers to efficiently overview these critical subtopics of plant science and technology and thus realize the concept and, hopefully, inspire the ideas of future experiments and the exploration of the knowledge and, eventually, lead to better development of future crops by scientists, plant biologists, and crop breeders.

The book editor, Dr. Jen-Tsung Chen, appreciates all contributors for their valuable chapters and the staff of Wiley for their instruction and assistance.

1

Mitigating Heat Stress Response in CRISPR/Cas-Mediated Edited Crops by Altering the Expression Pattern of Noncoding DNA

Emanpreet Kaur, Louie Cris Lopos, and Andriy Bilichak

Agriculture and Agri-Food Canada, Morden Research and Development Centre, Morden, Manitoba, Canada

1.1 Introduction

Heat stress, a significant constraint in crop production, is becoming increasingly threatening with the onset of climate change and the increased frequency of extreme heat waves. Plants are highly vulnerable to temperature fluctuations beyond their optimum range for growth, development, and reproduction (Govindaraj et al., 2018). Previously, elevated temperatures, especially heat waves, have already substantially reduced the regional yields of rice, wheat, and corn in Asia, North America, India, and Europe (Hassan et al., 2021). The Intergovernmental Panel on Climate Change's Sixth Assessment Report (IPCC's AR6) indicated that the global surface temperature rose by 1.09 °C from 1850–1900 to 2011–2020, with an anticipated rise of at least 0.2 °C per decade without substantial mitigation (IPCC, 2021). This could potentially compromise our capacity to attain food security in the future; the current predictive models suggest that yield losses among staple crops could be as high as 8% per 1 °C increase in global temperature (Zhao et al., 2017; Lee et al., 2024). While adaptive agricultural practices such as altering planting dates, improving irrigation availability and efficiency, utilizing shade structures, growing in controlled environments, and applying osmoprotectants could alleviate the effects of heat stress on crop yield, these approaches are not always cheap and flexible, especially at higher production levels. The deployment of thermotolerant cultivars that can withstand elevated temperatures without yield or agronomic penalties is still seen as the most sustainable solution to feeding the ever-growing human population amidst rising global temperatures.

Breeding for thermotolerance is associated with complexities originating from this trait's polygenic nature and the complex genetic regulatory network that controls its manifestation (Yeh et al., 2012). Currently, we are progressing in explaining the mechanisms and regulatory control of thermotolerance in model species such as *Arabidopsis* (*Arabidopsis thaliana*), but given the diversity of thermotolerance phenotypes among different plant species, this may not be sufficient (Yeh et al., 2012). Elucidating species-specific thermotolerant mechanisms will greatly benefit the development of thermotolerant cultivars.

Gene-editing platforms, particularly clustered regularly interspaced palindromic repeats (CRISPRs)/CRISPR-associated protein 9 (Cas9) and their derivatives, have been beneficial in generating novel thermotolerant phenotypes and functional elucidation. In plant breeding, the most

Genome and Epigenome Editing for Stress-Tolerant Crops, First Edition. Edited by Jen-Tsung Chen.

mainstream gene-editing approach is to mutate the coding region of a candidate protein-coding gene in a targeted manner to generate phenotypic variability to select upon. To a lesser extent, the mutation usually results in abolished gene function (knockout), gain-of-function, or neofunctionalization. Unfortunately, negative pleiotropic effects may also result when targeting genes involved in complex gene networks.

An emerging approach in genome editing for crop improvement is targeting noncoding DNA, particularly *cis*-regulatory elements (CREs). These noncoding DNA segments regulate gene expression when interacted with *trans*-regulatory elements. Contrary to the mainstream approach of modifying gene products, this approach aims to quantitatively modify the levels and/or the temporal–spatial gene expression to achieve phenotypic diversity while at the same time minimizing or avoiding pleiotropic effects (Swinnen et al., 2016). The transcriptional rewiring associated with the selection of CRE variants during the domestication of our staple crop species (Swinnen et al., 2016) and the diverse phenotypic changes induced when editing CREs (reviewed in Saeed et al. (2022)) greatly emphasize the phenotypic contributions of CREs in particular and noncoding DNA in general, making a strong case for the utility of this approach in accelerating crop improvement.

This chapter will review approaches for improving thermotolerance by modifying CREs and other functional, noncoding DNA elements using genome-editing tools, particularly those derived from CRISPR/Cas9. Since thermotolerance is a quantitative polygenic trait tightly regulated by a complex regulatory network, we believe this approach could greatly benefit its development.

1.2 Impact of Climate Change and Heat Stress on Crop Productivity

1.2.1 A Physiological Impact of Heat Stress on Plant Growth and Development

Heat stress is a significant environmental factor affecting crop yield and quality. The definition of heat stress depends on the natural habitat and is species-specific (Yeh et al., 2012). Mild temperature increase generally induces plant development and early flowering and alters immunity (Hua, 2013; Verhage et al., 2014; Capovilla et al., 2015; Gangappa et al., 2017). Plants can adapt to suboptimal temperatures through thermomorphogenesis – a range of morphological adaptations, including hypocotyl elongation, upward leaf movement (thermonasty), petiole elongation, reduced stomatal density, and formation of smaller and thinner leaves (Quint et al., 2016; Casal & Balasubramanian, 2019). Eventually, plants can cool themselves through open rosette structures and transpiration (Crawford et al., 2012; Park et al., 2019).

Extreme temperatures can cause irreversible plant damage, significantly affecting crop development and profitability and seriously threatening national and global food security (Lesk et al., 2022). Temperature stress disrupts photosynthesis, water metabolism, nutrient cycling, protein synthesis, reproduction, and the functionality of various enzymes, phytohormones, pollen development, and signaling molecules, resulting in sizeable reductions in yields (Crafts-Brandner & Salvucci, 2002; Zinn et al., 2010; Mishra et al., 2023). Molecular mechanisms of thermosensing and signaling have been primarily explored in *Arabidopsis* with the basic Helix-Loop-Helix (bHLH) transcription factor (TF) PHYTOCHROME INTERACTING FACTOR 4 (PIF4) considered as the core of temperature signaling pathways (Gangappa et al., 2017; Casal & Balasubramanian, 2019). One of the key downstream targets of the PIF4 pathway involved in promoting growth is *YUCCA8* (*YUC8*), which encodes a rate-limiting enzyme in auxin biosynthesis and is critical for thermomorphogenesis (Franklin et al., 2011; Lee et al., 2014).

At the same time, the PIF4-independent signaling pathways (PIF4 and PIF7 alike) were discovered to directly stimulate auxin biosynthesis and trigger thermomorphogenesis in a brassinosteroid-dependent manner (Chung et al., 2020; Fiorucci et al., 2020; Vu et al., 2021). The active form of the photoreceptor phytochrome B (phyB) senses elevated temperature and is converted from the active Pfr form to the inactive Pr form. The nuclear export of the phyB-Pr upon warmth releases PIF4 inhibition, initiating thermomorphogenesis (Legris et al., 2016; Qiu et al., 2019). Additionally, other mechanisms of temperature sensing were reported in *Arabidopsis*. The translation of *PIF7* mRNA is enhanced by elevated temperature through the relaxation of the *PIF7* mRNA hairpin structure, leading to the accumulation of PIF7 protein (Chung et al., 2020). Also, the transcriptional repressor EARLY FLOWERING 3 (ELF3) aggregates into inactive condensates during warming, contributing to sensing elevated temperatures (Jung et al., 2020).

Heat can damage photosynthetic machinery by inhibiting enzymes like RUBISCO and denaturing chloroplastic proteins. This leads to stomatal closure, limiting CO_2 uptake and resulting in the buildup of reactive oxygen species (ROS), which cause cellular damage, affecting lipids, proteins, and nucleic acids (Allakhverdiev et al., 2008; Wang et al., 2018). Heat stress significantly impacts nutrient cycling and plant nutrition. High temperatures can impair root function, reducing the plant's ability to uptake essential nutrients like nitrogen, phosphorus, and potassium. For instance, in tomato plants, heat stress has been found to decrease levels of nutrient-uptake and -assimilation proteins in roots, leading to reduced nutrient absorption (Giri et al., 2017). Additionally, soil warming can accelerate nutrient cycling, resulting in nutrient losses due to faster decomposition rates and reducing nutrient availability in the long term. High temperatures also interfere with plant nutrient translocation, further complicating metabolic processes essential for growth and productivity (Mishra et al., 2023). At the cellular level, heat stress can cause protein denaturation, membrane destabilization, and oxidative stress. Persistent heat stress may result in cell death and tissue necrosis, compromising the overall health and productivity of the plant (Haider et al., 2021). These changes lead to the accumulation of ROS, which can damage cellular components like lipids, proteins, and nucleic acids. Enzymes lose efficiency as temperatures exceed their optimal range; continued stress triggers phytohormonal imbalances, such as altered abscisic acid levels, which affect plant responses. These imbalances activate complex signaling cascades involving calcium, ROS, and other molecules, further challenging the plant's ability to cope with stress (Potters et al., 2007; Saidi et al., 2009; Devireddy et al., 2021).

1.2.2 Heat Shock Proteins (HSPs) and Their Roles

HSPs are a crucial component of a plant's response to heat stress, playing key roles in maintaining protein integrity and enabling cellular homeostasis under stressful conditions. These molecular chaperones assist in the folding, refolding, transport, and degradation of proteins, preventing their denaturation and aggregation caused by heat stress (Feder & Hofmann, 1999). They are often regulated by heat shock factors (HSFs), which are activated during extreme temperatures. In wheat, *HSFs* such as *HSFA2* and *HSFA6* are activated during heat stress, modulating downstream responses to improve heat tolerance (Xue et al., 2014). Once induced, HSPs bind to misfolded proteins, aid in refolding, and prevent the formation of toxic aggregates, playing a significant role in cellular survival under heat stress (Wu, 1995). Throughout the day, varying temperatures trigger the production and accumulation of different HSPs in plants. In the mid-morning, HSP20s are synthesized to prevent proteins from misfolding and aggregating before the temperature peaks by noon. Later in the day, as temperatures rise further, the production shifts to HSP60s, HSP70s, HSP90s, and HSP100s, which help resolubilize and reactivate proteins that may have become inactive or misfolded due to heat stress, ensuring their proper functioning (Finka et al., 2016). For instance, overexpressing certain HSPs in some crops like rice (HSP18), barley (HSP70, HSP90, and HSP100), *Arabidopsis* (HSP101), and cotton (HSP70-26) improved resilience, making them valuable targets in breeding programs for heat tolerance (Queitsch et al., 2000; Kuang et al., 2017; Chaudhary et al., 2019; Zhiyong et al., 2021). Modifying the expression of HSFs and other directly related noncoding DNA regions that regulate HSPs could be a strategic approach to enhance the overexpression of key HSPs, ultimately developing heat-resilient plants.

1.2.3 Epigenetic Regulation of Stress Response Pathways (Histone Posttranslational Modifications, DNA Methylation, and ncRNA Expression)

Recent evidence suggests that epigenetic mechanisms of gene expression and regulation are actively involved in thermosensing and heat stress tolerance in plants. Epigenetics, in a broader sense, refers to "chromatin modifications" through chemical modifications of histone proteins or DNA wrapped around them that do not change the base sequence (Deichmann, 2016). The regulation of gene expression occurs through different pathways, including DNA methylation, small RNAs, ATP-dependent chromatin remodeling, histone variants, histone modifications, histone chaperones, and long noncoding RNAs (lncRNAs). Some of these pathways regulate the expression of high-temperature-responsive genes to prevent heat-related damage and promote subsequent adaptation (reviewed in Perrella et al. (2022)). For example, the expression of many thermoresponsive genes is regulated by the repressive histone variant H2A.Z deposited in nucleosomes of temperature-regulated loci primarily at the +1 site (Kumar & Wigge, 2010). The deposition and eviction of repressive H2A.Z in exchange for permissive H2A variant are performed by the Snf2 ATPase remodeling complexes, SWR1-C and INO80-EIN6 ENHANCER (EEN), respectively (Xue et al., 2021). A component of SWR1-C, ACTIN-RELATED PROTEIN 6 (ARP6), is a mediator of temperature responses in *Arabidopsis,* and *arp6* mutants demonstrate elongated hypocotyls already at low temperatures, indicative of a constitutive warm temperature phenotype (Kumar & Wigge, 2010). INO80 and EEN are directly associated with PIF4 in activating the transcription of auxin-related genes under elevated temperatures through the H2A.Z eviction. The contribution of H2A.Z to stress-induced gene activation was further supported by the fact that its eviction was compromised in both *pif4* and *ino80* mutants. Overall, the H2A.Z-containing nucleosomes are not temperature sensors per se, but rather their presence depends on transcriptional regulators allowing for environment-dependent chromatin reorganization and release of stress-responsive gene expression (Cortijo et al., 2017).

In addition to chromatin remodeling, histone methylation and histone acetylation marks have been implemented to regulate the gene expression of growth-promoting genes in response to heat stress. For example, histone H3K4me2 (an activating mark) is demethylated in response to the binding of FLOWERING CONTROL LOCUS A (FCA) to PIF4-activated growth-promoting genes, like *YUC8*, therefore eventually suppressing high-temperature-induced hypocotyl elongation (Lee et al., 2014). Therefore, *fca* mutants with a continuous expression of the *YUC8* gene display hyper-elongated hypocotyls when exposed to 28 °C. At the same time, the H3K4me3 (considered an activating mark) demethylation of double mutant Jumonji C (JmjC) JMJ14 with its cofactor-producing enzyme cytosolic isocitrate dehydrogenases (cICDHs) showed suppression of several auxin-related genes, including *YUC8*, and resulted in reduced thermomorphogenesis capacity (Cui et al., 2021). These results suggest that distinctive histone demethylases can have either negative (FCA) or positive (JMJ14, JMJ15, and JMJ18) effects on the genes involved in thermomorphogenesis. Histone methylation can also affect alternative splicing (AS), and H3K36me3 enrichment was associated with differentially spliced (DiS) events upon temperature changes (Pajoro et al., 2017). Mutants of the histone H3K36 methyltransferases *SET DOMAIN-CONTAINING GROUP 8* (*SDG8*) and *SDG26* had reduced DiS events, suggesting that H3K36me3 is indeed required for AS (Sidaway-Lee et al., 2014). Histone acetylation affects gene expression through "opening" (acetylation) and "closing" (deacetylation) of specific chromatin regions in response to environmental stimuli (Asensi-Fabado et al., 2017). Histone acetyltransferases (HATs) and histone deacetylases (HDACs) catalyze this process, which ultimately modulate chromatin accessibility for TFs, other regulatory proteins, and the transcription machinery (Asensi-Fabado et al., 2017). Histone deacetylation is involved in H2A.Z eviction since the *HISTONE DEACETYLASE 9* (*hda9*) mutant displayed high levels of H2A.Z occupancy at the *YUC8* locus and a reduced ability for PIF4 binding at warm temperatures (van der Woude et al., 2019). This translates into an impaired response of the *hda9* mutants to hypocotyl elongation and a compact rosette morphology in response to elevated temperatures (van der Woude et al., 2019). Nevertheless, the role of other HDACs in the heat stress response is more elaborate since *hda15* mutant seedlings displayed longer hypocotyls, while *hda9* and *hda19* hypocotyls were shorter than the wild type at 27 °C (Shen et al., 2019). It remains to be shown whether HDA15 and HDA19 also regulate gene expression through H2A.Z occupancy.

Overall, histone and chromatin-modifying enzymes are guided to specific regions of the chromatin by heat-responsive factors such as TFs and lncRNAs, which act as a molecular signal to direct the epigenetic machinery to genes or genomic regions (Deng et al., 2018; Ueda & Seki, 2020). The interplay between these epigenetic regulators and their recruiters forms a sophisticated system for adapting to high-temperature conditions. For instance, in rice, the histone modification H3K9me2 (H3K9 methylation 2) on the *Oryza sativa FERTILIZATION-INDEPENDENT ENDOSPERM1* (*OsFIE1*) gene is sensitive to moderate heat stress (34 °C). This modification appears to regulate *OsFIE1* expression, which is important for rice seed development under elevated temperatures (Folsom et al., 2014), highlighting how histone modifications can directly impact important agronomic traits in response to heat stress. By altering the accessibility of DNA to TFs and other regulatory proteins, these epigenetics mechanisms provide a dynamic way for plants to rapidly adjust their gene expression profiles in response to heat stress.

DNA methylation is a dynamic process that regulates gene expression in response to environmental stress (Boyko et al., 2010; Li et al., 2016). DNA methylation involves the addition of methyl groups to cytosine bases in DNA, forming CG, CHG, and CHH patterns (where H represents A, T, or C), which play a crucial role in regulating plant responses to heat stress (Cokus et al., 2008; Popova et al., 2013; Liu et al., 2015, 2017; Talarico et al., 2024). *Arabidopsis* plants exposed to heat shock (HS) (42 °C for different durations – 6, 12, or 24 hours) demonstrated global DNA demethylation with many specific genes affected primarily related to stress response (e.g. HSFs, HSPs, fructose-bisphosphate aldolase 6, and 60S RIBOSOMAL PROTEIN L4-1) (Korotko et al., 2021). Similarly, heat-sensitive cotton also showed global DNA demethylation in response to heat stress (Min et al., 2014). In contrast, the genome-wide methylation greatly increases in heat-sensitive *Brassica napus* under extreme heat exposure (Gao et al., 2014). Further research on DNA methylation-deficient mutants revealed the importance of the RNA-directed DNA methylation (RdDM) pathway in plant heat tolerance (Popova et al., 2013). In *Arabidopsis*, different components of this pathway have varying effects on heat sensitivity. Plants lacking certain methyltransferases, such as domains rearranged methyltransferase 1/2 (*DRM1/DRM2*) and chromomethylase 3 (*CMT3*), exhibit reduced heat sensitivity. However, mutants deficient in other RdDM pathway components such as nuclear RNA polymerase D 2 (*nrpd2*), Dicer-like 3 (*dcl3*), RNA-dependent RNA polymerase 2 (*rdr2*), and argonaute 4 (*ago4*) lead to increased heat sensitivity. This suggests a complex relationship between DNA methylation and heat stress responses. Similarly, a copia-type, heat-responsive retrotransposon

ONSEN has been found to influence the heat responsiveness of nearby genes when it inserts into new locations (Ito et al., 2011). Additionally, the expression of certain genes, such as *CALMODULIN-LIKE 41* (*CML41*, *At3g50770*), is enhanced by heat stress and shows reduced methylation near transposable elements (TEs) (Naydenov et al., 2015). These findings suggest that the RdDM pathway modulates heat stress responses by regulating DNA methylation patterns near transposons and TEs, thereby affecting the expression of adjacent genes. Interestingly, plants lacking *CMT2*, which is involved in a specific type of DNA methylation (CHH), show improved heat tolerance, indicating that different methylation pathways may have distinct roles in heat stress responses (Shen et al., 2014).

Noncoding RNAs (ncRNAs), including micro RNAs (miRNAs), small-interfering RNAs (siRNAs), and lncRNAs, play crucial roles in regulating gene expression under heat stress, especially miRNAs and siRNAs. NcRNAs are functional RNA molecules transcribed from DNA but not translated into proteins. These ncRNAs form complex regulatory networks that enable plants to fine-tune their gene expression in response to environmental stresses, including heat. They act at various levels of gene regulation, from transcriptional control to posttranscriptional modification, allowing for rapid and flexible responses to changing conditions (Hirayama and Shinozaki, 2010; de Lima et al., 2012; Khraiwesh et al., 2012; Jha et al., 2023).

MiRNAs are short, typically 21–25 nucleotides long, ncRNAs that regulate gene expression post-transcriptionally (Voinnet, 2009). miRNAs are typically generated through a process involving the RNase III enzyme (Dicer-like [DCL] protein), which cleaves the miRNA primary transcripts into the precursor molecules, characterized by a distinctive stem-loop structure (de Lima et al., 2012). Once formed, mature miRNAs, along with protein factors, make an RNA-induced silencing complex (RISC) to target mRNAs to modulate the expression of the target genes posttranscriptionally. They function by base pairing with complementary sequences on target mRNAs, usually leading to mRNA degradation or translational repression (Baulcombe, 2004; Voinnet, 2009). This posttranscriptional regulation allows miRNAs to adjust gene expression in response to various cellular needs, including development, differentiation, stress responses, and environmental cues (Rogers & Chen, 2013). For instance, miR398 is a well-studied example of rapidly induced by heat stress in *Arabidopsis* and other plants. It targets genes encoding copper/zinc superoxide dismutases (CSD1 and CSD2) and their copper chaperone (CCS), which are involved in ROS scavenging. Downregulation of these targets by miR398 enhances heat tolerance (Zhu et al., 2011; Guan et al., 2013). DNA methylation can affect the expression of the miRNA-coded genes, which can regulate the genes involved in abiotic stress response. For example, methylation-sensitive amplification polymorphisms analysis of differentially methylated regions in response to temperature stress in *Populus simonii* identified seven miRNA-coded genes significantly affected by stress application (Ci et al., 2015). Three of those miRNAs, miR156i/j, miR390c, and miR396e/g, target the products of *ACYL-CoA OXIDASE* (*ACOX1* and *ACOX3*), *PHOSPHOLIPID/GLYCEROL ACYLTRANSFERASE FAMILY PROTEINS* (*LPCAT1* and *LPCAT2*), and *ISOCITRATE DEHYDROGENASE* (*IDH*), which are involved in lipid metabolism and peroxisome biogenesis.

Research has revealed complex interactions between heat stress responses and phytohormone signaling pathways in plants, mediated by various miRNAs. For example, microRNA miR167, which regulates *AUXIN RESPONSE FACTOR 8*, *ARF8* (a gene involved in the development of floral organs, gynoecium and stamen), shows a peculiar response to high temperatures. While miR167 decreases under heat stress, one specific variant, miR167h, exhibits a significant increase (Wu et al., 2006; Kruszka et al., 2014). Furthermore, other auxin signaling-related miRNAs such as miR390 and miR393 have been implicated in heat stress responses, indicating a broader involvement of auxin-related miRNAs in plant- thermotolerance mechanisms (Vidal et al., 2010; Xin et al., 2010; Guan et al., 2013; Hivrale et al., 2016).

Studies have shown that some miRNAs, typically associated with regulating plant developmental processes, also have important functions in how plants respond to heat stress. For instance, miR171 is a key regulator of gene expression for certain members of the *GRAS* family, specifically *SCARECROW-LIKE6-III* (*SCL6-III*) and *SCL6-IV*. These genes play crucial roles in various plant developmental processes, including forming radial patterns in root and shoot structures (Kamiya et al., 2003). By modulating the expression of these *GRAS* family genes, miR171 indirectly influences important aspects of plant architecture and growth. In *Arabidopsis*, heat stress increased miR171 levels, resulting in a greater suppression of *GRAS* gene expression (Mahale et al., 2014). Interestingly, a contrasting pattern was observed in *Populus* species, where pto-miR171 and ptc-miR171 showed reduced expression under heat stress conditions (Lu et al., 2008; Chen et al., 2012). This difference highlights the species-specific nature of miRNA responses to environmental stressors like heat. Further, wheat research has revealed that an additional target for miR164 is the heat shock protein HSP17. Under heat stress conditions, the expression of *HSP17* increased, suggesting a complex interplay between miR164 and its targets in response to elevated temperatures (Kumar et al., 2015). Furthermore, miR396 and its target gene *HaWRKY6*, as well as miR397 and its target gene *LACCASE*, have been implicated in heat stress adaptation (McCaig et al., 2005; Giacomelli et al., 2012).

siRNAs and lncRNAs are important ncRNAs that play crucial roles in plant gene regulation and stress responses. siRNAs are typically 21–24 nucleotides long and are derived from double-stranded RNA (dsRNA) precursors. The DCL enzymes cleave these precursors into siRNAs, which then associate with the RISC to target complementary mRNA for degradation or inhibit translation, playing a key role in regulating gene expression (Axtell, 2013). They can be classified into *trans*-acting siRNAs (ta-siRNAs), natural antisense transcript siRNAs (nat-siRNAs), and heterochromatic siRNAs. siRNAs are involved in various abiotic stress responses, including heat stress, where they regulate gene expression through mRNA degradation or translational repression (Sunkar & Zhu, 2007; Axtell, 2013). For example, three siRNAs were found to be downregulated by heat stress and upregulated by cold stress in wheat (Yao et al., 2010) and an miR173-cleaved ta-siRNA (*TAS1*) targets *HEAT-INDUCED TAS1 TARGET1* (*HTT1*) and *HTT2*, involved in thermotolerance in *Arabidopsis* (Khraiwesh et al., 2012; Li et al., 2014).

LncRNAs, longer than 200 nucleotides, can be categorized as antisense or intronic based on their genomic locations (Wierzbicki, 2012). They regulate gene expression at various levels, including chromatin modification, where they can recruit chromatin-modifying complexes to specific genomic loci to alter the epigenetic state of chromatin. During transcription, lncRNAs can interact with TFs or the transcriptional machinery to enhance or repress transcription. In posttranscriptional processing, lncRNAs can influence mRNA splicing, stability, and translation. LncRNAs have been implicated in plant development and stress responses, including heat stress (Budak et al., 2020; Xin et al., 2011). For example, Xin et al. (2011) characterized long-nonprotein-coding RNAs (npcRNAs) in wheat with powdery mildew infection and heat stress, four of which were identified as miRNA precursors (TalnRNA5, TapmlnRNA8, TapmlnRNA19, and TahlnRNA27), in which TahlnRNA27 was upregulated under heat stress. LncRNAs have been found to be differentially expressed under heat stress conditions in various plant species, suggesting their important roles in plant thermotolerance. However, while their involvement in heat stress responses is evident, a complete and detailed understanding of the specific mechanisms by which lncRNAs contribute to plant thermotolerance is still lacking. Further comprehensive studies are needed to fully unravel the complex functions of these ncRNAs in plant heat stress response pathways (Zhao et al., 2020).

Overall, the mechanisms mentioned earlier can lead to an epigenetic memory, which refers to the ability of plants to retain information about previous stress exposures and pass it on to subsequent generations. This mechanism plays a crucial role in plant adaptation, including heat stress (Kinoshita & Seki, 2014; Crisp et al., 2016; He & Li, 2018; Ramakrishnan et al., 2022).

1.2.4 Other Protective Mechanisms and Molecular Responses

Transcriptomic and proteomic adjustments are also prevalent responses of plants to heat stress exposure. AS is a widespread mechanism in plants that enhances the diversity of both transcripts and proteins (Syed et al., 2012). Research has demonstrated that numerous genes involved in plant stress responses undergo AS, which allows plants to adjust their gene expression in response to various environmental challenges (Qin et al., 2007; Matsukura et al., 2010; Guerra et al., 2015; Tognacca et al., 2022). For instance, in maize, the expression of *DEHYDRATION-RESPONSIVE ELEMENT BINDING 2B* (*DREB2B*), a key gene in stress response pathways, is regulated through AS when the plant experiences heat stress (Qin et al., 2007), and in rice, one of the three isoforms of gene *OsHSFA2d, OsHSFA2dI*, is specifically induced by HS. The gene is involved in the heat stress-induced unfolded protein response (Cheng et al., 2015), highlighting how plants can use AS to modulate their molecular responses to elevated temperatures.

Ubiquitination, also known as ubiquitylation, is a form of posttranslational modification. It involves the attachment of a small protein called ubiquitin to a target protein within the cell. This process requires three types of enzymes working in sequence: ubiquitin-activating enzyme (E1), ubiquitin-conjugating enzyme (E2), and ubiquitin ligase (E3) (Vierstra, 2009). The E3 enzymes play a crucial role in determining which proteins are targeted for ubiquitination, thus providing specificity to the process. Proteins that undergo ubiquitination can experience various outcomes. Ubiquitination is often associated with protein degradation, but it can also serve non-proteolytic functions (Hershko & Ciechanover, 1998). For example, in rice, the rice RING domain E3 ligase, *HEAT AND COLD INDUCED 1* (*OsHCI1*) has been identified as a key player in heat response. When exposed to high temperatures, *OsHCI1* adds a single ubiquitin molecule to the target proteins, known as monoubiquitination. This monoubiquitination triggers the movement of these proteins from the nucleus to the cytoplasm, which is thought to contribute to the plant's ability to withstand heat stress. Furthermore, *Arabidopsis* plants with overexpression of *OsHCI1*-enhenced yellow fluorescent protein (*EYFP*) displayed improved resistance to high temperatures (Lim et al., 2013).

1.3 Strategies for Mitigating Heat Stress Response Through Gene Editing and Derivative Technologies

1.3.1 Overview of Noncoding DNA

Noncoding DNA refers to genome regions that do not code for proteins. While once considered "junk DNA," it is now understood that noncoding DNA plays crucial roles in regulating gene expression and maintaining cellular function. Noncoding DNA includes sequences that control the activity of genes, influence chromatin structure, and regulate various cellular processes (Simna & Han, 2022). There is a growing consensus that the two key factors determining gene expression level are *cis*-regulatory modules (CRMs) and *trans*-acting factors (TAFs) (Schmitz et al., 2022). The former include gene-proximal promoters and distal enhancers that are part of CREs. Promoters are located near the transcription start site (TSS) of genes and are necessary for transcription initiation. The strength of a promoter could be influenced by several key factors, such as essential core promoter components, with the TATA box being a notable example, the proportion of guanine and cytosine nucleotides (GC content) within the promoter sequence, and the presence and arrangement of TF-binding sites near the promoter. Interaction between CREs and TAFs in a cell-specific manner determines the timing and level of gene expression. The plant core promoters (PCPs) around the TSSs with a minimum sequence region of 50–100 bp are rich in CREs and can drive the basal level of target gene transcription (Schmitz et al., 2022). The regulatory element can also affect gene expression under stress conditions. For example, in rice, a regulatory element called HS185 was identified in the *HSP70* promoter, which is capable of conferring heat stress responsiveness (Zhang et al., 2012).

Altering the promoters, enhancers, and insulator sequences can modulate the expression levels of heat-responsive genes, potentially enhancing the plant's ability to withstand temperature stress. Enhancers are distal regulatory elements that can enhance the transcription of genes, often by looping to the promoter region to bring TFs and other regulatory proteins in proximity to the transcriptional machinery. Unlike promoters, enhancers have some unique characteristics, such as they function regardless of their directional alignment, are effective whether located upstream or downstream of the minimal promoter, and can influence gene expression from considerable distances, often several kilobases away from the target gene (Blaise et al., 2016; Jores et al., 2020). For instance, in *Arabidopsis*, four candidate transcriptional activators (HsfA1a, HsfA1b, HsfA1d, and HsfA1e) were found to function as a *cis*-acting element in the HS-responsive expression of *DREB2A*, a key TF involved in both drought and heat stress responses (Yoshida et al., 2011).

Insulators, also known as boundary elements, are the genetic elements that function as boundaries within the genome to block the interaction between enhancers and promoters, thus preventing inappropriate activation or repression of genes. They can also act as barriers to the spread of heterochromatin (Gaszner & Felsenfeld, 2006). Several plant-derived insulator sequences across various species. One example of a potential plant insulator is a gypsy-like sequence in the *Arabidopsis* genome, which demonstrated enhancer-blocking effects between the 35S enhancer and the Pip (petal- and stamen-specific *PISTILLATA* promoter) (Singer & Cox, 2013). In rice, an insulator-like sequence named HS185 was identified in the *HSP70* gene promoter, capable of blocking the enhancer activity of the cauliflower mosaic virus 35S promoter (Zhang et al., 2012).

Matrix or scaffold attachment regions (MARs/SARs) are a well-studied class of plant-putative insulator elements. These AT-rich, noncoding sequences bind to the nuclear matrix in vitro and can form chromatin loops (Cockerill & Garrard, 1986; Dolgova & Dolgov, 2019). When used to flank transgenes, MARs can reduce position effects, leading to increased and more consistent transgene expression (Singer et al., 2012; Dolgova & Dolgov, 2019). For instance, MARs with insulator-like properties include a 2-kb region named transformation booster sequence (TBS) from *Petunia hybrida*, acting as an enhancer-blocking insulator when inserted between the CaMV 35S promoter/enhancer in *A. thaliana* (Hily et al., 2009) and Rb7 3′ MAR of tobacco (Singer et al., 2012).

1.3.2 The CRISPR/Cas Technology – Advantages and Limitations for Editing of the Noncoding DNA

CRISPR/Cas9 is a programmable, sequence-specific genome-editing system derived from the type II adaptive prokaryotic immunity. It is widely used in animal and plant research due to its cost-effectiveness, simplicity, and high efficiency. This genome-editing platform comprises the Cas9 nuclease, which induces a double-stranded break in the targeted genomic region, and the artificial single-guide RNA (sgRNA), which complexes with the Cas9 nuclease to guide it to the target site through Watson–Crick base pairing (Doudna & Charpentier, 2014). The sgRNA is engineered as a fusion of the naturally occurring RNA duplex of the protospacer-containing CRISPR RNA (crRNA), which dictates the sequence targeted by the sgRNA-Cas9 ribonucleoprotein, and the *trans*-activating crRNA (tracrRNA), which Cas9 recognizes (Doudna & Charpentier, 2014). One of the main limitations of this technology is the need to have a protospacer adjacent motif (PAM) immediately following the target sequence (Doudna & Charpentier, 2014). This PAM requirement not only restricts the diversity of genomic targets that can be edited by the CRISPR/Cas9 platform but also prevents the self-targeting of the T-DNA harboring the sgRNA transcriptional unit. Recently, Cas9 nucleases with differing PAM requirements have been explored, and more impressively, Cas9 nucleases with relaxed or almost PAM-less requirements have been developed (Nishimasu et al., 2018; Walton et al., 2020; Xu et al., 2021).

Specialized genome-editing platforms have been developed based on the CRISPR/Cas9 systems. For instance, base editors, which consist of a deactivated, noncatalytic Cas9 mutant nuclease (dCas9) fused to a nucleotide-modifying enzyme, either adenosine deaminase or cytidine deaminase, could impart targeted transition mutations (C·G to T·A and A·T to G·C) to the genome (Rees and Liu, 2018). The dCas9-sgRNA ribonucleoprotein transports cytidine or adenosine deaminase enzymes to a specific genomic site. There, the enzymes convert cytidine (C) or adenosine (A) bases into uracil (U) or inosine, which are subsequently transformed into thymine (T) and guanine (G) during DNA replication (Azameti & Dauda, 2021). Albeit powerful, this platform has a significant off-target activity, due to either the wide catalytic windows of the deaminase enzymes or the low specificity of the dCas9-sgRNA ribonucleoprotein. Additionally, cytidine base editors were documented to induce genome-wide cytidine deamination in rice (Jin et al., 2019). Efforts are underway to develop base editors with more specificity and the capacity for imparting transversion mutations (Azameti & Dauda, 2021). Like BE, epigenetic editors (EEs) use dCas9 as a targeting module. Fused to this dCas9 are effectors that modify epigenetic marks such as DNA methylases/demethylases (to modify DNA methylation) and HAT (facilitates the acetylation of histone H3 at lysine 27) (Shin et al., 2022; Qi et al., 2023). These tools have been used to modulate gene expression in several plant species (reviewed in Qi et al. (2023)).

The Prime-editing platform consists of a programmable nuclease-polymerase chimeric Prime-editor (PE) protein and an engineered prime-editing guide RNA (epegRNA), capable of introducing any user-defined short edits. The epegRNA complexes with the PE protein specify the genomic locus to be edited and the edits to be installed (Chen & Liu, 2023). The PE chimeric protein consists of the Cas9 nickase mutant (nCas9), which cuts one strand of the DNA instead of inducing double-stranded breaks. This is fused to a reverse transcriptase, typically the Moloney Murine Leukemia Virus (M-MLV), which installs the edits into the primed cut strand. The epegRNA, on the other hand, consists of the typical CRISPR/Cas9 sgRNA, extended to have a primer-binding site and the reverse-transcriptase template (Chen & Liu, 2023). The applicability of prime editing in modifying plant genomes has been limited by its low editing efficiency (Zong et al., 2022). Several improvements have been made to increase the expression of the components, make the PE chimeric protein more catalytic and localized to the nucleus, prevent the degradation of epegRNAs, remove the RNAse activity of the reverse transcriptase, and manipulate cellular processes that are involved in DNA repair, yet the editing efficiency of the platform in plants still remains challenging (Huang & Liu, 2023). Moreover, to improve the prime-editing platform's capacity to introduce large, theoretically even gene-sized edits, a derivative platform called PrimeRoot editing was developed (Sun et al., 2024). This platform employs dual PE-epegRNA ribonucleoproteins to install the recognition sequence of site-specific recombinases (RS), flanking the targeted genomic locus. Then, a site-specific recombinase will facilitate the recombination of the targeted genomic locus, which is now flanked with RS, and a donor vector containing the edits to be installed, similarly flanked with RS

(Sun et al., 2024). This approach was successful in the targeted integration of up to 11.1 kb of DNA segments into the rice genome, but its success is dependent on two layers: the capacity of PE to insert RS and the variable activity of different site-specific recombinases; this platform could further benefit from optimization (Sun et al., 2024).

Last is the CRISPR/dCas9+ transposase fusion and the CRISPR/Cas-associated transposase (CAST) editing platform. In essence, these platforms aim to achieve targeted integration of large DNA segments by amalgamating the RNA programmability of CRISPR/Cas systems and the integrative capacity of transposon-associated proteins. In doing so, the need for double-stranded breaks that are typical for the CRISPR/Cas9 system is abolished, preventing the consequent undesirable mutations. Efforts have been made to fuse dCas9 (the RNA-guided targeting unit) with a transposase (Chen & Wang, 2019) to increase the frequency of transposition into the user-defined target site. In plants, the rice transposase protein PONG was fused to Cas9 and 12 and successfully applied for targeted insertion of varying DNA segment sizes into *A. thaliana* and Soybean (Liu et al., 2024). Recently, it was discovered that prokaryotic Tn7-like transposon families are associated with minimal CRISPR/Cas systems as an RNA-guided targeting system that facilitates transposition (Peters et al., 2017). CAST is naturally occurring and can be engineered for gene-editing applications. For instance, a heterologous expression of a high-fidelity CAST from the prokaryote *Vibrio cholerae* resulted in a targeted, site-specific transposition of a donor DNA segment into plasmids (Klompe et al., 2019) and human cells (Lampe et al., 2024). This platform has not yet been used for plants.

Gene editing of noncoding DNA is a promising approach to developing crops with better agronomic characteristics and resiliency. Unfortunately, this approach is still in its infancy. The identification of noncoding DNA that can be targeted is mostly limited by the lack of systemic, in-depth exploration of their regulatory functions (Chen et al., 2023). This is mainly due to the wide array of regulatory noncoding DNA present in plant genomes and their degeneracy, the regulatory interactions involving these sequences within various highly complex regulatory networks and architectures, and the variability of these regulatory networks from one species to another.

CRISPR/Cas9 is a highly robust genome-editing tool, explaining its widespread use. While the development of CRISPR/Cas9 variants with relaxed or alternative PAM requirements expanded the range of targetable genomic sequences, the actual ability of the Cas9-sgRNA ribonucleoprotein to successfully generate double-stranded breaks is still greatly influenced by the epigenetic profile of the targeted genomic locus (Verkuijl & Rots, 2019; Lopos et al., 2023) and the activity of the individual sgRNA (Chari et al., 2017; Dong et al., 2022). Moreover, another important consideration is the minimization of CRISPR/Cas9 off-targets, which may introduce unwanted pleiotropic effects. The progress of genome sequencing has made multiple plant genomic sequences publicly available and facilitated the development of CRISPR/Cas9 off-target prediction tools (reviewed in Hahn & Nekrasov (2019)) and approaches (Hajiahmadi et al., 2019). However, there are usually substantial variations between cultivars, especially in noncoding DNA sequences. Hence, cultivar-specific genome sequences must also be available and integrated into the CRISPR/Cas9 off-target prediction tools.

While CRISPR/Cas9 editing is robust, it also generates random insertion/deletion mutation that is undesirable in applications requiring precise gene editing. The frequency of homology-directed repair in plants using CRISPR/Cas9 is low as the predominant mechanism to resolve double-stranded breaks is the nonhomologous end joining (NHEJ) repair pathway. CRISPR/Cas9-derived platforms such as Prime-editing and PrimeRoot editing could create a relatively more precise edit compared to CRISPR/Cas9 but the editing efficiency of these platforms is still quite low (Li et al., 2022a; Sun et al., 2024). On the other hand, base editing is greatly limited by its off-target activities and bystander editing corresponding to its wide catalytic window (Mishra et al., 2020) while the CRISPR/CAST is still in its infancy and the CRISPR/dCas9 transposase fusion also suffers from low insertion efficiency.

Finally, the delivery of gene-editing reagents into the plant cells and the regeneration of new plants from the edited cells also pose a problem. Biolistic and *Agrobacterium*-mediated transformations, the main methods to deliver gene-editing reagents into plant cells, require a tissue culture step to regenerate new plants. Unfortunately, many crop species and cultivars are still recalcitrant to tissue culture. To alleviate this, a strategy was developed in which developmental regulator (DR) genes are also codelivered along with the gene-editing reagents, to improve plant regeneration in tissue culture. Ectopic overexpression of a chimeric DR gene, the *GROWTH REGULATING FACTOR 4–GRF INTERACTING FACTOR 1 (GRF4–GIF1)*, resulted in improved regeneration efficiency in traditional tissue-culture recalcitrant crops such as wheat (Debernardi et al., 2020) without negative pleiotropic effects commonly associated with overexpressing other DR genes like *BABY BOOM*.

1.3.3 The Editing and Engineering of *cis*-Regulatory Elements (CREs)

While the genes responsible for mediating heat stress responses are often highly conserved between genotypes due to their survival implications, the heat-responsive CREs modulating these genes are rich in genetic variations, usually underlying the quantitative variability of transcriptomic heat stress responses. For instance, Waters et al. (2017) found that *cis*-regulatory variations were more common than *trans*-regulatory variations in differentially expressed transcripts upon exposure of 14-day-old maize seedlings to heat stress. Similarly, Zhou et al. (2021) identified different heat- and cold-responsive CRE variants in different maize inbred lines that underlie differentially expressed genes (DEGs) upon exposure to heat or cold stress. They used these CRE variations to approximate the possible heat or cold stress responses of different genotypes. The implications of CRE variations to the heat stress transcriptomic profile, and ultimately, to the heat stress response, emphasize their potential as gene-editing targets for achieving thermotolerance. The thermotolerance contributions of putatively causative CRE variants that were identified through transcriptomic studies, which may occur as presence/absence, single-nucleotide polymorphisms (SNPs), insertions/deletions (indels), copy number variations, and TE insertions, can be functionally verified through reporter assays, transgenesis, transient expression approaches, and association studies (Hernandez-Garcia & Finer, 2014; Zhou et al., 2021). Functionally verified causative CREs can then be edited into heat-susceptible but desirable cultivars using CRISPR/Cas9 and its derivative technologies. This strategy has been previously used to improve important agronomic traits in tomatoes. In this case, Li et al. (2022b) used CRISPR/Cas9 to mutagenize the M9 SNP in a CRE of the *KLUH* gene promoter, which was associated with large fruit sizes in tomatoes. The resulting mutants showed quantitatively varying fruit sizes, with some mutants showing larger fruits than the wild type (Li et al., 2022b).

Aside from editing CREs, it is also possible to engineer their regulatory activities by utilizing CRISPR/Cas9-derived platforms. Crucial to this operation is the availability of bioinformatic data pertaining to which CREs are performing regulatory functions. Once this is at hand, Saeed et al. (2022) suggest the following approaches: CRE disruption, CRE deletion, CRE addition, and promoter swapping/reconstitution. In the CRE disruption strategy, insertion/deletion (indel) mutations are targeted along the length of the identified CRE to disrupt its regulatory function. This is achieved using CRISPR/Cas9, which induces a targeted double-stranded break. This break is then resolved by the error-prone NHEJ pathway, resulting in indels. In the CRE deletion strategy, sgRNAs flanking the CREs to be removed are identified and used to guide the Cas9 nuclease to induce targeted double-stranded breaks, resulting in large nucleotide deletions that excise the CRE. A reverse approach can be utilized to add CREs with the desired regulatory function to promoters or nearby regulatory regions. Using CRISPR/Cas9, double-stranded breaks flanking the targeted genomic sequence are generated, and a donor template containing the desired CRE that is flanked by homology arms is provided. Alternatively, PrimeRoot editing can be employed to flank recombinase recognition sequences into the genomic sequence to be replaced by the desired CRE. A template containing the desired CRE flanked with the same recombinase recognition sequences is then cotransformed to facilitate homologous recombination. If the CRE is small, Prime-editing can be utilized instead of PrimeRoot editing. Lastly, in promoter swapping and reconstitution, an identified promoter can be excised and replaced with another or parts of different promoters containing various CREs can be fused.

CRE engineering has been successfully utilized to modify various traits in diverse crop species. For instance, Čermák et al. (2015) inserted the constitutive *CaMV35s* promoter between the endogenous promoter of the *anthocyanin mutant 1* (*ANT1*) gene and its coding sequence through *Agrobacterium*-mediated transformation of tomato cotyledon using a geminivirus replicon expressing CRISPR/Cas9:sgRNA transcriptional unit and the *CaMV35s* promoter donor sequence. This resulted in the heritable constitutive overproduction of the anthocyanin pigments in edited tomato lines. Similarly, to confer ethylene insensitivity and increased yield amidst abiotic stresses in maize, Shi et al. (2017) utilized both the promoter swapping and reconstitution approach, using CRISPR/Cas9, to elevate the expression levels of the maize *ARGOS8* gene, a negative regulator of ethylene response. They inserted and entirely swapped the endogenous *ARGOS8* promoter with the mildly ubiquitous *GOS2* promoter, resulting in increased grain yield relative to wild type when subjected to drought stress treatment during flowering stages. On the other hand, using CRISPR/Cas9, Liu et al. (2021) disrupted the promoter of the *CLE* gene and its compensatory gene, *FCP1*, in maize. These genes are negative regulators of grain yield by repressing meristematic proliferation through the conserved *CLAVATA (CLV)–WUSCHEL (WUS)* feedback signaling pathway. Some of the promoter-edited lines showcased improved yield-related traits such as ear length, kernel row number, cob diameter, and ear diameter.

Currently, these strategies have not yet been implemented for the development of thermotolerance in crops. Nonetheless, possible CRE targets can be identified based on the available literature. These potential targets are summarized in Table 1.1. For instance, the heat stress-responsive CRE variants previously identified by Waters et al. (2017), Zhou et al. (2021), and Liang et al. (2022) can be transferred to other maize genotypes to confer thermotolerance using the CRE editing approach through Prime-editing or PrimeRoot editing. In rice, Wu et al. (2022) identified an insertion of a 228-bp *miniature inverted-repeat transposable element* (*MITE*) in the promoter of the HS TF gene *HEAT-TOLERANCE GENE ON CHROMOSOME 3* (*HTG3*) that attenuated this gene's expression in response to heat stress, thereby causing heat sensitivity in rice. This particular *MITE* insertion can potentially be excised from the *HTG3* promoter through PrimeRoot editing or CRISPR/Cas9 platforms. In potatoes, Trapero-Mozos et al. (2018) discovered that an extra six TA repeats (10 TA repeats in total, 3–4 are present in wild-type alleles) present 495 bp upstream the genic start codon of the potato *HSc70* gene resulted in the elevated expression of this allelic variant at high temperatures, conferring higher tuber yield during heat-stressed conditions without yield penalties under nonstressed conditions. These extra TA repeats could potentially be added into the promoter of the other potato *HSc70* allelic variants using Prime-editing. Lastly, in grapevines, Liu et al. (2023) identified that the expression of the *HSFA2* gene is higher in the thermotolerant *Vitis davidii* "Tangwei" (*VdHSFA2*) compared to the heat-sensitive *Vitis vinifera*

Table 1.1 Summary of the potential CRE targets in different crops for the development of thermotolerance.

Potential CRE targets	Phenotypic effects of the potential targets	Crop species	Possible editing approaches to confer thermotolerance	References
Heat stress-responsive CREs	CRE variations leading to differential transcriptomic profiles and levels of heat stress response	Maize (*Zea mays*)	Identification and functional analysis of CRE variants in relation to heat stress response. Manipulation of the validated heat stress-responsive CRE variants using CRISPR/Cas9 and its derivative editing platforms	Liang et al. (2022), Waters et al. (2017), and Zhou et al. (2022)
228-bp *MITE* insertion within the *HEAT-TOLERANCE GENE ON CHROMOSOME 3 (HTG3)* promoter	Lowered the expression level of the *HTG3* gene, thus leading to heat stress sensitivity	Rice (*Oryza sativa*)	Excision using PrimeRoot editing or generating double-stranded breaks flanking the 228-bp *MITE* insertion through CRISPR/Cas9	Wu et al. (2022)
TA repetitive elements within the promoter of the *HSc70* gene	The additional six (6) TA repetitive elements within the promoter of an *HSc70* allelic variant conferred higher gene expression level leading to higher tuber yields during heat stress conditions	Potato (*Solanum tuberosum*)	Adding additional TA repetitive elements into the promoter of other *HSc70* alleles using Prime-editing	Trapero-Mozos et al. (2018)
Promoters of the *HSFA2* gene orthologs	The promoter of the *HSFA2* gene from the thermotolerant *Vitis davidii* "Tangwei" (*VdHSFA2*) drove higher gene expression in both normal and elevated temperatures when compared to the promoter of the same gene from the heat-sensitive *Vitis vinifera* "Jingxiu" (*VvHSFA2*)	Grapevine (*Vitis davidii* and *V. vinifera*)	*VdHSFA2* promoter could be inserted immediately upstream of the *VvHSFA2* gene through PrimeRoot editing or CRISPR/Cas9 editing Alternatively, causative CRE variations could be identified through promoter deletion/mutation analysis and association and promulgated into the *VvHSFA2* promoter using Prime-editing approaches	Liu et al. (2023)

"Jingxiu" (*VvHSFA2*). Transient reporter expression assay revealed that the promoter of *VdHSFA2* drove a higher expression level relative to *VvHSFA2* in both heat-stressed and nonheat-stressed conditions. The promoter insertion approach through PrimeRoot or CRISPR/Cas9 editing could potentially be utilized to insert the endogenous promoter of the *VdHSFA2* into the *VvHSFA2* gene to drive higher expression. Alternatively, causative CREs within the *VdHSFA2* promoter can be identified through promoter mutation assays and transferred to the *VvHSFA2* promoter through Prime-editing.

1.3.4 Integration of Omics Data and Artificial Intelligence (AI) Processing for Target Identification

Heat tolerance is a quantitative trait regulated by many quantitative trait loci (QTLs). For example, the first genetic studies of wheat heat tolerance were done in 1991, and the genes responsible for heat tolerance were mapped on chromosomes 3A, 3B, 4A, 4B, and 6A (Sun & Quick, 1991). Since then, over 130 QLTs have been mapped on almost every wheat chromosome, which involved in heat tolerance response (reviewed in Farhad et al. (2023)). Many of them were consistently located on chromosome 3B (Mason et al., 2010). Nevertheless, further fine-mapping is needed to decipher the genes and allelic variations responsible for differences in heat stress response. Genome-wide association study (GWAS) could further be used to dissect the complex genetic loci of heat tolerance in wheat (Khan et al., 2022; Fu et al., 2024). For example, the combination of GWAS, gene expression, haplotype, and gene function analysis within the physical intervals of the 15 QTL clusters identified pointed toward two new gene candidates – *TaWRKY74-B* and *TaSnRK3.15-B* (Fu et al., 2024). This was in line with the known role of WRKY TFs and *TaSnRK* genes in wheat growth and stress responses (Zhu et al., 2013; Jiang et al., 2022).

The overall end goal of *cis*-regulatory region modification is to alter gene expression, leading to a change in the target phenotype to introduce quantitative trait variation (QTV). Our knowledge of the transcriptional regulation of gene expression in plants has been based on the global transcriptome profile of the bulk tissues (e.g. root, leaves, embryo, and endosperm) through microarray and RNA-sequencing (RNA-seq) technologies. Due to composite gene expression profiles, these technologies limit our understanding of gene expression in specific cell types. The average transcript signal obtained from these analyses might not always correlate with the *cis*-regulatory regions responsible for the gene expression. Therefore, it becomes challenging to perform association studies on a large scale between particular DNA regulatory elements and their contribution to gene expression.

The rapidly evolving discipline of single-cell RNA-sequencing (scRNA-seq), combined with chromatin accessibility, can provide a global overview of the specific cell types' response to heat stress and the CREs involved (Giacomello, 2021). The method relies on the isolation of intact nuclei or protoplasts from target plant tissues through mechanical disruption of cells or enzymatic lysis, preparation of a next-generation sequencing (NGS) library for transcriptome or chromatin analysis with subsequent short read sequencing, and bioinformatic processing of the data. This allows the dissection of the tissue heterogeneity. Nevertheless, the information regarding the physical position of the cells is lost due to dissociation in single cells. This issue prompted the development of spatial transcriptomics techniques, allowing for in situ analysis of the gene expression profile at almost single-cell level (Crosetto et al., 2015). Such analysis is still challenging and could be divided into targeted (Chen et al., 2015; Eng et al., 2019) and untargeted (Stahl et al., 2016; Stickels et al., 2021). Whereas the former method allows expression quantification at the subcellular level for a selected number of genes, the latter can explore whole-tissue gene expression unbiasedly, although not always at the cellular level. In general, an intact tissue section is placed on a specially prepared slide that has 100 μm spots of poly-dT probes to ensure the binding of poly-A tails of mRNAs (Stahl et al., 2016; Zormpas et al., 2023). Spatial barcodes on the spots allow the physical connection of the expression data to the tissue image. Following the imaging step, the tissue is permeabilized, and RNA hybridizes to the spot closest to it, thus reflecting the physical location in the tissue. Eventually, the RNA is reverse-transcribed into cDNA, sequenced following NGS library preparation, and resolved to specific locations on the tissue image. Following its original implementation in 2016 (Stahl et al., 2016), resolution and sequencing methods for spatial transcriptomics have rapidly improved. An increase in resolution of the transcriptome profile was achieved by decreasing spot sizes on the slide to 55, 10, and 0.5 μm (10× Genomics Visium, Slide-Seq/V2, and Stereo-seq, respectively) (Rodriques et al., 2019; Stickels et al., 2021; Chen et al., 2022). The Stereo-seq technique offers a subcellular resolution using a DNA nanoball patterned array and in situ RNA capture (Chen et al., 2022).

Although still in its infancy, the scRNA-seq proved to be powerful in dissecting the transcriptomic profile of the germinating barley (*Hordeum vulgare*) grain, proliferating tomato (*Solanum lycopersicum*) and *Arabidopsis* (*A. thaliana*) calli (Song et al., 2023; Yin et al., 2024), rice (*O. sativa*) root (Zhang et al., 2021), and soybean (*Glycine max*) development through constructing a multiomic single-cell atlas (Zhang et al., 2024a, b). The combination of single-nucleus RNA sequencing (snRNA-seq) and single-cell sequencing of the assay for transposase-accessible chromatin (scATAC-seq) enables a direct correlation between transcriptomes and chromatin accessibility from the complex tissues at single-cell resolution (Hie et al., 2020; Vandereyken et al., 2023; Zhang et al., 2024a, b). This approach, for example, allowed the identification of accessible chromatin regions (ACRs) in 10 soybean tissues with 103 distinct cell types (Zhang et al., 2024a, b). ACRs are nucleosome-depleted regions that are preferred places of TFs binding when colocalized with CREs (Klemm et al., 2019). Eventually, the combination of epigenomics (chromatin accessibility in *cis*-regulatory regions) and transcriptomics data at the single-cell resolution can guide the gene-editing efforts to fine-tune the expression of the target genes for improved abiotic stress tolerance. Considering that cereals are the most vulnerable to high-temperature stress during the reproductive phase (Barnabas et al., 2008; Shi et al., 2022), which affects spikelet development due to insufficient assimilated supply and grain production, spatial transcriptomics of developing spikelets under normal and HS conditions combined with scATAC-seq could help with the strategy design to engineer improved tolerance to elevated temperatures during reproductive tissue development. In the long run, developing the crop multiomic atlas under normal growth conditions and in response to HS will be important to reveal the precise connection between CREs and gene expression.

Integration of the multiomics data for gene-editing target identification requires advanced processing techniques like AI algorithms (reviewed in Murmu et al. (2024)). For example, machine learning (ML) programs (e.g. support vector machine – SVM) were used to find essential genes associated with drought resistance in *Arabidopsis* (Liang et al., 2011) and key SNPs in maize (Shikha et al., 2017). Regulation of gene expression *in trans* through miRNAs could indicate the plant stress response level (Asefpour Vakilian, 2020). Regression models, such as decision trees (DTs), SVMs, and Naïve Bayes (NB), were effective in predicting the plant stress by measuring the plant miRNAs' concentration using gold nanoparticle-based biosensors. SVM with Gaussian kernel could predict plant stress at a high correlation level ($R^2 = 0.96$), considering miRNA concentrations as input features. Another study used SVM to predict miRNAs associated with four specific abiotic stresses: cold, drought, heat, and salt (Pradhan et al., 2023b). Overall, prediction accuracies for the independent datasets were over 80% for all abiotic stresses. The same group further reported the application of SVM to predict abiotic stress-responsive lncRNAs with an overall accuracy of over 68% (Pradhan et al., 2023a). ML models were also successfully used to predict gene expression responses to cold and heat. It was possible to achieve an accurate cross-species likelihood of gene activity following cold treatment by training the models using multispecies datasets (Meng et al., 2021). Furthermore, predictive models were developed to determine the relationship between thermal stress-responsive gene expression and TSSs in maize (Zhou et al., 2022). The prediction accuracy could be improved using models trained on data from more maize genotypes to accommodate different maize cultivars.

To identify CREs within PCPs, a deep-learning assisted analysis was done on a large-scale PCP dataset of 18,329 *Arabidopsis*, 34,415 maize, and 27,094 sorghum core promoters (Deng et al., 2023). The promoter strengths were measured in the previous study using six transient transfection systems (tobacco leaves with enhancer in the dark, tobacco leaves without enhancer in the dark, tobacco leaves with enhancer in the light, tobacco leaves without enhancer in the light, maize protoplasts with enhancer in the dark, and maize protoplasts without enhancer in the dark) (Jores et al., 2021). The algorithms developed to identify the CREs contained in a given PCP (iCREPCP) were able to find the base-resolution position of each CRE and its contribution to promoter strength with the mean training R^2 ranged from 0.490 to 0.782, and low variances, implying their feasibility. The model proved useful in the *trans*-species discovery of critical CREs for the rice *IPA1* gene – a key master regulator of plant architecture with experimentally verified promoter deletion sequence responsible for the trait (Song et al., 2022). Despite some limitations (promoter strength measured in vitro via tobacco leaves or maize protoplasts and short analysis region of −165 to +5), the model could predict CRE gene-editing targets that modulate the expression level of stress-responsive genes. Furthermore, incorporating nucleotide-resolution epigenetic signatures like chromatin accessibility and histone modifications, together with TF-binding motifs and sequence conservation, can aid in developing the models to assign predictive aggregate scores of the effect of each region in the promoter on gene expression (Zhou et al., 2023). The predictive scores were used to develop a CRISPR/Cas12a promoter editing (CAPE) system to introduce QTV for rice starch content, grain size, and plant height. Overall, combining the iCREPCP and CAPE methods into a single prediction model could further increase the precision of CRE predictions in plants.

1.4 Conclusions and Future Perspectives

Biological data generation, ML advancements, and gene editing have opened up new possibilities for understanding complex biological information (reviewed in Murmu et al. (2024)). However, integrating ML into plant molecular studies presents challenges due to the context-specific nature of ML approaches. Each ML algorithm has strengths and weaknesses, affecting predictive efficiency under different conditions. An ML model designed for one dataset may not generalize to others due to biological and technical variations. Omics datasets offer valuable information but often contain noise and sparsity, making it difficult to identify biological features (Joyce & Palsson, 2006). Researchers use resampling strategies like oversampling and undersampling to address imbalanced datasets (Maimon & Rokach, 2005). Techniques like dropout help prevent overfitting in deep learning models, enhancing their generalization ability to new data (Jiang et al., 2023). Handling big data in plant system biology involves addressing issues like sparsity, multicollinearity, and overfitting. Interpreting complex ML models remains challenging due to their "black box" nature. Despite these challenges, successful applications of AI in plant omics (Gao et al., 2023) highlight the need for robust algorithms and experimental validation. Collaboration between omics researchers, data scientists, and agricultural experts is crucial for maximizing AI's potential in plant defense and practical applications. The development of AI-assisted tools to search for the *cis*-regulatory regions responsible for abiotic stress response in plants will be detrimental to the successful application of gene editing in fine-tuning gene expression to create resilient crops. Future gene-editing applications may involve promoting the targeted crossover (CO) in elite hybrids to selectively bring together the best combination of stress-responsive genes and CREs without a linkage drug. ML tools can further assist with selecting the most susceptible target regions to CO (Demirci et al., 2018).

References

Allakhverdiev, S. I., Kreslavski, V. D., Klimov, V. V., Los, D. A., Carpentier, R., & Mohanty, P. (2008). Heat stress: an overview of molecular responses in photosynthesis. *Photosynthesis Research*, *98*(1–3), 541–550. https://doi.org/10.1007/s11120-008-9331-0

Asefpour Vakilian, K. (2020). Machine learning improves our knowledge about miRNA functions towards plant abiotic stresses. *Scientific Reports*, *10*(1), 3041. https://doi.org/10.1038/s41598-020-59981-6

Asensi-Fabado, M. A., Amtmann, A., & Perrella, G. (2017). Plant responses to abiotic stress: the chromatin context of transcriptional regulation. *Biochimica et Biophysica Acta, Gene Regulatory Mechanisms*, *1860*(1), 106–122. https://doi.org/10.1016/j.bbagrm.2016.07.015

Axtell, M. J. (2013). Classification and comparison of small RNAs from plants. *Annual Review of Plant Biology*, *64*, 137–159. https://doi.org/10.1146/annurev-arplant-050312-120043

Azameti, M. K., & Dauda, W. P. (2021). Base editing in plants: applications, challenges, and future prospects [review]. *Frontiers in Plant Science*, *12*. https://doi.org/10.3389/fpls.2021.664997

Barnabas, B., Jager, K., & Feher, A. (2008). The effect of drought and heat stress on reproductive processes in cereals. *Plant, Cell & Environment*, *31*(1), 11–38. https://doi.org/10.1111/j.1365-3040.2007.01727.x

Baulcombe, D. (2004). RNA silencing in plants. *Nature*, *431*(7006), 356–363. https://doi.org/10.1038/nature02874

Blaise, W., Johan, Z., Rurika, O., & Maike, S. (2016). Plant enhancers: a call for discovery. *Trends in Plant Science*, *21*(11), 974–987. https://doi.org/10.1016/j.tplants.2016.07.013

Boyko, A., Blevins, T., Yao, Y., Golubov, A., Bilichak, A., Ilnytskyy, Y., Hollunder, J., Meins, F. Jr., & Kovalchuk, I. (2010). Transgenerational adaptation of Arabidopsis to stress requires DNA methylation and the function of Dicer-like proteins. *PLoS One*, *5*(3), e9514. https://doi.org/10.1371/journal.pone.0009514

Budak, H., Kaya, S. B., & Cagirici, H. B. (2020). Long non-coding RNA in plants in the era of reference sequences. *Frontiers in Plant Science*, *11*, 276. https://doi.org/10.3389/fpls.2020.00276

Capovilla, G., Schmid, M., & Pose, D. (2015). Control of flowering by ambient temperature. *Journal of Experimental Botany*, *66*(1), 59–69. https://doi.org/10.1093/jxb/eru416

Casal, J. J., & Balasubramanian, S. (2019). Thermomorphogenesis. *Annual Review of Plant Biology*, *70*, 321–346. https://doi.org/10.1146/annurev-arplant-050718-095919

Čermák, T., Baltes, N. J., Čegan, R., Zhang, Y., & Voytas, D. F. (2015). High-frequency, precise modification of the tomato genome. *Genome Biology*, *16*(1), 232. https://doi.org/10.1186/s13059-015-0796-9

Chari, R., Yeo, N. C., Chavez, A., & Church, G. M. (2017). sgRNA Scorer 2.0: a species-independent model to predict CRISPR/Cas9 activity. *ACS Synthetic Biology*, *6*(5), 902–904.

Chaudhary, R., Baranwal, V. K., Kumar, R., Sircar, D., & Chauhan, H. (2019). Genome-wide identification and expression analysis of Hsp70, Hsp90, and Hsp100 heat shock protein genes in barley under stress conditions and reproductive development. *Functional & Integrative Genomics*, *19*(6), 1007–1022. https://doi.org/10.1007/s10142-019-00695-y

Chen, P. J., & Liu, D. R. (2023). Prime editing for precise and highly versatile genome manipulation. *Nature Reviews Genetics, 24*(3), 161–177. https://doi.org/10.1038/s41576-022-00541-1

Chen, S. P., & Wang, H. H. (2019). An engineered Cas-transposon system for programmable and site-directed DNA transpositions. *The CRISPR Journal, 2*(6), 376–394.

Chen, L., Ren, Y., Zhang, Y., Xu, J., Sun, F., Zhang, Z., & Wang, Y. (2012). Genome-wide identification and expression analysis of heat-responsive and novel microRNAs in *Populus tomentosa. Gene, 504*(2), 160–165. https://doi.org/10.1016/j.gene.2012.05.034

Chen, K. H., Boettiger, A. N., Moffitt, J. R., Wang, S., & Zhuang, X. (2015). RNA imaging. Spatially resolved, highly multiplexed RNA profiling in single cells. *Science, 348*(6233), aaa6090. https://doi.org/10.1126/science.aaa6090

Chen, A., Liao, S., Cheng, M., Ma, K., Wu, L., Lai, Y., Qiu, X., Yang, J., Xu, J., Hao, S., Wang, X., Lu, H., Chen, X., Liu, X., Huang, X., Li, Z., Hong, Y., Jiang, Y., Peng, J., et al. (2022). Spatiotemporal transcriptomic atlas of mouse organogenesis using DNA nanoball-patterned arrays. *Cell, 185*(10). 1777–1792.e1721. https://doi.org/10.1016/j.cell.2022.04.003

Chen, Y.-H., Lu, J., Yang, X., Huang, L.-C., Zhang, C.-Q., Liu, Q.-Q., & Li, Q.-F. (2023). Gene editing of non-coding regulatory DNA and its application in crop improvement. *Journal of Experimental Botany, 74*(19), 6158–6175. https://doi.org/10.1093/jxb/erad313

Cheng, Q., Zhou, Y., Liu, Z., Zhang, L., Song, G., Guo, Z., Wang, W., Qu, X., Zhu, Y., & Yang, D. (2015). An alternatively spliced heat shock transcription factor, Os2dI, functions in the heat stress-induced unfolded protein response in rice. *Plant Biology, 17*(2), 419–429. https://doi.org/10.1111/plb.12267

Chung, B. Y. W., Balcerowicz, M., Di Antonio, M., Jaeger, K. E., Geng, F., Franaszek, K., Marriott, P., Brierley, I., Firth, A. E., & Wigge, P. A. (2020). An RNA thermoswitch regulates daytime growth in Arabidopsis. *Nature Plants, 6*(5), 522–532. https://doi.org/10.1038/s41477-020-0633-3

Ci, D., Song, Y., Tian, M., & Zhang, D. (2015). Methylation of miRNA genes in the response to temperature stress in *Populus simonii. Frontiers in Plant Science, 6*, 921. https://doi.org/10.3389/fpls.2015.00921

Cockerill, P. N., & Garrard, W. T. (1986). Chromosomal loop anchorage of the kappa immunoglobulin gene occurs next to the enhancer in a region containing topoisomerase II sites. *Cell, 44*(2), 273–282. https://doi.org/10.1016/0092-8674(86)90761-0

Cokus, S. J., Feng, S., Zhang, X., Chen, Z., Merriman, B., Haudenschild, C. D., Pradhan, S., Nelson, S. F., Pellegrini, M., & Jacobsen, S. E. (2008). Shotgun bisulphite sequencing of the Arabidopsis genome reveals DNA methylation patterning. *Nature, 452*(7184), 215–219. https://doi.org/10.1038/nature06745

Cortijo, S., Charoensawan, V., Brestovitsky, A., Buning, R., Ravarani, C., Rhodes, D., van Noort, J., Jaeger, K. E., & Wigge, P. A. (2017). Transcriptional regulation of the ambient temperature response by H2A.Z nucleosomes and HSF1 transcription factors in Arabidopsis. *Molecular Plant, 10*(10), 1258–1273. https://doi.org/10.1016/j.molp.2017.08.014

Crafts-Brandner, S. J., & Salvucci, M. E. (2002). Sensitivity of photosynthesis in a C4 plant, maize, to heat stress. *Plant Physiology, 129*(4), 1773–1780.

Crawford, A. J., McLachlan, D. H., Hetherington, A. M., & Franklin, K. A. (2012). High temperature exposure increases plant cooling capacity. *Current Biology, 22*(10), R396–R397. https://doi.org/10.1016/j.cub.2012.03.044

Crisp, P. A., Ganguly, D., Eichten, S. R., Borevitz, J. O., & Pogson, B. J. (2016). Reconsidering plant memory: intersections between stress recovery, RNA turnover, and epigenetics. *Science Advances, 2*(2), e1501340. https://doi.org/10.1126/sciadv.1501340

Crosetto, N., Bienko, M., & van Oudenaarden, A. (2015). Spatially resolved transcriptomics and beyond. *Nature Reviews Genetics, 16*(1), 57–66. https://doi.org/10.1038/nrg3832

Cui, X., Zheng, Y., Lu, Y., Issakidis-Bourguet, E., & Zhou, D. X. (2021). Metabolic control of histone demethylase activity involved in plant response to high temperature. *Plant Physiology*, *185*(4), 1813–1828. https://doi.org/10.1093/plphys/kiab020

Debernardi, J. M., Tricoli, D. M., Ercoli, M. F., Hayta, S., Ronald, P., Palatnik, J. F., & Dubcovsky, J. (2020). A GRF-GIF chimeric protein improves the regeneration efficiency of transgenic plants. *Nature Biotechnology*, *38*(11), 1274–1279. https://doi.org/10.1038/s41587-020-0703-0

Deichmann, U. (2016). Epigenetics: the origins and evolution of a fashionable topic. *Developmental Biology*, *416*(1), 249–254. https://doi.org/10.1016/j.ydbio.2016.06.005

Demirci, S., Peters, S. A., de Ridder, D., & van Dijk, A. D. J. (2018). DNA sequence and shape are predictive for meiotic crossovers throughout the plant kingdom. *The Plant Journal*, *95*(4), 686–699. https://doi.org/10.1111/tpj.13979

Deng, X., Qiu, Q., He, K., & Cao, X. (2018). The seekers: how epigenetic modifying enzymes find their hidden genomic targets in Arabidopsis. *Current Opinion in Plant Biology*, *45*, 75–81. https://doi.org/10.1016/j.pbi.2018.05.006

Deng, K., Zhang, Q., Hong, Y., Yan, J., & Hu, X. (2023). iCREPCP: a deep learning-based web server for identifying base-resolution cis-regulatory elements within plant core promoters. *Plant Communications*, *4*(1), 100455. https://doi.org/10.1016/j.xplc.2022.100455

Devireddy, A. R., Tschaplinski, T. J., Tuskan, G. A., Muchero, W., & Chen, J.-G. (2021). Role of reactive oxygen species and hormones in plant responses to temperature changes. *International Journal of Molecular Sciences*, *22*(16), 8843. https://www.mdpi.com/1422-0067/22/16/8843

Dolgova, A. S., & Dolgov, S. V. (2019). Matrix attachment regions as a tool to influence plant transgene expression. *3. Biotech*, *9*(5), 176. https://doi.org/10.1007/s13205-019-1709-5

Dong, C., Gou, Y., & Lian, J. (2022). SgRNA engineering for improved genome editing and expanded functional assays. *Current Opinion in Biotechnology*, *75*, 102697.

Doudna, J. A., & Charpentier, E. (2014). The new frontier of genome engineering with CRISPR-Cas9. *Science*, *346*(6213), 1258096. https://doi.org/10.1126/science.1258096

Eng, C. L., Lawson, M., Zhu, Q., Dries, R., Koulena, N., Takei, Y., Yun, J., Cronin, C., Karp, C., Yuan, G. C., & Cai, L. (2019). Transcriptome-scale super-resolved imaging in tissues by RNA seqFISH. *Nature*, *568*(7751), 235–239. https://doi.org/10.1038/s41586-019-1049-y

Farhad, M., Kumar, U., Tomar, V., Bhati, P. K., Navaneetha Krishnan, J., Kishowar-E-Mustarin, Barek, V., Brestic, M., & Hossain, A. (2023). Heat stress in wheat: a global challenge to feed billions in the current era of the changing climate [review]. *Frontiers in Sustainable Food Systems*, *7*. https://doi.org/10.3389/fsufs.2023.1203721

Feder, M. E., & Hofmann, G. E. (1999). Heat-shock proteins, molecular chaperones, and the stress response: evolutionary and ecological physiology. *Annual Review of Physiology*, *61*, 243–282. https://doi.org/10.1146/annurev.physiol.61.1.243

Finka, A., Mattoo, R. U. H., & Goloubinoff, P. (2016). Experimental milestones in the discovery of molecular chaperones as polypeptide unfolding enzymes. *Annual Review of Biochemistry*, *85*, 715–742. https://doi.org/10.1146/annurev-biochem-060815-014124

Fiorucci, A. S., Galvão, V. C., Ince, Y., Boccaccini, A., Goyal, A., Allenbach Petrolati, L., Trevisan, M., & Fankhauser, C. (2020). PHYTOCHROME INTERACTING FACTOR 7 is important for early responses to elevated temperature in Arabidopsis seedlings. *The New Phytologist*, *226*(1), 50–58. https://doi.org/10.1111/nph.16316

Folsom, J. J., Begcy, K., Hao, X., Wang, D., & Walia, H. (2014). Rice fertilization-independent Endosperm1 regulates seed size under heat stress by controlling early endosperm development. *Plant Physiology*, *165*(1), 238–248. https://doi.org/10.1104/pp.113.232413

Franklin, K. A., Lee, S. H., Patel, D., Kumar, S. V., Spartz, A. K., Gu, C., Ye, S., Yu, P., Breen, G., Cohen, J. D., Wigge, P. A., & Gray, W. M. (2011). Phytochrome-interacting factor 4 (PIF4) regulates auxin biosynthesis at high temperature. *Proceedings of the National Academy of Sciences of the United States of America*, *108*(50), 20231–20235. https://doi.org/10.1073/pnas.1110682108

Fu, C., Zhou, Y., Liu, A., Chen, R., Yin, L., Li, C., & Mao, H. (2024). Genome-wide association study for seedling heat tolerance under two temperature conditions in bread wheat (*Triticum aestivum* L.). *BMC Plant Biology*, *24*(1), 430. https://doi.org/10.1186/s12870-024-05116-2

Gangappa, S. N., Berriri, S., & Kumar, S. V. (2017). PIF4 coordinates thermosensory growth and immunity in Arabidopsis. *Current Biology*, *27*(2), 243–249. https://doi.org/10.1016/j.cub.2016.11.012

Gao, G., Li, J., Li, H., Li, F., Xu, K., Yan, G., Chen, B., Qiao, J., & Wu, X. (2014). Comparison of the heat stress induced variations in DNA methylation between heat-tolerant and heat-sensitive rapeseed seedlings. *Breeding Science*, *64*(2), 125–133. https://doi.org/10.1270/jsbbs.64.125

Gao, S., Chen, S., Yang, M., Wu, J., Chen, S., & Li, H. (2023). Mining salt stress-related genes in *Spartina alterniflora* via analyzing co-evolution signal across 365 plant species using phylogenetic profiling. *aBIOTECH*, *4*(4), 291–302. https://doi.org/10.1007/s42994-023-00125-5

Gaszner, M., & Felsenfeld, G. (2006). Insulators: exploiting transcriptional and epigenetic mechanisms. *Nature Reviews Genetics*, *7*(9), 703–713. https://doi.org/10.1038/nrg1925

Giacomelli, J. I., Weigel, D., Chan, R. L., & Manavella, P. A. (2012). Role of recently evolved miRNA regulation of sunflower HaWRKY6 in response to temperature damage. *New Phytologist*, *195*(4), 766–773. https://doi.org/10.1111/j.1469-8137.2012.04259.x

Giacomello, S. (2021). A new era for plant science: spatial single-cell transcriptomics. *Current Opinion in Plant Biology*, *60*, 102041. https://doi.org/10.1016/j.pbi.2021.102041

Giri, A., Heckathorn, S., Mishra, S., & Krause, C. (2017). Heat stress decreases levels of nutrient-uptake and -assimilation proteins in tomato roots. *Plants*, *6*(1), 6. https://www.mdpi.com/2223-7747/6/1/6

Govindaraj, M., Pattanashetti, S. K., Patne, N., Kanatti, A. A., & Ciftci, Y. (2018). Breeding cultivars for heat stress tolerance in staple food crops. *Next Generation Plant Breeding*, *76480*(3).

Guan, Q., Lu, X., Zeng, H., Zhang, Y., & Zhu, J. (2013). Heat stress induction of mi398 triggers a regulatory loop that is critical for thermotolerance in Arabidopsis. *The Plant Journal*, *74*(5), 840–851. https://doi.org/10.1111/tpj.12169

Guerra, D., Crosatti, C., Khoshro, H. H., Mastrangelo, A. M., Mica, E., & Mazzucotelli, E. (2015). Post-transcriptional and post-translational regulations of drought and heat response in plants: a spider's web of mechanisms [review]. *Frontiers in Plant Science*, *6*. https://doi.org/10.3389/fpls.2015.00057

Hahn, F., & Nekrasov, V. (2019). CRISPR/Cas precision: do we need to worry about off-targeting in plants? *Plant Cell Reports*, *38*(4), 437–441. https://doi.org/10.1007/s00299-018-2355-9

Haider, S., Iqbal, J., Naseer, S., Yaseen, T., Shaukat, M., Bibi, H., Ahmad, Y., Daud, H., Abbasi, N. L., & Mahmood, T. (2021). Molecular mechanisms of plant tolerance to heat stress: current landscape and future perspectives. *Plant Cell Reports*, *40*(12), 2247–2271. https://doi.org/10.1007/s00299-021-02696-3

Hajiahmadi, Z., Movahedi, A., Wei, H., Li, D., Orooji, Y., Ruan, H., & Zhuge, Q. (2019). Strategies to increase on-target and reduce off-target effects of the CRISPR/Cas9 system in plants. *International Journal of Molecular Sciences*, *20*(15), 3719.

Hassan, M. U., Chattha, M. U., Khan, I., Chattha, M. B., Barbanti, L., Aamer, M., Iqbal, M. M., Nawaz, M., Mahmood, A., Ali, A., & Aslam, M. T. (2021). Heat stress in cultivated plants: nature, impact, mechanisms, and mitigation strategies – a review. *Plant Biosystems – An International Journal Dealing with All Aspects of Plant Biology*, *155*(2), 211–234. https://doi.org/10.1080/11263504.2020.1727987

He, Y., & Li, Z. (2018). Epigenetic environmental memories in plants: establishment, maintenance, and reprogramming. *Trends in Genetics*, *34*(11), 856–866. https://doi.org/10.1016/j.tig.2018.07.006

Hernandez-Garcia, C. M., & Finer, J. J. (2014). Identification and validation of promoters and cis-acting regulatory elements. *Plant Science*, *217–218*, 109–119. https://doi.org/10.1016/j.plantsci.2013.12.007

Hershko, A., & Ciechanover, A. (1998). The ubiquitin system. *Annual Review of Biochemistry*, *67*, 425–479. https://doi.org/10.1146/annurev.biochem.67.1.425

Hie, B., Peters, J., Nyquist, S. K., Shalek, A. K., Berger, B., & Bryson, B. D. (2020). Computational methods for single-cell RNA sequencing. *Annual Review of Biomedical Data Science*, *3*, 339–364. https://doi.org/10.1146/annurev-biodatasci-012220-100601

Hily, J. M., Singer, S. D., Yang, Y., & Liu, Z. (2009). A transformation booster sequence (TBS) from *Petunia hybrida* functions as an enhancer-blocking insulator in *Arabidopsis thaliana*. *Plant Cell Reports*, *28*(7), 1095–1104. https://doi.org/10.1007/s00299-009-0700-8

Hirayama, T., & Shinozaki, K. (2010). Research on plant abiotic stress responses in the post-genome era: past, present and future. *The Plant Journal*, *61*(6), 1041–1052. https://doi.org/10.1111/j.1365-313X.2010.04124.x

Hivrale, V., Zheng, Y., Puli, C. O. R., Jagadeeswaran, G., Gowdu, K., Kakani, V. G., Barakat, A., & Sunkar, R. (2016). Characterization of drought- and heat-responsive microRNAs in switchgrass. *Plant Science*, *242*, 214–223. https://doi.org/10.1016/j.plantsci.2015.07.018

Hua, J. (2013). Modulation of plant immunity by light, circadian rhythm, and temperature. *Current Opinion in Plant Biology*, *16*(4), 406–413. https://doi.org/10.1016/j.pbi.2013.06.017

Huang, Z., & Liu, G. (2023). Current advancement in the application of prime editing. *Frontiers in Bioengineering and Biotechnology*, *11*, 1039315.

Intergovernmental Panel on Climate Change (IPCC). (2021). Climate change 2021: The physical science basis. Contribution of Working Group I to the Sixth Assessment Report of the Intergovernmental Panel on Climate Change. Cambridge University Press. Retrieved October 2, 2024, from https://www.ipcc.ch/

Ito, H., Gaubert, H., Bucher, E., Mirouze, M., Vaillant, I., & Paszkowski, J. (2011). An siRNA pathway prevents transgenerational retrotransposition in plants subjected to stress. *Nature*, *472*(7341), 115–119. https://doi.org/10.1038/nature09861

Jha, U. C., Nayyar, H., Roychowdhury, R., Prasad, P. V. V., Parida, S. K., & Siddique, K. H. M. (2023). Non-coding RNAs (ncRNAs) in plant: master regulators for adapting to extreme temperature conditions. *Plant Physiology and Biochemistry*, *205*, 108164. https://doi.org/10.1016/j.plaphy.2023.108164

Jiang, B., Liu, Y., Niu, H., He, Y., Ma, D., & Li, Y. (2022). Mining the roles of wheat (*Triticum aestivum*) SnRK genes in biotic and abiotic responses [original research]. *Frontiers in Plant Science*, *13*. https://doi.org/10.3389/fpls.2022.934226

Jiang, P., Xue, Y., & Neri, F. (2023). Continuously evolving dropout with multi-objective evolutionary optimisation. *Engineering Applications of Artificial Intelligence*, *124*, 106504. https://doi.org/10.1016/j.engappai.2023.106504

Jin, S., Zong, Y., Gao, Q., Zhu, Z., Wang, Y., Qin, P., Liang, C., Wang, D., Qiu, J.-L., Zhang, F., & Gao, C. (2019). Cytosine, but not adenine, base editors induce genome-wide off-target mutations in rice. *Science*, *364*(6437), 292–295. https://doi.org/10.1126/science.aaw7166

Jores, T., Tonnies, J., Dorrity, M. W., Cuperus, J. T., Fields, S., & Queitsch, C. (2020). Identification of plant enhancers and their constituent elements by STARR-seq in tobacco leaves [OPEN]. *The Plant Cell*, *32*(7), 2120–2131. https://doi.org/10.1105/tpc.20.00155

Jores, T., Tonnies, J., Wrightsman, T., Buckler, E. S., Cuperus, J. T., Fields, S., & Queitsch, C. (2021). Synthetic promoter designs enabled by a comprehensive analysis of plant core promoters. *Nature Plants*, *7*(6), 842–855. https://doi.org/10.1038/s41477-021-00932-y

Joyce, A. R., & Palsson, B. Ø. (2006). The model organism as a system: integrating 'omics' data sets. *Nature Reviews Molecular Cell Biology*, *7*(3), 198–210. https://doi.org/10.1038/nrm1857

Jung, J. H., Barbosa, A. D., Hutin, S., Kumita, J. R., Gao, M., Derwort, D., Silva, C. S., Lai, X., Pierre, E., Geng, F., Kim, S. B., Baek, S., Zubieta, C., Jaeger, K. E., & Wigge, P. A. (2020). A prion-like domain in ELF3 functions as a thermosensor in Arabidopsis. *Nature*, *585*(7824), 256–260. https://doi.org/10.1038/s41586-020-2644-7

Kamiya, N., Itoh, J.-I., Morikami, A., Nagato, Y., & Matsuoka, M. (2003). The SCARECROW gene's role in asymmetric cell divisions in rice plants. *The Plant Journal*, *36*(1), 45–54. https://doi.org/10.1046/j.1365-313X.2003.01856.x

Khan, M. I., Kainat, Z., Maqbool, S., Mehwish, A., Ahmad, S., Suleman, H. M., Mahmood, Z., Ali, M., Aziz, A., Rasheed, A., & Li, H. (2022). Genome-wide association for heat tolerance at seedling stage in historical spring wheat cultivars [original research]. *Frontiers in Plant Science*, *13*. https://doi.org/10.3389/fpls.2022.972481

Khraiwesh, B., Zhu, J. K., & Zhu, J. (2012). Role of miRNAs and siRNAs in biotic and abiotic stress responses of plants. *Biochimica et Biophysica Acta*, *1819*(2), 137–148. https://doi.org/10.1016/j.bbagrm.2011.05.001

Kinoshita, T., & Seki, M. (2014). Epigenetic memory for stress response and adaptation in plants. *Plant and Cell Physiology*, *55*(11), 1859–1863. https://doi.org/10.1093/pcp/pcu125

Klemm, S. L., Shipony, Z., & Greenleaf, W. J. (2019). Chromatin accessibility and the regulatory epigenome. *Nature Reviews Genetics*, *20*(4), 207–220. https://doi.org/10.1038/s41576-018-0089-8

Klompe, S. E., Vo, P. L., Halpin-Healy, T. S., & Sternberg, S. H. (2019). Transposon-encoded CRISPR–Cas systems direct RNA-guided DNA integration. *Nature*, *571*(7764), 219–225.

Korotko, U., Chwiałkowska, K., Sańko-Sawczenko, I., & Kwasniewski, M. (2021). DNA demethylation in response to heat stress in *Arabidopsis thaliana*. *International Journal of Molecular Sciences*, *22*(4). https://doi.org/10.3390/ijms22041555

Kruszka, K., Pacak, A., Swida-Barteczka, A., Nuc, P., Alaba, S., Wroblewska, Z., Karlowski, W., Jarmolowski, A., & Szweykowska-Kulinska, Z. (2014). Transcriptionally and post-transcriptionally regulated microRNAs in heat stress response in barley. *Journal of Experimental Botany*, *65*(20), 6123–6135. https://doi.org/10.1093/jxb/eru353

Kuang, J., Liu, J., Mei, J., Wang, C., Hu, H., Zhang, Y., Sun, M., Ning, X., Xiao, L., & Yang, L. (2017). A class II small heat shock protein OsHsp18.0 plays positive roles in both biotic and abiotic defense responses in rice. *Scientific Reports*, *7*(1), 11333. https://doi.org/10.1038/s41598-017-11882-x

Kumar, S. V., & Wigge, P. A. (2010). H2A.Z-containing nucleosomes mediate the thermosensory response in Arabidopsis. *Cell*, *140*(1), 136–147. https://doi.org/10.1016/j.cell.2009.11.006

Kumar, R. R., Pathak, H., Sharma, S. K., Kala, Y. K., Nirjal, M. K., Singh, G. P., Goswami, S., & Rai, R. D. (2015). Novel and conserved heat-responsive microRNAs in wheat (*Triticum aestivum* L.). *Functional & Integrative Genomics*, *15*(3), 323–348. https://doi.org/10.1007/s10142-014-0421-0

Lampe, G. D., King, R. T., Halpin-Healy, T. S., Klompe, S. E., Hogan, M. I., Vo, P. L. H., Tang, S., Chavez, A., & Sternberg, S. H. (2024). Targeted DNA integration in human cells without double-strand breaks using CRISPR-associated transposases. *Nature Biotechnology*, *42*(1), 87–98.

Lee, H. J., Jung, J. H., Cortes Llorca, L., Kim, S. G., Lee, S., Baldwin, I. T., & Park, C. M. (2014). FCA mediates thermal adaptation of stem growth by attenuating auxin action in Arabidopsis. *Nature Communications*, *5*, 5473. https://doi.org/10.1038/ncomms6473

Lee, Z., Lim, J. A., Harikrishna, J. A., Islam, T., Abd Rahim, M. H., & Yaacob, J. S. (2024). Regulation of plant responses to temperature stress: a key factor in food security and for mitigating effects of climate change. *International Journal of Plant Production*, *18*(2), 141–159. https://doi.org/10.1007/s42106-024-00282-7

Legris, M., Klose, C., Burgie, E. S., Rojas, C. C., Neme, M., Hiltbrunner, A., Wigge, P. A., Schäfer, E., Vierstra, R. D., & Casal, J. J. (2016). Phytochrome B integrates light and temperature signals in Arabidopsis. *Science*, *354*(6314), 897–900. https://doi.org/10.1126/science.aaf5656

Lesk, C., Anderson, W., Rigden, A., Coast, O., Jägermeyr, J., McDermid, S., Davis, K. F., & Konar, M. (2022). Compound heat and moisture extreme impacts on global crop yields under climate change. *Nature Reviews Earth & Environment*, *3*(12), 872–889. https://doi.org/10.1038/s43017-022-00368-8

Li, S., Liu, J., Liu, Z., Li, X., Wu, F., & He, Y. (2014). Heat-induced TAS1 TARGET1 mediates thermotolerance via heat stress transcription factor A1a-directed pathways in Arabidopsis. *Plant Cell*, *26*(4), 1764–1780. https://doi.org/10.1105/tpc.114.124883

Li, J., Huang, Q., Sun, M., Zhang, T., Li, H., Chen, B., Xu, K., Gao, G., Li, F., Yan, G., Qiao, J., Cai, Y., & Wu, X. (2016). Global DNA methylation variations after short-term heat shock treatment in cultured microspores of *Brassica napus* cv. Topas. *Scientific Report*, *6*, 38401. https://doi.org/10.1038/srep38401

Li, J., Chen, L., Liang, J., Xu, R., Jiang, Y., Li, Y., Ding, J., Li, M., Qin, R., & Wei, P. (2022a). Development of a highly efficient prime editor 2 system in plants. *Genome Biology*, *23*(1), 161.

Li, Q., Feng, Q., Snouffer, A., Zhang, B., Rodríguez, G. R., & van der Knaap, E. (2022b). Increasing fruit weight by editing a cis-regulatory element in tomato KLUH promoter using CRISPR/Cas9 [original research]. *Frontiers in Plant Science*, *13*. https://doi.org/10.3389/fpls.2022.879642

Liang, Y., Zhang, F., Wang, J., Joshi, T., Wang, Y., & Xu, D. (2011). Prediction of drought-resistant genes in *Arabidopsis thaliana* using SVM-RFE. *PLoS One*, *6*(7), e21750. https://doi.org/10.1371/journal.pone.0021750

Liang, Z., Myers, Z. A., Petrella, D., Engelhorn, J., Hartwig, T., & Springer, N. M. (2022). Mapping responsive genomic elements to heat stress in a maize diversity panel. *Genome Biology*, *23*(1), 234. https://doi.org/10.1186/s13059-022-02807-7

Lim, S. D., Cho, H. Y., Park, Y. C., Ham, D. J., Lee, J. K., & Jang, C. S. (2013). The rice RING finger E3 ligase, OsHCI1, drives nuclear export of multiple substrate proteins and its heterogeneous overexpression enhances acquired thermotolerance. *Journal of Experimental Botany*, *64*(10), 2899–2914. https://doi.org/10.1093/jxb/ert143

de Lima, J. C., Loss-Morais, G., & Margis, R. (2012). MicroRNAs play critical roles during plant development and in response to abiotic stresses. *Genetics and Molecular Biology*, *35*(4 Suppl), 1069–1077. https://doi.org/10.1590/s1415-47572012000600023

Liu, J., Feng, L., Li, J., & He, Z. (2015). Genetic and epigenetic control of plant heat responses. *Frontiers in Plant Science*, *6*, 267. https://doi.org/10.3389/fpls.2015.00267

Liu, T., Li, Y., Duan, W., Huang, F., & Hou, X. (2017). Cold acclimation alters DNA methylation patterns and confers tolerance to heat and increases growth rate in Brassica rapa. *Journal of Experimental Botany*, *68*(5), 1213–1224. https://doi.org/10.1093/jxb/erw496

Liu, L., Gallagher, J., Arevalo, E. D., Chen, R., Skopelitis, T., Wu, Q., Bartlett, M., & Jackson, D. (2021). Enhancing grain-yield-related traits by CRISPR-Cas9 promoter editing of maize CLE genes. *Nature Plants*, *7*(3), 287–294. https://doi.org/10.1038/s41477-021-00858-5

Liu, X., Chen, H., Li, S., Lecourieux, D., Duan, W., Fan, P., Liang, Z., & Wang, L. (2023). Natural variations of HSFA2 enhance thermotolerance in grapevine. *Horticulture Research*, *10*(1), uhac250. https://doi.org/10.1093/hr/uhac250

Liu, P., Panda, K., Edwards, S. A., Swanson, R., Yi, H., Pandesha, P., Hung, Y.-H., Klaas, G., Ye, X., & Collins, M. V. (2024). Transposase-assisted target-site integration for efficient plant genome engineering. *Nature*, 1–8.

Lopos, L. C., Bykova, N. V., Robinson, J., Brown, S., Ward, K., & Bilichak, A. (2023). Diversity of transgene integration and gene-editing events in wheat (*Triticum aestivum* L.) transgenic plants generated using agrobacterium-mediated transformation. *Frontiers in Genome Editing*, *5*, 1265103.

Lu, S., Sun, Y.-H., & Chiang, V. L. (2008). Stress-responsive microRNAs in Populus. *The Plant Journal*, *55*(1), 131–151. https://doi.org/10.1111/j.1365-313X.2008.03497.x

Mahale, B. M., Fakrudin, B., Ghosh, S., & Krishnaraj, P. U. (2014). LNA mediated in situ hybridization of miR171 and miR397a in leaf and ambient root tissues revealed expressional homogeneity in response to shoot heat shock in *Arabidopsis thaliana*. *Journal of Plant Biochemistry and Biotechnology*, *23*(1), 93–103. https://doi.org/10.1007/s13562-013-0191-0

Maimon, O., & Rokach, L. (2005). *Data Mining and Knowledge Discovery Handbook, Vol. Vol. 2*, Springer.

Mason, R. E., Mondal, S., Beecher, F. W., Pacheco, A., Jampala, B., Ibrahim, A. M. H., & Hays, D. B. (2010). QTL associated with heat susceptibility index in wheat (*Triticum aestivum* L.) under short-term reproductive stage heat stress. *Euphytica*, *174*(3), 423–436. https://doi.org/10.1007/s10681-010-0151-x

Matsukura, S., Mizoi, J., Yoshida, T., Todaka, D., Ito, Y., Maruyama, K., Shinozaki, K., & Yamaguchi-Shinozaki, K. (2010). Comprehensive analysis of rice DREB2-type genes that encode transcription factors involved in the expression of abiotic stress-responsive genes. *Molecular Genetics and Genomics*, *283*(2), 185–196. https://doi.org/10.1007/s00438-009-0506-y

McCaig, B. C., Meagher, R. B., & Dean, J. F. D. (2005). Gene structure and molecular analysis of the laccase-like multicopper oxidase (LMCO) gene family in *Arabidopsis thaliana*. *Planta*, *221*(5), 619–636. https://doi.org/10.1007/s00425-004-1472-6

Meng, X., Liang, Z., Dai, X., Zhang, Y., Mahboub, S., Ngu, D. W., Roston, R. L., & Schnable, J. C. (2021). Predicting transcriptional responses to cold stress across plant species. *Proceedings of the National Academy of Sciences*, *118*(10), e2026330118. https://doi.org/10.1073/pnas.2026330118

Min, L., Li, Y., Hu, Q., Zhu, L., Gao, W., Wu, Y., Ding, Y., Liu, S., Yang, X., & Zhang, X. (2014). Sugar and auxin signaling pathways respond to high-temperature stress during anther development as revealed by transcript profiling analysis in cotton. *Plant Physiology*, *164*(3), 1293–1308. https://doi.org/10.1104/pp.113.232314

Mishra, R., Joshi, R. K., & Zhao, K. (2020). Base editing in crops: current advances, limitations and future implications. *Plant Biotechnology Journal*, *18*(1), 20–31.

Mishra, S., Spaccarotella, K., Gido, J., Samanta, I., & Chowdhary, G. (2023). Effects of heat stress on plant-nutrient relations: an update on nutrient uptake, transport, and assimilation. *International Journal of Molecular Sciences*, *24*(21), 15670.

Murmu, S., Sinha, D., Chaurasia, H., Sharma, S., Das, R., Jha, G. K., & Archak, S. (2024). A review of artificial intelligence-assisted omics techniques in plant defense: current trends and future directions [review]. *Frontiers in Plant Science*, *15*. https://doi.org/10.3389/fpls.2024.1292054

Naydenov, M., Baev, V., Apostolova, E., Gospodinova, N., Sablok, G., Gozmanova, M., & Yahubyan, G. (2015). High-temperature effect on genes engaged in DNA methylation and affected by DNA methylation in Arabidopsis. *Plant Physiology and Biochemistry*, *87*, 102–108. https://doi.org/10.1016/j.plaphy.2014.12.022

Nishimasu, H., Shi, X., Ishiguro, S., Gao, L., Hirano, S., Okazaki, S., Noda, T., Abudayyeh, O. O., Gootenberg, J. S., & Mori, H. (2018). Engineered CRISPR-Cas9 nuclease with expanded targeting space. *Science*, *361*(6408), 1259–1262.

Pajoro, A., Severing, E., Angenent, G. C., & Immink, R. G. H. (2017). Histone H3 lysine 36 methylation affects temperature-induced alternative splicing and flowering in plants. *Genome Biology*, *18*(1), 102. https://doi.org/10.1186/s13059-017-1235-x

Park, Y. J., Lee, H. J., Gil, K. E., Kim, J. Y., Lee, J. H., Lee, H., Cho, H. T., Vu, L. D., De Smet, I., & Park, C. M. (2019). Developmental programming of thermonastic leaf movement. *Plant Physiology*, *180*(2), 1185–1197. https://doi.org/10.1104/pp.19.00139

Perrella, G., Bäurle, I., & van Zanten, M. (2022). Epigenetic regulation of thermomorphogenesis and heat stress tolerance. *New Phytologist*, *234*(4), 1144–1160. https://doi.org/10.1111/nph.17970

Peters, J. E., Makarova, K. S., Shmakov, S., & Koonin, E. V. (2017). Recruitment of CRISPR-Cas systems by Tn7-like transposons. *Proceedings of the National Academy of Sciences*, *114*(35), E7358–E7366.

Popova, O. V., Dinh, H. Q., Aufsatz, W., & Jonak, C. (2013). The RdDM pathway is required for basal heat tolerance in Arabidopsis. *Molecular Plant*, *6*(2), 396–410. https://doi.org/10.1093/mp/sst023

Potters, G., Pasternak, T. P., Guisez, Y., Palme, K. J., & Jansen, M. A. K. (2007). Stress-induced morphogenic responses: growing out of trouble? *Trends in Plant Science*, *12*(3), 98–105. https://doi.org/10.1016/j.tplants.2007.01.004

Pradhan, U. K., Meher, P. K., Naha, S., Rao, A. R., & Gupta, A. (2023a). ASLncR: a novel computational tool for prediction of abiotic stress-responsive long non-coding RNAs in plants. *Functional & Integrative Genomics*, *23*(2), 113. https://doi.org/10.1007/s10142-023-01040-0

Pradhan, U. K., Meher, P. K., Naha, S., Rao, A. R., Kumar, U., Pal, S., & Gupta, A. (2023b). ASmiR: a machine learning framework for prediction of abiotic stress-specific miRNAs in plants. *Functional & Integrative Genomics*, *23*(2), 92. https://doi.org/10.1007/s10142-023-01014-2

Qi, Q., Hu, B., Jiang, W., Wang, Y., Yan, J., Ma, F., Guan, Q., & Xu, J. (2023). Advances in plant epigenome editing research and its application in plants. *International Journal of Molecular Sciences*, *24*(4), 3442. https://www.mdpi.com/1422-0067/24/4/3442

Qin, F., Kakimoto, M., Sakuma, Y., Maruyama, K., Osakabe, Y., Tran, L.-S. P., Shinozaki, K., & Yamaguchi-Shinozaki, K. (2007). Regulation and functional analysis of ZmDREB2A in response to drought and heat stresses in *Zea mays* L. *The Plant Journal*, *50*(1), 54–69. https://doi.org/10.1111/j.1365-313X.2007.03034.x

Qiu, Y., Li, M., Kim, R. J., Moore, C. M., & Chen, M. (2019). Daytime temperature is sensed by phytochrome B in Arabidopsis through a transcriptional activator HEMERA. *Nature Communications*, *10*(1), 140. https://doi.org/10.1038/s41467-018-08059-z

Queitsch, C., Hong, S. W., Vierling, E., & Lindquist, S. (2000). Heat shock protein 101 plays a crucial role in thermotolerance in Arabidopsis. *Plant Cell*, *12*(4), 479–492. https://doi.org/10.1105/tpc.12.4.479

Quint, M., Delker, C., Franklin, K. A., Wigge, P. A., Halliday, K. J., & van Zanten, M. (2016). Molecular and genetic control of plant thermomorphogenesis. *Nature Plants*, *2*, 15190. https://doi.org/10.1038/nplants.2015.190

Ramakrishnan, M., Zhang, Z., Mullasseri, S., Kalendar, R., Ahmad, Z., Sharma, A., Liu, G., Zhou, M., & Wei, Q. (2022). Epigenetic stress memory: a new approach to study cold and heat stress responses in plants. *Frontiers in Plant Science*, *13*, 1075279. https://doi.org/10.3389/fpls.2022.1075279

Rees, H. A., & Liu, D. R. (2018). Base editing: precision chemistry on the genome and transcriptome of living cells. *Nature Reviews Genetics*, *19*(12), 770–788.

Rodriques, S. G., Stickels, R. R., Goeva, A., Martin, C. A., Murray, E., Vanderburg, C. R., Welch, J., Chen, L. M., Chen, F., & Macosko, E. Z. (2019). Slide-seq: a scalable technology for measuring genome-wide expression at high spatial resolution. *Science*, *363*(6434), 1463–1467. https://doi.org/10.1126/science.aaw1219

Rogers, K., & Chen, X. (2013). Biogenesis, turnover, and mode of action of plant microRNAs. *Plant Cell*, *25*(7), 2383–2399. https://doi.org/10.1105/tpc.113.113159

Saeed, S., Usman, B., Shim, S.-H., Khan, S. U., Nizamuddin, S., Saeed, S., Shoaib, Y., Jeon, J.-S., & Jung, K.-H. (2022). CRISPR/Cas-mediated editing of cis-regulatory elements for crop improvement. *Plant Science*, *324*, 111435. https://doi.org/10.1016/j.plantsci.2022.111435

Saidi, Y., Finka, A., Muriset, M., Bromberg, Z., Weiss, Y. G., Maathuis, F. J. M., & Goloubinoff, P. (2009). The heat shock response in Moss plants is regulated by specific calcium-permeable channels in the plasma membrane. *The Plant Cell*, *21*(9), 2829–2843. https://doi.org/10.1105/tpc.108.065318

Schmitz, R. J., Grotewold, E., & Stam, M. (2022). Cis-regulatory sequences in plants: their importance, discovery, and future challenges [review]. *Plant Cell*, *34*(2), 718–741. https://doi.org/10.1093/plcell/koab281

Shen, X., De Jonge, J., Forsberg, S. K., Pettersson, M. E., Sheng, Z., Hennig, L., & Carlborg, Ö. (2014). Natural CMT2 variation is associated with genome-wide methylation changes and temperature seasonality. *PLoS Genetics*, *10*(12), e1004842. https://doi.org/10.1371/journal.pgen.1004842

Shen, Y., Lei, T., Cui, X., Liu, X., Zhou, S., Zheng, Y., Guerard, F., Issakidis-Bourguet, E., & Zhou, D. X. (2019). Arabidopsis histone deacetylase HDA15 directly represses plant response to elevated ambient temperature. *The Plant Journal*, *100*(5), 991–1006. https://doi.org/10.1111/tpj.14492

Shi, J., Gao, H., Wang, H., Lafitte, H. R., Archibald, R. L., Yang, M., Hakimi, S. M., Mo, H., & Habben, J. E. (2017). ARGOS 8 variants generated by CRISPR-Cas9 improve maize grain yield under field drought stress conditions. *Plant Biotechnology Journal*, *15*(2), 207–216.

Shi, W., Yang, J., Kumar, R., Zhang, X., Impa, S. M., Xiao, G., & Jagadish, S. V. K. (2022). Heat stress during gametogenesis irreversibly damages female reproductive organ in rice. *Rice (N Y)*, *15*(1), 32. https://doi.org/10.1186/s12284-022-00578-0

Shikha, M., Kanika, A., Rao, A. R., Mallikarjuna, M. G., Gupta, H. S., & Nepolean, T. (2017). Genomic selection for drought tolerance using genome-wide SNPs in maize. *Frontiers in Plant Science*, *8*, 550. https://doi.org/10.3389/fpls.2017.00550

Shin, H., Choi, W. L., Lim, J. Y., & Huh, J. H. (2022). Epigenome editing: targeted manipulation of epigenetic modifications in plants. *Genes & Genomics*, *44*(3), 307–315. https://doi.org/10.1007/s13258-021-01199-5

Sidaway-Lee, K., Costa, M. J., Rand, D. A., Finkenstadt, B., & Penfield, S. (2014). Direct measurement of transcription rates reveals multiple mechanisms for configuration of the Arabidopsis ambient temperature response. *Genome Biology*, *15*(3), R45. https://doi.org/10.1186/gb-2014-15-3-r45

Simna, S. P., & Han, Z. (2022). Prospects of non-coding elements in genomic DNA based gene therapy. *Current Gene Therapy*, *22*(2), 89–103. https://doi.org/10.2174/1566523221666210419090357

Singer, S. D., & Cox, K. D. (2013). A gypsy-like sequence from *Arabidopsis thaliana* exhibits enhancer-blocking activity in transgenic plants. *Journal of Plant Biochemistry and Biotechnology*, *22*(1), 35–42. https://doi.org/10.1007/s13562-012-0108-3

Singer, S. D., Liu, Z., & Cox, K. D. (2012). Minimizing the unpredictability of transgene expression in plants: the role of genetic insulators. *Plant Cell Reports*, *31*(1), 13–25. https://doi.org/10.1007/s00299-011-1167-y

Song, X., Meng, X., Guo, H., Cheng, Q., Jing, Y., Chen, M., Liu, G., Wang, B., Wang, Y., Li, J., & Yu, H. (2022). Targeting a gene regulatory element enhances rice grain yield by decoupling panicle number and size. *Nature Biotechnology*, *40*(9), 1403–1411. https://doi.org/10.1038/s41587-022-01281-7

Song, X., Guo, P., Xia, K., Wang, M., Liu, Y., Chen, L., Zhang, J., Xu, M., Liu, N., Yue, Z., Xu, X., Gu, Y., Li, G., Liu, M., Fang, L., Deng, X. W., & Li, B. (2023). Spatial transcriptomics reveals light-induced chlorenchyma cells involved in promoting shoot regeneration in tomato callus. *Proceedings of the National Academy of Sciences*, *120*(38), e2310163120. https://doi.org/10.1073/pnas.2310163120

Stahl, P. L., Salmen, F., Vickovic, S., Lundmark, A., Navarro, J. F., Magnusson, J., Giacomello, S., Asp, M., Westholm, J. O., Huss, M., Mollbrink, A., Linnarsson, S., Codeluppi, S., Borg, A., Ponten, F., Costea, P. I., Sahlen, P., Mulder, J., Bergmann, O., et al. (2016). Visualization and analysis of gene expression in tissue sections by spatial transcriptomics. *Science*, *353*(6294), 78–82. https://doi.org/10.1126/science.aaf2403

Stickels, R. R., Murray, E., Kumar, P., Li, J., Marshall, J. L., Di Bella, D. J., Arlotta, P., Macosko, E. Z., & Chen, F. (2021). Highly sensitive spatial transcriptomics at near-cellular resolution with Slide-seqV2. *Nature Biotechnology*, *39*(3), 313–319.

Sun, Q., & Quick, J. (1991). Chromosomal locations of genes for heat tolerance in tetraploid wheat. *Cereal Research Communications*, 431–437.

Sun, C., Lei, Y., Li, B., Gao, Q., Li, Y., Cao, W., Yang, C., Li, H., Wang, Z., & Li, Y. (2024). Precise integration of large DNA sequences in plant genomes using PrimeRoot editors. *Nature Biotechnology*, *42*(2), 316–327.

Sunkar, R., & Zhu, J.-K. (2007). Micro RNAs and short-interfering RNAs in plants. *Journal of Integrative Plant Biology*, *49*(6), 817–826. https://doi.org/10.1111/j.1744-7909.2007.00499.x

Swinnen, G., Goossens, A., & Pauwels, L. (2016). Lessons from domestication: targeting cis-regulatory elements for crop improvement. *Trends in Plant Science*, *21*(6), 506–515. https://doi.org/10.1016/j.tplants.2016.01.014

Syed, N. H., Kalyna, M., Marquez, Y., Barta, A., & Brown, J. W. (2012). Alternative splicing in plants – coming of age. *Trends in Plant Science*, *17*(10), 616–623. https://doi.org/10.1016/j.tplants.2012.06.001

Talarico, E., Zambelli, A., Araniti, F., Greco, E., Chiappetta, A., & Bruno, L. (2024). Unravelling the epigenetic code: DNA methylation in plants and its role in stress response. *Epigenomes*, *8*(3). https://doi.org/10.3390/epigenomes8030030

Tognacca, R. S., Rodríguez, F. S., Aballay, F. E., Cartagena, C. M., Servi, L., & Petrillo, E. (2022). Alternative splicing in plants: current knowledge and future directions for assessing the biological relevance of splice variants. *Journal of Experimental Botany*, *74*(7), 2251–2272. https://doi.org/10.1093/jxb/erac431

Trapero-Mozos, A., Morris, W. L., Ducreux, L. J., McLean, K., Stephens, J., Torrance, L., Bryan, G. J., Hancock, R. D., & Taylor, M. A. (2018). Engineering heat tolerance in potato by temperature-dependent expression of a specific allele of HEAT-SHOCK COGNATE 70. *Plant Biotechnology Journal*, *16*(1), 197–207.

Ueda, M., & Seki, M. (2020). Histone modifications form epigenetic regulatory networks to regulate abiotic stress response. *Plant Physiology*, *182*(1), 15–26. https://doi.org/10.1104/pp.19.00988

Vandereyken, K., Sifrim, A., Thienpont, B., & Voet, T. (2023). Methods and applications for single-cell and spatial multi-omics. *Nature Reviews Genetics*, *24*(8), 494–515. https://doi.org/10.1038/s41576-023-00580-2

Verhage, L., Angenent, G. C., & Immink, R. G. (2014). Research on floral timing by ambient temperature comes into blossom. *Trends in Plant Science*, *19*(9), 583–591. https://doi.org/10.1016/j.tplants.2014.03.009

Verkuijl, S. A. N., & Rots, M. G. (2019). The influence of eukaryotic chromatin state on CRISPR-Cas9 editing efficiencies. *Current Opinion in Biotechnology*, *55*, 68–73. https://doi.org/10.1016/j.copbio.2018.07.005

Vidal, E. A., Araus, V., Lu, C., Parry, G., Green, P. J., Coruzzi, G. M., & Gutiérrez, R. A. (2010). Nitrate-responsive miR393/*AFB3* regulatory module controls root system architecture in *Arabidopsis thaliana*. *Proceedings of the National Academy of Sciences*, *107*(9), 4477–4482. https://doi.org/10.1073/pnas.0909571107

Vierstra, R. D. (2009). The ubiquitin-26S proteasome system at the nexus of plant biology. *Nature Reviews Molecular Cell Biology*, *10*(6), 385–397. https://doi.org/10.1038/nrm2688

Voinnet, O. (2009). Origin, biogenesis, and activity of plant microRNAs. *Cell*, *136*(4), 669–687. https://doi.org/10.1016/j.cell.2009.01.046

Vu, L. D., Xu, X., Zhu, T., Pan, L., van Zanten, M., de Jong, D., Wang, Y., Vanremoortele, T., Locke, A. M., van de Cotte, B., De Winne, N., Stes, E., Russinova, E., De Jaeger, G., Van Damme, D., Uauy, C., Gevaert, K., & De Smet, I. (2021). The membrane-localized protein kinase MAP4K4/TOT3 regulates thermomorphogenesis. *Nature Communications*, *12*(1), 2842. https://doi.org/10.1038/s41467-021-23112-0

Walton, R. T., Christie, K. A., Whittaker, M. N., & Kleinstiver, B. P. (2020). Unconstrained genome targeting with near-PAMless engineered CRISPR-Cas9 variants. *Science*, *368*(6488), 290–296.

Wang, Q.-L., Chen, J.-H., He, N.-Y., & Guo, F.-Q. (2018). Metabolic reprogramming in chloroplasts under heat stress in plants. *International Journal of Molecular Sciences*, *19*(3), 849. https://www.mdpi.com/1422-0067/19/3/849

Waters, A. J., Makarevitch, I., Noshay, J., Burghardt, L. T., Hirsch, C. N., Hirsch, C. D., & Springer, N. M. (2017). Natural variation for gene expression responses to abiotic stress in maize. *The Plant Journal*, *89*(4), 706–717.

Wierzbicki, A. T. (2012). The role of long non-coding RNA in transcriptional gene silencing. *Current Opinion in Plant Biology*, *15*(5), 517–522. https://doi.org/10.1016/j.pbi.2012.08.008

van der Woude, L. C., Perrella, G., Snoek, B. L., van Hoogdalem, M., Novak, O., van Verk, M. C., van Kooten, H. N., Zorn, L. E., Tonckens, R., Dongus, J. A., Praat, M., Stouten, E. A., Proveniers, M. C. G., Vellutini, E., Patitaki, E., Shapulatov, U., Kohlen, W., Balasubramanian, S., Ljung, K., et al. (2019). Histone deacetylase 9 stimulates auxin-dependent thermomorphogenesis in *Arabidopsis thaliana* by mediating H2A.Z depletion. *Proceedings of the National Academy of Sciences of the United States of America*, *116*(50), 25343–25354. https://doi.org/10.1073/pnas.1911694116

Wu, C. (1995). Heat shock transcription factors: structure and regulation. *Annual Review of Cell and Developmental Biology*, *11*, 441–469. https://doi.org/10.1146/annurev.cb.11.110195.002301

Wu, M.-F., Tian, Q., & Reed, J. W. (2006). Arabidopsis microRNA167 controls patterns of ARF6 and ARF8 expression, and regulates both female and male reproduction. *Development*, *133*(21), 4211–4218. https://doi.org/10.1242/dev.02602

Wu, N., Yao, Y., Xiang, D., Du, H., Geng, Z., Yang, W., Li, X., Xie, T., Dong, F., & Xiong, L. (2022). A MITE variation-associated heat-inducible isoform of a heat-shock factor confers heat tolerance through regulation of *JASMONATE ZIM-DOMAIN* genes in rice. *New Phytologist*, *234*(4), 1315–1331. https://doi.org/10.1111/nph.18068

Xin, M., Wang, Y., Yao, Y., Xie, C., Peng, H., Ni, Z., & Sun, Q. (2010). Diverse set of microRNAs are responsive to powdery mildew infection and heat stress in wheat (*Triticum aestivum* L.). *BMC Plant Biology*, *10*(1), 123. https://doi.org/10.1186/1471-2229-10-123

Xin, M., Wang, Y., Yao, Y., Song, N., Hu, Z., Qin, D., Xie, C., Peng, H., Ni, Z., & Sun, Q. (2011). Identification and characterization of wheat long non-protein coding RNAs responsive to powdery mildew infection and heat stress by using microarray analysis and SBS sequencing. *BMC Plant Biology*, *11*(1), 61. https://doi.org/10.1186/1471-2229-11-61

Xu, Z., Kuang, Y., Ren, B., Yan, D., Yan, F., Spetz, C., Sun, W., Wang, G., Zhou, X., & Zhou, H. (2021). SpRY greatly expands the genome editing scope in rice with highly flexible PAM recognition. *Genome Biology*, *22*(1), 6. https://doi.org/10.1186/s13059-020-02231-9

Xue, G.-P., Drenth, J., & McIntyre, C. L. (2014). TaHsfA6f is a transcriptional activator that regulates a suite of heat stress protection genes in wheat (*Triticum aestivum* L.) including previously unknown Hsf targets. *Journal of Experimental Botany*, *66*(3), 1025–1039. https://doi.org/10.1093/jxb/eru462

Xue, M., Zhang, H., Zhao, F., Zhao, T., Li, H., & Jiang, D. (2021). The INO80 chromatin remodeling complex promotes thermomorphogenesis by connecting H2A.Z eviction and active transcription in Arabidopsis. *Molecular Plant*, *14*(11), 1799–1813. https://doi.org/10.1016/j.molp.2021.07.001

Yao, Y., Ni, Z., Peng, H., Sun, F., Xin, M., Sunkar, R., Zhu, J. K., & Sun, Q. (2010). Non-coding small RNAs responsive to abiotic stress in wheat (*Triticum aestivum* L.). *Functional & Integrative Genomics*, *10*(2), 187–190. https://doi.org/10.1007/s10142-010-0163-6

Yeh, C. H., Kaplinsky, N. J., Hu, C., & Charng, Y. Y. (2012). Some like it hot, some like it warm: phenotyping to explore thermotolerance diversity. *Plant Science*, *195*, 10–23. https://doi.org/10.1016/j.plantsci.2012.06.004

Yin, R., Chen, R., Xia, K., & Xu, X. (2024). A single-cell transcriptome atlas reveals the trajectory of early cell fate transition during callus induction in Arabidopsis. *Plant Communications*, *5*(8). https://doi.org/10.1016/j.xplc.2024.100941

Yoshida, T., Ohama, N., Nakajima, J., Kidokoro, S., Mizoi, J., Nakashima, K., Maruyama, K., Kim, J. M., Seki, M., Todaka, D., Osakabe, Y., Sakuma, Y., Schöffl, F., Shinozaki, K., & Yamaguchi-Shinozaki, K. (2011). Arabidopsis HsfA1 transcription factors function as the main positive regulators in heat shock-responsive gene expression. *Molecular Genetics and Genomics*, *286*(5–6), 321–332. https://doi.org/10.1007/s00438-011-0647-7

Zhang, Y. M., Zheng, Y. M., Xiao, N., Wang, L. N., Zhang, Y., Fang, R. X., & Chen, X. Y. (2012). Functional analysis of the HS185 regulatory element in the rice HSP70 promoter. *Molecular Biology Reports*, *39*(2), 1649–1657. https://doi.org/10.1007/s11033-011-0904-1

Zhang, T.-Q., Chen, Y., Liu, Y., Lin, W.-H., & Wang, J.-W. (2021). Single-cell transcriptome atlas and chromatin accessibility landscape reveal differentiation trajectories in the rice root. *Nature Communications*, *12*(1), 2053. https://doi.org/10.1038/s41467-021-22352-4

Zhang, X., Luo, Z., Marand, A. P., Yan, H., Jang, H., Bang, S., Mendieta, J. P., Minow, M. A. A., & Schmitz, R. J. (2024a). A spatially resolved multiomic single-cell atlas of soybean development. *bioRxiv*. 2024.2007.2003.601616. https://doi.org/10.1101/2024.07.03.601616

Zhang, X., Marand, A. P., Yan, H., & Schmitz, R. J. (2024b). Scifi-ATAC-seq: massive-scale single-cell chromatin accessibility sequencing using combinatorial fluidic indexing. *Genome Biology*, *25*(1), 90. https://doi.org/10.1186/s13059-024-03235-5

Zhao, C., Liu, B., Piao, S., Wang, X., Lobell, D. B., Huang, Y., Huang, M., Yao, Y., Bassu, S., Ciais, P., Durand, J.-L., Elliott, J., Ewert, F., Janssens, I. A., Li, T., Lin, E., Liu, Q., Martre, P., Müller, C., et al. (2017). Temperature increase reduces global yields of major crops in four independent estimates. *Proceedings of the National Academy of Sciences*, *114*(35), 9326–9331. https://doi.org/10.1073/pnas.1701762114

Zhao, J., Lu, Z., Wang, L., & Jin, B. (2020). Plant responses to heat stress: physiology, transcription, noncoding RNAs, and epigenetics. *International Journal of Molecular Sciences*, *22*(1). https://doi.org/10.3390/ijms22010117

Zhiyong, N., Na, L., Yuehua, Y., Chenxi, B., Quanjia, C., & Yanying, Q. (2021). The cotton 70-kDa heat shock protein GhHSP70-26 plays a positive role in the drought stress response. *Environmental and Experimental Botany*, *191*, 104628. https://doi.org/10.1016/j.envexpbot.2021.104628

Zhou, P., Enders, T. A., Myers, Z. A., Magnusson, E., Crisp, P. A., Noshay, J. M., Gomez-Cano, F., Liang, Z., Grotewold, E., Greenham, K., & Springer, N. M. (2021). Prediction of conserved and variable heat and cold stress response in maize using cis-regulatory information. *The Plant Cell*, *34*(1), 514–534. https://doi.org/10.1093/plcell/koab267

Zhou, J., Liu, G., Zhao, Y., Zhang, R., Tang, X., Li, L., Jia, X., Guo, Y., Wu, Y., Han, Y., Bao, Y., He, Y., Han, Q., Yang, H., Zheng, X., Qi, Y., Zhang, T., & Zhang, Y. (2023). An efficient CRISPR-Cas12a promoter editing system for crop improvement. *Nature Plants*, *9*(4), 588–604. https://doi.org/10.1038/s41477-023-01384-2

Zhu, C., Ding, Y., & Liu, H. (2011). MiR398 and plant stress responses. *Physiologia Plantarum*, *143*(1), 1–9. https://doi.org/10.1111/j.1399-3054.2011.01477.x

Zhu, X., Liu, S., Meng, C., Qin, L., Kong, L., & Xia, G. (2013). WRKY transcription factors in wheat and their induction by biotic and abiotic stress. *Plant Molecular Biology Reporter*, *31*(5), 1053–1067. https://doi.org/10.1007/s11105-013-0565-4

Zinn, K. E., Tunc-Ozdemir, M., & Harper, J. F. (2010). Temperature stress and plant sexual reproduction: uncovering the weakest links. *Journal of Experimental Botany*, *61*(7), 1959–1968.

Zong, Y., Liu, Y., Xue, C., Li, B., Li, X., Wang, Y., Li, J., Liu, G., Huang, X., Cao, X., & Gao, C. (2022). An engineered prime editor with enhanced editing efficiency in plants. *Nature Biotechnology*, *40*(9), 1394–1402. https://doi.org/10.1038/s41587-022-01254-w

Zormpas, E., Queen, R., Comber, A., & Cockell, S. J. (2023). Mapping the transcriptome: realizing the full potential of spatial data analysis. *Cell*, *186*(26), 5677–5689. https://doi.org/10.1016/j.cell.2023.11.003

2

Design Future Crops with Stress Resilience by CRISPR/Cas Reprogramming Noncoding RNAs

Hidam Bishworjit Singh[1], Niraj Kumar[2,3], Riwandahun Marwein[2,3], Ajay Kumar Keot[2,3], Manash Pratim Sarmah[4], and Dhanawantari L. Singha[5]

[1] *Department of Biotechnology, Gauhati University, Guwahati, Assam, India*
[2] *Biological Sciences and Technology Division, CSIR-North East Institute of Science and Technology (NEIST), Jorhat, Assam, India*
[3] *Academy of Scientific and Innovative Research (AcSIR), Ghaziabad, India*
[4] *Department of Zoology, University of Science & Technology, Baridua, Meghalaya, India*
[5] *Rabindranath Tagore University, Hojai, Assam, India*

2.1 Introduction

RNA stands for ribose nucleic acid, a class of nucleic acid that is chemically similar to another class of nucleic acid deoxyribose nucleic acid (DNA) but has less stability than DNA. They are formed by DNA-dependent RNA polymerase from DNA (Minchin & Lodge, 2019). There are mainly three types of RNA viz. mRNA (transcribed from DNA and carries information to make protein), tRNA (carries amino acid to the ribosome during protein synthesis), and rRNA (site for protein synthesis). However, there are other groups of noncoding RNA (ncRNA) such as microRNA (miRNA), small nuclear RNA (snRNA), and small interfering RNA (siRNA), which have been known for their involvement in the regulation of various cellular processes (Wang & Farhana, 2024).

ncRNAs are a diverse group of RNA transcribed by the action of RNA polymerase having a size range of 20–200 long nucleotides but they do not code protein. ncRNA have been classified into housekeeping ncRNA (constitutive or regular ncRNA) and regulatory ncRNA. Housekeeping ncRNAs are constitutively expressed within the cells and they include ribosomal RNAs (rRNAs), transfer RNAs (tRNAs), snRNAs, small nucleolar RNAs (snoRNAs), and small Cajal body-specific RNAs (scaRNAs). However, regulatory ncRNA can be divided into two classes based on the length of nucleotides. They are of long noncoding RNAs (lncRNAs) (>200 nt) and small noncoding RNAs (sncRNAs) (19–25 nts). Among them, lncRNAs include *cis*-natural antisense (cis-NATs), *trans*-natural antisense transcripts (*trans*-NATs), and pseudogenes, sense or antisense (based on the origin of strand), divergent or convergent (based on the orientation of transcription), and as intronic or intergenic (based on the location on chromosome) (Waititu et al., 2020; Yadav et al., 2024). This class of regulatory ncRNAs regulates the expression of genes either by interacting with transcriptional repressors or activators to repress or activate gene regulatory pathways. Further, lncRNAs interact with splicing complexes or factors to control the RNA splicing process and also regulate gene expression patterns by regulating gene promoters by recruiting chromatin remodeling complex, etc. (Waititu et al., 2020), whereas sRNA includes piRNAs (piwi-interacting RNAs; 24–32 nts),

Genome and Epigenome Editing for Stress-Tolerant Crops, First Edition. Edited by Jen-Tsung Chen.

miRNAs (19–25 nts), and siRNAs (20–24 nts) (Waititu et al., 2020; Yadav et al., 2024). miRNA inhibits the particular transcript expression either by cleaving the transcript or by inhibiting translation by binding to the 5′ untranslated region of the transcript to inhibit ribosome recruitment, whereas siRNA works by inducing posttranscriptional gene silencing and also promotes gene silencing through DNA methylations (Yadav et al., 2024).

Plants are sessile organisms that are affected by various biotic and abiotic stresses including nutrient deficiency, drought, pathogen infection, UV-B radiation, cold, mechanical stress, and salinity stress. In this situation, the plant regulates the expression of miRNA to tackle the stresses (Khraiwesh et al., 2012). During the stresses, miRNA can transmit from cell to cell distantly and act as transmitters and communicate between cells and the environment for the growth metabolism and development of plants. The exogenous miRNAs, miR156 and miR399, and root-derived Nb-miR164 have been shown to have mobile activity to act their functions (Ma et al., 2022). During biotic stress or pathogenic attack, miRNAs also help in the regulation of plant immune responses. For example, miRNA regulates the expression of the R gene by cleaving it to reduce the expression of the R protein, which recognizes the effector protein secreted by pathogens (Song et al., 2021).

Similarly, siRNAs also help the plants resist against environmental stresses. For example, 24-nt siRNA corresponding to the *SRO5* gene targets the *P5CDH* gene and downregulated its expression leading to the accumulation of proline during salt stress (Basel et al., 2012). Similarly, during biotic stress, a particular type of siRNA, *nat-siRNAATGB2* repress the expression pentatricopeptide repeats protein-like (PPRL) gene, a negative regulator of RPS2 resistance pathway (Katiyar-Agarwal et al., 2006).

lncRNAs are also involved in the management of stresses in plants. Several differentially expressed biotic and abiotic stress-responsive plant lncRNAs were identified in many plant species (Yu et al., 2019). For example, the elevated expression of lncRNA DRIR induces tolerance to drought and salt stresses (Qin et al., 2017). Further, during abiotic stress like *Phytophthora infestans* infection lncRNA16397 of tomato has been reported to function in cis to induce expression of GLUTAREDOXIN 22, thereby leading to enhanced pathogen resistance (Chen et al., 2017).

Climate change can be driven by both natural dynamics such as solar cycles or volcanic eruptions and human activities. In the past two centuries, deforestation along with continuous emission of greenhouse gases from agricultural practices mainly driven climate changes globally. These changes in climate trigger a series of environmental stresses in developing countries thereby leading to plant productivity, health, and nutritional quality, reducing global food security. But plant science community developed a lot of practices for developing climate-resilient crops to increase crop diversity and productivity through genetic improvement and development of new crops (Gao & Cui, 2023). Further, climate change influences various physiological processes in plants at the biochemical level, thereby altering the composition and concentration of metabolites within plants (Terán et al., 2024). Besides, outbreaks of pathogens due to climate changes that are favorable for pathogens to vulnerable hosts are also reported, thereby leading to affecting agricultural productivity. Further, changes in climate also alter the plant–microbe interaction, which is crucial for the proper development and growth of host plants (Summanwar et al., 2020; Singh et al., 2023). Overall, climate change affects crop productivity and biodiversity, which are crucial for sustainable crop production (Torquebiau et al., 2016).

Although the development of high-yielding cultivars tolerant to environmental stresses has been achieved through plant breeding techniques and genome modifications, both of the techniques have some limitations. Traditional breeding techniques still remain a relatively slow process. Further, through the application of genetic engineering techniques, the time required for the development of new is greatly reduced but this technique cannot completely eliminate the need for a plant breeding approach (Summanwar et al., 2020). But recently, genome-editing techniques through the application of sequence-specific nucleases like zinc-finger nucleases (ZFNs), transcription activator-like effector nucleases (TALENs), and clustered regularly interspersed short palindromic repeats/CRISPR-associated protein 9 (CRISPRs/Cas9) have been applied in the generation of disease-resistant crops (Summanwar et al., 2020). Emerging evidence on the identification of the role of ncRNAs as an important role in various molecular processes suggested that these ncRNAs are important regulators in various developmental and cellular processes in the process of plant immunity and adaptation to environmental stresses (Summanwar et al., 2020).

2.2 Noncoding RNAs Associated with Drought in Plants

Drought is one of the most significant abiotic stresses for crops. It can severely reduce crop yield and thus create challenges for global food security. Water deficiency in fields creates complex molecular changes in crop plants, which modulates end metabolites and results in phenotypic change. In response to low water conditions, plants reprogram their molecular dynamics to overcome the stress. In plants, drought-responsive ncRNAs are characterized in different crop plants, which are reported to play key roles in drought adaptation (Gelaw & Sanan-Mishra, 2021). Additionally, a ncRNA can function negatively in drought response like the miRNA, miR1916 functions as a negative regulator of drought in tomato and tobacco (Chen et al. 2019a), This category of ncRNA genes in crops could be knocked out by the CRISPR/Cas gene-editing method to develop drought-resistant crops.

2.3 Noncoding RNA Associated with Salt Stress Tolerance in Plant

Salinization condition in crop fields is a major challenge globally and reduce agricultural yield, and the area affected by salinity is projected to increase every year. Higher salt in soil can adversely impose toxicity on plants by developing conditions like oxidative and hyperosmotic stress in plant cells leading to death (Das et al., 2023). Plants respond to salt stress by reprogramming gene transcription and metabolism as adaptive machinery. The ncRNA genes are one major survival strategy and play a decisive role in plants' ability to minimize salt stress-induced conditions (Zhang et al. 2022a, b). Utilizing modern gene-editing tools based on CRISPR/Cas, these ncRNA could be targeted to develop salt stress-resilient future crops. Importantly, the role of lncRNA in salt stress is studied in different crop plants and emerged as a key player in plant salinity stress response. They can fine-tune the expression profile of the protein-coding genes of the salt stress pathway (Das et al., 2023). For example, in rice, the miRNA, OsMIR528, was found to positively regulate salt stress, whereas in cotton a lncRNA, i.e. lncRNA973, mediates salt stress response (Zhou et al., 2017; Zhang et al., 2019). In soybean plants, the long noncoding lncRNA77580-mediated salt stress response (Niu et al., 2021). These ncRNA could be modified by base change or enhanced expression by CRISPR tools, CRISPR activators, or CRISPR base editors. Additionally, their target sites could also be modified by CRISPR tools. This reprogramming of ncRNA in crop plants will lead to the development of salt stress-resilient varieties.

2.4 Important to Edit Noncoding RNA by CRISPR/Cas to Reprogram Them

CRISPR/Cas has evolved as a powerful gene-editing method in plants as it can efficiently and precisely target theoretically any locus in plant genomic DNA. In recent years, using CRISPR/Cas technology, the ncRNA (miRNA) genes have been edited in plants including rice, wheat, and tomato (Deng et al., 2022). The general pipeline for the CRISPR/Cas gene editing of ncRNA genes in plants is represented in Figure 2.1. Plant ncRNA plays a significant role in regulating genes and in controlling many biological processes in plants. Many pathway studies in crops including the major cereal plant and rice unveil the regulatory role of ncRNA. These pathways could contribute to agronomic traits like yield enhancement, height control, or nutrient uptake in rice. For example, in rice, the noncoding miRNA, miR156 targets *OsSPL14* (Squamosa Promoter Binding Protein Like) to control its expression, which ultimately affects plant architecture and yield (Kaur et al., 2020). Similarly, noncoding miRNA (miR319) determines wheat plant architecture (height,

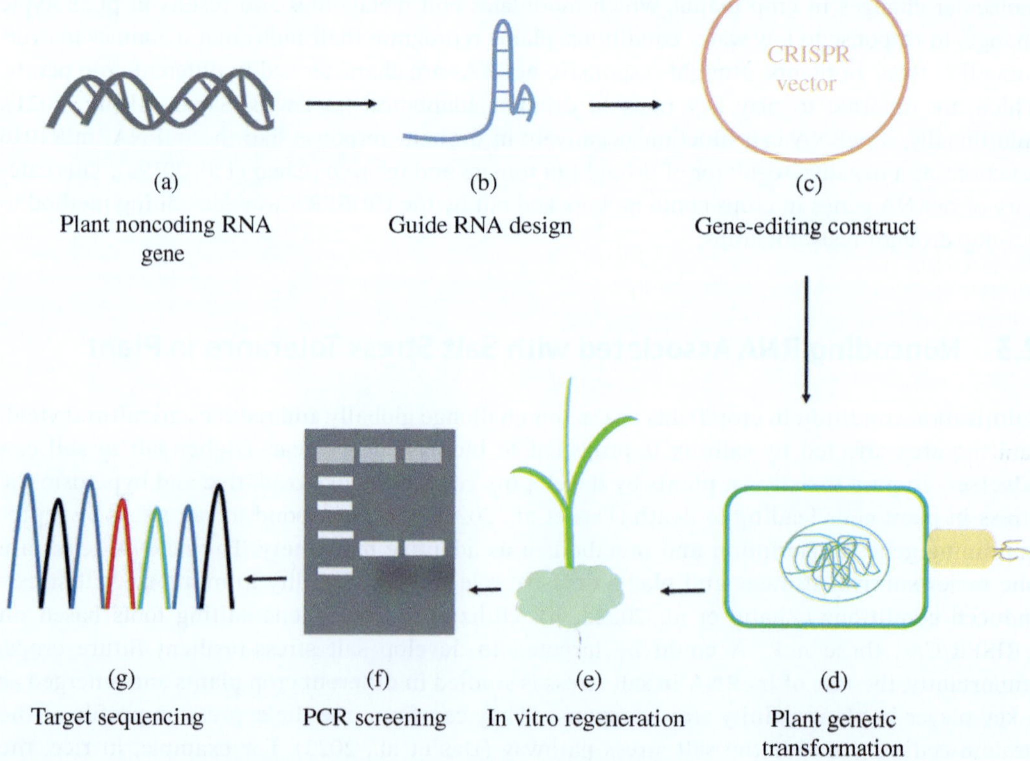

Figure 2.1 Pipeline for CRISPR/Cas gene editing of noncoding RNA genes in plants. (a) In the very first step, the noncoding RNA gene is selected to edit. (b) Guide RNA is designed to target the selected gene using guide RNA design tools. (c) CRISPR plasmid containing Cas enzyme, guide RNA, and plant selectable marker is created. (d) *Agrobacterium*-mediated plant transformation to deliver gene-editing components into plant cells (other methods of gene transfer like particle bombardment could also be used). (e) Selection of plant cells (tissue) containing gene-editing cassette through antibiotic selection method containing. (f) Molecular confirmation of transformation of *Cas* and *guide RNA* through PCR and (g) Sanger sequencing of the guide RNA target region for editing confirmation.

spikes, culm thickness, tiller number, and flag leaf) by regulating its target gene *TaGAMYB3* (Jian et al., 2022). In another example, a rice miRNA namely miR396 represses *GROWTH-REGULATING FACTORs* (*OsGRFs*), *OsGRF4*, and *OsGRF6* (Zhang et al., 2020). This miRNA-mediated control of transcription results in an increase in grain size and branching of panicles. Furthermore, in tomato, the miRNA miR172 precisely modulates the transcription of *SlAP2a* transcription factor gene to affect plant development and the ripening process (Chung et al., 2020). The role of lncRNAs in plant immunity is being studied in plants like *Arabidopsis*, wheat, rice, tomato, and maize. For example, cold-induced lncRNAs *COOLAIR*, *COLDWARP*, and *COLDAIR* silence transcription of *FLOWERING LOCUS C (FLC)* in *Arabidopsis*. Tomato lncRNAs, *lncRNA16397*, and *lncRNA3373* are found to regulate defense response against *P. infestans* infection (Huang et al., 2023). These demonstrate regulatory mechanisms of ncRNA in plants and build the platform for precisely manipulating them by utilizing modern gene-editing technology, CRISPR/Cas. The different tools developed based on CRISPR such as CRISPR base editor, prime editors, CRISPR activator, or CRISPR interference could be used to edit the ncRNA genes. Knowing the accurate functions of plant ncRNA genes offers opportunities to manipulate them for enhancement of yield, nutritional biofortification, nutrient uptake efficiencies, etc. Another reason that demands editing of ncRNA is to assess their function or their target that affects plant growth and development pathway. This necessitates knockout, knockdown, or base modification studies. The most precise and simple methods of gene editing, i.e. CRISPR/Cas, remain fundamental to achieving the goal of ncRNA manipulation.

2.5 Gene-Editing Mechanism in Plants with CRISPR/Cas Technology

CRISPR/Cas genome-editing tool has been extensively exploited for translational research in plants for developing plants of superior agronomic traits. The tool is advantageous over conventional breeding methods because of its simplicity, feasibility, inexpensive, speedy outcome, and most importantly its precision in target modifications (Khatodia et al., 2016). The molecular mechanism of CRISPR/Cas genome editing is represented in Figure 2.2. The function of the system depends on two components, CRISPR-associated Cas protein and a single-guide RNA (sgRNA) entity. The system works through an RNA-guided endonuclease target recognition and cleavage originated from bacterial adaptive immunity against invading bacteriophages. CRISPR locus is a sequence of arrays that consist of direct repeat palindromic sequence interspaced by a similar size nonrepeating spacer sequence. Adjacent to these arrays are multiple conserved gene sequences called Cas genes (Nussenzweig & Marraffini, 2020). The molecular mechanism of CRISPR/Cas immunity is divided into three stages, viz., adaption or spacer integration, crRNA maturation, and interference. Subsequent to foreign DNA invasion, the adaptation stage begins in which the foreign genetic material is recognized, cut into smaller pieces termed spacer or protospacer, and integrated into the host CRISPR array, which can serve as a memory sequence for future infections. The selection of the protospacer depends on a short protospacer adjacent motif (PAM) sequence located adjacent to the protospacer and is unique for every CRISPR/Cas system. For example, Cas9 recognizes NGG PAM downstream of the protospacer. The spacer sequence is integrated between the AT-rich leader and the first CRISPR repeat in the CRISPR array. During the maturation stage, the CRISPR array is transcribed into pre-crRNA transcript, which is further processed into mature small RNA called CRISPR RNA or crRNA. In the interference stage, mature crRNAs are used as guides to scan for invading DNA with the help of the Cas protein to degrade foreign genetic material to provide immunity (Barrangou & Marraffini, 2014; Hille & Charpentier, 2016). On the basis of the Cas effector

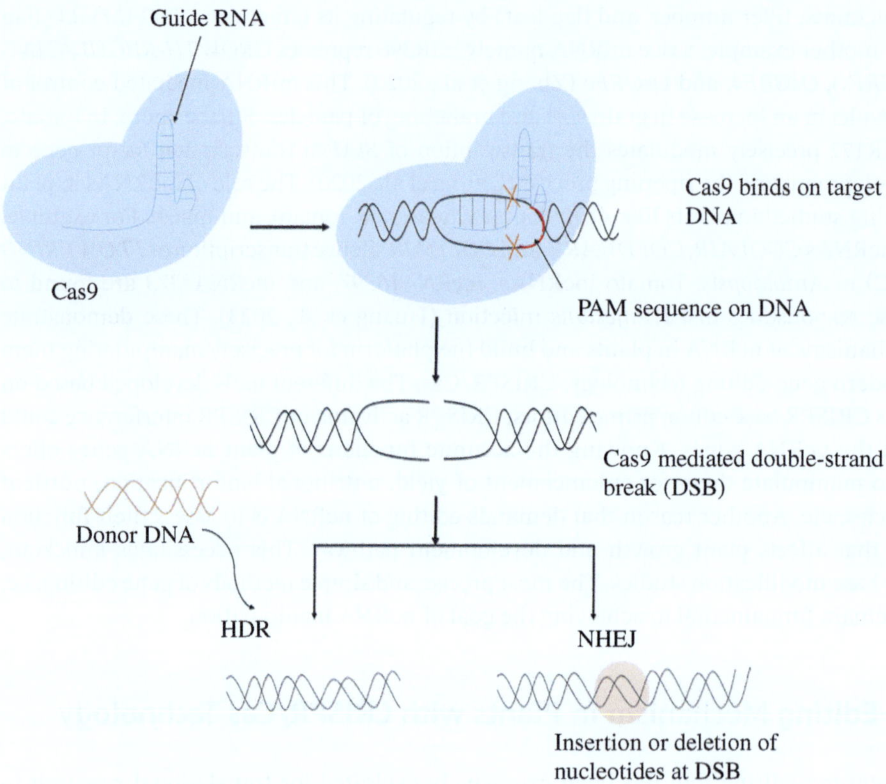

Figure 2.2 CRISPR/Cas9 gene-editing mechanism.

protein interference module, the system is divided into two classes, Class 1 and Class 2, which are further divided into six types and nineteen subtypes on the basis of signature proteins. Types I, III, and IV are categorized within the Class 1 system, whereas Class 2 consists of Types II, V, and VI utilizing only one effector module. The major difference between the two classes is that in Class 1 the effector module which is responsible for target sequence identification and nuclease cleavage activity consists of multiple Cas proteins complex termed the CRISPR-associated complex for antiviral defense (Cascade). In Class 2, however, a single large multi-functional Cas protein is involved in target recognition, pre-crRNA processing, and interference (Mosterd et al., 2021). Cas3, Cas10, and Cas8 are the signature proteins of Types I, III, and IV, respectively, whereas Cas9, Cas12, and Cas13 belong to respective Types II, V, and VI. Cas1 and Cas2 are universally present in all the types. Identified from several microorganisms such as *Streptococcus pyogenes* (SpCas9), *S. aureus* (SaCas9), *S. thermophilus* (StCas9), and *Franscisella novicida* (FnCas9), Type V Cas9 protein is widely used for gene editing in various systems including plant and thus have a well-studied protein structure. It is bilobed consisting of two structural lobes, REC, recognition lobe, and NUC, nuclease lobe. HNH and RuvC are the two catalytic domains within the NUC lobe. REC lobe functions in identifying the target DNA region. After recognition, the HNH domain is responsible for cutting of sgRNA complementary strand, whereas the RuvC domain cleaves the opposite nontarget strand (Mosterd et al., 2021; Allemailem et al., 2022).

Cas9 enzyme binds to sgRNA forming a complex. This complex loads on the double-stranded DNA molecule and moves and upon recognizing a guide RNA complementary target adjacent to

PAM (5′ NGG 3′), Cas9 unwinds the DNA and with its nuclease activity induces double-strand break (DSB). The DSB activates the DNA repair mechanism. DNA break can be repaired by either homology-directed repair (HDR) or nonhomologous end joining (NHEJ) pathway. In the presence of a donor DNA (template for homology-mediated repair), precise base modification or targeted gene insertion can be achieved. Whereas in the case of NHEJ, random insertion or deletion (Indels) occurs at the site of the break, this mutation often brings frameshift mutation in the protein-coding gene.

2.6 Functions of Different Cas Proteins

Cas1 and Cas2 complexes are universal Cas proteins required for spacer acquisition and endonuclease activity, but only the catalytic activity of Cas1 is essential for spacer binding and integration. In addition to Cas1–Cas2, type II A also required Csn2, Cas9, and *trans*-activating CRISPR RNA (tracrRNA) for spacer acquisition (McGinn & Marraffini, 2019). Cas6 endoribonuclease performs the cleavage and processing of precursor crRNA (pre-crRNA) into mature crRNA in both Type I and III systems. Cas6 binds to the individual repeat sequence and cleaves at a specific 5′ end site releasing a single spacer sequence, which then forms an interference complex with other Cas proteins. In comparison to previous types, the Type II system undergoes two cleavage steps, which are first within the repeats followed by spacer cleavage to yield mature crRNA. RNase III host protein and noncoding tracrRNA are two important components required for cleavage. The tracrRNA annealed to the individual repeat sequence and then cleaved by RNase III releasing a mature tracrRNA–crRNA duplex. The duplex forms a ribonucleoprotein complex with Cas9 nuclease, which helps in the scanning of foreign genetic elements during the interference stage (Hille & Charpentier, 2016; Nussenzweig & Marraffini, 2020). Application in genetic engineering replaces the tracrRNA–crRNA duplex into sgRNA of 20 nucleotides in length and adjacent to the PAM NGG sequence. During the interference stage, multiple Cas proteins termed Cascade (a CRISPR-associated complex for antiviral defense) are involved in the recognition and degradation of invading DNA. These proteins include Cse (1 and 2), Cas5, Cas7, and Cas6e in type I-E *Escherichia coli*, and Csy (1, 2, and 3), Cas6f in type I–F system. The crRNA–Cascade complex scanned for PAM sequence in the foreign genome and upon recognition unwinds the dsDNA and subsequent annealing of crRNA and protospacer creates an R-loop-like structure. Formation of this structure activates Cas3 (having both helicase and nuclease domains), which creates a nick in the nontarget strand followed by unwinding and degradation of both strands. In the type III system, crRNA and Csm/cmr complex scanned for complementary targets and Cas10 acts as nuclease activity degrading both DNA and RNA (Rath et al., 2015). Type II system employs a sgRNA, which is the tracrRNA–crRNA duplex and single Cas effector protein Cas9 for interfering with the invading DNA. The sgRNA–Cas9 complex functions as an RNA-guided endonuclease scanned for the PAM motif, which is NGG in the case of Cas9, facilitating the annealing of sgRNA with the protospacer. Binding activates both HNH and RuvC nuclease domains of Cas9 with the HNH domain cleaving the annealed strand and RuvC the opposite strand. The resulting DSB is corrected by host NHEJ and HDR mechanisms. NHEJ creates indel (insertion or deletion) mutations of varying length changing the reading frame and causing frameshift mutations in the target region inducing functional gene knockout and inactivation, whereas HDR is suitable for gene knock-in or replacement with donor homology template (Mengstie & Wondimu, 2021).

2.7 Different Cas9 Variants for Genome Editing in Plants

2.7.1 CRISPR/Cas12a

The most versatile application of the CRISPR/Cas system apart from Cas9 is Cas12a/Cpf1 discovered by MIT University researchers. The nuclease function of Cpf1 was identified from the bacterium *Prevotella* and *Francisella1* with varied PAM sequence requirements, i.e. TTTN or TTN, where N is either A or T or C or G or TTTV where V is A or C or G bases. Cpf1 has three homologs, viz. FnCas12a from *Francisella novicida*, *AsCas12a* from *Acidaminococcus* sp., and *LbCas12a* from *Lachnospiraceae bacterium* (Bandyopadhyay et al., 2020). The differences between Cas9 and Cpf1 include PAM sequence requirement, the protospacer sequence for Cpf1 is 5′ upstream of the PAM sequence and cleavage occurs at 18–24 nucleotide bases from the PAM generating sticky ends overhangs as compared to blunt end of Cas9, in the maturation stage Cpf1 have its own ability to process the mature crRNA, whereas Cas9 requires tracrRNA for processing. Cpf1 cleaves the target DNA with a single crRNA, which is shorter than sgRNA for Cas9 (43 nt versus ~100 nt), thus making it cost-effective when chemically synthesized. Apart from DNA cleavage, Cpf1 takes the function of RNaseIII activity to process the pre-crRNA that can be exploited for multiplex genome editing with tandemly arrayed pre-crRNA expressing constructs, and lastly, Cpf1 has been reported to cause less off-target activity in mammalian cells as compared to Cas9 (Zaidi et al., 2017). Furthermore, Cpf1 is bilobed in structure connected by wedge (WED) domain and is divided into (i) NUC lobe composed of RuvC (I–III), bridge helix (BH), and NUC domain and (ii) REC lobe consisting of REC1 and REC2 domains. The RuvC domain is responsible for the cleavage of both target and nontarget DNA strands (Paul & Montoya, 2020).

2.7.2 dCas System

The deactivated nuclease domain of Cas9 or Cpf1 is called dCas9 or dCpf1 and is mainly used for precise transcriptional regulations. The inactivated dCas9 is fused with an activator or repressor effector protein, guided by sgRNA to the target genomic region in which binding of either the effectors in the regulatory elements results in activation (CRISPRa) or repression (CRISPRi) of target gene expression. Thus, this system can be used for gene functional screening, exploration of regulatory elements, and delivering specific effector proteins like GFP and DNA demethylase to the target location (Tang et al., 2017; Cai et al., 2023). dCas9 can further be reengineered for use as a base editor or epigenetic modifier. Cytosine or adenine deaminase enzyme is fused with dCas where the protein guided by sgRNA locally denatures and unwinds the DNA facilitating binding of base editor to the noncomplementary strand and modified base upstream of PAM sequence (Moradpour & Abdulah, 2020).

2.7.3 Cas9 Nickase (nCas9)

This system is created by modifying the nuclease domain of Cas9 such as amino acid alanine substitutions at D10 catalytic site in RuvC or H840A in the HNH domain converting nuclease to nickase activity. A single-strand nick is generated in the target DNA, which is repaired by high-fidelity homologous recombination (Fauser et al., 2014). The major concern using native Cas9 nuclease is the uncontrolled off-target effects, especially with plants having large genomes and several duplications. Therefore, paired Cas9 nickase can be an alternative strategy to induce DSBs with heritable indel mutations and less off-target effects (Trevino & Zhang, 2014). nCas9 has also been used for base editing in place of dCas9 by fusing with a specific deaminase enzyme (Chen et al. 2019b).

2.7.4 Base and Prime Editing CRISPR Tools

In plants, precise genome editing without inducing DSB and achieving low off-targets can be achieved using base editing which is of two types: cytosine base editor (CBE) and adenine base editor (ABE). Fusion of cytosine deaminase to either dCas9 or nCas9 induces base substitution C → G to T → A. Additionally, coupling of uracil DNA glycosylase inhibitor to CBE gives higher editing efficiency. In the case of ABE, an adenosine deaminase is fused with nCas9 substituting A → T to G → C. Different variants of deaminase enzymes such as ABE-P1S when fused with nCas9 give double editing efficiency in rice targeting seven genes (Rehman et al., 2022; Kaya, 2024). Using different Cas9 variants that recognize different PAMs has also been proven to improve base editing efficiency in plants. Advanced base editing technology like saturated targeted endogenous mutagenesis editors (STEMs), which is basically a combination of both base editors and simultaneous substitutions, has been efficiently used to produce herbicide tolerance in plants (Sukegawa et al., 2021). Since base editing is only applicable for four types of transitions, i.e. purine to purine and pyrimidine to pyrimidine, prime editing (PE) is another tool expanding the scope of genome editing without the requirement of DSB. In PE, nCas9 is fused with a modified reverse transcriptase and a multifunctional primer editing guide RNA (pegRNA). The pegRNA recognizes the target and a nick is created on the nontarget strand upstream of the PAM exposing 3′OH ends. The primer-binding site (PBS) in the pegRNA binds with the overhangs 3′OH ends and the RT template makes the desired base edits and through cellular repair mechanism changes are introduced. PE has a scope not just as base substitution but also as insertion and deletion mutations (Mengstie & Wondimu, 2021; Kaya, 2024).

2.8 Success in CRISPR/Cas-Editing Noncoding RNA Genes in Plants

Due to their determining role in plants' diverse processes, the ncRNAs have been chosen to edit using the most commonly used gene-editing method, i.e. CRISPR/Cas9, for functional analysis in recent years. The CRISPR/Cas9-mediated gene-editing method is the most studied and characterized and thus widely opted for gene editing in plants. Reprogramming the ncRNAs through mutations like knockout, knockdown, or base modification has wide scope to develop stress-resilient crops to maintain or enhance yield production in climate change scenarios. Rice, tomato, maize, potato, soybean, etc. are members of the most important crops on the planet earth and are extensively cultivated globally. After *Arabidopsis*, the characterized ncRNAs belong to these major crops. From the literature study, it is evident that CRISPR/Cas9-mediated gene editing of ncRNA genes is being done in major crops. The outcome of different studies suggests that the CRISPR/Cas9 system is suitable for creating mutations in ncRNA genes in crops. Table 2.1 summarizes the recent progress of CRISPR/Cas9-mediated editing of ncRNA genes in different crops in recent years.

2.9 Challenges Associated with Editing Noncoding RNA in Plants

While editing ncRNA genes by the CRISPR/Cas method presents an opportunity for plant genetic modification, it also suffers challenges. The protein-coding genes in plants are often knockdown, knockout, or modified (nucleotide base modification) for a purposeful goal. A similar approach could be employed for editing plant ncRNA genes too. But there exist complexities,

Table 2.1 Summary of reports published in recent years where noncoding RNA in different plants is edited by the CRISPR/Cas9 method.

Noncoding RNA (S)	Plant	Target gene/pathway	Biological role	References
MIR396e and *MIR396f*	Rice	Rice *GROWTH-REGULATING FACTORs*	Higher grain size and panicle branching	Zhang et al. (2020)
lncD09 and lncA07	Cotton	Jasmonic acid	Increased susceptibility to insect infestation	Zhang et al. (2022a, b)
MIR156a-MIR156c, *MIR156k*, and *MIR156l*	Rice	Upregulation of *IPA1*	Enhanced seed dormancy	Miao et al. (2019)
miR482b and *miR482c*	Tomato	defense pathway	Reduced susceptibility to *Phytophthora infestans*	Hong et al. (2021)
lncRNA1459	Tomato	Ethylene production pathway	Repression of tomato ripening process	Li et al. (2018)
MIR168	Rice	—	Growth and stress responses	Zhou et al. (2022)
lncRNA77580	Soybean	Protein-coding genes involved in salt stress	Salt stress response	Niu et al. (2021)
MIR160a, MIR160b, and *MIR390a*	Potato	Auxin response factors (ARFs)	Auxin-mediated pathway	Lukan et al. (2022)
PILNCR2	Maize	*ZmPHT1* (inorganic phosphate [Pi] transporters)	Phosphate tolerance	Wang et al. (2023)
lncWOX11a	Poplar	*PeWOX11a*	Adventitious root formation	Ran et al. (2023)

unlike protein-coding genes the ncRNA genes are quite short in length and sometimes they may exist within an intron of protein-coding host genes. For example, in many cases, an miRNA is located in the introns of protein-coding host genes (Barrangou et al., 2015). Editing these kinds of noncoding genes may also affect the host genes and so create a challenge. Additionally, the ncRNA genes (miRNAs, snRNAs, and snoRNAs) are generally small in length in nucleotides, and designing efficient guide RNAs for CRISPR/Cas-based modifications in such a limited window is another critical factor. In short-length DNA sequences, getting a guide RNA for Cas protein having no off-target sites might be difficult. In the CRISPR/Cas9 system, the guide RNA requires an NGG sequence motif on DNA as PAM and the probability of having PAM in an AT-rich genome will be lesser. Similarly, the Cas12a binds adjacent to the "TTTV" PAM motif. An miRNA gene sequence of 18–25 nucleotides will make it difficult to locate suitable PAM and guide RNA (Deng et al., 2022). Furthermore, it is critical to analyze the targets of a ncRNA, reports demonstrated a single ncRNA can target multiple genes to regulate their expression. Knockout mutation of ncRNA in these circumstances will lead to undesired off-target effects (Lucero et al., 2020; Bhogireddy et al., 2021). Therefore, all ncRNA (such as miRNA) may not be modified by the CRISPR/Cas system. Moreover, evidence indicates that creating small indel mutations in ncRNA genes might lead to the biogenesis of new variants, which can create unknown effects. Besides, ncRNA regulates their target genes at the transcription or translation, this circumstance requires detailed comprehensive studies to well characterize the regulation machinery of ncRNAs, before selecting them for gene-editing applications to develop stress-resilient crop varieties.

2.10 Conclusion and Future Directions

Biotic and abiotic stress lowers agricultural crop productivity. It is assumed that these stress conditions will occur more in the coming years due to global warming and climate change. Drought, salinity, and disease incidence are critical factors that may limit crop yield in the future. The development of stress resilience crop varieties is crucial for world food security. In recent years, it has been evident that ncRNAs in plants play an important role in stress adaptation and resistance. The ncRNA is a subset of cellular RNAs, which does not undergo translation to form a protein. Their length varies from more than 200 bp long (lncRNA) to a small length of about 21–25 nucleotides (miRNA). In general, they regulate the transcription of protein-coding genes by binding to complementary sequences. As regulators of gene expression, they are involved in pathways that determine important traits in agricultural crop plants including resistance to stress or increase in yields. With the progress made in molecular technologies and gene-editing methods (CRISPR/Cas), it is possible to edit ncRNA genes or their target genes. This reprogramming of ncRNA for the development of stress resilience crops is hopeful. Though, CRISPR/Cas gene editing of ncRNA for stress resilience is in the starting phase yet looks like a solution for the need for future climate resilience crop varieties.

References

Allemailem, K. S., Alsahli, M. A., Almatroudi, A., Alrumaihi, F., Alkhaleefah, F. K., Rahmani, A. H., & Khan, A. A. (2022). Current updates of CRISPR/Cas9-mediated genome editing and targeting within tumor cells: an innovative strategy of cancer management. *Cancer Communications, 42*(12), 1257–1287. https://doi.org/10.1002/cac2.12366

Bandyopadhyay, A., Kancharla, N., Javalkote, V. S., Dasgupta, S., & Brutnell, T. P. (2020). CRISPR-Cas12a (Cpf1): a versatile tool in the plant genome editing tool box for agricultural advancement. *Frontiers in Plant Science, 11*. https://doi.org/10.3389/fpls.2020.584151

Barrangou, R., & Marraffini, L. A. (2014). CRISPR-Cas systems: prokaryotes upgrade to adaptive immunity. *Molecular Cell, 54*(2), 234–244. https://doi.org/10.1016/j.molcel.2014.03.011

Barrangou, R., Birmingham, A., Wiemann, S., Beijersbergen, R. L., Hornung, V., & van Smith, A. B. (2015). Advances in CRISPR-Cas9 genome engineering: lessons learned from RNA interference. *Nucleic Acids Research, 43*(7), 3407–3419. https://doi.org/10.1093/nar/gkv226

Basel, K., Jian-Kang, Z., & Jianhua, Z. (2012). Role of miRNAs and siRNAs in biotic and abiotic stress responses of plants. *Biochimica et Biophysica Acta (BBA) - Gene Regulatory Mechanisms, 1819*(2), 137–148. https://doi.org/10.1016/j.bbagrm.2011.05.001

Bhogireddy, S., Mangrauthia, S. K., Kumar, R., Pandey, A. K., Singh, S., Jain, A., Budak, H., Varshney, R. K., & Kudapa, H. (2021). Regulatory non-coding RNAs: a new frontier in regulation of plant biology. *Functional & Integrative Genomics, 21*(3–4), 313–330. https://doi.org/10.1007/s10142-021-00787-8

Cai, R., Lv, R., Shi, X., Yang, G., & Jin, J. (2023). CRISPR/dCas9 tools: epigenetic mechanism and application in gene transcriptional regulation. *International Journal of Molecular Sciences, 24*(19), 14865. https://doi.org/10.3390/ijms241914865

Chen, G., Cui, J., Wang, L., Zhu, Y., Lu, Z., & Jin, B. (2017). Genome-wide identification of circular RNAs in *Arabidopsis thaliana*. *Frontiers in Plant Science, 8*. https://doi.org/10.3389/fpls.2017.01678

Chen, L., Meng, J., & Luan, Y. (2019a). miR1916 plays a role as a negative regulator in drought stress resistance in tomato and tobacco. *Biochemical and Biophysical Research Communications, 508*(2), 597–602. https://doi.org/10.1016/j.bbrc.2018.11.165

Chen, K., Wang, Y., Zhang, R., Zhang, H., & Gao, C. (2019b). CRISPR/Cas genome editing and precision plant breeding in agriculture. *Annual Review of Plant Biology*, *70*(1), 667–697. https://doi.org/10.1146/annurev-arplant-050718-100049

Chung, M.-Y., Nath, U. K., Vrebalov, J., Gapper, N., Lee, J. M., Lee, D.-J., Kim, C. K., & Giovannoni, J. (2020). Ectopic expression of miRNA172 in tomato (*Solanum lycopersicum*) reveals novel function in fruit development through regulation of an AP2 transcription factor. *BMC Plant Biology*, *20*(1), 283. https://doi.org/10.1186/s12870-020-02489-y

Das, P., Agarwala, N., Gill, S. S., & Varshney, R. K. (2023). Emerging role of plant long non coding RNAs (lncRNAs) in salinity stress response. *Plant Stress*, *10*, 100265. https://doi.org/10.1016/j.stress.2023.100265

Deng, F., Zeng, F., Shen, Q., Abbas, A., Cheng, J., Jiang, W., Chen, G., Shah, A. N., Holford, P., Tanveer, M., Zhang, D., & Chen, Z.-H. (2022). Molecular evolution and functional modification of plant miRNAs with CRISPR. *Trends in Plant Science*, *27*(9), 890–907. https://doi.org/10.1016/j.tplants.2022.01.009

Fauser, F., Schiml, S., & Puchta, H. (2014). Both CRISPR Cas-based nucleases and nickases can be used efficiently for genome engineering in *Arabidopsis thaliana*. *The Plant Journal*, *79*(2), 348–359. https://doi.org/10.1111/tpj.12554

Gao, L., & Cui, X. (2023). Climate change and food security: plant science roles. *Molecular Plant*, *16*(10), 1481–1483. https://doi.org/10.1016/j.molp.2023.09.019

Gelaw, T. A., & Sanan-Mishra, N. (2021). Non-coding RNAs in response to drought stress. *International Journal of Molecular Sciences*, *22*(22), 12519. https://doi.org/10.3390/ijms222212519

Hille, F., & Charpentier, E. (2016). CRISPR-Cas: biology, mechanisms and relevance. *Philosophical Transactions of the Royal Society B: Biological Sciences*, *371*(1707), 20150496. https://doi.org/10.1098/rstb.2015.0496

Hong, Y., Meng, J., He, X., Zhang, Y., Liu, Y., Zhang, C., Qi, H., & Luan, Y. (2021). Editing miR482b and miR482c Simultaneously by CRISPR/Cas9 Enhanced Tomato Resistance to *Phytophthora infestans*. *Phytopathology®*, *111*(6), 1008–1016. https://doi.org/10.1094/PHYTO-08-20-0360-R

Huang, J., Zhou, W., Zhang, X., & Li, Y. (2023). Roles of long non-coding RNAs in plant immunity. *PLoS Pathogens*, *19*(5), e1011340. https://doi.org/10.1371/journal.ppat.1011340

Jian, C., Hao, P., Hao, C., Liu, S., Mao, H., Song, Q., Zhou, Y., Yin, S., Hou, J., Zhang, W., Zhao, H., Zhang, X., & Li, T. (2022). The miR319/TaGAMYB3 module regulates plant architecture and improves grain yield in common wheat (*Triticum aestivum*). *New Phytologist*, *235*(4), 1515–1530. https://doi.org/10.1111/nph.18216

Katiyar-Agarwal, S., Morgan, R., Dahlbeck, D., Borsani, O., Villegas, A., Zhu, J.-K., Staskawicz, B. J., & Jin, H. (2006). A pathogen-inducible endogenous siRNA in plant immunity. *Proceedings of the National Academy of Sciences*, *103*(47), 18002–18007. https://doi.org/10.1073/pnas.0608258103

Kaur, R., Bhunia, R. K., & Rajam, M. V. (2020). MicroRNAs as potential targets for improving rice yield via plant architecture modulation: recent studies and future perspectives. *Journal of Biosciences*, *45*(1), 116. https://doi.org/10.1007/s12038-020-00084-9

Kaya, H. B. (2024). Base editing and prime editing. In: *A Roadmap for Plant Genome Editing*, Springer Nature Switzerland, 17–39. https://doi.org/10.1007/978-3-031-46150-7_2

Khatodia, S., Bhatotia, K., Passricha, N., Khurana, S. M. P., & Tuteja, N. (2016). The CRISPR/Cas genome-editing tool: application in improvement of crops. *Frontiers in Plant Science*, *7*. https://doi.org/10.3389/fpls.2016.00506

Khraiwesh, B., Zhu, J.-K., & Zhu, J. (2012). Role of miRNAs and siRNAs in biotic and abiotic stress responses of plants. *Biochimica et Biophysica Acta (BBA) – Gene Regulatory Mechanisms*, *1819*(2), 137–148. https://doi.org/10.1016/j.bbagrm.2011.05.001

Li, R., Fu, D., Zhu, B., Luo, Y., & Zhu, H. (2018). CRISPR/Cas9-mediated mutagenesis of *lncRNA1459* alters tomato fruit ripening. *The Plant Journal*, 94(3), 513–524. https://doi.org/10.1111/tpj.13872

Lucero, L., Fonouni-Farde, C., Crespi, M., & Ariel, F. (2020). Long noncoding RNAs shape transcription in plants. *Transcription, 11*(3–4), 160–171. https://doi.org/10.1080/21541264.2020.1764312

Lukan, T., Veillet, F., Križnik, M., Coll, A., Mahkovec Povalej, T., Pogačar, K., Stare, K., Chauvin, L., Chauvin, J.-E., & Gruden, K. (2022). CRISPR/Cas9-mediated fine-tuning of miRNA expression in tetraploid potato. *Horticulture Research, 9*. https://doi.org/10.1093/hr/uhac147

Ma, X., Zhao, F., & Zhou, B. (2022). The characters of non-coding RNAs and their biological roles in plant development and abiotic stress response. *International Journal of Molecular Sciences, 23*(8), 4124. https://doi.org/10.3390/ijms23084124

McGinn, J., & Marraffini, L. A. (2019). Molecular mechanisms of CRISPR-Cas spacer acquisition. *Nature Reviews Microbiology, 17*(1), 7–12. https://doi.org/10.1038/s41579-018-0071-7

Mengstie, M. A., & Wondimu, B. Z. (2021). Mechanism and applications of CRISPR/Cas-9-mediated genome editing. *Biologics: Targets and Therapy, 15*, 353–361. https://doi.org/10.2147/BTT.S326422

Miao, C., Wang, Z., Zhang, L., Yao, J., Hua, K., Liu, X., Shi, H., & Zhu, J.-K. (2019). The grain yield modulator miR156 regulates seed dormancy through the gibberellin pathway in rice. *Nature Communications, 10*(1), 3822. https://doi.org/10.1038/s41467-019-11830-5

Minchin, S., & Lodge, J. (2019). Understanding biochemistry: structure and function of nucleic acids. *Essays in Biochemistry, 63*(4), 433–456. https://doi.org/10.1042/EBC20180038

Moradpour, M., & Abdulah, S. N. A. (2020). CRISPR/dCas9 platforms in plants: strategies and applications beyond genome editing. *Plant Biotechnology Journal, 18*(1), 32–44. https://doi.org/10.1111/pbi.13232

Mosterd, C., Rousseau, G. M., & Moineau, S. (2021). A short overview of the CRISPR-Cas adaptation stage. *Canadian Journal of Microbiology, 67*(1), 1–12. https://doi.org/10.1139/cjm-2020-0212

Niu, F., Jiang, Q., Sun, X., Hu, Z., Wang, L., & Zhang, H. (2021). Large DNA fragment deletion in lncRNA77580 regulates neighboring gene expression in soybean (*Glycine max*). *Functional Plant Biology, 48*(11), 1139–1147. https://doi.org/10.1071/FP20400

Nussenzweig, P. M., & Marraffini, L. A. (2020). Molecular mechanisms of CRISPR-Cas immunity in bacteria. *Annual Review of Genetics, 54*(1), 93–120. https://doi.org/10.1146/annurev-genet-022120-112523

Paul, B., & Montoya, G. (2020). CRISPR-Cas12a: functional overview and applications. *Biomedical Journal, 43*(1), 8–17. https://doi.org/10.1016/j.bj.2019.10.005

Qin, T., Zhao, H., Cui, P., Albesher, N., & Xiong, L. (2017). A nucleus-localized long non-coding RNA enhances drought and salt stress tolerance. *Plant Physiology, 175*(3), 1321–1336. https://doi.org/10.1104/pp.17.00574

Ran, N., Liu, S., Qi, H., Wang, J., Shen, T., Xu, W., & Xu, M. (2023). Long non-coding RNA lncWOX11a suppresses adventitious root formation of poplar by regulating the expression of PeWOX11a. *International Journal of Molecular Sciences, 24*(6), 5766. https://doi.org/10.3390/ijms24065766

Rath, D., Amlinger, L., Rath, A., & Lundgren, M. (2015). The CRISPR-Cas immune system: Biology, mechanisms and applications. *Biochimie, 117*, 119–128. https://doi.org/10.1016/j.biochi.2015.03.025

Rehman, R. S., Zafar, S. A., Ali, M., Pasha, A. N., Naveed, M. S., Waseem, M., Ahmad, M., & Raza, A. (2022). CRISPR-Cas mediated genome editing: a paradigm shift towards sustainable agriculture and biotechnology. *Asian Plant Research Journal, 27–49*. https://doi.org/10.9734/aprj/2022/v9i130197

Singh, B. K., Delgado-Baquerizo, M., Egidi, E., Guirado, E., Leach, J. E., Liu, H., & Trivedi, P. (2023). Climate change impacts on plant pathogens, food security and paths forward. *Nature Reviews Microbiology, 21*(10), 640–656. https://doi.org/10.1038/s41579-023-00900-7

Song, L., Fang, Y., Chen, L., Wang, J., & Chen, X. (2021). Role of non-coding RNAs in plant immunity. *Plant Communications*, *2*(3), 100180. https://doi.org/10.1016/j.xplc.2021.100180

Sukegawa, S., Saika, H., & Toki, S. (2021). Plant genome editing: ever more precise and wide reaching. *The Plant Journal*, *106*(5), 1208–1218. https://doi.org/10.1111/tpj.15233

Summanwar, A., Basu, U., Rahman, H., & Kav, N. N. V. (2020). Non-coding RNAs as emerging targets for crop improvement. *Plant Science*, *297*, 110521. https://doi.org/10.1016/j.plantsci.2020.110521

Tang, X., Lowder, L. G., Zhang, T., Malzahn, A. A., Zheng, X., Voytas, D. F., Zhong, Z., Chen, Y., Ren, Q., Li, Q., Kirkland, E. R., Zhang, Y., & Qi, Y. (2017). A CRISPR-Cpf1 system for efficient genome editing and transcriptional repression in plants. *Nature Plants*, *3*(3), 17018. https://doi.org/10.1038/nplants.2017.18

Terán, F., Vives-Peris, V., Gómez-Cadenas, A., & Pérez-Clemente, R. M. (2024). Facing climate change: plant stress mitigation strategies in agriculture. *Physiologia Plantarum*, *176*(4). https://doi.org/10.1111/ppl.14484

Torquebiau, E., Tissier, J., & Grosclaude, J.-Y. (2016). How climate change reshuffles the cards for agriculture. In: *Climate Change and Agriculture Worldwide*, Springer, Netherlands, 1–16. https://doi.org/10.1007/978-94-017-7462-8_1

Trevino, A. E., & Zhang, F. (2014). Genome editing using Cas9 nickases. In: *Methods in Enzymology*, Academic Press, 161–174. https://doi.org/10.1016/B978-0-12-801185-0.00008-8

Waititu, J. K., Zhang, C., Liu, J., & Wang, H. (2020). Plant non-coding RNAs: origin, biogenesis, mode of action and their roles in abiotic stress. *International Journal of Molecular Sciences*, *21*(21), 8401. https://doi.org/10.3390/ijms21218401

Wang, D., & Farhana, A. (2024). Biochemistry, RNA structure. In: *StatPearls*, StatPearls Publishing. http://www.ncbi.nlm.nih.gov/books/NBK558999/

Wang, Y., Wang, Z., Du, Q., Wang, K., Zou, C., & Li, W.-X. (2023). The long non-coding RNA PILNCR2 increases low phosphate tolerance in maize by interfering with miRNA399-guided cleavage of ZmPHT1s. *Molecular Plant*, *16*(7), 1146–1159. https://doi.org/10.1016/j.molp.2023.05.009

Yadav, A., Mathan, J., Dubey, A. K., & Singh, A. (2024). The emerging role of non-coding RNAs (ncRNAs) in plant growth, development, and stress response signaling. *Non-Coding RNA*, *10*(1), 13. https://doi.org/10.3390/ncrna10010013

Yu, Y., Zhang, Y., Chen, X., & Chen, Y. (2019). Plant noncoding RNAs: hidden players in development and stress responses. *Annual Review of Cell and Developmental Biology*, *35*(1), 407–431. https://doi.org/10.1146/annurev-cellbio-100818-125218

Zaidi, S. S.-A., Mahfouz, M. M., & Mansoor, S. (2017). CRISPR-Cpf1: a new tool for plant genome editing. *Trends in Plant Science*, *22*(7), 550–553. https://doi.org/10.1016/j.tplants.2017.05.001

Zhang, X., Dong, J., Deng, F., Wang, W., Cheng, Y., Song, L., Hu, M., Shen, J., Xu, Q., & Shen, F. (2019). The long non-coding RNA lncRNA973 is involved in cotton response to salt stress. *BMC Plant Biology*, *19*(1), 459. https://doi.org/10.1186/s12870-019-2088-0

Zhang, J., Zhou, Z., Bai, J., Tao, X., Wang, L., Zhang, H., & Zhu, J.-K. (2020). Disruption of MIR396e and MIR396f improves rice yield under nitrogen-deficient conditions. *National Science Review*, *7*(1), 102–112. https://doi.org/10.1093/nsr/nwz142

Zhang, J., Li, J., Saeed, S., Batchelor, W. D., Alariqi, M., Meng, Q., Zhu, F., Zou, J., Xu, Z., Si, H., Wang, Q., Zhang, X., Zhu, H., Jin, S., & Yuan, D. (2022a). Identification and functional analysis of lncRNA by CRISPR/Cas9 during the cotton response to Sap-sucking insect infestation. *Frontiers in Plant Science*, *13*. https://doi.org/10.3389/fpls.2022.784511

Zhang, Y., Zhou, Y., Zhu, W., Liu, J., & Cheng, F. (2022b). Non-coding RNAs fine-tune the balance between plant growth and abiotic stress tolerance. *Frontiers in Plant Science, 13*. https://doi.org/10.3389/fpls.2022.965745

Zhou, J., Deng, K., Cheng, Y., Zhong, Z., Tian, L., Tang, X., Tang, A., Zheng, X., Zhang, T., Qi, Y., & Zhang, Y. (2017). CRISPR-Cas9 based genome editing reveals new insights into MicroRNA function and regulation in rice. *Frontiers in Plant Science, 8*. https://doi.org/10.3389/fpls.2017.01598

Zhou, J., Zhang, R., Jia, X., Tang, X., Guo, Y., Yang, H., Zheng, X., Qian, Q., Qi, Y., & Zhang, Y. (2022). CRISPR-Cas9 mediated *OsMIR168a* knockout reveals its pleiotropy in rice. *Plant Biotechnology Journal, 20*(2), 310–322. https://doi.org/10.1111/pbi.13713

3

Improving Plant Abiotic Stress Tolerance by Modulating LncRNAs Using CRISPR/Cas9 Technology

Muhammad Waseem[1,2,3,4], Sana Basharat[1,2,3,4], Iffat Shaheen[5], Liu Pingwu[1,2,3,4], Muhammad Shareef[6], and Nimra Shahid[6]

[1] *College of Tropical Agriculture and Forestry, Hainan University, Haikou, Hainan, China*
[2] *Key Laboratory of Tropical Horticultural Crop Quality Regulation, College of Horticulture, Hainan University, Haikou, Hainan, China*
[3] *Hainan Yazhou Bay Seed Laboratory, Sanya Nanfan Research Institute of Hainan University, Sanya, China*
[4] *Fang Zhiyuan Academician Team Innovation Center of Hainan Province, Sanya, China*
[5] *Department of Agriculture, Bahauddin Zakariya University, Multan, Pakistan*
[6] *Department of Botany, University of Narowal, Narowal, Punjab, Pakistan*

3.1 Introduction

Plants being sessile organisms face heterogeneous and fluctuating environmental stresses throughout their life and constantly adapt to maintain growth and ensure survival and reproduction. As a result, plants have evolved plenty of adaptive mechanisms to sense and respond to stress and modulate multiple genes expression at various levels like epigenetic, transcriptional, and posttranscriptional. Additionally, like protein-coding RNAs (mRNAs), long noncoding RNAs (lncRNAs) provide an additional layer of adaptation and mediating environmental responses (Traubenik et al., 2024). The lncRNAs are a diverse class of ncRNAs with more than 200 nucleotides in length and are classified into several categories based on genomic context. These include those (i) transcribed from intergenic regions and act in *trans*-manner – long intergenic ncRNAs (lincRNAs); exons of strands of sense strands act in *cis*- or *trans* context – natural antisense transcripts (NATs); originated from intronic regions – intronic ncRNAs (incRNAs); from protein-coding gene on sense strand – sense lncRNA; nonsequential back-splicing of introns and/or exons – circular lncRNAs (circRNAs) (Waseem et al., 2020). Conversely, unlike mRNAs, lncRNAs are less conserved between species, have lower abundance, and are confined to specific tissues or cells (Waseem et al., 2020). lncRNAs regulate plenty of biological processes like chromatin modifications, alternative splicing, and mRNA inhibition to regulate the expression of coding genes (Jha et al., 2020a). Additionally, plant lncRNAs have been reported to regulate various plant growth, development, and responses to biotic/abiotic stresses (Waseem et al., 2020).

Tolerance to environmental stress is a complex biological trait mediated by an intricate network of genes. In plant abiotic stress, crosstalk involves components of metabolic, regulatory, and signaling networks that affect plant developmental processes and affect final productivity (Misra et al., 2023). Recent advances in RNA-seq technology and meta-analysis annotated plenty of lncRNAs in several plant species, at different developmental stages, mutants, and

Genome and Epigenome Editing for Stress-Tolerant Crops, First Edition. Edited by Jen-Tsung Chen.

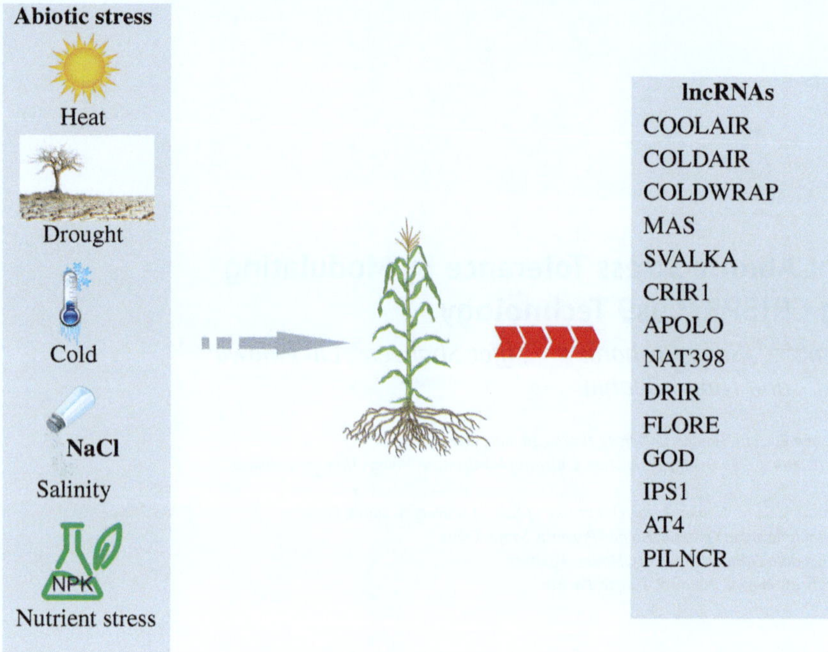

Figure 3.1 lncRNA-mediated abiotic stress adaptations in plants. Cold-responsive lncRNAs include *COLDAIR*, *COOLAIR*, *COLDWRAP*, *SVALKA*, *MAS*, and *CRIR1*. *FLORE*, *DRIR*, and *GOD* respond to drought in ABA-dependent and -independent manner. *APS1* and *PILNCR* lncRNAs are involved in inorganic phosphorus starvation responses. *Source:* Rawpixel.com/Adobe Stock Photos.

stress conditions. A significant portion of plant lncRNAs remains functionally uncharacterized, leaving a wealth of potential biological roles waiting to be uncovered. Unlike animals, to date, the characterization of lncRNAs in various molecular pathways is comparative at an early stage. However, only a few popular studies on lncRNAs in the development of plants with enhanced agronomic traits and their applications in crop-breeding programs are shown in Figure 3.1.

Clustered regulatory interspaced short palindromic repeats/CRISPR-associated proteins (CRISPRs/Cas) have revolutionized the field of molecular biology and emerged as an important biotechnological approach. To date, CRISPR-Cas9 has been confined to editing coding genes in numerous plant species like *Arabidopsis*, tomato, rice, and tobacco; however, studies targeting lncRNAs are scarce. The CRISPR-Cas9 system will benefit the lncRNAs functional characterization and would be critical to modulating their expression to develop stress-tolerant crops.

3.2 The CRISPR-Cas Technology

Crop-breeding programs assisted in the development of plenty of crops with desirable traits including stress tolerance without compromising productivity. Crossbreeding is one of the effective ways for sustainable breeding in traditional crop-breeding programs and has been widely used to modify agronomic crops. Conversely, mutation breeding and transgenic breeding are modern breeding approaches for crop improvement (Manghwar et al., 2019). These breeding approaches have limitations in terms of breeding time. Genome-editing technique improves targeted traits requiring less time, efficiency, and accuracy. Compared with crop breeding, mutation breeding, and transgenic breeding, genome breeding is fast and with short breeding time (Figure 3.2).

In the past decade, the advancement of site-specific nucleases such as transcription activator-like effector nucleases (TALENs) and zinc finger nucleases (ZFNs) has significantly transformed traditional plant breeding to generate highly effective crop genetic variations (Kim et al., 1996; Christian et al., 2010; Miller et al., 2011). ZFNs and TALENZs are programmable DNA binding

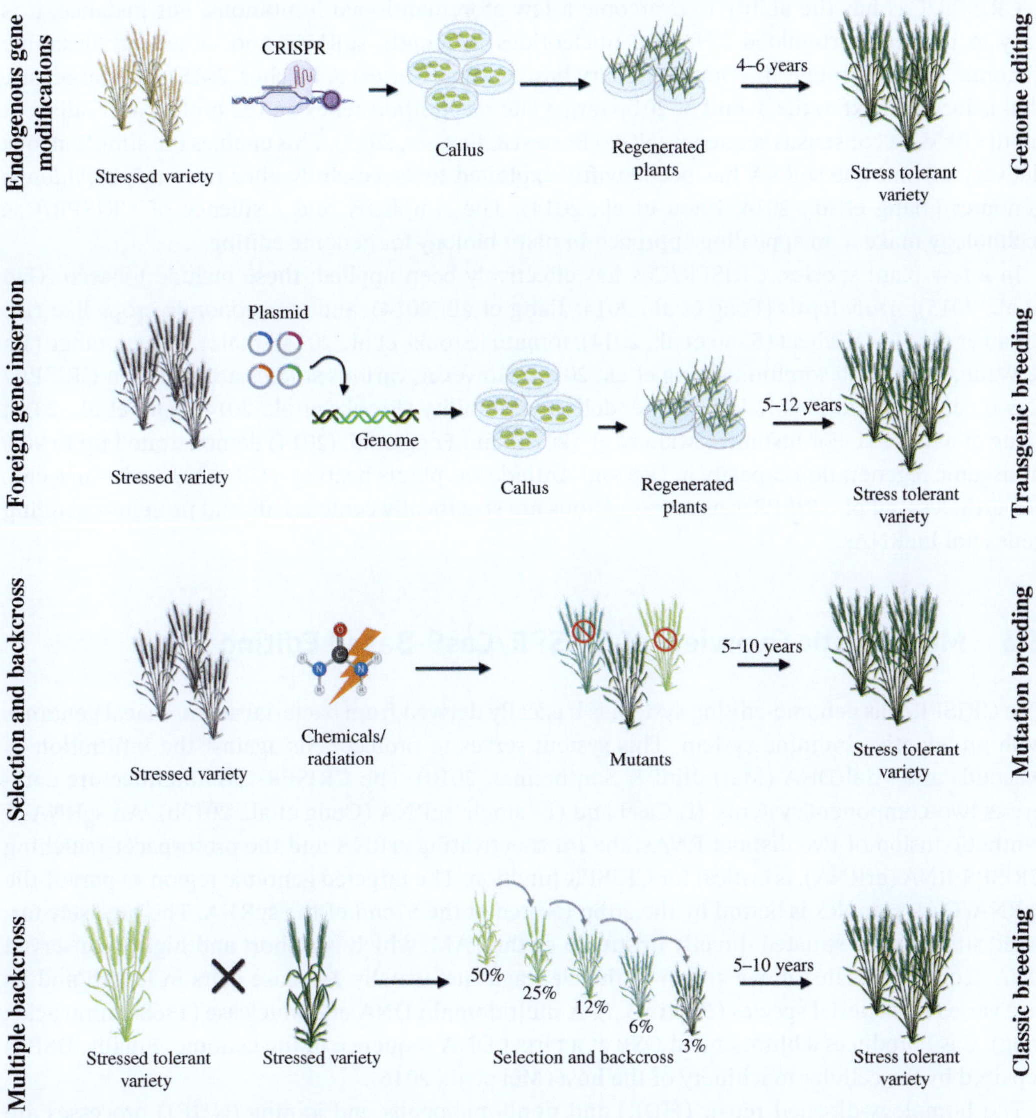

Figure 3.2 Comparison of crop-breeding approaches. Genome editing: precisely modifying target genes or regulatory elements for improving desirable traits. Genome breeding is fast and short time (4–6 years) to improve agronomic traits. Transgenic breeding: improve desirable traits in the time 8–12 years viz. exogenous transformation of genes or regulatory elements associated with desirable traits. Mutation breeding: involves the application of chemical or physical irradiation to improve traits in 8–10 years to generate novel genetic variability. Crossbreeding: crossbreeding involves a cross between a recipient with donor having a desirable trait and selecting outstanding progeny with the desired trait takes 8–10 years.

modules by generating double-strand breaks (DSBs) that cause genome alterations, which were subsequently repaired by homologous and nonhomologous recombination (Reyon et al., 2012; Belhaj et al., 2015). CRISPR/Cas has emerged as a pivotal genome-editing technique (Horvath & Barrangou, 2010; Cong et al., 2013a). However, these techniques have certain limitations including cost, design, low efficiency, and success rate. CRISPR loci comprise short spacers that vary in length separated by short repeats, transcribed into ncRNA which subsequently complexed with Cas and finally guide to cleave the DNA at the target site (Mali et al., 2013; Belhaj et al., 2015).

CRISPR/Cas has the ability to overcome a few aforementioned limitations. For instance, it is easy to design short-guided RNA (20 nucleotides in length, sgRNA) and target any desirable genomic region through its complementary base paring (Bortesi & Fischer, 2015). The target site that is located next to the 3′ end of 20 bp target site recognition relies on the protospacer adjacent motif (PAM), a consensus sequence, *NGG* (Bortesi & Fischer, 2015). This enables the simultaneous delivery of multiple sgRNA has been swiftly exploited to successfully alter rice and *Arabidopsis* genomes (Jiang et al., 2014; Zhou et al., 2014). The simplicity and resilience of CRISPR/Cas technology make it an appealing approach in plant biology for genome editing.

In a few plant species, CRISPR/Cas has effectively been applied; these include tobacco (Gao et al., 2015), *Arabidopsis* (Feng et al., 2014; Jiang et al., 2014), and in agronomic crops like rice (Shan et al., 2014), wheat (Shan et al., 2014), tomato (Brooks et al., 2014), maize, sweet orange (Jia & Wang, 2014), and sorghum (Jiang et al., 2013). However, various studies are based on CRISPR/Cas-mediated mutations following Mendelian heritability (Brooks et al., 2014; Feng et al., 2014; Jiang et al., 2014). For instance, Miao et al. (2013) and Feng et al. (2014) demonstrated up to 90% transgenic regeneration capacity in rice and *Arabidopsis* plants bearing a CRISPR/Cas9 transgene. Nevertheless, all of CRISPR/Cas investigations are specifically centered around proteins encoding genes not lncRNAs.

3.3 Mechanistic Overview of CRISPR/Cas9-Based Editing

The CRISPR/Cas genome-editing system is basically derived from bacterial and archaeal genomes with an adaptive immune system. This system serves to protect cells against the infiltration of plasmids and viral DNA (Marraffini & Sontheimer, 2010). The CRISPR-Cas9 architecture comprises two-component systems: (i) Cas9 and (ii) single sgRNA (Cong et al., 2013b). An sgRNA, a synthetic fusion of two distinct RNAs: the *trans*-activating crRNA and the protospacer-matching CRISPR RNA (crRNA), is critical for CRISPR function. The targeted genomic region as part of the sgRNA/Cas9 complex is bound by the 20 bp located at the 5′ end of the sgRNA. The precisely targeted site must be situated directly upstream of the PAM, which is a short and highly conserved DNA sequence located downstream of the cleavage site, usually 2–3 base pairs in length and its size varies in bacterial species (Figure 3.3). A multidomain DNA endonuclease (1368 amino acids long), Cas9 produces a blunt-ended DSB at a target DNA sequence in the genome. Finally, DSB is repaired by the cellular machinery of the host (Mei et al., 2016).

The homology-directed repair (HDR) and nonhomologous end joining (NHEJ) processes are responsible for repairing DSBs generated by the Cas9 proteins (Liu et al., 2020). HDR is a highly precise technique that utilizes a DNA template that exhibits homology and is active during the cell cycle, which requires a substantial quantity of donor-DNA specifying targeted nucleotide sequence. It executes the targeted gene substitution or insertion by introducing a donor DNA template with sequence similarity at the anticipated DSB position (Yang et al., 2022). In an enzymatic reaction

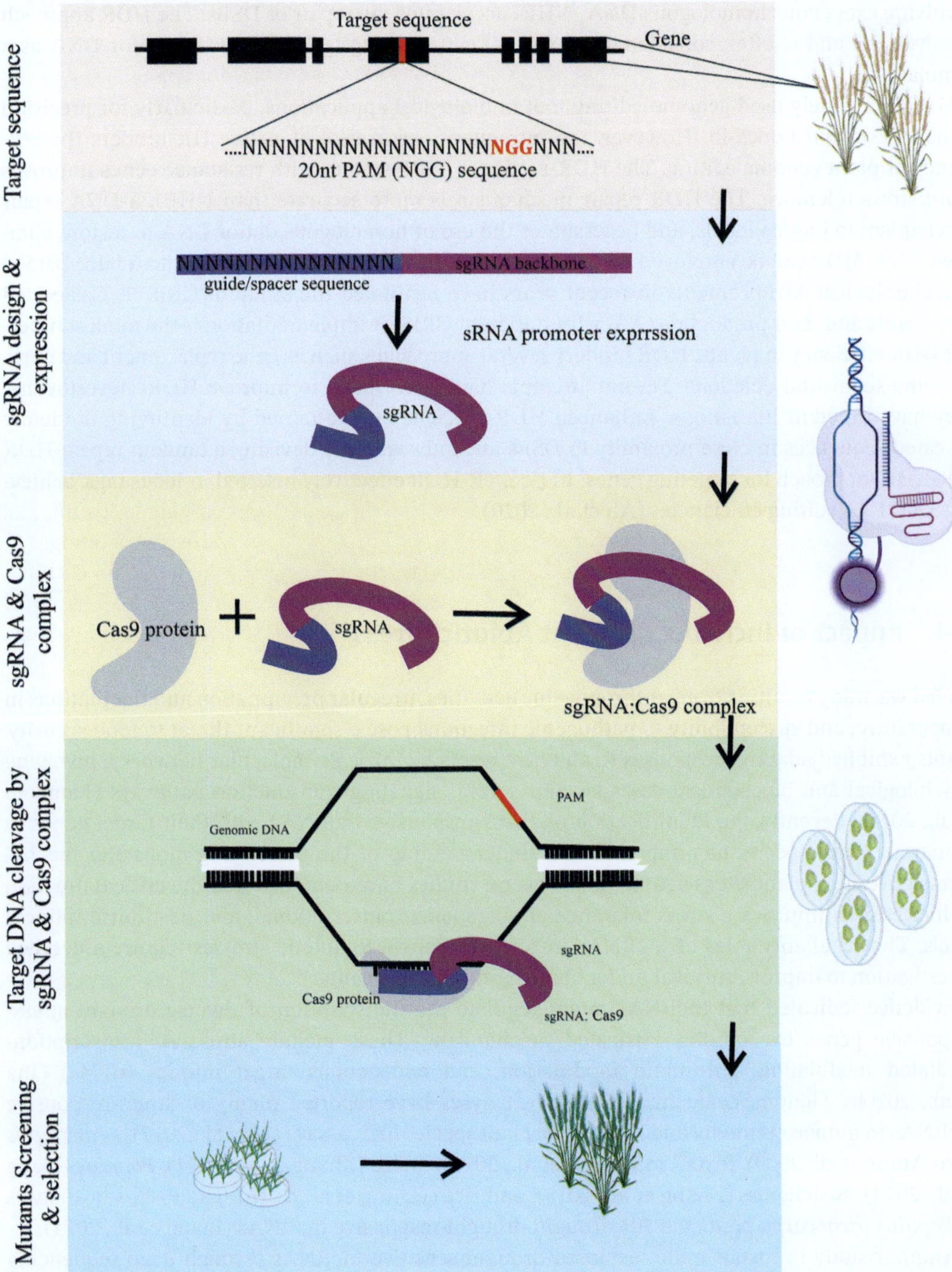

Figure 3.3 Flow chart to generate mutagenized plant by CRISPRs-Cas (clustered regulatory interspaced short palindromic repeats/CRISPR-associated). CRISPR-Cas-mediated genome editing on single-guided RNA (sgRNA), Cas9 protein, forms a complex with complementary sequences of the target site in DNA next to PAM (protospacer adjacent motif; NGG). The complex sgRNA-Cas9 generates a mutation in the genomic target sequence through a double-strand DNA break.

involving exogenous homologous DNA, NHEJ accelerates the repair of DSBs. The HDR approach involves HR and is often considered an optimal gene-editing technique using donor DNA as a template.

HDR is a widely used genome-editing tool in biological applications, particularly for precision mutagenesis and knock-in. However, the infrequent occurrence of native HR hinders the efficiency of plant genome editing. The HDR-mediated/Cas9 system with resistance genes improves plant stress tolerance. The HDR repair mechanism is more accurate than NHEJ, a DNA repair mechanism to induce indels, and necessitates the use of homologous donor DNA to restore damaged DNA. HDR can be employed for targeted mutations at desirable sequence sites (Jain, 2015).

Technological advancements in recent years have facilitated the study of CRISPR/Cas-edited crop plants and their production. A key limitation of CRISPR implementation is the modest inherent HDR efficiency in plants. HDR hinders several approaches such as gene replacement and chromosome segmental deletions. Several attempts have been made to improve HDR; nevertheless, they have inherent limitations. Enhanced HDR efficiency was attained by identifying tandemly repeated sequences in close proximity to DSBs and subsequently devising a tandem repeat-HDR (TR-HDR) approach for targeting genes. In rice, TR-HDR effectively inserted in-locus tags, achieving 3.4–11.4% editing efficiencies (Ali et al., 2020).

3.4 Impact of lncRNAs on Plant Abiotic Stress

Global warming resulting from anthropogenic activities, irregular precipitation and fluctuations in temperature, and susceptibility to pathogenic infections pose a significant threat to food security. Plants exhibited adaptive responses to abiotic stresses by intricate molecular networks, involving physiological and biochemical cascades and several signaling transduction pathways (Lamaoui et al., 2018). Recently, the identification of stress-responsive lncRNAs with their target genes in numerous plant species has improved our understanding of the underlying molecular mechanisms governing plant stress adaptations. Several studies have demonstrated the critical function of lncRNAs in improving stress tolerance at epigenetic, transcriptional, and posttranscriptional levels. The regulatory roles of lncRNAs in plant responses to abiotic stresses require a detailed investigation to improve survival under challenging environments.

Evidence indicated that lncRNAs might regulate the transcription of diverse drought stress-responsive genes by complex intricated mechanisms. These include antisense transcription-mediated modulation, chromatin modulation, and endogenous target mimics (eTMs) (Jha et al., 2020b). Genome-scale transcriptome analyses have reported plenty of stress-responsive lncRNAs in numerous model and agronomic plant species like cassava (Li et al., 2017), *Arabidopsis* (Ben Amor et al., 2009), foxtail millet (Qi et al., 2013), maize (Zhang et al., 2014), *Populus* (Liang et al., 2014), switchgrass (Zhang et al., 2018), and rice (Chung et al., 2016). The RNA-seq analysis of *Populus trichocarpa* predicted 504 drought drought-responsive lncRNAs (Liang et al., 2014).

Another study in foxtail millet explored drought-sensitive lncRNAs through deep sequencing and uncovered about 584 lncRNA, including NAR (2) and lincRNAs (17) lncRNAs (Qi et al., 2013). In cassava, 318 lncRNAs and 153 NAT lncRNAs induced by cold and drought stress were predicted in strand-specific RNA-seq analysis, respectively (Li et al., 2017). Chung et al. (2016) predicted 98 lncRNAs within the entire rice genome context. For instance, two (2) drought-responsive NAT lncRNAs, such as *Os02g0250700-01* and *Os02g0180800-01*, target the late embryogenesis abundant protein gene and cinnamoyl-CoA reductase genes, respectively.

Furthermore, lncRNA may act as a target mimic of protein-coding genes, defined as eTM. For example, *lincRNA340* serves as eTM for miR169, which regulates the expression of the drought-responsive *NUCLEAR FACTOR Y (NF-Y)* (Figure 3.4) (Li et al., 2017). lncRNAs can modulate the transcriptional expression of several genes. The lncRNAs also functioned as transcriptional regulators in stress responses like drought being explored in model plants like *Arabidopsis*. A nuclear-localized lncRNA (755 nt in length), *DROUGHT INDUCED lncRNA (DRIR)*, modulates plenty of drought-responsive transcripts, encompassing those associated with annexin (*ANNAT7*), ABA signaling (*ABI5, RD29A/B*, and *P5CS1*), and aquaporin genes (*NIP1, TIP4*). Furthermore, DRIR may affect the activity of transcription factor *NAM/ATAF/CUC 3 (NAC3)* and *FUCOSYLTRANSFERASE 4 (FUT4)*, leading to salinity and drought tolerance (Qin et al., 2017). These findings indicate that DRIR may act as a positive regulator of drought tolerance. In conclusion, the integration of drought-responsive lncRNAs may act as a target mimic of miRNAs that subsequently regulate gene expression and transcription regulation of diverse drought-responsive genes. Furthermore, these lncRNAs may serve as a critical regulatory hub at both transcriptional and posttranscriptional levels for drought-responsive signal transduction pathways.

Plant growth and development are significantly affected by the presence of specific nutrients in the soil. Soil nutrient absorption is a crucial physiological process involving several membrane transporters that promote nutritional homeostasis (Tegeder & Masclaux-Daubresse, 2018). Phosphorus (Pi) is a vital macronutrient that facilitates plant development and growth, serves as a growth limiting factor, and provides inorganic Pi for ATP synthesis (Adnan et al., 2017). Nonetheless, the intricate regulatory mechanisms governing Pi homeostasis in plants are limited.

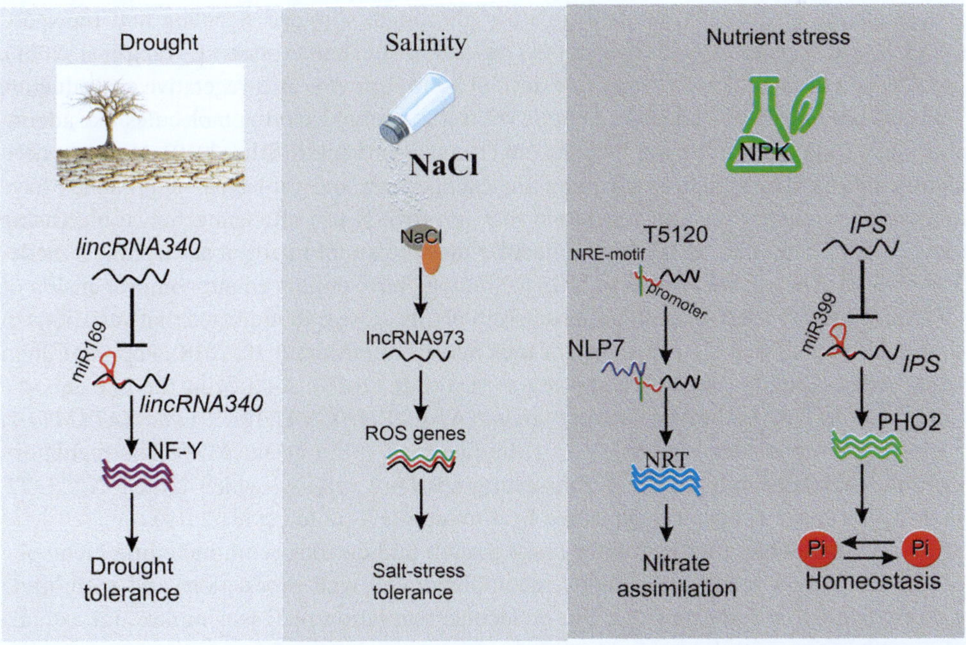

Figure 3.4 lncRNA-mediated stress responses in the plant. Under drought stress, a lncRNA, *lincRNA340*, acts as a target mimic for miR169, which triggers nuclear factor Y (NF-Y) genes. A salinity-responsive lncRNA, lncRNA973, serves as a positive regulator of reactive oxygen species (ROS). To maintain phosphate homeostasis at the soil–root interface, IPS1 regulates the expression of PHO2 by decoy miR399 in the root. Likewise, NLP7-mediated lncRNA *T5120* is functioned in nitrate assimilation in the plant root system.

Moreover, plants have developed various physiological and molecular mechanisms to enhance Pi utilization efficiency and availability in soil (Wang et al., 2018). The lncRNAs also function in regulating phosphate availability in plants has been extensively studied in several plant species including *Arabidopsis* and rice (Jabnoune et al., 2013; Yuan et al., 2017). Strand-specific RNA sequencing investigations in *Arabidopsis* under phosphorus-sufficient (P+) and phosphorus-deficient (P−) circumstances identified around 1212 lncRNAs, with about 309 lncRNAs exhibiting differential expression in both scenarios (Yuan et al., 2017). *Npc536* is a lncRNA in *Arabidopsis* that is upregulated in plant parts like leaves and roots due to drought and phosphate deficiency, influencing root system development under saline stress conditions (Ben Amor et al., 2009).

Phosphate acquisition, transport, and homeostasis are influenced by *PHOSPHATE 2* (*PHO2*) in *Arabidopsis*, with its expression modulated by miR399 (Huang et al., 2013). *PHO2* expression is essential for regulating phosphate homeostasis in settings with abundant phosphate supply. Under Pi deprivation, elevated levels of miR399 block *PHO2*, leading to its cleavage and enhancing the Pi translocation and uptake from root to shoot. In *Arabidopsis*, *IPS* possesses a 23-nucleotide long binding affinity for miR399, acting as a target mimic for miR399, which regulates the accumulation of the target gene *PHO2* during phosphate deficit and enhances its absorption (Franco-Zorrilla et al., 2007). In this situation, lncRNA *IPS* interacts with miR399, thereby diminishing its action on *PHO2* when phosphorus is plentiful in the soil. The elevated levels of *PHO2* impede *PHOSPHATE TRANSPORTER 1 (PHR1)* (Bari et al., 2006), hence averting Pi toxicity in roots through the promotion of sufficient Pi uptake. In phosphorus-deficient conditions, miR399 induces the degradation of *PHO2* mRNA, hence activating PHR and enhancing the accumulation of inorganic phosphate in the root (Figure 3.4). In *Medicago truncatula*, three lncRNAs are designated as *PHOSPHATE DEFICIENCY* induced long noncoding RNAs (PDILs) for their role in regulating phosphate shortage signaling. PDIL1 inhibits *MtPHO2* degradation viz miR399, hence regulating phosphate shortage signaling and transport, whereas PDIL2/3 participates in the transcriptional regulation of Pi homeostasis (T. Wang et al., 2017).

Nitrogen (N) is a crucial element for plant development, growth, and vegetative reproductive success. N is crucial for the biogenesis of peptide subunits, energy-carrying molecules like adenosine triphosphate (ATP), and nitrogen metabolism (Hawkesford & Griffiths, 2019). Nitrogen deficiency significantly impairs plant yield; nevertheless, genetic and molecular approaches have pinpointed various transporters and regulators that improve N-use efficiency, hence increasing plant productivity (Chen et al., 2016). The molecular mechanism of nitrogen absorption is inadequately understood. The advancements in NGS technology have enhanced our comprehension of the role of N-responsive lncRNAs in N-homeostasis in plants. Research indicated that variations in nitrogen availability modified the expression of lncRNAs in rice (Shin et al., 2018), Populus (Chen et al., 2016), barley (Chen et al., 2022), and maize (Lv et al., 2016). A study in the rice genome under N-deprivation clarified the functions of two cis-NAT-lncRNAs, *NATAMT1*.1 and *NATAMT1*.2, generated from the sense strand of *AMT1* in N-homeostasis (Shin et al., 2018). The regulatory mechanism of N-responsive lncRNA, a *trans*-acting siRNA 3 (*TAS3*), which targets *NITRATE TRANSPORTER*, is comprehensively explained in *Arabidopsis* (Fukuda et al., 2019).

Boron (B) is a vital element necessary for plant growth and development, including metabolic processes like nitrate assimilation, phenolic metabolism, cell wall production, and membrane integrity. Recent research indicated that the molecular regulation of B is complex and extends beyond the coding gene, with lncRNAs identified as novel regulators of gene expression modulation. A comprehensive genome-wide analysis of lncRNAs to B-toxicity identified roughly ~8000 lncRNAs (Unver & Tombuloglu, 2020), whereas the investigation of B-deficiency in *Poncirus trifoliata* uncovered 2101 lncRNAs, including 729 that were upregulated and 721 that were downregulated (Zhou et al., 2019).

Temperature, whether low or high, is a crucial limiting factor in plant growth and output. High-throughput RNA sequencing facilitates the identification of numerous QTLs and diverse regulatory genes associated with cold tolerance in plants (Jha et al., 2017). Similarly, numerous lncRNAs have been identified as candidate genes associated with cold tolerance, as extensively recorded across different plants. In *Arabidopsis*, weather-dependent reproductive process-related lncRNAs, such as *COLDWRAP*, *COLDAIR*, and *COOLAIR*, are involved in flower induction (Heo & Sung, 2011; Song et al., 2012; D.-H. Kim & Sung, 2017).

Vernalization is a distinct mechanism in flowering plants that initiates flowering under ideal spring conditions (Kim et al., 2009). In *Arabidopsis*, *FLC* functions as a flowering repressor. A vernalization-induced lncRNA, *COOLAIR*, is a NAT-lncRNA of *FLC*, involved in FLC suppression (Castaings et al., 2014). Another lncRNA, *COLDAIR*, derived from the first intron of the FLC gene inhibits the expression of the *FLC* locus via *PRC2*-mediated chromatin remodeling (Heo & Sung, 2011). The lncRNA *COLDWRAP*, originating from the *FLC* promoter, may contribute to the prolonged repression of FLC during vernalization (D. H. Kim et al., 2009). The functional mechanisms of cold-induced cryptic antisense *CBF1* (*asCBF1*) and *SVALKA* enhanced our comprehension of cold acclimatization processes in plants (Kindgren et al., 2018). In reaction to cold stress, the transcription of the cold-responsive *MAF4* induced by *MAS*, which recruits the WDR5a complex for epigenetic change at the MAF4 locus, leading to its activation and the consequent suppression of flowering (Zhao et al., 2018).

3.5 CRISPR/Cas lncRNA Editing Improves Plant Abiotic Stress Tolerance

CRISPR/Cas9 genome editing can be used to modify plant lncRNAs to help them withstand abiotic stresses. Unlike lncRNAs, CRISPR/Cas facilitated great progress in manipulating protein-coding genes. This is probably because functional characterization of lncRNAs is fraught with challenges and no effective method is available to study lncRNAs in plants. Here, this chapter provides summary and critical analysis of CRISPR/Cas-mediated lncRNA-editing findings in plants.

Ramírez Gonzales et al. (2021) demonstrated that an antisense lncRNA encoded by a *CYCLING DOF FACTOR 1* (*CDF1*) locus called *FLORE* is able to enhance tolerance to abiotic stress by its overexpression. CRISPR-Cas9 mutants indicate that *FLORE* increases drought tolerance by reducing water loss by controlling stomatal guard cell dynamics. In *Arabidopsis*, a lncRNA, *DANA1 (for Drought Associated long Noncoding RNA 1)* as a positive regulator of drought together with *DANA1-INTERACTING PROTEIN 1 (DIP1)*, a member of L1p/L10e family. CRISPR/Cas-knockout *DANA1* mutants encourage the accumulation of ABA contents by promoting HDA9-mediated histone deacetylation of *CYP707A1* and *CYP707A2*, thereby enhancing drought tolerance (Cai et al., 2024).

According to Du et al. (2018), inorganic phosphate deficiency-induced lncRNA, *Pi-deficiency-induced lncRNA 2* (*PILNCR2*) could inhibit the cleavage of *ZmPHO2* by *ZmmiRNA399*. Wang et al. (2023) further demonstrated that *PILNCR2* loss-of-function mutants suppress inorganic phosphate tolerance in maize. Taken together, these findings suggest the role of lncRNAs in inorganic phosphate homeostasis in plants. Niu et al. (2021) identified a nuclear-localized salt stress-responsive intergenic lncRNA located on chromosome 20 (*lncRNA77580*) in soybean. They found that deletion mutation in *lncRNA77580* by dual-single guide RNA/Cas9 significantly altered the expression of several salt stress-responsive genes. Another study by Jia et al. (2023) identified a lncRNA *BplncSIR1* that was highly induced by salt stress in birch. CRISPR-edited (knockout) *BplncSIR1* improves growth and confers salinity tolerance in birch by regulating *BplNAC2*, controlling stomatal aperture, and improving reactive oxygen scavenging (ROS) capability (Jia et al., 2024).

3.6 Future Directions and Perspective

Despite the great potential of CRISPR/Cas-mediated lncRNA editing to generate stress-tolerant plants, several significant challenges need to be resolved. A major challenge is the precision in lncRNA identification and functional characterization in plants. Despite similarities between protein-coding genes and lncRNAs, the key difference is in their cross-species conservation. lncRNAs have expression pattern specificity that is specific to certain plant tissues/organs or stress conditions. This variability in expression makes it more challenging to anticipate the functional characterization of identified lncRNAs. Applications of advanced sequencing technologies like NGS, including RNA-seq and chromatin interaction analysis, have significantly aided in the discovery of new lncRNAs. Nevertheless, the functional analysis of lncRNAs in plants is still in its infancy, requiring more detailed investigation to comprehend their various functions in challenging environmental conditions.

A significant barrier to CRISPR/Cas9 editing is the possibility of off-target consequences, which may lead to unpredicted alterations in the genome and phenotypes. Although much effort has been made to minimize off-target effects by identifying and introducing Cas9 variants and improving gRNA design. To prevent undesired outcomes, it is pivotal to prioritize the accuracy of lncRNA editing.

The second significant barrier in hindering the commercialization of lncRNA-edited plants and the legislative regulatory frameworks governing the utilization of CRISPR/Cas-edited crops vary greatly among nations. The issue of public acceptance of genetically modified crops is a significant concern that necessitates consideration. Establishing public trust and acceptance of CRISPR/Cas technology requires transparency and publicizing its advantages and potential challenges.

In the future, integrating CRISPR/Cas technology with other omics approaches, including proteomics, transcriptomics, and metabolomics, could improve our understanding of lncRNA functions in the context of stress responses, for editing and identifying new lncRNA targets, and also aid in generating crops that have improved abiotic stress resilience. Moreover, another important emerging aspect is CRISPR/Cas-based transcriptional activators (CRISPRa) and repressors (CRISPRi), providing new ways for altering lncRNA expression without modulating the fundamental DNA sequence. CRISPR/Cas-based transcriptional can be used to temporarily activate or suppress lncRNAs in specific environmental conditions, providing a more dynamic and flexible way of conferring stress tolerance.

An effective approach entails utilizing multiplexed CRISPR/Cas systems to simultaneously modify a few lncRNAs with their corresponding targets. The interactive and repetitive nature of the stress regulatory network could be utilized to achieve synergistic effects that enhance the stress tolerance of crop plants. With the emerging understanding of epigenetic control of lncRNAs, it might become possible to utilize CRISPR/Cas systems for epigenetic regulations associated with lncRNAs. This strategy has the potential to add an extra layer of lncRNA expression regulation and present novel methods for improving plant resistance to abiotic stressors.

3.7 Conclusion

CRISPR/Cas-mediated lncRNA editing is a potential approach to developing abiotic stress-tolerant crops. Researchers generated stress-tolerant plants by utilizing the diverse regulatory capabilities of lncRNAs to modulate the expression of stress-responsive genes. Nevertheless, notable hurdles to overcome, including the precise determination of the lncRNA functions and the possibility of off-target impacts. Although there are a few other obstacles, the perspective for editing plants lncRNA is promising, particularly due to the ongoing progress of CRISPR/Cas techniques and integration with omics approaches. By employing appropriate techniques and conducting meticulous study, the use of CRISPR/Cas-mediated lncRNA editing holds great potential in advancing crop cultivation that can withstand progressively challenging environmental conditions.

References

Adnan, M., Shah, Z., Fahad, S., Arif, M., Alam, M., Khan, I. A., Mian, I. A., et al. (2017). Phosphate-solubilizing bacteria nullify the antagonistic effect of soil calcification on bioavailability of phosphorus in alkaline soils. *Scientific Reports*, *7*(1), 16131.

Ali, Z., Shami, A., Sedeek, K., Kamel, R., Alhabsi, A., Tehseen, M., Hassan, N., Butt, H., Kababji, A., & Hamdan, S. M. (2020). Fusion of the Cas9 endonuclease and the VirD2 relaxase facilitates homology-directed repair for precise genome engineering in rice. *Communications Biology*, *3*, 44.

Bari, R., Pant, B. D., Stitt, M., & Scheible, W.-R. (2006). PHO2, microRNA399, and PHR1 define a phosphate-signaling pathway in plants. *Plant Physiology*, *141*(3), 988–999.

Belhaj, K., Chaparro-Garcia, A., Kamoun, S., Patron, N. J., & Nekrasov, V. (2015). Editing plant genomes with CRISPR/Cas9. *Current Opinion in Biotechnology*, *32*, 76–84.

Ben Amor, B., Wirth, S., Merchan, F., Laporte, P., d'Aubenton-Carafa, Y., Hirsch, J., Maizel, A., Mallory, A., Lucas, A., Deragon, J. M., Vaucheret, H., Thermes, C., Crespi, M., & Crespi, M. (2009). Novel long non-protein coding RNAs involved in *Arabidopsis* differentiation and stress responses. *Genome Research*, *19*(1), 57–69.

Bortesi, L., & Fischer, R. (2015). The CRISPR/Cas9 system for plant genome editing and beyond. *Biotechnology Advances*, *33*, 41–52.

Brooks, C., Nekrasov, V., Lippman, Z. B., & Eck, J. (2014). Efficient gene editing in tomato in the first generation using the clustered regularly interspaced short palindromic repeats/CRISPR-associated9 system. *Plant Physiology*, *166*, 1292–1297.

Cai, J., Zhang, Y., He, R., Jiang, L., Qu, Z., Gu, J., Yang, J., Legascue, M. F., Wang, Z.-Y., & Ariel, F. (2024). LncRNA DANA1 promotes drought tolerance and histone deacetylation of drought responsive genes in *Arabidopsis*. *EMBO Reports*, *25*, 796–812.

Castaings, L., Bergonzi, S., Albani, M. C., Kemi, U., Savolainen, O., & Coupland, G. (2014). Evolutionary conservation of cold-induced antisense RNAs of FLOWERING LOCUS C in Arabidopsis thaliana perennial relatives. *Nature Communications*, *5*(1), 4457.

Chen, M., Chenlu, W., Hai, B., Hui, C., & Yanwei, W. (2016). Genome-wide identification and characterization of novel lncRNAs in Populus under nitrogen deficiency. *Molecular Genetics and Genomics*, 291, 1663–1680.

Chen, Z., Li, L., Halford, N. G., Hongwei, X., Huang, L., Gao, R., Ruiju, L., & Liu, C. (2022). Advances in barley breeding for improving nitrogen use efficiency. *Agronomy*, *12*(7), 1682.

Christian, M., Cermak, T., Doyle, E. L., Schmidt, C., Zhang, F., & Hummel, A. (2010). Targeting DNA double-strand breaks with TAL effector nucleases. *Genetics*, *186*, 757–761.

Chung, P. J., Jung, H., Jeong, D.-H., Ha, S.-H., Choi, Y. D., & Kim, J.-K. (2016). Transcriptome profiling of drought responsive noncoding RNAs and their target genes in rice. *BMC Genomics*, *17*, 563–563.

Cong, L., Ran, F. A., Cox, D., Lin, S., Barretto, R., & Habib, N. (2013a). Multiplex genome engineering using CRISPR/Cas systems. *Science*, *339*, 819–823.

Cong, L., Ran, F. A., Cox, D., Lin, S., Barretto, R., Habib, N., Hsu, P. D., Wu, X., Jiang, W., & Marraffini, L. A. (2013b). Multiplex genome engineering using CRISPR/Cas systems. *Science*, *339*, 819–823.

Du, Q., Wang, K., Zou, C., Xu, C., & Li, W.-X. (2018). The PILNCR1-miR399 regulatory module is important for low phosphate tolerance in maize. *Plant Physiology*, *177*, 1743–1753.

Feng, Z., Mao, Y., Xu, N., Zhang, B., Wei, P., & Yang, D. L. (2014). Multigeneration analysis reveals the inheritance, specificity, and patterns of CRISPR/Cas-induced gene modifications in Arabidopsis. *Proceedings of the National Academy of Sciences of the United States of America*, *111*, 4632–4637.

Franco-Zorrilla, J. M., Valli, A., Todesco, M., Mateos, I., Puga, M. I., Rubio-Somoza, I., Leyva, A., Weigel, D., García, J. A., & Paz-Ares, J. (2007). Target mimicry provides a new mechanism for regulation of microRNA activity. *Nature Genetics*, *39*(8), 1033–1037.

Fukuda, M., Nishida, S., Kakei, Y., Shimada, Y., & Fujiwara, T. (2019). Genome-wide analysis of long intergenic noncoding RNAs responding to low-nutrient conditions in Arabidopsis thaliana: possible involvement of trans-acting siRNA3 in response to low nitrogen. *Plant and Cell Physiology*, *60*(9), 1961–1973.

Gao, J., Wang, G., Ma, S., Xie, X., Wu, X., & Zhang, X. (2015). CRISPR/Cas9mediated targeted mutagenesis in *Nicotiana tabacum*. *Plant Molecular Biology*, *87*, 99–110.

Hawkesford, M. J., & Griffiths, S. (2019). Exploiting genetic variation in nitrogen use efficiency for cereal crop improvement. *Current Opinion in Plant Biology*, *49*, 35–42.

Heo, J. B., & Sung, S. (2011). Vernalization-mediated epigenetic silencing by a long intronic noncoding RNA. *Science*, *331*(6013), 76–79.

Horvath, P., & Barrangou, R. (2010). CRISPR/Cas, the immune system of bacteria and archaea. *Science*, *327*, 167–170.

Huang, T.-K., Han, C.-L., Lin, S.-I., Chen, Y.-J., Tsai, Y.-C., Chen, Y.-R., Chen, J.-W., et al. (2013). Identification of downstream components of ubiquitin-conjugating enzyme PHOSPHATE2 by quantitative membrane proteomics in Arabidopsis roots. *The Plant Cell*, *25*(10), 4044–4060.

Jabnoune, M., Secco, D., Lecampion, C., Robaglia, C., Shu, Q., & Poirier, Y. (2013). A rice cis-natural antisense RNA acts as a translational enhancer for its cognate mRNA and contributes to phosphate homeostasis and plant fitness. *The Plant Cell*, *25*(10), 4166–4182.

Jain, M. (2015). *Function Genomics of Abiotic Stress Tolerance in Plants: A CRISPR Approach*, Frontiers Media SA.

Jha, U. C., Bohra, A., Parida, S. K., & Jha, R. (2017). Integrated "omics" approaches to sustain global productivity of major grain legumes under heat stress. *Plant Breeding*, *136*(4), 437–459.

Jha, U. C., Nayyar, H., Jha, R., Khurshid, M., Zhou, M., Mantri, N., & Siddique, K. H. M. (2020a). Long non-coding RNAs: emerging players regulating plant abiotic stress response and adaptation. *BMC Plant Biology*, *20*, 1–20.

Jha, U. C., Nayyar, H., Jha, R., Khurshid, M., Zhou, M., Mantri, N., & Siddique, K. H. M. (2020b). Long non-coding RNAs: emerging players regulating plant abiotic stress response and adaptation. *BMC Plant Biology*, *20*, 466.

Jia, H., & Wang, N. (2014). Targeted genome editing of sweet orange using Cas9/sgRNA. *PLoS One*, *9*, e93806.

Jia, Y., Zhao, H., Niu, Y., & Wang, Y. (2023). Identification of birch lncRNAs and mRNAs responding to salt stress and characterization of functions of lncRNA. *Horticulture Research*, *10*, uhac277.

Jia, Y., Zhao, H., Niu, Y., & Wang, Y. (2024). Long noncoding RNA from *Betula platyphylla*, Bplncsir1, confers salt tolerance by regulating BpNAC2 to mediate reactive oxygen species scavenging and stomatal movement. *Plant Biotechnology Journal, 22*, 48–65.

Jiang, W., Zhou, H., & BI, H., Fromm, M., Yang, B. & Weeks, D. P. (2013). Demonstration of CRISPR/Cas9/sgRNA-mediated targeted gene modification in Arabidopsis, tobacco, sorghum and rice. *Nucleic Acids Research, 41*, 188.

Jiang, W., Yang, B., & Weeks, D. P. (2014). Efficient CRISPR/Cas9-mediated gene editing in Arabidopsis thaliana and inheritance of modified genes in the T2 and T3 generations. *PLoS One, 9*, e99225.

Kim, D.-H., & Sung, S. (2017). Vernalization-triggered intragenic chromatin loop formation by long noncoding RNAs. *Developmental Cell, 40*(3), 302–312.

Kim, Y. G., Cha, J., & Chandrasegaran, S. (1996). Hybrid restriction enzymes: zinc finger fusions to Fok I cleavage domain. *Proceedings of the National Academy of Sciences of the United States of America, 93*, 1156–1160.

Kim, D.-H., Doyle, M. R., Sung, S., & Amasino, R. M. (2009). Vernalization: winter and the timing of flowering in plants. *Annual Review of Cell and Developmental, 25*(1), 277–299.

Kindgren, P., Ard, R., Ivanov, M., & Marquardt, S. (2018). Transcriptional read-through of the long non-coding RNA SVALKA governs plant cold acclimation. *Nature Communications, 9*(1), 4561.

Lamaoui, M., Jemo, M., Datla, R., & Bekkaoui, F. (2018). Heat and drought stresses in crops and approaches for their mitigation. *Frontiers in Chemistry, 6*, 26–26.

Li, S., Yu, X., Lei, N., Cheng, Z., Zhao, P., He, Y., Wang, W., & Peng, M. (2017). Genome-wide identification and functional prediction of cold and/or drought-responsive lncRNAs in cassava. *Scientific Reports, 7*, 45981.

Liang, D., Zhang, Z., Wu, H., Huang, C., Shuai, P., Ye, C.-Y., Tang, S., Wang, Y., Yang, L., & Wang, J. (2014). *Single-Base-Resolution Methylomes of Populus trichocarpa Reveal the Association Between DNA Methylation and Drought Stress*, Springer, 1–11.

Liu, L., Zhang, J., Xu, J., Li, Y., Guo, L., Wang, Z., Zhang, X., Zhao, B., Guo, Y.-D., & Zhang, N. (2020). CRISPR/Cas9 targeted mutagenesis of SlLBD40, a lateral organ boundaries domain transcription factor, enhances drought tolerance in tomato. *Plant Science, 301*, 110683.

Lv, Y., Liang, Z., Ge, M., Qi, W., Zhang, T., Lin, F., Peng, Z., & Zhao, H. (2016). Genome-wide identification and functional prediction of nitrogen-responsive intergenic and intronic long non-coding RNAs in maize (Zea mays L.). *BMC Genomics, 17*, 1–15.

Mali, P., Yang, L., Esvelt, K. M., Aach, J., Guell, M., & Dicarlo, J. E. (2013). RNA-guided human genome engineering via Cas9. *Science, 339*, 823–826.

Manghwar, H., Lindsey, K., Zhang, X. L., & Jin, S. X. (2019). CRISPR/Cas system: recent advances and future prospects for genome editing. *Trends in Plant Science, 24*, 1102–1125.

Marraffini, L. A., & Sontheimer, E. J. (2010). Self versus non-self discrimination during CRISPR RNA-directed immunity. *Nature, 463*, 568–571.

Mei, Y., Wang, Y., Chen, H., Sun, Z. S., & Ju, X.-D. (2016). Recent progress in CRISPR/Cas9 technology. *Journal of Genetics and Genomics, 43*, 63–75.

Miao, J., Guo, D., Zhang, J., Huang, Q., Qin, G., & Zhang, X. (2013). Targeted mutagenesis in rice using CRISPR-cas system. *Cell Research, 23*, 1233–1236.

Miller, J. C., Tan, S., Qiao, G., Barlow, K. A., Wang, J., & Xia, D. F. (2011). A TALE nuclease architecture for efficient genome editing. *Nature Biotechnology, 29*, 143–148.

Misra, V., Mall, A. K., Pandey, H., Srivastava, S., & Sharma, A. (2023). Advancements and prospects of CRISPR/Cas9 technologies for abiotic and biotic stresses in sugar beet. *Frontiers in Genetics, 14*, 1235855.

Niu, F., Jiang, Q., Sun, X., Hu, Z., Wang, L., Zhang, H., & Bozhkov, P. (2021). Large DNA fragment deletion in lncRNA77580 regulates neighboring gene expression in soybean (*Glycine max*). *Functional Plant Biology, 48*, 1139–1147.

Qi, X., Xie, S., Liu, Y., Yi, F., & Yu, J. (2013). Genome-wide annotation of genes and noncoding RNAs of foxtail millet in response to simulated drought stress by deep sequencing. *Plant Molecular Biology, 83*, 459–473.

Qin, T., Zhao, H., Cui, P., Albesher, N., & Xiong, L. (2017). A nucleus-localized long non-coding RNA enhances drought and salt stress tolerance. *Plant Physiology, 175*(3), 1321–1336. https://doi.org/10.1104/pp.17.00574. Epub 2017 Sep 8. PMID: 28887353; PMCID: PMC5664461

Ramírez Gonzales, L., Shi, L., Bergonzi, S. B., Oortwijn, M., Franco-Zorrilla, J. M., Solano-Tavira, R., Visser, R. G. F., Abelenda, J. A., & Bachem, C. W. B. (2021). Potato cycling DOF FACTOR 1 and its lncRNA counterpart StFLORE link tuber development and drought response. *The Plant Journal, 105*, 855–869.

Reyon, D., Tsai, S. Q., Khayter, C., Foden, J. A., Sander, J. D., & Joung, J. K. (2012). Flash assembly of TALENs for high-throughput genome editing. *Nature Biotechnology, 30*, 460–465.

Shan, Q., Wang, Y., Li, J., & Gao, C. (2014). Genome editing in rice and wheat using the CRISPR/Cas system. *Nature Protocols, 9*, 2395–2410.

Shin, S.-Y., Jeong, J. S., Lim, J. Y., Kim, T., Park, J. H., Kim, J.-K., & Shin, C. (2018). Transcriptomic analyses of rice (Oryza sativa) genes and non-coding RNAs under nitrogen starvation using multiple omics technologies. *BMC Genomics, 19*, 1–20.

Song, J., Angel, A., Howard, M., & Dean, C. (2012). Vernalization–a cold-induced epigenetic switch. *Journal of Cell Science, 125*(16), 3723–3731.

Tegeder, M., & Masclaux-Daubresse, C. (2018). Source and sink mechanisms of nitrogen transport and use. *New Phytologist, 217*(1), 35–53.

Traubenik, S., Charon, C., & Blein, T. (2024). From environmental responses to adaptation: the roles of plant lncRNAs. *Plant Physiology, 195*, 232–244.

Unver, T., & Tombuloglu, H. (2020). Barley long non-coding RNAs (lncRNA) responsive to excess boron. *Genomics, 112*(2), 1947–1955.

Wang, T., Zhao, M., Zhang, X., Liu, M., Yang, C., Chen, Y., Chen, R., Wen, J., Mysore, K. S., & Zhang, W. H. (2017). Novel phosphate deficiency-responsive long non-coding RNAs in the legume model plant Medicago truncatula. *Journal of Experimental Botany, 68*, 5937–5948.

Wang, F., Deng, M., Xu, J., Zhu, X., & Mao, C. (2018). Molecular mechanisms of phosphate transport and signaling in higher plants. In: *Seminars in Cell & Developmental Biology, Vol. 74*, Academic Press, 114–122.

Wang, Y., Wang, Z., Du, Q., Wang, K., Zou, C., & Li, W.-X. (2023). The long non-coding RNA PILNCR2 increases low phosphate tolerance in maize by interfering with miRNA399-guided cleavage of ZmPHT1s. *Molecular Plant, 16*, 1146–1159.

Waseem, M., Liu, Y., & Xia, R. (2020). Long non-coding RNAs, the dark matter: an emerging regulatory component in plants. *International Journal of Molecular Sciences, 22*, 86.

Yang, S. H., Kim, E., Park, H., & Koo, Y. (2022). Selection of the high efficient sgRNA for CRISPR-Cas9 to edit herbicide related genes, PDS, ALS, and EPSPS in tomato. *Applied Biological Chemistry, 65*, 13.

Yuan, Y., Gao, M., Zhang, M., Zheng, H., Zhou, X., Guo, Y., Zhao, Y., Kong, F., & Li, S. (2017). QTL mapping for phosphorus efficiency and morphological traits at seedling and maturity stages in wheat. *Frontiers in Plant Science, 8*, 614.

Zhang, W., Han, Z., Guo, Q., Liu, Y., Zheng, Y., Wu, F., & Jin, W. (2014). Identification of maize long non-coding RNAs responsive to drought stress. *PLoS One, 9*, e98958.

Zhang, C., Tang, G., Peng, X., Sun, F., Liu, S., & XI, Y. (2018). Long non-coding RNAs of switchgrass (*Panicum virgatum* L.) in multiple dehydration stresses. *BMC Plant Biology*, *18*, 79–79.

Zhao, X., Li, J., Lian, B., Hanqing, G., Li, Y., & Qi, Y. (2018). Global identification of Arabidopsis lncRNAs reveals the regulation of MAF4 by a natural antisense RNA. *Nature Communications*, *9*(1), 5056.

Zhou, H., Liu, B., Weeks, D. P., Spalding, M. H., & Yang, B. (2014). Large chromosomal deletions and heritable small genetic changes induced by CRISPR/Cas9 in rice. *Nucleic Acids Research*, *42*, 10903–10914.

Zhou, G.-F., Zhang, L.-P., Bi-Xian Li, O., Sheng, Q.-J. W., Yao, F.-X., Guan, G., & Liu, G.-D. (2019). Genome-wide identification of long non-coding RNA in trifoliate orange (Poncirus trifoliata (L.) Raf) leaves in response to boron deficiency. *International Journal of Molecular Sciences*, *20*(21), 5419.

4

CRISPR/Cas-Modified Long Noncoding RNAs for Regulating Plant Abiotic Responses

Diliane Harumi Yaguinuma[1], Jardel de Oliveira[1], Fernanda Freitas de Oliveira[2], and Tiago Benedito dos Santos[3]

[1] *Department of Agronomy - Postgraduate Program in Agronomy - Plant Production, São Paulo Western University, Presidente Prudente-SP, Brazil*
[2] *Genetics Department, Federal University of Parana, Curitiba-PR, Brazil*
[3] *Department of Agronomy - Program in Agronomy - Plant Production, São Paulo Western University, Presidente Prudente-SP, Brazil*

4.1 Introduction

Plants constantly face challenging adverse conditions that hinder growth and reduce yields. To survive as stationary organisms, they have evolved sophisticated strategies to detect, resist, and adapt to biotic stressors – such as viruses, fungi, and pathogens – and abiotic stressors, including temperature fluctuations, drought, and salinity (Imran et al., 2021). Abiotic stresses, particularly drought and temperature changes, are becoming constants due to climate change, making it essential to understand the molecular basis of plant responses (dos Santos et al., 2022).

To cope with temperature changes, plants have developed mechanisms involving signal transduction pathways that activate stress-specific transcription factors (TFs), genes, and antioxidants. Hormones regulate these pathways by altering levels in response to stress, and they interact with reactive oxygen species (ROS), which further influence hormone biosynthesis (Devireddy et al., 2021). WRKY-TFs is considered a large TF family, which participates in regulating stress-sensitive genes, helping plants balance growth and stress responses (Khoso et al., 2022).

In terms of drought response, the genes (*PtDi19-2* and *PtDi19-7*) in poplar improve significantly drought tolerance via ABA-dependent pathways (Wu et al., 2022). Similarly, *HvSAP8* and *HvSAP16* in *Hordeum vulgare* enhance drought tolerance, increasing grain weight and shoot number (Baidyussen et al., 2021). In *Oryza sativa*, *OsERF83* regulates genes related to drought tolerance, while *OsERF115/AP2EREBP110* promotes both heat and drought resistance (Park et al., 2021). The biosynthesis of the osmoprotectant proline in *Stipa purpurea*, through the *SpP5CS* gene, increases drought resistance by consequently increasing proline levels (Yang et al., 2021).

Among the various abiotic stresses, salt stress uses mechanisms like the SOS signaling pathway, involving *SOS1*, *SOS2*, and *SOS3*, to regulate Na^+ concentrations and osmotic pressure in plants (Shi et al., 2000; Qiu et al., 2002; Quintero et al., 2002). Additionally, glycosyl inositol phosphorylceramide (GIPC) sphingolipids may act as salt stress sensors, and *AtANNEXIN 4* plays a role in increasing cytoplasmic calcium levels under salt stress (Jiang et al., 2019; Ma et al., 2019).

Genome and Epigenome Editing for Stress-Tolerant Crops, First Edition. Edited by Jen-Tsung Chen.

Lastly, long noncoding RNAs (lncRNAs) have emerged as critical regulators in plant responses to abiotic stress, playing key roles in environmental adaptation and stress memory (Yang et al., 2023a). As a molecule with multiple biological functions, their involvement in regulating gene expression makes them promising targets for enhancing stress tolerance in crops. With the advancement of CRISPR/Cas9 genome editing, it is now possible to precisely manipulate lncRNAs or their associated regulatory pathways to improve plant resilience. Nonetheless, by targeting specific lncRNAs that influence drought and stress responses, CRISPR can help develop crops with enhanced tolerance, offering a powerful tool for ensuring yield stability in changing climates in the world.

4.1.1 Understanding a Little About Long Noncoding RNA (lncRNA)

lncRNA can function as decoys and is considered a transcription product with no relevant biological function, being classified as noise (Mercer et al., 2009; Wu et al., 2020). According to the literature, lncRNAs are classified as RNA that can have a function independent of its protein-coding potential and that is produced by another mechanism (Wu et al., 2020; Wierzbicki et al., 2021). The above is true with the first uses of the term lncRNA (Shamovsky & Nudler, 2006) and avoids the requirement for an arbitrary cutoff (200 nucleotides – nt). This class of RNAs is described as transcripts that can exceed more than 200 nt in length and cannot be translated into protein (Kapranov et al., 2007). Variability with respect to length, lncRNAs can vary considerably (Long et al., 2017). Conversely, this variation in length can affect the amplification efficiency and mechanism during the construction process of a given library, respectively (Fang & Fullwood, 2016).

The lncRNAs can be transcribed by RNA pol: II, III, IV, and V and also control distinct target regulation through multiple ways, which can include, for example, chromatin remodeling, transcriptional repression, and RNA splicing (Ariel et al., 2015; reviewed by Waititu et al., 2020). Research has shown that lncRNAs can be classified in several ways, by considering their different aspects (Quinn & Chang, 2016; Wierzbicki et al., 2021). It is believed that lncRNAs are classified into seven different classes according to their biogenesis (Budak et al., 2020). This classification may involve length, location in relation to protein-coding genes (PCGs), their association with other DNA elements (known function), biogenesis/stability, sublocation or subcellular origin, function, association with specific biological processes, presence in repetitions, and conservation of sequence and structure (Laurent et al., 2015). It should be noted that a lncRNA can fit into more than one of these categories. It is important to make clear that this class of RNA can perform several important functions, such as plant growth, development, and responses to different abiotic stresses at the transcriptional and posttranscriptional levels (D'Ario et al., 2017; Bhogireddy et al., 2021; Figure 4.1).

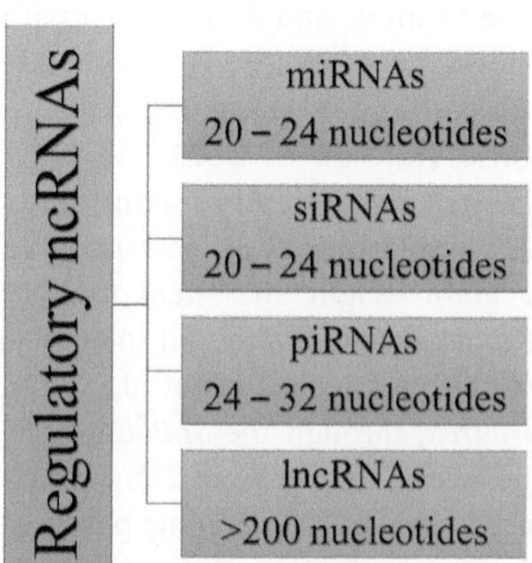

Figure 4.1 Classification of the regulatory ncRNAs consists of microRNAs (miRNAs), short interfering RNAs (siRNAs), piwi-interacting RNAs (piRNAs), and long noncoding RNAs (lncRNAs).

The action of lncRNAs is much more studied in animal systems than in plants (Mercer et al., 2009; Rinn & Chang, 2012; Wierzbicki et al., 2021). However, there is evidence that, in both systems, lncRNAs can act in recruitment and/or histone modification, thereby promoting chromatin reorganization (as reviewed by De Lucia &

Dean, 2011; Bhogireddy et al., 2021). Some of these RNAs can even encode small peptides that are rapidly degraded by proteasomes (Amor et al., 2009). Also, some miRNAs can target lncRNAs by producing phased small interfering RNAs (phasiRNAs). The lncRNAs can also be involved in the origin of miRNAs or in regulating the expression of miRNAs (Meng et al., 2021).

Studies performed out to date demonstrate that the expression of lncRNAs is generally more cell- and tissue-specific than PCGs (Derrien et al., 2012; Rinn & Chang, 2012). For example, in *Brassica rapa*, two lncRNAs have been described to mainly target miR160 and are involved in pollen development (Huang et al., 2018). Wu et al. (2019) reported that lncRNA XR_514690.2 in *Prunus mume* negatively regulates ppe-miR172d and also positively regulates *AP2* (*APETALA2*), which is believed to be related to flower development. Thus, this differential expression between different stages of differentiation indicates a function of refinement in cell fate. This spatiotemporal-specific expression may be linked to the establishment of well-defined gene expression barriers and cell-type-specific regulatory genetic programs (Ebert & Sharp, 2012; Wang et al., 2013). The advancement of new technologies (accumulation of extensive data sets of lncRNAs) has played a crucial role in new scientific discoveries (Pinkney et al., 2020). Numerical estimation of lncRNA has been a challenge among databases. For example, curated and manually annotated databases such as GENCODE estimate that these transcripts can reach approximately 19,000 genes that are transcribed into lncRNA (Frankish et al., 2023). However, automatically annotated databases such as NONCODE have estimates of 100,000 lncRNAs (Guigó, 2023). It is important to emphasize that the importance of lncRNAs has been widely recognized in recent decades and that their study has become a rapidly growing field of research. Nodule primordia of *Medicago truncculata* was the first study in plants where the first lncRNA was identified (EARLY NODULIN 40-*ENOD40*) (Crespi et al., 1994). lncRNAs have been identified in several plant species: *Arabidopsis thaliana* (Sun et al., 2020; Liu et al., 2020), *O. sativa* (Chen et al., 2018, 2021), *Zea mays* (Li et al., 2014), *Arachis hypogaea* L. (Ma et al., 2020b), *Gossypium hirsutum* (Zhao et al., 2018; Zhang et al., 2021), and *H. vulgare* L.JN8 (Li et al., 2023).

Previously, Li et al. (2014) identified 20,163 potential lncRNA candidates in maize. Of these, more than 90% were classified as small RNA precursors, and half of them were expressed in a tissue-specific manner. Zhang et al. (2014) in analysis of corn subjected to drought identified 1724 lncRNAs, and of these, 664 transcripts were responsive to stress. Khoei et al. (2021) investigated the participation of *Glycine max* lncRNAs in response to two nematodes (*Heterodera glycines* and *Rotylenchulus reniformis*) and identified 21 lncRNAs that appear to be involved in the plant's response to the pathogen.

Although few studies report the function of these molecules in plants, several of them are related to the impact of lncRNAs on vernalization processes (Yamaguchi & Abe, 2012; Liu et al., 2018a, b, 2022; Shea et al., 2019). In these conditions, where there is long exposure to cold during winter, the expression of an antisense transcript called Cold-Assisted Intronic Noncoding RNA (*COOLAIR*) is induced (Hawkes et al., 2016; Kim et al., 2017). An example of how lncRNAs are involved in environmental responses can be identified in the regulation of the flowering mechanism in *Arabidopsis* plants (*FLOWERING LOCUS C – FLC*). Furthermore, FLC can also be regulated by other antisense lncRNAs called Cold-Induced Long Antisense Intragenic RNA (COOLAIR) (Fatica & Bozzoni, 2014). It was also identified that an important region for the stable silencing of FLC by vernalization overlaps the promoter of lncRNAs COLDAIR (Sung et al., 2006; Kim et al., 2017; Jampala et al., 2021). "*INDUCED BY PHOSPHATE STARVATION 1*" is a member of the *TPS1/mt4* gene family that acts as a target mimic of miR399 in the fine regulation of expression of *PHO2* (phosphate transporter) in *Arabidopsis*, *Solanum lycopersicum*, and *M. truncatula* but does not encode protein (Aung et al., 2006; Franco-Zorrilla et al., 2007; reviewed by Wang et al. (2023)).

Arguably, with advances in new biological technologies, such as deep RNA sequencing (RNA-seq), researchers can study diverse organisms in depth (Moran et al., 2012; Budak et al., 2020). Zhang et al. (2021) studied the response mechanisms of lncRNAs to low-phosphorus stress in two soybean genotypes. In total, 4166 novel lncRNAs were identified (Zhang et al., 2021).

Additionally, we must take into consideration that the importance of lncRNAs (previously described as junk sequence) has been widely recognized and discussed in recent decades, and their study has become a rapidly growing field of research in all areas.

4.2 Long Noncoding RNA (lncRNA) from Plants May Also be Related to Stress

Plants are considered sessile species and, throughout their evolution, unfortunately, they were unable to escape adversity and stress (Chaudhry & Sidhu, 2022; Bibi & Rahman, 2023). However, in order to guarantee their survival, they have developed tolerance mechanisms when exposed to contrasting environmental conditions (Zhang et al., 2023). It is also known that some plant lncRNAs are involved in the abiotic stress mechanism, and this involvement is done through chromatin remodeling (Matzke & Mosher, 2014; Waititu et al., 2020). In recent decades, many lncRNAs have been identified and are involved in the mechanism related to abiotic stresses (Jha et al., 2020; Yang et al., 2023a). lncRNAs play numerous roles in plants, such as in development, in responses to certain hormones, in different stresses, and can also be associated with epigenetic events (Jha et al., 2020; Patra et al., 2023; Tremblay & Qüesta, 2023). We cannot fail to mention that genomic information through the advent of new next-generation sequencing (NGS) technologies is triggered, and there has been a great advance in the discovery of new lncRNAs in plants. Several studies of lncRNAs under abiotic stress have been described in plant species: *A. thaliana* (Aung et al., 2006; Amor et al., 2009; Yuan et al., 2016), *Musa itinerans* (Liu et al., 2018a, b), *Brassica juncea* (Bhatia et al., 2020), *B. rapa* ssp. Chinensis (Wang et al., 2019a), *Brassica napus* L. (Tan et al., 2020), *Capsicum annuum* (Baruah et al., 2021), *Solanum tuberosum* (Ramírez Gonzales et al., 2021), *G. hirsutum* (Deng et al., 2018; Zhang et al., 2021), and *Sorgum bicolor* (Zou et al., 2024). We report some studies that used lncRNAs under abiotic stress conditions (monocotyledons and dicotyledons) in Table 4.1.

Table 4.1 lncRNAs involved in abiotic stress response in monocotyledons and dicotyledons.

Name of lncRNAs	Type of stress	References
IPS1	↑ phosphate deficiency – *Arabidopsis thaliana*	Aung et al. (2006), Franco-Zorrilla et al. (2007), and Wu et al. (2013)
TalnRNA5	↑ heat stress – *Triticum aestivum*	Yao et al. (2010) and Xin et al. (2011)
TahlnRNA27	↑ heat stress – *Triticum aestivum*	Yao et al. (2010) and Xin et al. (2011)
TalnRNA21	↑ heat stress – *Triticum aestivum*	Yao et al. (2010) and Xin et al. (2011)
TahlnRNA3	↑ heat stress – *Triticum aestivum*	Yao et al. (2010) and Xin et al. (2011)
TahlnRNA14	↑ heat stress – *Triticum aestivum*	Yao et al. (2010) and Xin et al. (2011)
TahlnRNA19	↑ heat stress – *Triticum aestivum*	Yao et al. (2010) and Xin et al. (2011)
TahlnRNA36	↑ heat stress – *Triticum aestivum*	Yao et al. (2010) and Xin et al. (2011)
TahlnRNA41	↑ heat stress – *Triticum aestivum*	Yao et al. (2010) and Xin et al. (2011)
TahlnRNA42	↑ heat stress – *Triticum aestivum*	Yao et al. (2010) and Xin et al. (2011)

Table 4.1 (Continued)

Name of lncRNAs	Type of stress	References
TahlnRNA47	↑ heat stress – *Triticum aestivum*	Yao et al. (2010) and Xin et al. (2011)
TahlnRNA52	↑ heat stress – *Triticum aestivum*	Yao et al. (2010) and Xin et al. (2011)
lincRNA1128	↓ drought stress – *Populus trichocarpa*	Shuai et al. (2014)
lincRNA1393	↓ drought stress – *Populus trichocarpa*	Shuai et al. (2014)
lincRNA3018	↓ drought stress – *Populus trichocarpa*	Shuai et al. (2014)
lincRNA2752	↑ drought stress – *Populus trichocarpa*	Shuai et al. (2014)
lincRNA1795	↓ drought stress – *Populus trichocarpa*	Shuai et al. (2014)
lincRNA20	↑ drought stress – *Populus trichocarpa*	Shuai et al. (2014)
lincRNA2623	↓ drought stress – *Populus trichocarpa*	Shuai et al. (2014)
lincRNA2623	↓ drought stress – *Populus trichocarpa*	Shuai et al. (2014)
lincRNA2762	↓ drought stress – *Populus trichocarpa*	Shuai et al. (2014)
lincRNA2198	↑ drought stress – *Populus trichocarpa*	Shuai et al. (2014)
lincRNA2131	↑ drought stress – *Populus trichocarpa*	Shuai et al. (2014)
lincRNA2085	↑ drought stress – *Populus trichocarpa*	Shuai et al. (2014)
lincRNA2962	↑ drought stress – *Populus trichocarpa*	Shuai et al. (2014)
lincRNA1534	↑ drought stress – *Populus trichocarpa*	Shuai et al. (2014)
lincRNA1039	↑ drought stress – *Populus trichocarpa*	Shuai et al. (2014)
lincRNA2962	↓ drought stress – *Populus trichocarpa*	Shuai et al. (2014)
lncRNAs	↑ drought stress – *Zea mays*	Zhang et al. (2014)
lncRNAs	↑ light stress – *Arabidopsis thaliana*	Wang et al. (2014a, b)
lncRNAs	↑ light stress – *Arabidopsis thaliana*	Wang et al. (2014a, b)
COOLAIR	↑ cold stress – *Arabidopsis thaliana*	Csorba et al. (2014)
lncRNAs	↑ and ↓phosphate deficiency – *Oryza sativa*	Xu et al. (2016)
pti-MIR5471	↑ phosphate deficiency – *Phaeodactylum tricornutum*	Cruz de Carvalho et al. (2016)
pti-MIR5472	↑ phosphate deficiency – *Phaeodactylum tricornutum*	Cruz de Carvalho et al. (2016)
lincRNAs	↑ and ↓ nitrogen deficiency stress – *Populus tomentosa*	Chen et al. (2016)
lincRNAs	↑ and ↓ cold and heat stress – *Brassica rapa*	Song et al. (2016)
lncRNAs	↑ salt stress – *Vitis vinifera* L.	Deng et al. (2018)
VIT_200s0179n00030	↑ cold stress – *Vitis vinifera* L.	Wang et al. (2019a, b)
VIT_207s0141n00070	↑ cold stress – *Vitis vinifera* L.	Wang et al. (2019a, b)
VIT_207s0005n00480	↑ cold stress – *Vitis vinifera* L.	Wang et al. (2019a, b)
VIT_201s0010n00070	↓ cold stress – *Vitis vinifera* L.	Wang et al. (2019a, b)
VIT_208s0007n00270	↓ cold stress – *Vitis vinifera* L.	Wang et al. (2019a, b)
VIT_209s0002n00020	↓ cold stress – *Vitis vinifera* L.	Wang et al. (2019a, b)
lncRNAs	↑ salt stress – *Cicer arietinum* L.	Kumar et al. (2021)
lncRNAs	↑ salt stress – *Solanum pennellii*	Li et al. (2022)

According to the table, the symbol ↑ means that the expression is induced. The arrow symbol ↓ refers to the low expression of lncRNAs under abiotic stress conditions.

We report some studies for a greater understanding. It was observed that the lncRNA IPS1 can mimic miRNA399 in *A. thaliana* subjected to P (phosphorus) deficiency conditions, its action blocks the interaction of its target (PHOSPHATE 2 – PHO2) (Franco-Zorrilla et al., 2007; Wu et al., 2013). Xu et al. (2016) can attest to the similar role of lncRNAs in *O. sativa* P deficiency. In *Manihot esculenta*, lncNAT14 was active in plants subjected to cold stress after cooling acclimation (CCA) (Li et al., 2017). These authors were able to observe that under cold-induced stress, lncRNAs were expressed and activated similarly under control conditions (Li et al., 2017). Previously, in *H. vulgare* L., 535 differentially expressed lncRNAs were identified as being responsive to drought stress induced by the agent polyethylene glycol (PEG) (Qiu et al., 2019). To adapt to the deteriorating survival environment, in other plants, as well, there is a description of some lncRNAs involved in the mechanism of response to drought (Shuai et al., 2014; Zhang et al., 2014; Imaduwage & Hewadikaram, 2024; Zou et al., 2024). It was observed in *S. lycopersicum* that lncRNAs may be involved in reproductive development under drought stress (Lamin-Samu et al., 2022). In this study, a total of 67,770 lncRNAs were identified, of which 3053 lncRNAs responded to drought stress.

Among abiotic stresses, saline stress is one of the main restrictions of crops (dos Santos et al., 2022; Ondrasek et al., 2022). *Spirodela polyrhiza* presented a total of 2815 lncRNAs under salt stress conditions, of which 185 effectively participate in differential expression in response to salinity (Fu et al., 2020). In *Solanum pennelli*, lncRNAs were identified and their involvement in salt tolerance was identified (Li et al., 2022). According to these authors, 133 lncRNAs were upregulated in roots, and 100 lncRNAs were downregulated in tomato plants after salt treatment (Li et al., 2022). One *Phyllostachys edulis* lncRNA (*PelncRNA1*) was differentially expressed after UV-B treatment (Yu et al., 2023). A total of 7709 novel lncRNAs were identified in *Raphanus sativus* L., and 363 of these lncRNAs are involved in differential expression in response to salt stress (Sun et al., 2023). Notably, the predicted target genes of these salt-responsive lncRNAs can trigger and be involved in several processes: transcription regulation, scavenging of ROS, and ionic homeostasis, among other mechanisms (Sun et al., 2023). In agreement with this state, the expression levels of *PelncRNA1* and its target genes increased during UV-B treatment. Through high-throughput sequencing in *S. bicolor*, 11,955 lncRNAs were found in leaves (Zou et al., 2024). Many of these lncRNAs identified and that will be characterized may play important roles in the adaptation of plants to the adversities that will possibly affect humanity in the coming years.

4.2.1 CRISPR in Plant Genome Editing

Climatic changes significantly impact agricultural activities, resulting in reduced agricultural productivity, in addition to alterations in the frontiers of agroecosystems (Masson-Delmotte et al., 2021; Bibi & Rahman, 2023). Due to the continuous growth of the global population, it is expected that the food demand will rise between 70% and 100% by 2050, which will exert additional pressure on agricultural land. Developing nations depend heavily on agriculture and are particularly vulnerable to climatic change consequences and food insecurity (Kaushik et al., 2023). One of the UN's sustainable development goals (SDGs) is to achieve food security, making it essential to approach the factors that threaten agricultural production systems' sustainability to lessen the risks imposed by global climatic changes (Kaushik et al., 2023). It follows then that it is crucial to improve agricultural productivity and sustainability at a global level. Scientific advances and technological innovations in crop production and breeding are urgently needed its application to assure food safety in the future.

Taking everything into account, genome-editing techniques were developed to introduce precise modifications (punctual) in plants, allowing the achievement of specific desired characteristics (Saini et al., 2023). These techniques are driving a new era of plant improvements, distinguished by precision in the development of enhanced crops (Chen et al., 2019; Cardi et al., 2023; Lorenzo et al., 2023; Saini et al., 2023). The CRISPR system (clustered regularly interspaced short palindromic repeats) in Cas clusters (CRISPR-associated proteins) has stood out as one of the most advanced tools for plant genome editing. Its precision and efficiency have revolutionized genetic engineering (CRISPR/Cas), allowing specific gene modification to enhance the characteristics of crops (Shan et al., 2013; Chen et al., 2019; Liu & Doudna, 2020; Shahriar et al., 2021). Notably, the CRISPR-Cas system is a diverse set originating from the adaptive immunity of many bacteria and almost all archaea, each one of which consists of a combination of Cas effector proteins and CRISPR RNAs (crRNAs) (Liu & Doudna, 2020; Saini et al., 2023). These defense systems utilize small RNAs to recognize and silence, in a targeted way, foreign nucleic acids. CRISPR/Cas systems consist of Cas genes organized in CRISPR operons and matrices that contain directed sequences of the genome, called spacers, interleaved with identical repetitions (Terns & Terns, 2011; Wiedenheft et al., 2012; Liu & Doudna, 2020).

CRISPR/Cas-mediated immunity involves three main phases. In the adaptation stage, bacteria and archaea that have CRISPR loci answer to the invasion of viruses or plasmids by incorporating short fragments of foreign sequences, called protospacers, in the host chromosome, in the proximal end of the CRISPR matrix (Terns & Terns, 2011; Wiedenheft et al., 2012; Sternberg et al., 2016; Mosterd et al., 2021; Newsom et al., 2021). Therefore, in the expression and interference stages, the transition of the repetition spacer element in RNA CRISPR precursor (pre-crRNA) molecules, which are then processed by enzymatic cleavage, resulting in short crRNAs. These crRNAs can pair themselves with complementary sequences of protospacers in viral targets or invading plasmids (Deltcheva et al., 2011; Sashital et al., 2011; Newsom et al., 2021). The reconnaissance of these targets by the crRNAs guides the Cas proteins, which act together with the crRNAs to silence the foreign sequences (Brouns et al., 2008; Lintner et al., 2011; Zhang et al., 2012; Murugan et al., 2017; Liu & Doudna, 2020).

We can divide the CRISPR/Cas systems into three main types (Makarova et al., 2020). Types I and III share similar characteristics: both utilize Cas endonucleases to process the pre-crRNAs that, once mature, form complex multi-Cas capable of recognizing and then, cleaving nucleic acids complementary to the crRNAs. In contrast, the type II system follows a distinct mechanism, in which a crRNA transactivator (tracrRNA) guides the processing of pre-crRNA through ribonuclease RNase III in the presence of the Cas9 protein. Cas9 is considered mainly responsible for the silencing of foreign DNA guided by crRNA (Barrangou et al., 2007; Sapranauskas et al., 2011; Makarova et al., 2020).

Concurrently, the CRISPR/Cas system has shown itself effective in various agricultural species of great importance, such as *Z. mays* (Svitashev et al., 2016), *O. sativa* (Jiang et al., 2013; Wang et al., 2017), and *Triticum aestivum* (Shan et al., 2013; Wang et al., 2014), allowing the direct introduction of mutations in the elite germplasm. Some plant species utilized the CRISPR/Cas9 in *O. sativa* (Shan et al., 2015; Zhang et al., 2019), *T. aestivum* (Li et al., 2020), *Nicotiana benthamiana* and *A. thaliana* (Li et al., 2013), *S. tuberosum*, and *M. esculenta* (Chen et al., 2019; Zhu et al., 2020). For the first time in history, plant breeders managed to control the specific introduction of directed variations in the genetic sequence, which represents a revolutionary resource for rapid crop improvements. Since then, there has been continual progress in CRISPR/Cas systems, including the introduction of CRISPR/Cpf1, which expanded even more the capabilities and applications of this technology in agricultural biotechnology (Zetsche et al., 2015).

In recent years, the use of CRISPR/Cas technology in plants has allowed the identification and functional analysis of TFs associated with kinases similar to receptors in *O. sativa* (Chen et al., 2022) and genes involved in plant immunity and regulation of transposable elements (TEs) in *H. vulgare* L.

(Galli et al., 2022). However, CRISPR/Cas is capable of creating crops that are resistant to both ambiental stresses (Ali et al., 2022; Shelake et al., 2022) and diseases (Karmakar et al., 2022; Talakayalaet al., 2022). Another important advantage of CRISPR/Cas systems is the ability to direct multiple alleles at the same time, which allows the genetic editing of polyploid plants (Ahmad et al., 2021; Eid et al., 2021).

Elite crop varieties can have their genomes edited in a precise manner to generate new variants that present greater resistance to ambiental stresses and greater efficiency in the use of nutrients, without the limitations imposed by existing genetic variation. In addition, the continual edition of these elite varieties does not introduce potentially prejudicial alleles that could emerge from crossbreeding and recombinations, eliminating also the necessity of prolonged backcrossings to reconstruct the original genetic base (Scheben et al., 2017). Thanks to their reduced cost, simplicity of use, and high efficiency, CRISPR/Cas became the most widely adopted tool for genome editing in plants (Yin et al., 2017).

4.2.2 CRISPR in Long Noncoding (lncRNA) Plant RNA Editing

The creation of innovative techniques in enhancement is essential to optimize time, resources utilized, and the necessary space for the production and selection of genetically modified plants (Gao, 2021; Wanga et al., 2021). In this context, modern genome-editing tools present a significant potential to modify plant genomes, seeking an understanding of genetic functions and agricultural characteristics enhancements desirable (Zhang et al., 2018; Miladinovic et al., 2021).

Genome edition through CRISPR/Cas technology has been widely applied in the past decade, especially in plants' polypeptides codifying regions (Liu et al., 2021; Maharajan et al., 2022; Chincinska et al., 2023), and this innovative method began to be explored for the study of noncodifying regions of plant genome's biological functions (Deng et al., 2022). One of the advantages of the CRISPR/Cas system is the capability to direct multiple alleles at the same time, which allows genetic editing in polyploid plants (Ahmad et al., 2021; Eid et al., 2021).

The predominant group of regulatory noncodifying RNAs (ncRNAs) in plants is composed of endogenous transcriptions of varying sizes that do not generate polypeptides, but play a crucial role in genic expression control. Indeed, these noncodifying transcriptions in plants are categorized into various subgroups that include small interfering RNAs (siRNAs), microRNAs (miRNAs), lncRNAs, small RNAs derived from tRNA (tsRNAs, from 15 to 42 nt), and circular RNAs (circRNAs, with more than 100 nucleotides) (Waititu et al., 2020; Bhogireddy et al., 2021; Chao et al., 2022).

Updating, we can mention that the lncRNAs are a class of RNA molecules with more than 200 nt that, in general, are not translated into proteins (Trivedi & Hasan, 2020; Traubenik et al., 2024). They act as crucial regulators of genic expression and play different roles in various cellular processes. Some lncRNAs act as "sponges" when interacting with miRNAs, modulating their disponibility and, in that way, affecting the genetic expression. (Traubenik et al., 2024). This competitive interaction gives origin to complex regulatory networks that influence plant development, as well as their response to stress (Yang et al., 2023b). Some lncRNAs are involved in the mechanism of differentiation and cellular development in plants (Whittaker & Dean, 2017; Chen et al., 2020).

However, research focused on functional lncRNA studies in plants is still limited due to their low levels of expression, their low conservation in different plant species, and their expression specificity in different tissues (Basak & Nithin, 2015; Li et al., 2018). CRISPR/Cas technology has been pointed out as a promising tool to overcome these challenges in regulatory lncRNAs functional analysis in plants (Basso et al., 2019; Wu et al., 2020; Bhogireddy et al., 2021). There is still little information related to lncRNA manipulation mediated by CRISPR/Cas. That is due, in part, to the fact that lncRNA functional analysis faces many challenges and uncertainties that need to be approached in future research (Yu & Zhu, 2019).

Plant and animal genomes express a great number of lncRNAs, making it fundamental a system to classify and understand their functions. Furthermore, this perception requires a reevaluation of the quantity and kind of information needed to program the development of a particular organism (Mattick et al., 2023). It is of fundamental importance to understand the role of lncRNAs and their dynamic interactions with other macromolecules because it can achieve a wider vision of cellular biology, in the development of organisms and the interactions of genes and the environment. Those studies will open new perspectives on the complexity of genetic regulation and its importance to fundamental cellular processes, including the development and the response to ambiental stress (Mattick et al., 2023).

Recent studies have shown that lncRNAs have played crucial roles in the regulation of the signaling of ethylene in tomatoes, influencing important characteristics such as color, flavor, and texture of the fruit (Ma et al., 2020a). Furthermore, those lncRNAs have a vital role in the defense against viral and fungal infections (Sharma et al., 2022). Despite these discoveries, the use of CRISPR/Cas to investigate the functional role of those lncRNAs is still in its initial stage of exploration.

We can mention, for example, that lncRNAs (*G. max*) are involved in various crucial processes, such as the development of flowers, reproduction, nodule formation, immune responses, and adaptation to saline stress (Ghose & Rai, 2021). Niu et al. (2021), however, utilized a CRISPR/Cas9 tool to cause great DNA deletions in a lncRNA (lncRNA77580) to obtain results in functional characterization in other plant species. Studies in cotton lncRNAs using the CRISPR/Cas tool are in the initial stages of development (Bravo-Vázquez et al., 2024). In this interim, the application and the practice of genome editing through CRISPR/Cas in the functional characterization of those lncRNAs is still little explored and is in the early stages of development, evidencing the need for future investigations in this promising area.

The number of studied lncRNAs is growing rapidly, and the scientific literature is richer and richer in discoveries on those RNAs. Despite the diversity of reports, some common themes are arising that highlight the ubiquity and relevance of lncRNAs in the differentiation and development process. These themes include the association of the lncRNAs with proteins that modify chromatin, the expression of lncRNA in "potentiators" of development, and the formation of nucleated coacervates by RNA, which result in the separation of stages inside the cell (Mattick et al., 2023).

According to Bravo-Vázquez et al. (2024), the CRISPR/Cas technology has transformed the research on the biology of ncRNAs in plants, standing out as a powerful tool for functional characterization of ncRNAs and for the enhancement of plants mediated by ncRNAs. Even though there are important reports about the edition of lncRNA mediated by CRISPR/Cas in significant crop species, there is still a vast field to be explored in the next decade to clarify the persistent questions about the biologic meaning of lncRNAs in plants utilizing this genomic-editing technology. Therefore, future studies should utilize the CRISPR/Cas tool to deepen the comprehension of lncRNA functions in plants' secondary metabolism, for example, and in its preservation.

4.2.3 Perspectives

lncRNAs play a critical role in regulating various aspects of plant stress responses, both abiotic and biotic, including stress perception, signal transduction, and downstream pathways. These lncRNAs exert their influence through diverse mechanistic strategies, such as acting as enhancers, decoys, scaffolds, or guides, which are intricately integrated into plant stress-response networks. Their functions include regulating transcription machinery, modifying chromatin in promoter regions, modulating TF binding affinity in enhancer regions, facilitating RNA–DNA hybridization, and influencing chromatin topological structures (Jin et al., 2023).

Currently, the validation of lncRNA function is largely dependent on overexpression techniques. However, CRISPR/Cas technology offers a promising, flexible approach for studying lncRNAs by generating loss and gain of function mutants. Future research is expected to expand our understanding of lncRNAs in plant stress responses and develop new strategies to enhance crop stress tolerance. Some studies have already explored CRISPR/Cas-mediated manipulation of lncRNAs (Bravo-Vázquez et al., 2024).

4.3 Conclusion

In summary, studies on lncRNAs have grown exponentially in the past decade, and this has generated great expectations regarding their potential applications. Due to the regulatory role that lncRNAs exert on gene expression, they may represent important and promising targets. Furthermore, the ability of lncRNAs to interact with other molecules, such as proteins and RNAs, can also be better explored. Although there are still many challenges to be overcome in the field of lncRNAs, for example, the lack of specific bioinformatics tools for the study of these molecules, the need to better understand the molecular mechanisms involved, among other mechanisms, will be a challenge to be faced in the next years.

Given the above, lncRNAs appear to be promising and important molecules, acting directly or indirectly in the transcription and translation pathways, which makes them strong candidates for future studies to better understand this molecule. Unfortunately, very shortly, numerous abiotic stresses will occur, massively threatening crop development and yield. Understanding and functional characterization and mechanism of lncRNAs can provide new insights and strategies aimed at using them in breeding programs and improving their adaptability to other vegetable crops against adverse stresses.

References

Ahmad, S., Tang, L., Shahzad, R., Mawia, A. M., Rao, G. S., Jamil, S., Wei, C., Sheng, Z., Shao, G., Wei, X., Hu, P., Mahfouz, M. M., Hu, S., & Tang, S. (2021). CRISPR-based crop improvements: a way forward to achieve zero hunger. *Journal of Agricultural and Food Chemistry*, *69*(30), 8307–8323. https://doi.org/10.1021/acs.jafc.1c02653

Ali, Z., Rai, S. K., Jan, S., & Raina, K. (2022). An assessment on CRISPR Cas as a novel asset in mitigating drought stress. *Genetic Resources and Crop Evolution*, *69*(6), 2011–2027. https://doi.org/10.1007/s10722-022-01364-z

Amor, B. B., Wirth, S., Merchan, F., Laporte, P., d'Aubenton-Carafa, Y., Hirsch, J., Maizel, A., Mallory, A., Lucas, A., Deragon, J. M., Vaucheret, H., Thermes, C., & Crespi, M. (2009). Novel long non-protein coding RNAs involved in *Arabidopsis* differentiation and stress responses. *Genome Research*, *19*(1), 57–69. http://www.genome.org/cgi/doi/10.1101/gr.080275.108

Ariel, F., Romero-Barrios, N., Jégu, T., Benhamed, M., & Crespi, M. (2015). Battles and hijacks: noncoding transcription in plants. *Trends in Plant Science*, *20*(6), 362–371. https://doi.org/10.1016/j.tplants.2015.03.003

Aung, K., Lin, S. I., Wu, C. C., Huang, Y. T., Su, C. L., & Chiou, T. J. (2006). *pho2*, a phosphate overaccumulator, is caused by a nonsense mutation in a microRNA399 target gene. *Plant Physiology*, *141*(3), 1000–1011. https://doi.org/10.1104/pp.106.078063

Baidyussen, A., Jatayev, S., Khassanova, G., Amantayev, B., Sereda, G., Sereda, S., Gupta, N. K., Gupta, S., Schramm, C., Anderson, P., Jenkins, C. L. D., Soole, K. L., Langridge, P., & Shavrukov, Y. (2021). Expression of specific alleles of zinc-finger transcription factors, *HvSAP8* and *HvSAP16*, and corresponding SNP markers, are associated with drought tolerance in barley populations. *International Journal of Molecular Sciences*, *22*(22). https://doi.org/10.3390/ijms222212156

Barrangou, R., Fremaux, C., Deveau, H., Richards, M., Boyaval, P., Moineau, S., Romero, D. A., & Horvath, P. (2007). CRISPR provides acquired resistance against viruses in prokaryotes. *Science*, *315*(5819), 1709–1712. https://doi.org/10.1126/science.1138140

Baruah, P. M., Krishnatreya, D. B., Bordoloi, K. S., Gill, S. S., & Agarwala, N. (2021). Genome wide identification and characterization of abiotic stress responsive lncRNAs in *Capsicum annuum*. *Plant Physiology and Biochemistry*, *162*, 221–236. https://doi.org/10.1016/j.plaphy.2021.02.031

Basak, J., & Nithin, C. (2015). Targeting non-coding RNAs in plants with the CRISPR-Cas technology is a challenge yet worth accepting. *Frontiers in Plant Science*, *6*, 1001. https://doi.org/10.3389/fpls.2015.01001

Basso, M. F., Ferreira, P. C. G., Kobayashi, A. K., Harmon, F. G., Nepomuceno, A. L., Molinari, H. B. C., & Grossi-de-Sa, M. F. (2019). Micro RNA s and new biotechnological tools for its modulation and improving stress tolerance in plants. *Plant Biotechnology Journal*, *17*(8), 1482–1500. https://doi.org/10.1111/pbi.13116

Bhatia, G., Singh, A., Verma, D., Sharma, S., & Singh, K. (2020). Genome-wide investigation of regulatory roles of lncRNAs in response to heat and drought stress in *Brassica juncea* (Indian mustard). *Environmental and Experimental Botany*, *171*, 103922. https://doi.org/10.1016/j.envexpbot.2019.103922

Bhogireddy, S., Mangrauthia, S. K., Kumar, R., Pandey, A. K., Singh, S., Jain, A., Budak, H., Varshney, R. K., & Kudapa, H. (2021). Regulatory non-coding RNAs: a new frontier in regulation of plant biology. *Functional & Integrative Genomics*, *21*(3), 313–330. https://doi.org/10.1007/s10142-021-00787-8

Bibi, F., & Rahman, A. (2023). An overview of climate change impacts on agriculture and their mitigation strategies. *Agriculture*, *13*(8), 1508. https://doi.org/10.3390/agriculture13081508

Bravo-Vázquez, L. A., Méndez-García, A., Chamu-García, V., Rodríguez, A. L., Bandyopadhyay, A., & Paul, S. (2024). The applications of CRISPR/Cas-mediated microRNA and lncRNA editing in plant biology: shaping the future of plant non-coding RNA research. *Planta*, *259*(2). https://doi.org/10.1007/s00425-023-04303-z

Brouns, S. J., Jore, M. M., Lundgren, M., Westra, E. R., Slijkhuis, R. J., Snijders, A. P., Dickman, M. J., Makarova, K. S., Koonin, E. V., & Van Der Oost, J. (2008). Small CRISPR RNAs guide antiviral defense in prokaryotes. *Science*, *321*(5891), 960–964. https://doi.org/10.1126/science.1159689

Budak, H., Kaya, S. B., & Cagirici, H. B. (2020). Long non-coding RNA in plants in the era of reference sequences. *Frontiers in Plant Science*, *11*, 276. https://doi.org/10.3389/fpls.2020.00276

Cardi, T., Murovec, J., Bakhsh, A., Boniecka, J., Bruegmann, T., Bull, S. E., Eeckhaut, T., Fladung, M., Galovic, V., Linkiewicz, A., Lukan, T., Mafra, I., Michalski, K., Kavas, M., Nicolia, A., Nowakowska, J., Sági, L., Sarmiento, C., Yıldırım, K., Zlatković, M., Hensel, G., & Van, L. K. (2023). CRISPR/Cas-mediated plant genome editing: outstanding challenges a decade after implementation. *Trends in Plant Science*, *28*(10), 1144–1165. https://doi.org/10.1016/j.tplants.2023.05.012

Chao, H., Hu, Y., Zhao, L., Xin, S., Ni, Q., Zhang, P., & Chen, M. (2022). Biogenesis, functions, interactions, and resources of non-coding RNAs in plants. *International Journal of Molecular Sciences*, *23*(7), 3695. https://doi.org/10.3390/ijms23073695

Chaudhry, S., & Sidhu, G. P. S. (2022). Climate change regulated abiotic stress mechanisms in plants: a comprehensive review. *Plant Cell Reports*, *41*(1), 1–31. https://doi.org/10.1007/s00299-021-02759-5

Chen, M., Wang, C., Bao, H., Chen, H., & Wang, Y. (2016). Genome-wide identification and characterization of novel lncRNAs in *Populus* under nitrogen deficiency. *Molecular Genetics and Genomics*, *291*, 1663–1680. https://doi.org/10.1007/s00438-016-1210-3

Chen, L., Shi, S., Jiang, N., Khanzada, H., Wassan, G. M., Zhu, C., Peng, X., Jie, X., Chen, Y., Yu, Q., He, X., Fu, J., Chen, X., Hu, L., Ouyang, L., Sun, X., He, H., & Bian, J. (2018). Genome-wide analysis of long non-coding RNAs affecting roots development at an early stage in the rice response to cadmium stress. *BMC Genomics*, *19*, 1–10. https://doi.org/10.1186/s12864-018-4807-6

Chen, K., Wang, Y., Zhang, R., Zhang, H., & Gao, C. (2019). CRISPR/Cas genome editing and precision plant breeding in agriculture. *Annual Review of Plant Biology*, *70*(1), 667–697. https://doi.org/10.1146/annurev-arplant-050718-100049

Chen, L., Zhu, Q. H., & Kaufmann, K. (2020). Long non-coding RNAs in plants: emerging modulators of gene activity in development and stress responses. *Planta*, *252*(5), 92. https://doi.org/10.1007/s00425-020-03480-5

Chen, J., Zhong, Y., & Qi, X. (2021). LncRNA TCONS_00021861 is functionally associated with drought tolerance in rice (*Oryza sativa* L.) via competing endogenous RNA regulation. *BMC Plant Biology*, *21*, 1–12. https://doi.org/10.1186/s12870-021-03195-z

Chen, K., Ke, R., Du, M., Yi, Y., Chen, Y., Wang, X., Yao, L., Liu, H., Hou, X., Xiong, L., Yang, Y., & Xie, K. (2022). A FLASH pipeline for arrayed CRISPR library construction and the gene function discovery of rice receptor-like kinases. *Molecular Plant*, *15*(2), 243–257. https://doi.org/10.1016/j.molp.2021.09.015

Chincinska, I. A., Miklaszewska, M., & Sołtys-Kalina, D. (2023). Recent advances and challenges in potato improvement using CRISPR/Cas genome editing. *Planta*, *257*, 25. https://doi.org/10.1007/s00425-022-04054-3

Crespi, M. D., Jurkevitch, E., Poiret, M., d'Aubenton-Carafa, Y., Petrovics, G., Kondorosi, E., & Kondorosi, A. (1994). enod40, a gene expressed during nodule organogenesis, codes for a non-translatable RNA involved in plant growth. *The EMBO Journal*, *13*(21), 5099–5112. https://doi.org/10.1002/j.1460-2075.1994.tb06839.x

Cruz de Carvalho, M. H., Sun, H. X., Bowler, C., & Chua, N. H. (2016). Noncoding and coding transcriptome responses of a marine diatom to phosphate fluctuations. *New Phytologist*, *210*(2), 497–510. https://doi.org/10.1111/nph.13787

Csorba, T., Questa, J. I., Sun, Q., & Dean, C. (2014). Antisense *COOLAIR* mediates the coordinated switching of chromatin states at *FLC* during vernalization. *Proceedings of the National Academy of Sciences*, *111*(45), 16160–16165. https://doi.org/10.1073/pnas.1419030111

D'Ario, M., Griffiths-Jones, S., & Kim, M. (2017). Small RNAs: big impact on plant development. *Trends in Plant Science*, *22*(12), 1056–1068. https://doi.org/10.1016/j.tplants.2017.09.009

De Lucia, F., & Dean, C. (2011). Long non-coding RNAs and chromatin regulation. *Current Opinion in Plant Biology*, *14*(2), 168–173. https://doi.org/10.1016/j.pbi.2010.11.006

Deltcheva, E., Chylinski, K., Sharma, C. M., Gonzales, K., Chao, Y., Pirzada, Z. A., Eckert, M. R., Vogel, J., & Charpentier, E. (2011). CRISPR RNA maturation by *trans*-encoded small RNA and host factor RNase III. *Nature*, *471*(7340), 602–607. https://doi.org/10.1038/nature09886

Deng, F., Zhang, X., Wang, W., Yuan, R., & Shen, F. (2018). Identification of *Gossypium hirsutum* long non-coding RNAs (lncRNAs) under salt stress. *BMC Plant Biology*, *18*, 1–14. https://doi.org/10.1186/s12870-018-1238-0

Deng, F., Zeng, F., Shen, Q., Abbas, A., Cheng, J., Jiang, W., Chen, G., Shah, A. N., Holford, P., Tanveer, M., Zhang, D., & Chen, Z. H. (2022). Molecular evolution and functional modification of plant miRNAs with CRISPR. *Trends in Plant Science*, *27*(9), 890–907. https://doi.org/10.1016/j.tplants.2022.01.009

Derrien, T., Johnson, R., Bussotti, G., Tanzer, A., Djebali, S., Tilgner, H., Guernec, G., Martin, D., Merkel, A., Knowles, D. G., Lagarde, J., Veeravalli, L., Ruan, X., Ruan, Y., Lassmann, T., Carninci, P., Brown, J. B., Lipovich, L., Gonzalez, J. M., Thomas, M., Davis, C. A., Shiekhattar, R., Gingeras, T. R., Hubbard, T. J., Notredame, C., Harrow, J., & Guigó, R. (2012). The GENCODE v7 catalog of human long noncoding RNAs: analysis of their gene structure, evolution, and expression. *Genome Research*, *22*(9), 1775–1789. https://doi.org/10.1101/gr.132159.111

Devireddy, A. R., Tschaplinski, T. J., Tuskan, G. A., Muchero, W., & Chen, J. G. (2021). Role of reactive oxygen species and hormones in plant responses to temperature changes. *International Journal of Molecular Sciences*, *22*(16). https://doi.org/10.3390/ijms22168843

Dos Santos, T. B., Ribas, A. F., de Souza, S. G. H., Budzinski, I. G. F., & Domingues, D. S. (2022). Physiological responses to drought, salinity, and heat stress in plants: a review. *Stresses*, *2*(1), 113–135. https://doi.org/10.3390/stresses2010009

Ebert, M. S., & Sharp, P. A. (2012). Roles for microRNAs in conferring robustness to biological processes. *Cell*, *149*(3), 515–524. https://doi.org/10.1016/j.cell.2012.04.005

Eid, A., Mohan, C., Sanchez, S., Wang, D., & Altpeter, F. (2021). Multiallelic, targeted mutagenesis of magnesium chelatase with CRISPR/Cas9 provides a rapidly scorable phenotype in highly polyploid sugarcane. *Frontiers in Genome Editing*, *3*, 654996. https://doi.org/10.3389/fgeed.2021.654996

Fang, Y., & Fullwood, M. J. (2016). Roles, functions, and mechanisms of long non-coding RNAs in cancer. *Genomics, Proteomics & Bioinformatics*, *14*(1), 42–54. https://doi.org/10.1016/j.gpb.2015.09.006

Fatica, A., & Bozzoni, I. (2014). Long non-coding RNAs: new players in cell differentiation and development. *Nature Reviews Genetics*, *15*(1), 7–21. https://doi.org/10.1038/nrg3606

Franco-Zorrilla, J. M., Valli, A., Todesco, M., Mateos, I., Puga, M. I., Rubio-Somoza, L. A., Weigel, D., García, J. A., & Paz-Ares, J. (2007). Target mimicry provides a new mechanism for regulation of microRNA activity. *Nature Genetics*, *39*(8), 1033–1037. https://doi.org/10.1038/ng2079

Frankish, A., Carbonell-Sala, S., Diekhans, M., Jungreis, I., Loveland, J. E., Mudge, J. M., Sisu, C., Wright, J. C., Arnan, C., Barnes, I., Banerjee, A., Bennett, R., Berry, A., Bignell, A., Boix, C., Calvet, F. L., Cerdán-Vélez, D., Cunningham, F., Davidson, C., Donaldson, S., Dursun, C., Fatima, R., Giorgetti, S., Giron, C. G., Gonzalez, J. M., Hardy, M., Harrison, P. W., Hourlier, T., Hollis, Z., Hunt, T., James, B., Jiang, Y., Johnson, R., Kay, M., Lagarde, J., Martin, F. J., Martínez, G. L., Nair, S., Ni, P., Pozo, F., Ramalingam, V., Ruffier, M., Schmitt, B. M., Schreiber, J. M., Steed, E., Suner, M. M., Sumathipala, D., Sycheva, I., Uszczynska-Ratajczak, B., Wass, E., Yang, Y. T., Yates, A., Zafrulla, Z., Choudhary, J. S., Gerstein, M., Guigo, R., Hubbard, T. J. P., Kellis, M., Kundaje, A., Paten, B., Tress, M. L., & Flicek, P. (2023). GENCODE: reference annotation for the human and mouse genomes in 2023. *Nucleic Acids Research*, *51*(D1), D942–D949. https://doi.org/10.1093/nar/gkac1071

Fu, L., Ding, Z., Tan, D., Han, B., Sun, X., & Zhang, J. (2020). Genome-wide discovery and functional prediction of salt-responsive lncRNAs in duckweed. *BMC Genomics*, *21*, 1–14. https://doi.org/10.1186/s12864-020-6633-x

Galli, M., Martiny, E., Imani, J., Kumar, N., Koch, A., Steinbrenner, J., & Kogel, K. H. (2022). CRISPR/Sp Cas9-mediated double knockout of barley Microrchidia MORC1 and MORC6a reveals their strong involvement in plant immunity, transcriptional gene silencing and plant growth. *Plant Biotechnology Journal*, *20*(1), 89–102. https://doi.org/10.1111/pbi.13697

Gao, C. (2021). Genome engineering for crop improvement and future agriculture. *Cell*, *184*(6), 1621–1635. https://doi.org/10.1016/j.cell.2021.01.005

Ghose, K., & Rai, K. M. (2021). Long noncoding RNAs and their implication in novel trait development in soybean. *Long Noncoding RNAs in Plants*, 133–150. https://doi.org/10.1016/B978-0-12-821452-7.00009-X

Guigó, R. (2023). Genome annotation: From human genetics to biodiversity genomics. *Cell Genomics*, *3*(8). https://doi.org/10.1016/j.xgen.2023.100375

Hawkes, E. J., Hennelly, S. P., Novikova, I. V., Irwin, J. A., Dean, C., & Sanbonmatsu, K. Y. (2016). COOLAIR antisense RNAs form evolutionarily conserved elaborate secondary structures. *Cell Reports*, *16*(12), 3087–3096. https://doi.org/10.1016/j.celrep.2016.08.045

Huang, L., Dong, H., Zhou, D., Li, M., Liu, Y., Zhang, F., Feng, Y., Yu, D., Lin, S., & Cao, J. (2018). Systematic identification of long non-coding RNA s during pollen development and fertilization in *Brassica rapa*. *The Plant Journal*, *96*(1), 203–222. https://doi.org/10.1111/tpj.14016

Imaduwage, I. U. H., & Hewadikaram, M. (2024). Predicted roles of long non-coding RNAs in abiotic stress tolerance responses of plants. *Molecular Horticulture*, *4*(1), 20. https://doi.org/10.1186/s43897-024-00094-3

Imran, Q. M., Falak, N., Hussain, A., Mun, B. G., & Yun, B. W. (2021). Abiotic stress in plants, stress perception to molecular response and role of biotechnological tools in stress resistance. *Agronomy*, *11*(8). https://doi.org/10.3390/agronomy11081579

Jampala, P., Garhewal, A., & Lodha, M. (2021). Functions of long non-coding RNA in *Arabidopsis thaliana*. *Plant Signaling & Behavior*, *16*(9). https://doi.org/10.1080/15592324.2021.1925440

Jha, U. C., Nayyar, H., Jha, R., Khurshid, M., Zhou, M., Mantri, N., & Siddique, K. H. (2020). Long non-coding RNAs: emerging players regulating plant abiotic stress response and adaptation. *BMC Plant Biology*, *20*, 1–20. https://doi.org/10.1186/s12870-020-02595-x

Jiang, W. Z., Zhou, H. B., Bi, H. H., Fromm, M., Yang, B., & Weeks, D. P. (2013). Demonstration of CRISPR/Cas9/sgRNA-mediated targeted gene modification in Arabidopsis, tobacco, sorghum and rice. *Nucleic Acids Research*, *41*, e188. https://doi.org/10.1093/nar/gkt780

Jiang, Z., Zhou, X., Tao, M., Yuan, F., Liu, L., Wu, F., Wu, X., Xiang, Y., Niu, Y., Liu, F., Li, C., Ye, R., Byeon, B., Xue, Y., Zhao, H., Wang, H. N., Crawford, B. M., Johnson, D. M., Hu, C., Pei, C., Zhou, W., Swift, G. B., Zhang, H., Vo-Dinh, T., Hu, Z., Siedow, J. N., & Pei, Z.-M. (2019). Plant cell-surface GIPC sphingolipids sense salt to trigger Ca^{2+} influx. *Nature*, *572*(7769), 341–346. https://doi.org/10.1038/s41586-019-1449-z

Jin, X., Wang, Z., Li, X., Ai, Q., Wong, D. C. J., Zhang, F., Yang, J., Zhang, N., & Si, H. (2023). Current perspectives of lncRNAs in abiotic and biotic stress tolerance in plants. *Frontiers in Plant Science*, *14*. https://doi.org/10.3389/fpls.2023.1334620

Kapranov, P., Cheng, J., Dike, S., Nix, D. A., Duttagupta, R., Willingham, A. T., Stadler, P. F., Hertel, J., Hackermüller, J., Hofacker, I. L., Bell, I., Cheung, E., Drenkow, J., Dumais, E., Patel, S., Helt, G., Ganesh, M., Ghosh, S., Piccolboni, A., Sementchenko, V., Tammana, H., & Gingeras, T. R. (2007). RNA maps reveal new RNA classes and a possible function for pervasive transcription. *Science*, *316*(5830), 1484–1488. https://doi.org/10.1126/science.1138341

Karmakar, S., Das, P., Panda, D., Xie, K., Baig, M. J., & Molla, K. A. (2022). A detailed landscape of CRISPR-Cas-mediated plant disease and pest management. *Plant Science*, *323*, 111376. https://doi.org/10.1016/j.plantsci.2022.111376

Kaushik, B., Singh, K., Tiwari, D. K., & Singh, U. K. (2023). Impact of climate change on crop yield due to pests and crop diseases: future projections. *Microscopy and Microanalysis*. https://doi.org/10.1093/micmic/ozad067.022

Khoei, M. A., Karimi, M., Karamian, R., Amini, S., & Soorni, A. (2021). Identification of the complex interplay between nematode-related lncRNAs and their target genes in *Glycine max* L. *Frontiers in Plant Science*, *12*, 779597. https://doi.org/10.3389/fpls.2021.779597

Khoso, M. A., Hussain, A., Ritonga, F. N., Ali, Q., Channa, M. M., Alshegaihi, R. M., Meng, Q., Ali, M., Zaman, W., Brohi, R. D., Liu, F., & Manghwar, H. (2022). WRKY transcription factors (TFs): molecular switches to regulate drought, temperature, and salinity stresses in plants. *Frontiers in Plant Science*, *13*. https://doi.org/10.3389/fpls.2022.1039329

Kim, D. H., Xi, Y., & Sung, S. (2017). Modular function of long noncoding RNA, COLDAIR, in the vernalization response. *PLoS Genetics*, *13*(7), e1006939. https://doi.org/10.1371/journal.pgen.1006939

Kumar, N., Bharadwaj, C., Sahu, S., Shiv, A., Shrivastava, A. K., Reddy, S. P. P., Soren, K. R., Patil, B. S., Pal, M., Soni, A., Roorkiwal, M., & Varshney, R. K. (2021). Genome-wide identification and functional prediction of salt-stress related long non-coding RNAs (lncRNAs) in chickpea (*Cicer arietinum* L.). *Physiology and Molecular Biology of Plants*, *27*, 2605–2619. https://doi.org/10.1007/s12298-021-01093-0

Lamin-Samu, A. T., Zhuo, S., Ali, M., & Lu, G. (2022). Long non-coding RNA transcriptome landscape of anthers at different developmental stages in response to drought stress in tomato. *Genomics*, *114*(4), 110383. https://doi.org/10.1016/j.ygeno.2022.110383

Laurent, G. S., Wahlestedt, C., & Kapranov, P. (2015). The landscape of long noncoding RNA classification. *Trends in Genetics*, *31*(5), 239–251. https://doi.org/10.1016/j.tig.2015.03.007

Li, J. F., Norville, J. E., Aach, J., McCormack, M., Zhang, D., Bush, J., Church, G. M., & Sheen, J. (2013). Multiplex and homologous recombination-mediated genome editing in *Arabidopsis* and *Nicotiana benthamiana* using guide RNA and Cas9. *Nature Biotechnology*, *31*, 688–691. https://doi.org/10.1038/nbt.2654

Li, L., Eichten, S. R., Shimizu, R., Petsch, K., Yeh, C. T., Wu, W., Chettoor, A. M., Givan, S. A., Cole, R. A., Fowler, J., Evans Matthew, M. S., Scanlon, M. J., Yu, J., Schnable, P. S., Timmermans Marja, C. P., Springer, N. M., & Muehlbauer, G. J. (2014). Genome-wide discovery and characterization of maize long non-coding RNAs. *Genome Biology*, *15*, 1–15. https://doi.org/10.1186/gb-2014-15-2-r40

Li, S., Yu, X., Lei, N., Cheng, Z., Zhao, P., He, Y., Wang, W., & Peng, M. (2017). Genome-wide identification and functional prediction of cold and/or drought-responsive lncRNAs in cassava. *Scientific Reports*, *7*(1), 45981. https://doi.org/10.1038/srep45981

Li, R., Fu, D., Zhu, B., Luo, Y., & Zhu, H. (2018). CRISPR/Cas9-mediated mutagenesis of lncRNA1459 alters tomato fruit ripening. *The Plant Journal*, *94*(3), 513–524. https://doi.org/10.1111/tpj.13872

Li, J., Wang, Z., He, G., Ma, L., & Deng, X. W. (2020). CRISPR/Cas9-mediated disruption of TaNP1 genes results in complete male sterility in bread wheat. *Journal of Genetics and Genomics*, *47*(5), 263–272. https://doi.org/10.1016/j.jgg.2020.05.004

Li, N., Wang, Z., Wang, B., Wang, J., Xu, R., Yang, T., Huang, S., Wang, H., & Yu, Q. (2022). Identification and characterization of long non-coding RNA in tomato roots under salt stress. *Frontiers in Plant Science*, *13*, 834027. https://doi.org/10.3389/fpls.2022.834027

Li, L., Guo, N., Liu, T., Yang, S., Hu, X., Shi, S., & Li, S. (2023). Genome-wide identification and characterization of long non-coding RNA in barley roots in response to *Piriformospora indica* colonization. *Plant Science*, *330*, 111666. https://doi.org/10.1016/j.plantsci.2023.111666

Lintner, N. G., Kerou, M., Brumfield, S. K., Graham, S., Liu, H., Naismith, J. H., Sdano, M., Peng, N., She, Q., Copié, V., Young, M. J., White, M. F., & Lawrence, C. M. (2011). Structural and functional characterization of an archaeal clustered regularly interspaced short palindromic repeat (CRISPR)-associated complex for antiviral defense (CASCADE). *Journal of Biological Chemistry*, *286*(24), 21643–21656. https://doi.org/10.1074/jbc.M111.238485

Liu, T. Y., & Doudna, J. A. (2020). Chemistry of Class 1 CRISPR-Cas effectors: binding, editing, and regulation. *Journal of Biological Chemistry*, *295*(42), 14473–14487. https://doi.org/10.1074/jbc.REV120.007034

Liu, T., Wu, P., Wang, Q., Wang, W., Zhang, C., Sun, F., Liu, Z., Lee, Y., & Hou, X. (2018a). Comparative transcriptome discovery and elucidation of the mechanism of long noncoding RNAs during vernalization in *Brassica rapa*. *Plant Growth Regulation*, *85*, 27–39. https://doi.org/10.1007/s10725-018-0371-y

Liu, W., Cheng, C., Lin, Y., XuHan, X., & Lai, Z. (2018b). Genome-wide identification and characterization of mRNAs and lncRNAs involved in cold stress in the wild banana (*Musa itinerans*). *PLoS One*, *13*(7), e0200002. https://doi.org/10.1371/journal.pone.0200002

Liu, Z. W., Zhao, N., Su, Y. N., Chen, S. S., & He, X. J. (2020). Exogenously overexpressed intronic long noncoding RNAs activate host gene expression by affecting histone modification in *Arabidopsis*. *Scientific Reports*, *10*(1), 3094. https://doi.org/10.1038/s41598-020-59697-7

Liu, Q., Yang, F., Zhang, J., et al. (2021). Application of CRISPR/Cas9 in crop quality improvement. *International Journal of Molecular Sciences*, *22*, 4206. https://doi.org/10.3390/ijms22084206

Liu, X., Luo, M., Li, M., & Wei, J. (2022). Transcriptomic analysis reveals lncRNAs associated with flowering of *Angelica sinensis* during vernalization. *Current Issues in Molecular Biology*, *44*(5), 1867–1888. https://doi.org/10.3390/cimb44050128

Long, Y., Wang, X., Youmans, D. T., & Cech, T. R. (2017). How do lncRNAs regulate transcription? *Science Advances*, *3*(9), eaao2110. https://doi.org/10.1126/sciadv.aao2110

Lorenzo, C. D., Debray, K., Herwegh, D., Develtere, W., Impens, L., Schaumont, D., Vandeputte, W., Aesaert, S., Coussens, G., Boe, Y. D., Demuynck, K., Hautegem, T. V., Pauwels, L., Jacobs, T. B., Ruttink, T., Nelissen, H., & Inzé, D. (2023). BREEDIT: a multiplex genome editing strategy to improve complex quantitative traits in maize. *The Plant Cell*, *35*(1), 218–238. https://doi.org/10.1093/plcell/koac243

Ma, L., Ye, J., Yang, Y., Lin, H., Yue, L., Luo, J., Long, Y., Fu, H., Liu, X., Zhang, Y., Wang, Y., Chen, L., Kudla, J., Wang, Y., Han, S., Song, C. P., & Guo, Y. (2019). The SOS2-SCaBP8 complex generates and fine-tunes an AtANN4-dependent calcium signature under salt stress. *Developmental Cell*, *48*(5). 697–709.e5. https://doi.org/10.1016/j.devcel.2019.02.010

Ma, L., Mu, J., Grierson, D., Wang, Y., Gao, L., Zhao, X., Zhu, B., Luo, Y., Shi, K., Wang, Q., & Zuo, J. (2020a). Noncoding RNAs: functional regulatory factors in tomato fruit ripening. *Theoretical and Applied Genetics*, *133*, 1753–1762. https://doi.org/10.1007/s00122-020-03582-4

Ma, X., Zhang, X., Traore, S. M., Xin, Z., Ning, L., Li, K., Zhao, K., Li, Z., He, G., & Yin, D. (2020b). Genome-wide identification and analysis of long noncoding RNAs (lncRNAs) during seed development in peanut (*Arachis hypogaea* L.). *BMC Plant Biology*, *20*, 1–14. https://doi.org/10.1186/s12870-020-02405-4

Maharajan, T., Krishna, T. P. A., Rakkammal, K., et al. (2022). Application of CRISPR/Cas system in cereal improvement for biotic and abiotic stress tolerance. *Planta*, *256*, 106. https://doi.org/10.1007/s00425-022-04023-w

Makarova, K. S., Wolf, Y. I., Iranzo, J., Shmakov, S. A., Alkhnbashi, O. S., Brouns, S. J., Charpentier, E., Cheng, D., Haft, D. H., Horvath, P., Moineau, S., Mojica Francisco, J. M., Scott, D., Shah Shiraz, A., Siksnys, V., Terns, Č. V. M. P., White, M. F., Yakunin, A. F., Yan, W., Zhang, F., Garrett, R. A., Backofen, R., Oost, J. V. D., Barrangou, R., & Koonin, E. V. (2020). Evolutionary classification of CRISPR-Cas systems: a burst of class 2 and derived variants. *Nature Reviews Microbiology*, *18*(2), 67–83. https://doi.org/10.1038/s41579-019-0299-x

Masson-Delmotte, V., Zhai, P., Pirani, A., Connors, S. L., Péan, C., Berger, S., Caud, N., Chen, Y., Goldfarb, L., Gomis, M. I., Huang, M., Leitzell, K., Lonnoy, E., Matthews, J. B. R., Maycock, T. K.,

Waterfield, T., Yelekçi, O., Yu, R., & Zhou, B. (2021). Climate change 2021: the physical science basis. In: *Contribution of Working Group I to the Sixth Assessment Report of the Intergovernmental Panel on Climate Change*, *2*(1), 2391. https://doi.org/10.1017/9781009157896

Mattick, J. S., Amaral, P. P., Carninci, P., Carpenter, S., Chang, H. Y., Chen, L. L., Chen, R., Dean, C., Dinger, M. E., Fitzgerald, K. A., Gingeras, T. R., Guttman, M., Hirose, T., Huarte, M., Johnson, R., Kanduri, C., Kapranov, P., Lawrence, J. B., Lee, J. T., Mendell, J. T., Mercer, T. R., Moore, K. J., Nakagawa, S., Rinn, J. L., Spector, D. L., Ulitsky, I., Wan, Y., Wilusz, J. E., & Wu, M. (2023). Long non-coding RNAs: definitions, functions, challenges and recommendations. *Nature Reviews Molecular Cell Biology*, *24*(6), 430–447. https://doi.org/10.1038/s41580-022-00566-8

Matzke, M. A., & Mosher, R. A. (2014). RNA-directed DNA methylation: an epigenetic pathway of increasing complexity. *Nature Reviews Genetics*, *15*(6), 394–408. https://doi.org/10.1038/nrg3683

Meng, X., Li, A., Yu, B., & Li, S. (2021). Interplay between miRNAs and lncRNAs: Mode of action and biological roles in plant development and stress adaptation. *Computational and Structural Biotechnology Journal*, *19*, 2567–2574. https://doi.org/10.1016/j.csbj.2021.04.062

Mercer, T., Dinger, M., & Mattick, J. (2009). Long non-coding RNAs: insights into functions. *Nature Reviews Genetics*, *10*, 155–159. https://doi.org/10.1038/nrg2521

Miladinovic, D., Antunes, D., Yildirim, K., Bakhsh, A., Cvejić, S., Kondić-Špika, A., Jeromela, A. M., Opsahl-Sorteberg, H. G., Zambounis, A., & Hilioti, Z. (2021). Targeted plant improvement through genome editing: from laboratory to field. *Plant Cell Reports*, *40*, 935–951. https://doi.org/10.1007/s00299-020-02655-4

Moran, V. A., Perera, R. J., & Khalil, A. M. (2012). Emerging functional and mechanistic paradigms of mammalian long non-coding RNAs. *Nucleic Acids Research*, *40*(14), 6391–6400. https://doi.org/10.1093/nar/gks296

Mosterd, C., Rousseau, G. M., & Moineau, S. (2021). A short overview of the CRISPR-Cas adaptation stage. *Canadian Journal of Microbiology*, *67*(1), 1–12. https://doi.org/10.1139/cjm-2020-0212

Murugan, K., Babu, K., Sundaresan, R., Rajan, R., & Sashital, D. G. (2017). The revolution continues: newly discovered systems expand the CRISPR-Cas toolkit. *Molecular Cell*, *68*(1), 15–25. https://doi.org/10.1016/j.molcel.2017.09.007

Newsom, S., Parameshwaran, H. P., Martin, L., & Rajan, R. (2021). The CRISPR-Cas mechanism for adaptive immunity and alternate bacterial functions fuels diverse biotechnologies. *Frontiers in Cellular and Infection Microbiology*, *10*, 619763. https://doi.org/10.3389/fcimb.2020.619763

Niu, F., Jiang, Q., Sun, X., Hu, Z., Wang, L., Zhang, H., & Bozhkov, P. (2021). Large DNA fragment deletion in lncRNA77580 regulates neighboring gene expression in soybean (*Glycine max*). *Functional Plant Biology*, *48*(11), 1139–1147. https://doi.org/10.1071/FP20400

Ondrasek, G., Rathod, S., Manohara, K. K., Gireesh, C., Anantha, M. S., Sakhare, A. S., & Horvatinec, J. (2022). Salt stress in plants and mitigation approaches. *Plants*, *11*(6), 717. https://doi.org/10.3390/plants11060717

Park, S. I., Kwon, H. J., Cho, M. H., Song, J. S., Kim, B. G., Baek, J. H., Kim, S. L., Ji, H. S., Kwon, T. R., Kim, K. H., & Yoon, I. S. (2021). The OsERF115/AP2EREBP110 transcription factor is involved in the multiple stress tolerance to heat and drought in rice plants. *International Journal of Molecular Sciences*, *22*(13). https://doi.org/10.3390/ijms22137181

Patra, G. K., Gupta, D., Rout, G. R., & Panda, S. K. (2023). Role of long non coding RNA in plants under abiotic and biotic stresses. *Plant Physiology and Biochemistry*, *194*, 96–110. https://doi.org/10.1016/j.plaphy.2022.10.030

Pinkney, H. R., Wright, B. M., & Diermeier, S. D. (2020). The lncRNA toolkit: databases and in silico tools for lncRNA analysis. *Non-coding RNA*, *6*(4), 49. https://doi.org/10.3390/ncrna6040049

Qiu, Q. S., Guo, Y., Dietrich, M. A., Schumaker, K. S., & Zhu, J. K. (2002). Regulation of SOS1, a plasma membrane Na^+/H^+ exchanger in *Arabidopsis thaliana*, by SOS2 and SOS3. *Proceedings of the National Academy of Sciences, 99*(12), 8436–8441. https://doi.org/10.1073/pnas.122224699

Qiu, C. W., Zhao, J., Chen, Q., & Wu, F. (2019). Genome-wide characterization of drought stress responsive long non-coding RNAs in Tibetan wild barley. *Environmental and Experimental Botany, 164*, 124–134. https://doi.org/10.1016/j.envexpbot.2019.05.002

Quinn, J. J., & Chang, H. Y. (2016). Unique features of long non-coding RNA biogenesis and function. *Nature Reviews Genetics, 17*(1), 47–62. https://doi.org/10.1038/nrg.2015.10

Quintero, F. J., Ohta, M., Shi, H., Zhu, J. K., & Pardo, J. M. (2002). Reconstitution in yeast of the *Arabidopsis* SOS signaling pathway for Na^+ homeostasis. *Proceedings of the National Academy of Sciences, 99*(13), 9061–9066. https://doi.org/10.1073/pnas.132092099

Ramírez Gonzales, L., Shi, L., Bergonzi, S. B., Oortwijn, M., Franco-Zorrilla, J. M., Solano-Tavira, R., Visser, R. G. F., Abelenda, J. A., & Bachem, C. W. (2021). Potato CYCLING DOF FACTOR 1 and its lncRNA counterpart StFLORE link tuber development and drought response. *The Plant Journal, 105*(4), 855–869. https://doi.org/10.1111/tpj.15093

Rinn, J. L., & Chang, H. Y. (2012). Genome regulation by long noncoding RNAs. *Annual Review of Biochemistry, 81*(1), 145–166. https://doi.org/10.1146/annurev-biochem-051410-092902

Saini, H., Thakur, R., Gill, R., Tyagi, K., & Goswami, M. (2023). CRISPR/Cas9-gene editing approaches in plant breeding. *GM Crops & Food, 14*(1), 1–17. https://doi.org/10.1080/21645698.2023.2256930

Sapranauskas, R., Gasiunas, G., Fremaux, C., Barrangou, R., Horvath, P., & Siksnys, V. (2011). The Streptococcus thermophilus CRISPR/Cas system provides immunity in *Escherichia coli*. *Nucleic Acids Research, 39*(21), 9275–9282. https://doi.org/10.1093/nar/gkr606

Sashital, D. G., Jinek, M., & Doudna, J. A. (2011). An RNA-induced conformational change required for CRISPR RNA cleavage by the endoribonuclease Cse3. *Nature Structural & Molecular Biology, 18*(6), 680–687. https://doi.org/10.1038/nsmb.2043

Scheben, A., Wolter, F., Batley, J., Puchta, H., & Edwards, D. (2017). Towards CRISPR/Cas crops – bringing together genomics and genome editing. *New Phytologist, 216*(3), 682–698. https://doi.org/10.1111/nph.14702

Shahriar, S. A., Islam, M. N., Chun, C. N. W., Rahim, M. A., Paul, N. C., Uddain, J., & Siddiquee, S. (2021). Control of plant viral diseases by CRISPR/Cas9: resistance mechanisms, strategies and challenges in food crops. *Plants, 10*(7), 1264. https://doi.org/10.3390/plants10071264

Shamovsky, I., & Nudler, E. (2006). Gene control by large noncoding RNAs. *Science's STKE, 2006*(355), pe40. https://doi.org/10.1126/stke.3552006pe40

Shan, Q. W., Wang, Y. P., Li, J., Zhang, Y., Chen, K. L., Liang, Z., Zhang, K., Liu, J. X., Xi, J. J., Qiu, J. L., et al. (2013). Targeted genome modification of crop plants using a CRISPR-Cas system. *Nature Biotechnology, 31*, 686–688. https://doi.org/10.1038/nbt.2650

Shan, Q., Zhang, Y., Chen, K., Zhang, K., & Gao, C. (2015). Creation of fragrant rice by targeted knockout of the Os BADH 2 gene using TALEN technology. *Plant Biotechnology Journal, 13*(6), 791–800. https://doi.org/10.1111/pbi.12312

Sharma, Y., Sharma, A., Madhu, S., Singh, K., & Upadhyay, S. K. (2022). Long non-coding RNAs as emerging regulators of pathogen response in plants. *Noncoding RNA, 8*(1), 4. https://doi.org/10.3390/ncrna8010004

Shea, D. J., Nishida, N., Takada, S., Itabashi, E., Takahashi, S., Akter, A., Miyaji, N., Osabe, K., Mehraj, H., Shimizu, M., Seki, M., Kakizaki, T., Okazaki, K., Dennis, E. S., & Fujimoto, R. (2019). Long noncoding RNAs in *Brassica rapa* L. following vernalization. *Scientific Reports, 9*(1), 9302. https://doi.org/10.1038/s41598-019-45650-w

Shelake, R. M., Kadam, U. S., Kumar, R., Pramanik, D., Singh, A. K., & Kim, J. Y. (2022). Engineering drought and salinity tolerance traits in crops through CRISPR-mediated genome editing: targets, tools, challenges, and perspectives. *Plant Communications*, *3*(6). https://doi.org/10.1016/j.xplc.2022.100417

Shi, H., Ishitani, M., Kim, C., & Zhu, J. K. (2000). The *Arabidopsis thaliana* salt tolerance gene SOS1 encodes a putative Na^+/H^+ antiporter. *Proceedings of the National Academy of Sciences*, *97*(12), 6896–6901. https://doi.org/10.1073/pnas.120170197

Shuai, P., Liang, D., Tang, S., Zhang, Z., Ye, C. Y., Su, Y., Xia, X., & Yin, W. (2014). Genome-wide identification and functional prediction of novel and drought-responsive lincRNAs in *Populus trichocarpa*. *Journal of Experimental Botany*, *65*(17), 4975–4983. https://doi.org/10.1093/jxb/eru256

Song, X., Liu, G., Huang, Z., Duan, W., Tan, H., Li, Y., & Hou, X. (2016). Temperature expression patterns of genes and their coexpression with LncRNAs revealed by RNA-Seq in non-heading Chinese cabbage. *BMC Genomics*, *17*, 1–15. https://doi.org/10.1186/s12864-016-2625-2

Sternberg, S. H., Richter, H., Charpentier, E., & Qimron, U. (2016). Adaptation in CRISPR-Cas systems. *Molecular Cell*, *61*(6), 797–808. https://doi.org/10.1016/j.molcel.2016.01.030

Sun, Z., Huang, K., Han, Z., Wang, P., & Fang, Y. (2020). Genome-wide identification of *Arabidopsis* long noncoding RNAs in response to the blue light. *Scientific Reports*, *10*(1), 6229. https://doi.org/10.1038/s41598-020-63187-1

Sun, X., Tang, M., Xu, L., Luo, X., Shang, Y., Duan, W., Huang, Z., Jin, C., & Chen, G. (2023). Genome-wide identification of long non-coding RNAs and their potential functions in radish response to salt stress. *Frontiers in Genetics*, *14*, 1232363. https://doi.org/10.3389/fgene.2023.1232363

Sung, S., Schmitz, R. J., & Amasino, R. M. (2006). A PHD finger protein involved in both the vernalization and photoperiod pathways in *Arabidopsis*. *Genes & Development*, *20*(23), 3244–3248. http://www.genesdev.org/cgi/doi/10.1101/gad.1493306

Svitashev, S., Schwartz, C., Lenderts, B., Young, J. K., & Mark, C. A. (2016). Genome editing in maize directed by CRISPR-Cas9 ribonucleoprotein complexes. *Nature Communications*, *7*, 13274. https://doi.org/10.1038/ncomms13274

Talakayala, A., Ankanagari, S., & Garladinne, M. (2022). CRISPR-Cas genome editing system: a versatile tool for developing disease resistant crops. *Plant Stress*, *3*, 100056. https://doi.org/10.1016/j.stress.2022.100056

Tan, X., Li, S., Hu, L., & Zhang, C. (2020). Genome-wide analysis of long non-coding RNAs (lncRNAs) in two contrasting rapeseed (*Brassica napus* L.) genotypes subjected to drought stress and re-watering. *BMC Plant Biology*, *20*, 1–20. https://doi.org/10.1186/s12870-020-2286-9

Terns, M. P., & Terns, R. M. (2011). CRISPR-based adaptive immune systems. *Current Opinion in Microbiology*, *14*(3), 321–327. https://doi.org/10.1016/j.mib.2011.03.005

Traubenik, S., Charon, C., & Blein, T. (2024). From environmental responses to adaptation: the roles of plant lncRNAs. *Plant Physiology*, *195*(1), 232–244. https://doi.org/10.1093/plphys/kiae034

Tremblay, B. J., & Qüesta, J. I. (2023). Mechanisms of epigenetic regulation of transcription by lncRNAs in plants. *IUBMB Life*, *75*(5), 427–439. https://doi.org/10.1002/iub.2681

Trivedi, P. K., & Hasan, A. M. (2020). Updates on plant long non-coding RNAs (lncRNAs): the regulatory components. *Plant Cell, Tissue and Organ Culture*, *140*(2), 259–269. https://doi.org/10.1007/s11240-019-01726-z

Waititu, J. K., Zhang, C., Liu, J., & Wang, H. (2020). Plant non-coding RNAs: origin, biogenesis, mode of action and their roles in abiotic stress. *International Journal of Molecular Sciences*, *21*(21), 8401. https://doi.org/10.3390/ijms21218401

Wang, Y., Xu, Z., Jiang, J., Xu, C., Kang, J., Xiao, L., Wu, M., Xiong, J., Guo, X., & Liu, H. (2013). Endogenous miRNA sponge lincRNA-RoR regulates Oct4, Nanog, and Sox2 in human embryonic stem cell self-renewal. *Developmental Cell*, *25*(1). https://doi.org/69-80.10.1016/j.devcel.2013.03.002

Wang, Y. P., Cheng, X., Shan, Q. W., Zhang, Y., Liu, J. X., Gao, C. X., & Qiu, J. L. (2014a). Simultaneous editing of three homoeoalleles in hexaploid bread wheat confers heritable resistance to powdery mildew. *Nature Biotechnology*, *32*, 947–951. https://doi.org/10.1038/nbt.2969

Wang, H., Chung, P. J., Liu, J., Jang, I. C., Kean, M. J., Xu, J., & Chua, N. H. (2014b). Genome-wide identification of long noncoding natural antisense transcripts and their responses to light in Arabidopsis. *Genome Research*, *24*(3), 444–453. http://www.genome.org/cgi/doi/10.1101/gr.165555.113

Wang, M., Mao, Y., Lu, Y., Tao, X., & Zhu, J.-k. (2017). Multiplex gene editing in rice using the CRISPR-Cpf1 system. *Molecular Plant*, *10*, 1011–1013. https://doi.org/10.1016/j.molp.2017.03.001

Wang, A., Hu, J., Gao, C., Chen, G., Wang, B., Lin, C., Song, L., Ding, Y., & Zhou, G. (2019a). Genome-wide analysis of long non-coding RNAs unveils the regulatory roles in the heat tolerance of Chinese cabbage (*Brassica rapa* ssp. Chinensis). *Scientific Reports*, *9*(1), 5002. https://doi.org/10.1038/s41598-019-41428-2

Wang, P., Dai, L., Ai, J., Wang, Y., & Ren, F. (2019b). Identification and functional prediction of cold-related long non-coding RNA (lncRNA) in grapevine. *Scientific Reports*, *9*(1), 6638. https://doi.org/10.1038/s41598-019-43269-5

Wang, X., Yuan, D., Liu, Y., Liang, Y., He, J., Yang, X., Hang, R., Jia, H., Mo, B., Tian, F., Chen, X., & Liu, L. (2023). INDETERMINATE1 autonomously regulates phosphate homeostasis upstream of the miR399-ZmPHO2 signaling module in maize. *The Plant Cell*, *35*(6), 2208–2231. https://doi.org/10.1093/plcell/koad089

Wanga, M. A., Shimelis, H., Mashilo, J., & Laing, M. D. (2021). Opportunities and challenges of speed breeding: A review. *Plant Breeding*, *140*(2), 185–194. https://doi.org/10.1111/pbr.12909

Whittaker, C., & Dean, C. (2017). The FLC locus: a platform for discoveries in epigenetics and adaptation. *Annual Review of Cell and Developmental Biology*, *33*(1), 555–575. https://doi.org/10.1146/annurev-cellbio-100616-060546

Wiedenheft, B., Sternberg, S. H., & Doudna, J. A. (2012). RNA-guided genetic silencing systems in bacteria and archaea. *Nature*, *482*(7385), 331–338. https://doi.org/10.1038/nature10886

Wierzbicki, A. T., Blevins, T., & Swiezewski, S. (2021). Long noncoding RNAs in plants. *Annual Review of Plant Biology*, *72*(1), 245–271. https://doi.org/10.1146/annurev-arplant-093020-035446

Wu, H. J., Wang, Z. M., Wang, M., & Wang, X. J. (2013). Widespread long noncoding RNAs as endogenous target mimics for microRNAs in plants. *Plant Physiology*, *161*(4), 1875–1884. https://doi.org/10.1104/pp.113.215962

Wu, X., Shi, T., Iqbal, S., Zhang, Y., Liu, L., & Gao, Z. (2019). Genome-wide discovery and characterization of flower development related long non-coding RNAs in *Prunus mume*. *BMC Plant Biology*, *19*, 1–17. https://doi.org/10.1186/s12870-019-1672-7

Wu, L., Liu, S., Qi, H., Cai, H., & Xu, M. (2020). Research progress on plant long non-coding RNA. *Plants*, *9*(4), 408. https://doi.org/10.3390/plants9040408

Wu, C., Lin, M., Chen, F., Chen, J., Liu, S., Yan, H., & Xiang, Y. (2022). Homologous drought-induced 19 proteins, PtDi19-2 and PtDi19-7, enhance drought tolerance in transgenic plants. *International Journal of Molecular Sciences*, *23*(6). https://doi.org/10.3390/ijms23063371

Xin, M., Wang, Y., Yao, Y., Song, N., Hu, Z., Qin, D., Xie, C., Peng, H., Ni, Z., & Sun, Q. (2011). Identification and characterization of wheat long non-protein coding RNAs responsive to powdery mildew infection and heat stress by using microarray analysis and SBS sequencing. *BMC Plant Biology*, *11*, 1–13. https://doi.org/10.1186/1471-2229-11-61

Xu, X. W., Zhou, X. H., Wang, R. R., Peng, W. L., An, Y., & Chen, L. L. (2016). Functional analysis of long intergenic non-coding RNAs in phosphate-starved rice using competing endogenous RNA network. *Scientific Reports*, *6*(1), 20715. https://doi.org/10.1038/srep20715

Yamaguchi, A., & Abe, M. (2012). Regulation of reproductive development by non-coding RNA in *Arabidopsis*: to flower or not to flower. *Journal of Plant Research*, *125*, 693–704. https://doi.org/10.1007/s10265-012-0513-7

Yang, D., Ni, R., Yang, S., Pu, Y., Qian, M., Yang, Y., & Yang, Y. (2021). Functional characterization of the *Stipa purpurea* P5CS gene under drought stress conditions. *International Journal of Molecular Sciences*, *22*(17). https://doi.org/10.3390/ijms22179599

Yang, H., Cui, Y., Feng, Y., Hu, Y., Liu, L., & Duan, L. (2023a). Long non-coding RNAs of plants in response to abiotic stresses and their regulating roles in promoting environmental adaption. *Cells*, *12*(5), 729. https://doi.org/10.3390/cells12050729

Yang, W., Bai, Q., Li, Y., Chen, J., & Liu, C. (2023b). Epigenetic modifications: allusive clues of lncRNA functions in plants. *Computational and Structural Biotechnology Journal*, *21*, 1989–1994.

Yao, Y., Ni, Z., Peng, H., Sun, F., Xin, M., Sunkar, R., Zhu, J. K., & Sun, Q. (2010). Non-coding small RNAs responsive to abiotic stress in wheat (*Triticum aestivum* L.). *Functional & Integrative Genomics*, *10*, 187–190. https://doi.org/10.1007/s10142-010-0163-6

Yin, K., Gao, C., & Qiu, J. L. (2017). Progress and prospects in plant genome editing. *Nature Plants*, *3*(8), 1–6. https://doi.org/10.1038/nplants.2017.107

Yu, T., & Zhu, H. (2019). Long non-coding RNAs: rising regulators of plant reproductive development. *Agronomy*, *9*(2), 53. https://doi.org/10.3390/agronomy9020053

Yu, L., Ding, Y., & Zhou, M. (2023). A long non-coding RNA PelncRNA1 is involved in *Phyllostachys edulis* response to UV-B stress. *PeerJ*, *11*, e15243. https://doi.org/10.7717/peerj.15243

Yuan, J., Zhang, Y., Dong, J., Sun, Y., Lim, B. L., Liu, D., & Lu, Z. J. (2016). Systematic characterization of novel lncRNAs responding to phosphate starvation in *Arabidopsis thaliana*. *BMC Genomics*, *17*, 1–16. https://doi.org/10.1186/s12864-016-2929-2

Zetsche, B., Gootenberg, J. S., Abudayyeh, O. O., Slaymaker, I. M., Makarova, K. S., Essletzbichler, P., Volz, S. E., Joung, J., Oost, J. V. D., Regev, A., Koonin, E. V., & Zhang, F. (2015). Cpf1 is a single RNA-guided endonuclease of a class 2 CRISPR-Cas system. *Cell*, *163*(3), 759–771. https://doi.org/10.1016/j.cell.2015.09.038

Zhang, J., Rouillon, C., Kerou, M., Reeks, J., Brugger, K., Graham, S., Reimann, J., Cannone, G., Liu, H., Albers, S. V., Naismith, J. H., Spagnolo, L., & White, M. F. (2012). Structure and mechanism of the CMR complex for CRISPR-mediated antiviral immunity. *Molecular Cell*, *45*(3), 303–313. https://doi.org/10.1016/j.molcel.2011.12.013

Zhang, W., Han, Z., Guo, Q., Liu, Y., Zheng, Y., Wu, F., & Jin, W. (2014). Identification of maize long non-coding RNAs responsive to drought stress. *PLoS One*, *9*(6), e98958. https://doi.org/10.1371/journal.pone.0098958

Zhang, Y., Massel, K., Godwin, I. D., & Gao, C. (2018). Applications and potential of genome editing in crop improvement. *Genome Biology*, *19*(1), 210. https://doi.org/10.1186/s13059-018-1586-y

Zhang, A., Liu, Y., Wang, F., Li, T., Chen, Z., Kong, D., Bi, J., Zhang, F., Luo, X., Wang, J., Tang, J., Yu, X., Liu, G., & Luo, L. (2019). Enhanced rice salinity tolerance via CRISPR/Cas9-targeted mutagenesis of the OsRR22 gene. *Molecular Breeding*, *39*, 1–10. https://doi.org/10.1007/s11032-019-0954-y

Zhang, X., Shen, J., Xu, Q., Dong, J., Song, L., Wang, W., & Shen, F. (2021). Long noncoding RNA lncRNA354 functions as a competing endogenous RNA of miR160b to regulate ARF genes in response to salt stress in upland cotton. *Plant, Cell & Environment*, *44*(10), 3302–3321. https://doi.org/10.1111/pce.14133

Zhang, Y., Xu, J., Li, R., Ge, Y., Li, Y., & Li, R. (2023). Plants' response to abiotic stress: mechanisms and strategies. *International Journal of Molecular Sciences*, *24*(13), 10915. https://doi.org/10.3390/ijms241310915

Zhao, T., Tao, X., Feng, S., Wang, L., Hong, H., Ma, W., Shang, G., Guo, S., He, Y., Zhou, B., & Guan, X. (2018). LncRNAs in polyploid cotton interspecific hybrids are derived from transposon neofunctionalization. *Genome Biology*, *19*, 1–17. https://doi.org/10.1186/s13059-018-1574-2

Zhu, H., Li, C., & Gao, C. (2020). Applications of CRISPR-Cas in agriculture and plant biotechnology. *Nature Reviews Molecular Cell Biology*, *21*(11), 661–677. https://doi.org/10.1038/s41580-020-00288-9

Zou, C., Zhao, S., Yang, B., Chai, W., Zhu, L., Zhang, C., & Gai, Z. (2024). Genome-wide characterization of drought-responsive long non-coding RNAs in sorghum (*Sorghum bicolor*). *Plant Physiology and Biochemistry*, *214*, 108908. https://doi.org/10.1016/j.plaphy.2024.108908

5

Functional Analysis of Plant Noncoding Genomes Using CRISPR-Cas9-Mediated Approaches

Anna Rera, Privilege Chikove, Nyashadzashe Shengezi, Angeline Jurry, and Jyoti Prakash Sahoo

Faculty of Agriculture and Allied Sciences, C.V. Raman Global University, Bhubaneswar, India

5.1 Introduction

The plant genome is a vast repository of genetic information, with a significant portion comprising noncoding regions. These regions, which do not encode proteins, are integral to the regulation of gene expression and are essential for proper plant development and response to environmental challenges. Noncoding DNA includes regulatory elements such as promoters, enhancers, silencers, and long noncoding RNAs (lncRNAs). Promoters are sequences that facilitate the initiation of transcription by RNA polymerase, while enhancers are distal elements that increase the transcription of specific genes (Zhang et al., 2020). Silencers act to repress transcription, often in a tissue-specific manner. lncRNAs, which are transcribed but not translated into proteins, play diverse roles in gene regulation, including chromatin remodeling, transcriptional interference, and post-transcriptional regulation (Wang et al., 2021). These regulatory elements are critical in orchestrating the spatial and temporal expression of genes, allowing plants to adapt to varying environmental conditions, undergo developmental transitions, and maintain cellular homeostasis (Li et al., 2021). The dynamic regulation mediated by noncoding regions is fundamental to processes such as stress responses, where precise gene expression patterns determine a plant's ability to withstand biotic and abiotic stresses (Li et al., 2021). Despite their importance, the functional characterization of noncoding regions remains challenging due to their complex and often redundant nature. Advanced tools and technologies are needed to unravel their roles and mechanisms (Ouyang et al., 2020). CRISPR-Cas9 technology has emerged as a groundbreaking tool for genome editing, offering precision, efficiency, and versatility. Originally discovered as an adaptive immune mechanism in bacteria, clustered regularly interspaced short palindromic repeat (CRISPR) systems utilize RNA-guided nucleases to target and cleave foreign DNA. The most widely used system, CRISPR-Cas9, comprises two main components: the Cas9 nuclease and a guide RNA (gRNA) (Haeussler et al., 2020).

The gRNA is designed to be complementary to a specific DNA sequence within the target genome. When introduced into a plant cell along with Cas9, the gRNA directs Cas9 to the target site, where Cas9 induces a double-strand break (DSB). The cell's intrinsic repair mechanisms, such as nonhomologous end joining (NHEJ) or homology-directed repair (HDR), then repair

Genome and Epigenome Editing for Stress-Tolerant Crops, First Edition. Edited by Jen-Tsung Chen.

the break (Li et al., 2021). NHEJ often results in insertions or deletions (indels) that can disrupt gene function, while HDR can introduce precise genetic changes using a donor template (Montague et al., 2020). CRISPR-Cas9's versatility extends beyond creating knockouts (Ouyang et al., 2020). CRISPR interference (CRISPRi) utilizes a catalytically inactive Cas9 (dCas9) fused to repressor domains to block transcription of target genes, effectively silencing them. Conversely, CRISPR activation (CRISPRa) employs Cas9 fused to activator domains to upregulate gene expression. Additionally, CRISPR-mediated base editing enables the direct conversion of one DNA base pair to another without inducing DSBs, providing a method for precise nucleotide changes (Long et al., 2020). Recent advancements in genome-editing technologies, particularly the CRISPR-Cas9 system, have opened new avenues for exploring these noncoding regions. This chapter delves into the functional analysis of the plant noncoding genome using the CRISPR-Cas9-mediated approach, highlighting its significance, methodologies, and potential applications.

5.2 Noncoding Genome in Plants

5.2.1 Significance of Noncoding Elements in Gene Regulation

Noncoding elements play a crucial role in gene regulation, significantly influencing the expression of genes and the overall genomic architecture. These elements include noncoding RNAs (ncRNAs), enhancers, silencers, insulators, and promoters, which do not encode proteins but are essential for regulating various genetic functions. One of the most studied noncoding elements is enhancers, which can increase the transcription of genes located thousands of base pairs away. Enhancers function through binding transcription factors and facilitating the formation of a loop that brings enhancers into close proximity with promoters, thus increasing the efficiency of transcription initiation. This spatial organization underscores the complexity of genomic regulation where noncoding elements serve as key regulators of gene expression patterns essential for cellular function and organismal development (Long et al., 2016). ncRNAs, particularly microRNAs (miRNAs) and lncRNAs, also play significant roles in gene regulation. miRNAs typically bind to messenger RNAs (mRNAs) and inhibit their translation or lead to their degradation, thus fine-tuning gene expression post-transcriptionally. For instance, miRNAs are involved in various cellular processes, including development, differentiation, and stress response, highlighting their regulatory versatility (Bartel, 2004). lncRNAs, on the other hand, participate in chromatin remodeling, transcriptional interference, and the recruitment of transcription factors or chromatin-modifying complexes to specific genomic loci.

This demonstrates the multifunctionality of lncRNAs in gene regulation, ranging from transcriptional to posttranscriptional modifications (Rinn & Chang, 2012). Insulators are another class of noncoding elements that prevent the inappropriate interaction between enhancers and promoters, thereby maintaining the integrity of gene expression domains. They play a critical role in ensuring that enhancers activate the correct target genes and prevent erroneous gene activation that could lead to diseases such as cancer. For instance, the CTCF protein is a well-known insulator binding factor that contributes to the establishment of topologically associating domains (TADs), which are large genomic regions within which enhancers and promoters interact more frequently (Phillips & Corces, 2009). Collectively, these noncoding elements constitute an intricate network of regulatory controls that are essential for precise gene expression and genomic stability. Understanding their roles enhances our comprehension of genetic regulation and provides insights into the molecular mechanisms underlying various diseases, including developmental disorders and cancers (ENCODE Project Consortium, 2012). These insights are paving the way for novel therapeutic strategies targeting noncoding regions of the genome.

5.2.2 Types of Noncoding Elements

Noncoding elements in the genome, though they do not encode proteins, play critical roles in the regulation of gene expression. The types of noncoding elements may include promoters, enhancers, silencers, insulators, satellite DNA, and lncRNAs (ENCODE Project Consortium, 2012).

5.2.2.1 Promoters

Promoters are sequences of DNA located at the beginning of a gene. They act as binding sites for RNA polymerase and transcription factors, initiating transcription. Promoters typically contain specific motifs, such as the TATA box, that are recognized by the transcriptional machinery. The activity of promoters is crucial for determining the timing, location, and level of gene expression. For example, the core promoter elements are necessary for the assembly of the transcription initiation complex, while proximal promoter elements can modulate the efficiency of transcription initiation (Long et al., 2016). Variations in promoter sequences can lead to differences in gene expression, impacting cellular function and contributing to diseases when dysregulated.

5.2.2.2 Enhancers

Enhancers are regulatory DNA sequences that can be located far from the genes they regulate. They function by increasing the transcription of associated genes, often through the binding of transcription factors. Enhancers can operate independently of their position and orientation relative to the target gene. The ability of enhancers to influence gene expression involves the formation of DNA loops that bring enhancers into close proximity with promoters, facilitating the recruitment of the transcriptional machinery (Long et al., 2016). Enhancers are also highly context dependent, with their activity being regulated by cell type-specific transcription factors and chromatin states.

5.2.2.3 Long Noncoding RNAs (lncRNAs)

lncRNAs are transcripts longer than 200 nucleotides that do not encode proteins. They are involved in a wide range of regulatory functions at multiple levels of gene expression, including chromatin remodeling, transcriptional regulation, and posttranscriptional processing. lncRNAs can interact with DNA, RNA, and proteins, influencing the expression of genes by acting as scaffolds, guides, decoys, or enhancers. For instance, lncRNAs can recruit chromatin-modifying complexes to specific genomic loci, thereby modulating chromatin states and gene expression (Rinn & Chang, 2012). The versatility of lncRNAs in gene regulation makes them significant players in cellular differentiation, development, and disease pathogenesis. In addition to promoters, enhancers, and lncRNAs, the genome contains several other critical noncoding elements that play essential roles in gene regulation and genomic organization. These include silencers, insulators, miRNAs, and satellite DNA (Rinn & Chang, 2012).

5.2.2.4 Silencers

Silencers are DNA sequences that can repress the transcription of genes. They function by binding to repressor proteins that inhibit the assembly of the transcriptional machinery at the promoter, thus reducing or preventing gene expression. Silencers can be located upstream, downstream, or within the gene they regulate, and they often act in a position-independent manner. An example is the neural restrictive silencer element (NRSE), which binds the repressor protein REST to silence neural-specific genes in non-neural tissues (Ooi & Wood, 2007). The regulatory role of silencers is crucial for maintaining proper gene expression patterns and ensuring cell type-specific expression.

5.2.2.5 Insulators

Insulators are genomic elements that block the interaction between enhancers and promoters or separate different transcriptional domains within the genome. They function by forming boundaries that prevent the spread of active or repressive chromatin states. Insulators can also organize the genome into TADs, which are regions where the DNA interacts more frequently within itself than with other regions. The protein CCCTC-binding factor (CCTF) is a well-known insulator-binding protein that contributes to the formation of these boundaries (Phillips & Corces, 2009). Insulators play a vital role in maintaining the integrity of gene expression programs and preventing inappropriate regulatory interactions.

5.2.2.6 MicroRNAs (miRNAs)

miRNAs are small noncoding RNAs, typically 20–24 nucleotides long, which regulate gene expression posttranscriptionally. They function by base pairing with complementary sequences on target mRNAs, leading to mRNA degradation or inhibition of translation. miRNAs are involved in diverse biological processes, including development, differentiation, proliferation, and apoptosis. Their dysregulation is associated with various diseases, particularly cancers, where they can act as oncogenes or tumor suppressors (Bartel, 2004). The study of miRNAs has expanded our understanding of the complexity of gene regulation at the posttranscriptional level.

5.2.2.7 Satellite DNA

Satellite DNA consists of repetitive, noncoding sequences that are often found in centromeric and pericentromeric regions of chromosomes. These sequences do not encode proteins but play essential roles in chromosomal structure and function. Satellite DNA is involved in the formation of heterochromatin, which is a tightly packed form of DNA that is transcriptionally inactive. This organization helps maintain genome stability and proper chromosome segregation during cell division. Although traditionally considered "junk DNA," satellite DNA is now recognized for its structural and regulatory importance (Garrido-Ramos, 2017).

5.3 CRISPR-Cas9 Technology

CRISPR-Cas9 is based on the adaptive mechanism of bacteria and archaea, to protect themselves from invading viruses (Doudna & Charpentier, 2014). The principle is based on two components, which are CRISPR and CRISPR-associated (Cas) protein. CRISPR is a repeated palindromic short sequence associated with spacer sequences from previous viral infections. It consists of crRNA and tracrRNA: (Hsu et al., 2014). crRNA is a short RNA sequence with a guide sequence intended to match a specific DNA target. The crRNA directs the Cas9 enzyme to the correct site or target sequence on a genome. tracrRNA is a *trans*-activating CRISPR RNA (tracrRNA) that helps in crRNA maturation and folding. They both are joined together forming a guard RNA complex (gRNA) (Doudna & Charpentier, 2014).

Moreover, Cas9 is a nuclease that makes a double-stranded cut on the viral DNA. This short DNA fragment is then used as a protospacer and stored in a bacterial genome. These viral spacer fragments are placed between repeated palindromic sequences (Hsu et al., 2014). Upon reinfection, the Cas9 protein destroys the viral genome, depending on CRISPR RNA (crRNA) and a tracrRNA. crRNA being complementary to stored viral DNA fragment (viral spacer) and tracrRNA acting as a scaffold form gRNA. gRNA acts as a director and Cas9 as the scissors (Bannikov & Lavrov, 2017). Before the cut, the Cas9 scans for a specific sequence adjacent to the viral protospacer. This sequence is referred to as a protospacer adjacent motif (PAM). The protospacer is upstream of the PAM sequence, and then Cas9 will make a double-stranded cutoff of the target sequence (Shiimori et al., 2018). As the viral DNA lacks a repair system, this cut will incapacitate the virus. Through the engineering of chimeric single-guide RNA (sgRNA), scientific research offered the possibility of genetic manipulation. This relies on taking advantage of cellular DNA repair pathways. While there are several DNA repair pathways, the key ones used for gene editing are NHEJ and HDR (Zaboikin et al., 2017). NHEJ is exploited to render genes nonfunctional, while HDR is exploited to insert new genes or fragments of genetic material.

Plant adaptability to changing climate is very pivotal to food security, in this CRISPR can be used to address and manipulate genes responsible for stress response and resistance. CAS9 does not need any genetic engineering, making it flexible to be manipulated by various gRNA, this means cloning is unnecessary. Another advantage of CRISPR/Cas9 compared to ZFNs (Zinc-finger nucleases) and TALENs (Transcription activator-like effector nucleases) is the ease of multiplexing. Researchers can engineer large genomic deletions or inversions by targeting two widely spaced cleavage sites on the same chromosome. It is a highly used gene-editing tool due to its higher targeting efficiency and cost-effectiveness. By using CRISPR technology in agriculture, we have managed to increase product yield to achieve food security. This technique has also improved disease resistance in studied plants, which also in turn helped boost crop yield.

5.4 Applications of CRISPR-Cas9 in Plant Genomics

According to Schimmel et al. (2023), the accuracy of CRISPR-Cas9 technology guarantees a very dependable approach to genome-editing techniques that do not unintentionally cause unexpected changes within the genome. Since the discovery of the technology, countless attempts have been made using CRISPR-Cas9 to edit the genome in plants. Numerous studies were done on *Arabidopsis thaliana* before the technology of genome editing was applied to crops. There are various uses of CRISPR-Cas9 systems. It has been used in medical studies on diseases caused by viruses, cancer, genetic disorders, and pathogen identification (Schwank et al., 2013). This system has been successfully applied in mice to correct mutations in monogenic diseases, including the one that causes Duchenne muscular dystrophy (DMD) (Long et al., 2016).

5.4.1 Disease Resistance

With CRISPR-Cas9, remarkable genetic modifications have been made to improve metabolic pathways, providing resistance against biotic (fungal, bacterial, or viral pathogens) or abiotic (cold, drought, or salt) stresses, herbicide resistance, and many more. Horvath & Barrangou (2010) noted that CRISPR-Cas9 targets nucleic acid in a sequence-specific manner, hence providing acquired immunity against viruses and plasmids in many bacteria and most archaea. Numerous bacterial phylogenetic groupings contained CRISPR and genes coding for their related proteins, according to comparative genomic investigations (Makarova et al., 2006). It is likely that all bacteria have this adaptive immune system because it is necessary for bacterial survival.

5.4.2 Crop Improvement

CRISPR-Cas9 has the capability to enhance the yield and growth of crops by altering genes that regulate plant growth and yield. Crops with larger yields can be developed, for instance, by modifying the genes related to flowering time. The nutritional profile for crops is improved by specifically altering and targeting genes associated with nutrient synthesis and storage. For example, modifying rice's DNA to make it more vitamin A-rich. Potato plants that produce amylopectin starch, a highly desired commercial characteristic, developed after the granule-bound starch synthase (GBSS) gene was knocked out using CRISPR-Cas9 in a single transfection session. Rice, the potential of CRISPR/Cas9 technology to improve the quality of agricultural goods, is one of its most significant uses in terms of sustainability. The first effective CRISPR/Cas9 manipulation of rice, a key food supply for the world's population (Fukagawa & Ziska, 2019), was accomplished by Miao et al. (2013), who showed that the method could be applied for targeted alterations in rice.

Plant breeding is the scientific study of modifying a plant's traits to yield desirable qualities. Increase in yield, resistance to disease and pests, improved quality, resistance to abiotic stress, photosensitivity, synchronous maturity, and elimination of toxic substances are the main objectives of plant breeding. By directly introducing desired features into plants, CRISPR-Cas9 can breed more quickly than several generations of crossbreeding. By editing multiple traits at once, wild plants can be domesticated, rendering them suitable for cultivation and consumption. Following the discovery of Watson and Crick's DNA in the 1940s as a genetic material, a new biological age was made possible and various plant breeders embraced these discoveries well (Fukagawa & Ziska, 2019).

Presently with methods like genome sequencing and editing, the plant breeding industry has improved and has seen huge profits (Orton et al., 2020). Plant breeding methods are of two major types according to the techniques and tools that are available: conventional and unconventional breeding. Conventional breeding is the process in which natural approaches are used in developing new varieties. During the cross-pollination process, desirable features from various gene pools are combined to produce products that are closely related. Based on this, the end result of conventional breeding is the combination and matching of preexisting traits to produce crops that are desired (Acquaah, 2015). Conventional breeding commences with setting a clear objective, creating and assembling variation, selection, evaluation, release, multiplication, and distribution of new plants. Because conventional breeding techniques took a long period, modern breeding techniques were required (Fukagawa & Ziska, 2019).

Prefertilization and postfertilization of sexual obstacles, as well as incompatibility, were also significant concerns (Bhargava and Srivastava, 2019). Conventional breeding techniques have been integrated with other scientific disciplines to produce a particular and distinctive feature such as reduced harvest time, resistance to pests, weed resistance, and increased nutrient absorption (Lamichhane & Thapa, 2022). Utilizing genomic selection, enviromics, and high-throughput phenotyping (HTP) allowed for larger genetic gains. Additionally, according to Crossa et al. (2017), the "Modern Plant-Breeding Triangle" is made up of phenomics, enviromics, and genetics. By using bioinformatics techniques to identify the phenotypic markers corresponding to the desired trait, genotypic markers are employed in marker-assisted selection. This approach has shown to be far more precise and efficient than the traditional direct phenotypic selection method, which can be labor-intensive, time-consuming, and less accurate. In recent times, climate change has been quite drastic (Fukagawa & Ziska, 2019). The constant change in rainfall, soil acidity, increased temperature, etc. can significantly affect the growth and yield of crops. To identify the superior genes with desirable qualities, genomic selection predicts the genotypes and phenotypes of a training population using DNA-based markers (Fukagawa & Ziska, 2019).

5.5 Designing CRISPR-Cas9 Experiments for Noncoding Regions

Designing CRISPR-Cas9 experiments to investigate noncoding regions in plant genomes involves several critical steps: identifying functional noncoding elements, selecting appropriate target sites, and optimizing delivery methods. Each step requires careful consideration to ensure precise and effective genome editing (Lamichhane & Thapa, 2022).

5.5.1 Identification of Functional Noncoding Elements

Identifying functional noncoding elements is the first step in designing CRISPR-Cas9 experiments. This process often involves integrating data from multiple sources, such as genomic, transcriptomic, and epigenomic datasets. Techniques like chromatin immune precipitation followed by sequencing (ChIP-seq) can identify regions bound by transcription factors or marked by specific histone modifications, indicating potential regulatory elements (Ouyang et al., 2020). Additionally, RNA sequencing (RNA-seq) and assays for transposase-accessible chromatin using sequencing (ATAC-seq) help pinpoint active regions of the genome that may harbor functional noncoding elements (Zhang et al., 2020; Wang et al., 2021). Comparative genomics can also be employed to identify conserved noncoding regions across different species, which are likely to be functionally important (Li et al., 2021). Once candidate regions are identified, further functional validation can be performed using CRISPR-Cas9 to disrupt or modify these elements and observe the resulting phenotypic changes.

5.5.2 Selection of Target Sites

The selection of target sites within the identified noncoding elements is crucial for the success of CRISPR-Cas9 experiments. Effective targeting requires designing gRNAs that are specific to the chosen noncoding regions and minimizing potential off-target effects. Tools like CRISPR Design Tool and CRISPOR can help design and evaluate gRNAs for their specificity and efficiency (Haeussler et al., 2020; Montague et al., 2020). When selecting target sites, it is important to consider the functional motifs within the noncoding regions. For instance, if targeting a promoter or enhancer, the gRNAs should be designed to disrupt key transcription factor-binding sites or other regulatory motifs essential for the activity of these elements (Long et al., 2020). For lncRNAs, targeting the promoter or critical exons can help elucidate their roles in gene regulation (Wang et al., 2020).

5.5.3 Optimization of Delivery Methods

Optimizing the delivery of CRISPR-Cas9 components into plant cells is a critical step for successful genome editing. Various methods can be used depending on the plant species and the specific requirements of the experiment. Common delivery methods include *Agrobacterium*-mediated transformation, particle bombardment (ballistics), and protoplast transfection. *Agrobacterium-mediated transformation*: this method is widely used for many dicotyledonous plants. It involves using *Agrobacterium* tumefactions to transfer the CRISPR-Cas9 construct into plant cells. This method is efficient and can result in stable integration of the construct into the plant genome (Zhang et al., 2020). *Particle bombardment (ballistics)*: this method is suitable for both monocots and dicots, particle bombardment uses high-velocity microprojectiles coated with DNA to deliver CRISPR-Cas9 constructs into plant cells. This method can be particularly useful for species that are recalcitrant to *Agrobacterium*-mediated transformation (Liu et al., 2021). *Protoplast transfection*: this method involves the direct delivery of CRISPR-Cas9 constructs into plant protoplasts using polyethylene glycol (PEG) or electroporation. It allows for transient expression and is useful for rapid testing of gRNA efficiency and target site validation (Li et al., 2020).

5.6 Techniques for Characterizing CRISPR-Cas9-Mediated Mutations

This includes some methodology mostly used to authenticate the availability of particular mutations. Techniques for characterizing mutations include PCR amplification and gel electrophoresis, DNA sequencing, and functional assays (Fukagawa & Ziska, 2019).

5.6.1 PCR Amplification and Gel Electrophoresis

PCR conditions are optimized to a specific target region to be amplified. This PCR amplification requires specific primers to the target sequence. CRISPR target sequence primers are designed to flank the target, then PCR is performed to amplify the target. PCR is followed by PCR product purification. This removes unwanted products, for example, enzymes, primers, and unused dNTPs (Shiimori et al., 2018). *Enzymatic Purification* – involves enzymatic use, like Exonuclease I and shrimp alkaline phosphate (SAP). Exonuclease I is an (*Escherichia coli*) DNA exonuclease specifically used to remove single-stranded nucleotide 3′–5′ direction. In PCR product purification, this enzyme is used to degrade single-stranded primers by digesting them into dNTPs. This is done before Sanger DNA sequencing and single-nucleotide polymorphism (SNP) analysis. As for SAP, its main purpose is to remove the phosphate group (dephosphorylation) from unused dNTPs from the PCR. Both enzymes are inactivated by heating at 80 °C for 10 minutes so that we do not affect the sequencing reaction mix (Watanabe et al., 2010). *Gel Extraction* – use of gel electrophoresis to purify PCR products. It is done using a DNA ladder to extract your DNA of interest and by using different band sizes. This technique is also used for RNA and protein separation according to different molecular sizes. Therefore, the CRISPR-edited segment from the PCR can be isolated from a mixture of DNA samples for further studies using this technique (Bhattacharya & Van Meir, 2019)

5.6.2 Sanger DNA Sequencing

This technique is used to provide the exact DNA sequence of a provided DNA fragment (Crossley et al., 2020). *Components: DNA polymerase; a sequencing primer (usually one of the PCR primers); normal deoxynucleotides (dNTPs); chain-terminating dideoxynucleotides (ddNTPs), each labeled with a different fluorescent dye corresponding to A, T, C, or G.* Sanger's method uses PCR products as their template. DNA polymerase extends the primer adding dNTPs elongating the strand. ddNTPs have no hydroxyl group on their sugar C-3; therefore, whenever this type of nucleotide is added it terminates the extension of the primer, and the fragment is formed.

By using four different tubes each containing its type of ddNTP (ddATP, ddGTP, ddCTP, and ddTTP), different length fragments are produced in each tube relying on the number of bases when ddNTP is added by DNA polymerase enzyme as it is a random process (Crossley et al., 2020). Primers used are those from the PCR. The length of the fragments is estimated by performing acrylamide gel electrophoresis and autoradiography. After obtaining an auto radiogram, we arrange bases according to the size and attain the DNA sequence. Modern Sanger DNA sequence ddNTPs are attached to a fluorescence die, and each ddNTP has a different color tag. This now can allow us to perform a sequencing reaction in one tube rather than four tubes. A polyacrylamide gel electrophoresis can now also be performed in a single well. Advanced systems can separate these fluorescence-tagged fragments by using capillary electrophoresis. These are therefore equipped with a charged coupled device (CCD) for detecting fluorescence signals. As each fragment passes a detector, the fluorescent label is excited, and the resulting emission is recorded. This generates a series of peaks, each representing a nucleotide in the sequence (Eren et al., 2022)

5.6.3 Sequence Analysis and Next-Generation Sequencing (NGS)

To observe CRISPR mutations, the generated sequence is compared to the reference sequence. Overlapping peaks may be a sign of sample heterogeneity, such as populations of DNA that have been edited and unedited (Bloh et al., 2021). NGS technique goes beyond traditional sequencing, providing more accurate results, and offers ultrahigh throughput for DNA and RNA sequencing. It is used to study the order of nucleotides for DNA or RNA in a target sequence or the entire genome (Raza & Ahmad, 2016). Illumina Sequencing, a predominant NGS is a powerful tool for identifying mutations resulting from CRISPR/Cas9 genome editing. It gives insights into on-target, off-target, and specificity of CRISPR edits (Levy & Boone, 2019). PCR-amplified CRISPR-edited fragments are joined to adapters through the ligation process. These adapters are necessary for binding the DNA fragments to the sequencing platform and allow for their amplification and sequencing (Eren et al., 2022).

5.6.3.1 Cluster Generation, Sequencing by Synthesis (SBS), and Data Analysis

Flow cell loading: the prepared library is loaded onto a flow cell, which is a glass slide with lanes coated with oligonucleotides complementary to the adapters. *Bridge amplification*: the DNA fragments bind to the oligonucleotides on the flow cell surface. The fragments are then amplified through a process called bridge amplification, where they form clusters of clonal DNA molecules. Each cluster contains thousands of identical copies of the original DNA fragment. After the amplification, reverse strands are cleaved and washed away, leaving only the forward strands. Prime ends are then blocked to prevent unwanted priming. Primer binds to the complementary adapter sequence on each DNA fragment. Nucleotide incorporation is done to the flow cell, and each nucleotide is fluorescently labeled. As the DNA polymerase adds complementary nucleotides, they get excited emitting a fluorescent signal which is red sequentially and recording the results. For data analysis, the fluorescent color is converted into nucleotides each color corresponding to a specific nucleotide (A, T, C, or G). This process is done through base-calling algorithms (Almogy et al., 2022).

5.6.4 Functional Assays and Enzymatic Mismatch Cleavage (Gel-Based) Assays

Reporter assay: this measures the activity of particular genes or regulatory elements targeted by CRISP-Cas9. *Luciferase assay*: when associated with CRISPR-Cas9 technology, it is an effective mechanism for confirming gene changes and investigating their functional effects. In this technique, luciferase is linked to a target gene promoter. When expressed, it emits light which is then measured by a luminometer, and the light is directly proportional to gene expression. A decrease in light indicates reduced gene expression, while an increase in light indicates enhanced gene expression. Therefore, this helps in the validation of knocked-in or knocked-out genes by measuring luminescence. In contrast, when CRISPRi is used to repress gene expression, the reduction promoter activity is evaluated by decreased luciferase expression (Gentzel et al., 2020). *Phenotypic assay*: in most cases, the result phenotype of CRISPR-edited cells is known. Therefore, by observing the phenotypic changes in those edited cells, it can be determined which cells have been successfully edited and which cells have not (Watson et al., 2020).

Another straightforward way to detect on-target gene changes is to employ enzymatic assays followed by gel analysis. T7EI enzyme is utilized for this purpose. T7EI does not reveal the edited region's exact sequence composition, but it can indicate that a change in the genomic DNA sequence happened (Chiang et al., 2016). Therefore, by observing the phenotypic changes in those edited cells, we can determine which cells have been successfully edited and which cells have not (Watson et al., 2020). However, for further genomic analysis sequencing techniques must be used. Moreover, a design of CRISPR-Cas9 experiment for noncoding regions and functional analysis is indexed in Figure 5.1.

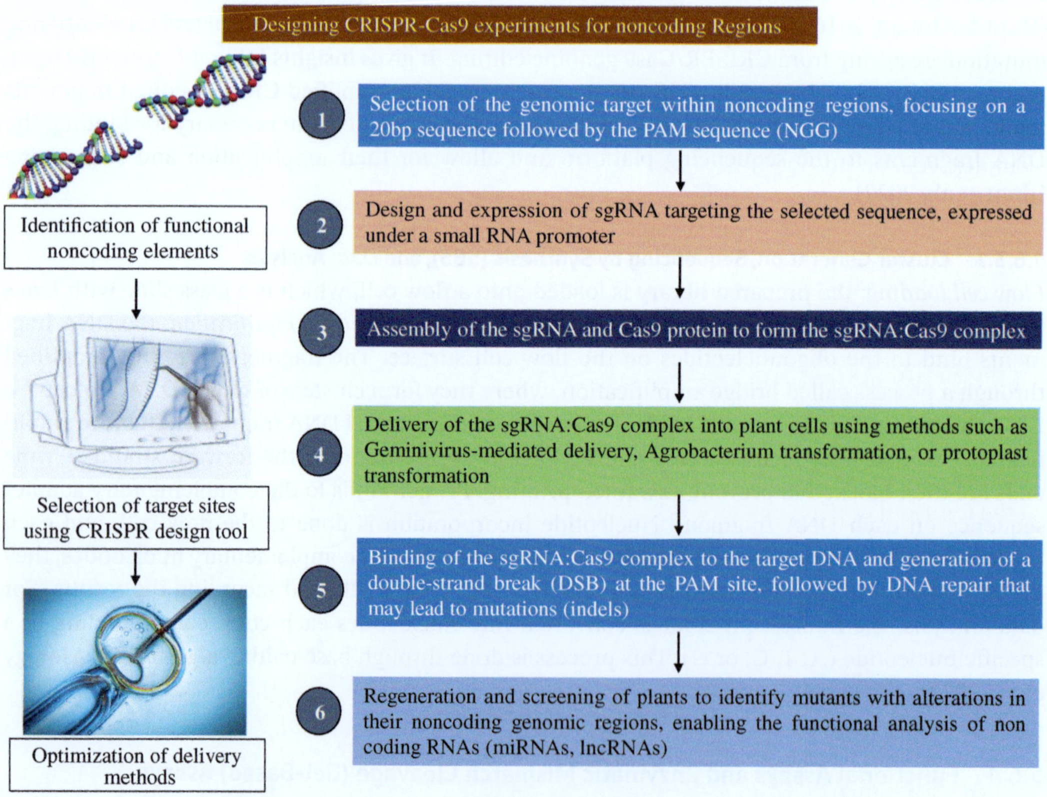

Figure 5.1 CRISPR-Cas9 experiment for noncoding regions and functional analysis. *Source:* arudolf/Adobe Stock Photos, nevodka.com/Adobe Stock Photos.

5.7 Future Prospects

The functional analysis of plant noncoding genomes using CRISPR-Cas9-mediated approaches has opened new frontiers in plant genomics. Noncoding regions, once considered "junk DNA," are now recognized as vital components of the genome, playing crucial roles in gene regulation, chromatin structure, and genome stability. Through the precise editing capabilities of CRISPR-Cas9, researchers can now systematically dissect these noncoding elements to elucidate their functions and contributions to plant development, stress responses, and overall genomic architecture. CRISPR-Cas9 has proven to be an indispensable tool for exploring the complexities of noncoding DNA. It allows for targeted mutagenesis and functional interrogation of regulatory elements such as promoters, enhancers, and lncRNAs (Eren et al., 2022). This has led to significant insights into how these elements control gene expression and influence phenotypic traits. For instance, by disrupting or modifying specific noncoding regions, researchers can observe changes in gene expression patterns and plant phenotypes, providing direct evidence of the roles these regions play (Eren et al., 2022).

Moreover, the application of CRISPR-Cas9 in plant genomics is not limited to model organisms; it extends to a wide range of agriculturally important species. Optimizing delivery methods, such as *Agrobacterium*-mediated transformation, particle bombardment, and protoplast transfection, has enabled efficient and stable gene editing across diverse plant species (Eren et al., 2022). This versatility is crucial for translating basic research findings into practical applications, such as developing crops with improved traits like disease resistance, stress tolerance, and enhanced nutritional profiles. The advancements in CRISPR-Cas9 technology also facilitate the creation of high-throughput functional genomics platforms. These platforms can systematically screen large genomic regions for functional elements, accelerating the discovery process and enabling a comprehensive understanding of the noncoding genome. Such capabilities are essential for addressing complex genetic questions and driving innovation in plant breeding and biotechnology (Eren et al., 2022).

5.8 Conclusion

CRISPR-Cas9-mediated approaches have revolutionized the functional analysis of plant noncoding genomes. By enabling precise and targeted manipulation of noncoding DNA, this technology has unveiled the hidden regulatory landscapes of plant genomes, providing new avenues for research and crop improvement. As CRISPR-Cas9 continues to evolve, it will undoubtedly remain at the forefront of plant genomic research, driving forward our understanding of plant biology and enhancing agricultural productivity and sustainability.

References

Acquaah, G. (2015). Conventional plant breeding principles and techniques. In: *Advances in Plant Breeding Strategies: Breeding, Biotechnology and Molecular Tools*, 115–158.

Almogy, G., Pratt, M., Oberstrass, F., Lee, L., Mazur, D., Beckett, N., et al. (2022). Cost-efficient whole genome-sequencing using novel mostly natural sequencing-by-synthesis chemistry and open fluidics platform. BioRxiv, 2022.05.29.493900. https://doi.org/10.1101/2022.05.29.493900

Bannikov, A. V., & Lavrov, A. V. (2017). CRISPR/CAS9, the king of genome editing tools. *Molecular Biology*, *51*(4), 514–525. https://doi.org/10.1134/S0026893317040033

Bartel, D. P. (2004). MicroRNAs: genomics, biogenesis, mechanism, and function. *Cell*, *116*(2), 281–297. https://doi.org/10.1016/S0092-8674(04)00045-5

Bhargava, A., & Srivastava, S. (2019). Toward participatory plant breeding. In: *Participatory Plant Breeding: Concept and Applications*, 69–86.

Bhattacharya, D., & Van Meir, E. G. (2019). A simple genotyping method to detect small CRISPR-Cas9 induced indels by agarose gel electrophoresis. *Scientific Reports*, *9*(1), 4437. https://doi.org/10.1038/s41598-019-39950-4

Bloh, K., Kanchana, R., Bialk, P., Banas, K., Zhang, Z., Yoo, B.-C., & Kmiec, E. B. (2021). Deconvolution of complex DNA repair (DECODR): establishing a novel deconvolution algorithm for comprehensive analysis of CRISPR-edited Sanger sequencing data. *The CRISPR Journal*, *4*(1), 120–131.

Chiang, T.-W. W., Le Sage, C., Larrieu, D., Demir, M., & Jackson, S. P. (2016). CRISPR-Cas9D10A nickase-based genotypic and phenotypic screening to enhance genome editing. *Scientific Reports*, *6*(1), 24356.

Crossa, J., Pérez-Rodríguez, P., Cuevas, J., Montesinos-López, O., Jarquín, D., De Los Campos, G., et al. (2017). Genomic selection in plant breeding: methods, models, and perspectives. *Trends in Plant Science*, *22*(11), 961–975.

Crossley, B. M., Bai, J., Glaser, A., Maes, R., Porter, E., Killian, M. L., Clement, T., & Toohey-Kurth, K. (2020). Guidelines for Sanger sequencing and molecular assay monitoring. *Journal of Veterinary Diagnostic Investigation*, *32*(6), 767–775.

Doudna, J. A., & Charpentier, E. (2014). The new frontier of genome engineering with CRISPR-Cas9. *Science*, *346*(6213), 1258096.

ENCODE Project Consortium (2012). An integrated encyclopedia of DNA elements in the human genome. *Nature*, *489*(7414), 57–74.

Eren, K., Taktakoğlu, N., & Pirim, I. (2022). DNA sequencing methods: from past to present. *Eurasian Journal of Medicine*, *54*(Suppl. 1), S47–S56.

Fukagawa, N. K., & Ziska, L. H. (2019). Rice: Importance for global nutrition. *Journal of Nutritional Science and Vitaminology*, *65*(Supplement), S2–S3. https://doi.org/10.3177/jnsv.65.S2

Garrido-Ramos, M. A. (2017). Satellite DNA: an evolving topic. *Genes*, *8*(9), 230. https://doi.org/10.3390/genes8090230

Gentzel, I. N., Park, C. H., Bellizzi, M., Xiao, G., Gadhave, K. R., Murphree, C., Yang, Q., LaMantia, J., Redinbaugh, M. G., & Balint-Kurti, P. (2020). A CRISPR/dCas9 toolkit for functional analysis of maize genes. *Plant Methods*, *16*, 1–9.

Haeussler, M., Schönig, K., Eckert, H., et al. (2020). Evaluation of off-target and on-target scoring algorithms and integration into the guide RNA selection tool CRISPOR. *Genome Biology*, *21*(1), 81. https://doi.org/10.1186/s13059-016-1012-2

Horvath, P., & Barrangou, R. (2010). CRISPR/Cas, the immune system of bacteria and archaea. *Science*, *327*(5962), 167–170.

Hsu, P. D., Lander, E. S., & Zhang, F. (2014). Development and applications of CRISPR-Cas9 for genome engineering. *Cell*, *157*(6), 1262–1278. https://doi.org/10.1016/j.cell.2014.05.010

Lamichhane, S., & Thapa, S. (2022). Advances from conventional to modern plant breeding methodologies. *Plant Breeding and Biotechnology*, *10*(1), 1–14.

Levy, S. E., & Boone, B. E. (2019). Next-generation sequencing strategies. *Cold Spring Harbor Perspectives in Medicine*, *9*(7), a025791.

Li, J. F., Zhang, D., & Sheen, J. (2020). Targeted plant genome editing via the Agrobacterium-mediated CRISPR-Cas9 system. *Nature Protocols*, *15*(1), 1314–1347.

Li, J., Zhang, J., Ren, M., et al. (2021). Comparative genomics reveals conservation and diversity of functional elements in the non-coding regions of grass genomes. *Genome Biology*, *22*(1), 36.

Liu, Q., Wang, C., Jiao, J., et al. (2021). High efficiency biolistic transformation of maize immature embryos. *Journal of Integrative Plant Biology*, *63*(8), 1452–1456.

Long, H. K., Prescott, S. L., & Wysocka, J. (2016). Ever-changing landscapes: transcriptional enhancers in development and evolution. *Cell*, *167*(5), 1170–1184.

Long, J., Walker, J., Sheppard, S., et al. (2020). Genome-wide analysis identifies regulatory elements responsible for enhancer activity and binding specificity. *Nature Genetics*, *52*(5), 485–495.

Makarova, K. S., Grishin, N. V., Shabalina, S. A., Wolf, Y. I., & Koonin, E. V. (2006). A putative RNA-interference-based immune system in prokaryotes: computational analysis of the predicted enzymatic machinery, functional analogies with eukaryotic RNAi, and hypothetical mechanisms of action. *Biology Direct*, *1*, 1–26.

Miao, J., Guo, D., Zhang, J., Huang, Q., Qin, G., Zhang, X., et al. (2013). Targeted mutagenesis in rice using CRISPR-Cas system. *Cell Research*, *23*(10), 1233–1236.

Montague, T. G., Cruz, J. M., Gagnon, J. A., et al. (2020). CHOPCHOP: a CRISPR/Cas9 and TALEN web tool for genome editing. *Nucleic Acids Research*, *48*(W1), W401–W407.

Ooi, L., & Wood, I. C. (2007). Regulation of gene expression in the nervous system by REST and its cofactors. *Cellular and Molecular Life Sciences*, *64*(4), 581–597.

Orton, G. S., Tabataba-Vakili, F., Eichstädt, G., Rogers, J., Hansen, C. J., Momary, T. W., et al. (2020). A survey of small-scale waves and wave-like phenomena in Jupiter's atmosphere detected by JunoCam. *Journal of Geophysical Research: Planets*, *125*(7), e2019JE006369.

Ouyang, X., Li, J., Li, Y., et al. (2020). Comprehensive identification and functional analysis of regulatory elements in the Rice genome using CRISPR/Cas9. *Plant Biotechnology Journal*, *18*(4), 863–876.

Phillips, J. E., & Corces, V. G. (2009). CTCF: master weaver of the genome. *Cell*, *137*(7), 1194–1211.

Raza, K., & Ahmad, S. (2016). Principle, Analysis, Application and Challenges of Next-Generation Sequencing: A Review. arXiv preprint arXiv:1606.05254.

Rinn, J. L., & Chang, H. Y. (2012). Genome regulation by long noncoding RNAs. *Annual Chapter of Biochemistry*, *81*, 145–166.

Schimmel, J., Munoz-Subirana, N., Kool, H., van Schendel, R., van der Vlies, S., Kamp, J. A., et al. (2023). Modulating mutational outcomes and improving precise gene editing at CRISPR-Cas9-induced breaks by chemical inhibition of end-joining pathways. *Cell Reports*, *42*(2), 112019. https://doi.org/10.1016/j.celrep.2023.112019

Schwank, G., Koo, B. K., Sasselli, V., Dekkers, J. F., Heo, I., Demircan, T., et al. (2013). Functional repair of CFTR by CRISPR/Cas9 in intestinal stem cell organoids of cystic fibrosis patients. *Cell stem cell*, *13*(6), 653–658.

Shiimori, M., Garrett, S. C., Graveley, B. R., & Terns, M. P. (2018). Cas4 nucleases define the PAM, length, and orientation of DNA fragments integrated at CRISPR loci. *Molecular Cell*, *70*(5), 814–824.

Wang, Y., Luo, X., Sun, F., et al. (2020). LncRNA-dependent mechanisms of enhancing or inhibiting gene transcription in plants. *Frontiers in Plant Science*, *11*, 1478.

Wang, Z., Wang, S., Wu, M., et al. (2021). ATAC-seq analysis reveals chromatin accessibility dynamics during soybean seed development. *Frontiers in Plant Science*, *12*, 709017.

Watanabe, K., Emoto, N., Sunohara, M., Kawakami, M., Kage, H., Nagase, T., Ohishi, N., & Takai, D. (2010). Treatment of PCR products with exonuclease I and heat-labile alkaline phosphatase improves the visibility of combined bisulfite restriction analysis. *Biochemical and Biophysical Research Communications*, *399*(3), 422–424.

Watson, C. J., Monstad-Rios, A. T., Bhimani, R. M., Gistelinck, C., Willaert, A., Coucke, P., Hsu, Y.-H., & Kwon, R. Y. (2020). Phenomics-based quantification of CRISPR-induced mosaicism in zebrafish. *Cell Systems*, *10*(3), 275–286.

Zaboikin, M., Zaboikina, T., Freter, C., & Srinivasakumar, N. (2017). Non-homologous end joining and homology directed DNA repair frequency of double-stranded breaks introduced by genome editing reagents. *Plos One*, *12*(1), e0169931. https://doi.org/10.1371/journal.pone.0169931

Zhang, Y., Liu, T., Meyer, C. A., et al. (2020). Model-based analysis of ChIP-Seq (MACS). *Nature Protocols*, *15*(3), 1650–1672.

6

Small RNA-Mediated Plant Protection Constructed by CRISPR/Cas Genome Editing

Alagu Manickavelu

Department of Genomic Science, Central University of Kerala, Kasaragod, Kerala, India

6.1 Introduction

Plants are continuously exposed to a range of biotic stress factors, such as viruses, bacteria, fungi, and insects. Over time, they have developed complex defense systems to recognize and combat these challenges. Conventional plant protection methods, including the use of chemical pesticides and selective breeding, have long been central to agricultural practices. However, these approaches present considerable drawbacks, such as negative environmental effects, the emergence of resistant pathogen strains, and the lengthy process involved in breeding programs, as noted by Agrios (2005).

Recent advances in molecular biology have transformed our understanding of plant defense mechanisms, enabling the development of more precise and sustainable approaches. RNA interference (RNAi) and clustered regularly interspaced palindromic repeats (CRISPRs)/CRISPR-associated (Cas) protein genome-editing technologies have emerged as powerful tools for engineering disease-resistant crops. RNAi, first described by Fire et al. (1998), involves silencing-specific genes through small RNAs (sRNAs). CRISPR/Cas, a technology developed based on the work of Jinek et al. (2012), allows for targeted modifications to the plant genome. Together, these technologies offer new possibilities for enhancing plant immunity and reducing reliance on chemical interventions.

sRNAs are short, noncoding RNA molecules that regulate gene expression at the posttranscriptional level. In plants, sRNAs play a crucial role in defense against biotic stresses by orchestrating gene silencing and modulating immune signaling pathways. As highlighted by Voinnet (2009), the two primary classes of sRNAs involved in plant defense are microRNAs (miRNAs) and small interfering RNAs (siRNAs). miRNAs, typically 20–24 nucleotides long, are derived from endogenous primary transcripts known as pri-miRNAs. These pri-miRNAs are processed by the enzyme Dicer-like 1 (DCL1) into precursor (pre-miRNAs) and mature miRNAs. Once incorporated into the RNA-induced silencing complex (RISC), miRNAs guide the complex to target complementary mRNAs, leading to their degradation or translational repression. In plant defense, miRNAs regulate genes involved in immune responses, such as those encoding transcription factors, hormone signaling components, and receptor-like kinases, as described by Jones-Rhoades & Bartel (2004). siRNAs, on the other hand, are typically produced from double-stranded RNA (dsRNA) precursors and function similarly by guiding

Genome and Epigenome Editing for Stress-Tolerant Crops, First Edition. Edited by Jen-Tsung Chen.

RISC to complementary RNA sequences, leading to their cleavage and degradation. siRNAs are particularly important in defending against viral pathogens by targeting viral RNAs for degradation, as detailed by Baulcombe (2004). In addition to miRNAs and siRNAs, other sRNA classes, such as *trans*-acting siRNAs (ta-siRNAs) and natural antisense siRNAs (nat-siRNAs), also contribute to plant defense, as discussed by Allen et al. (2005).

CRISPR and Cas proteins represent a groundbreaking technology for genome editing. Initially discovered as part of a bacterial immune defense system, CRISPR/Cas has been adapted for precise genetic modifications across a wide array of organisms, including plants (Barrangou et al., 2007). Due to its simplicity, versatility, and high efficiency, CRISPR/Cas has become the preferred tool for genetic editing, surpassing older methods like transcription activator-like effector nucleases (TALENs) and zinc finger nucleases (Cong et al., 2013). In terms of plant protection, this technology holds remarkable promise for engineering crops with stronger resistance to biotic stress. By specifically altering genes that play key roles in pathogen detection, signal transmission, and defense mechanisms, CRISPR/Cas can be leveraged to develop crops with enhanced disease resistance (Zaidi et al., 2017). Additionally, it allows for the modification of sRNA pathways, offering precise control over gene regulation in response to biotic challenges (Miao et al., 2018).

This chapter explores the intersection of sRNA biology and CRISPR/Cas genome editing, focusing on how these technologies can be integrated to develop next-generation crops with robust, engineered resistance to pathogens. The following sections will delve into the mechanisms by which sRNAs mediate plant defense, how CRISPR/Cas can be used to manipulate these pathways, and the potential applications of these technologies in sustainable agriculture.

6.2 Principles of Small RNA-Mediated Plant Defense Mechanisms

6.2.1 Types of Small RNAs in Plants

6.2.1.1 MicroRNAs (miRNAs)

miRNAs are short, approximately 21–24 nucleotides in length, and play crucial roles in gene regulation in plants. They are derived from long primary transcripts that form hairpin structures and are processed by DCL proteins. The mature miRNAs are then incorporated into the RISC, guiding the complex to complementary mRNA targets, leading to mRNA degradation or translational repression (Bartel, 2004).

6.2.1.2 Small Interfering RNAs (siRNAs)

siRNAs are another major class of sRNAs involved in gene silencing. siRNAs are typically 21–23 nucleotides long and are generated from dsRNA precursors by DCL proteins. They are involved in posttranscriptional gene silencing and can trigger transcriptional silencing through chromatin modifications. siRNAs are often associated with transposons and repetitive sequences, helping to maintain genome stability (Vaucheret, 2006).

6.2.1.3 Other Classes of Small RNAs

Besides miRNAs and siRNAs, plants produce other sRNAs, such as ta-siRNAs and nat-siRNAs. ta-siRNAs are derived from TAS genes and target mRNAs for silencing in trans, while nat-siRNAs arise from complementary antisense transcripts and regulate gene expression through RNA-directed DNA methylation (Axtell, 2013).

6.2.1.4 Biogenesis and Function

The biogenesis of sRNAs involves several steps, including transcription by RNA polymerase II or IV, processing by DCL proteins, and incorporation into RISC. The function of sRNAs is primarily to regulate gene expression by guiding RISC to specific target mRNAs, leading to their degradation or inhibition of translation. This regulatory mechanism is critical for controlling various physiological processes and responses to environmental stimuli (Jones-Rhoades et al., 2006). The overview of sRNA classes and their role in plant defense is shown in Table 6.1.

6.2.2 Small RNA Pathways in Response to Biotic Stress

6.2.2.1 Viral Pathogens

sRNAs play a significant role in plant defense against viral infections. During viral infection, plant cells produce sRNAs from viral RNA genomes, which are then processed into virus-derived

Table 6.1 Summary of small RNA classes and their functions in plant defense.

Small RNA class	Type	Biogenesis	Roles in plant defense	Examples
MicroRNAs (miRNAs)	Regulatory	Transcribed as primary miRNAs (pri-miRNAs) and processed into mature miRNAs by Drosha and Dicer proteins	Control gene expression by attaching to specific mRNAs, resulting in their degradation or inhibition of translation. Involved in regulating plant responses to various stresses, including pathogen attacks	miR159 (regulates MYB transcription factors)
Small interfering RNAs (siRNAs)	Silencing	Produced from double-stranded RNA (dsRNA) by Dicer proteins, leading to the formation of siRNA duplexes	Guide RNA-induced silencing complex (RISC) to cleave target RNAs with complementary sequences. Plays a role in silencing transposons and viral RNAs, and regulating defense gene expression	tasiRNAs (e.g. tasiR-ARF)
Trans-acting siRNAs (tasiRNAs)	Regulatory	Generated from long noncoding RNAs by Dicer-like enzymes, followed by amplification through RDR6 and further processing	Regulate gene expression by targeting specific mRNAs for degradation. Involved in systemic signaling of defense responses	tasiR-ARF (regulates auxin response factors)
Endogenous siRNAs (endo-siRNAs)	Silencing	Derived from endogenous sources such as transposons or repetitive elements	Contribute to silencing of transposons and repetitive elements, thereby maintaining genome stability and regulating defense pathways	21-nt and 24-nt siRNAs
Small RNAs derived from transposons (ts-siRNAs)	Silencing	Generated from transposon sequences	Involved in silencing of transposons and maintaining genomic integrity. Can also influence the expression of nearby genes	—

siRNAs (vsiRNAs). These vsiRNAs guide RISC in targeting viral RNAs for degradation, thus limiting viral replication and spread. This RNA silencing mechanism is a key component of the antiviral defense system in plants (Ding & Voinnet, 2007).

6.2.2.2 Bacterial Pathogens

In response to bacterial pathogens, plants utilize sRNAs to regulate defense-related genes and modulate immune responses. sRNAs can target bacterial effector genes or plant genes manipulated by pathogens to promote infection. For example, some miRNAs have been shown to regulate genes involved in pathogen recognition and signaling pathways, enhancing the plant's ability to respond to bacterial attacks (Liu et al., 2012).

6.2.2.3 Fungal Pathogens

sRNAs also play a role in defense against fungal pathogens. Similar to their role in viral and bacterial defense, sRNAs can target fungal RNAs or plant genes involved in the defense response. For instance, siRNAs derived from transposons or repetitive elements can help silence genes targeted by fungal effectors, contributing to the overall resistance of the plant to fungal infections (Li et al., 2015).

6.2.3 Role of Small RNAs in Systemic Acquired Resistance (SAR)

SAR is a form of long-distance signaling that provides broad-spectrum protection against a wide range of pathogens. sRNAs are involved in SAR by serving as signals that travel from the site of infection to distant parts of the plant. This long-distance signaling helps to activate defense mechanisms throughout the plant, even in tissues that were not directly exposed to the pathogen (Durrant & Dong, 2004). The contribution of sRNAs to SAR involves their movement through the plant's vascular system. For example, sRNAs generated at the site of infection can move through the phloem to distant tissues, where they regulate gene expression and enhance the plant's resistance to subsequent pathogen attacks. This systemic movement of sRNAs is crucial for coordinating a comprehensive defense response across the entire plant (Yoo et al., 2004). The exact mechanisms by which sRNAs contribute to SAR are still being elucidated. Research has shown that sRNAs can modulate the expression of genes involved in defense signaling pathways, such as salicylic acid (SA) and jasmonic acid (JA) pathways. By fine-tuning these signaling pathways, sRNAs help to establish and maintain the SAR response, ensuring effective protection against diverse pathogens (Petersen et al., 2007).

6.3 CRISPR/Cas-Mediated Engineering of Small RNA Pathways

6.3.1 Precision Editing of Small RNA Genes Using CRISPR/Cas

The CRISPR/Cas system has proven to be an effective tool for precise genome editing in plants, particularly for targeting sRNA genes. This method allows for specific alterations in the genome by creating double-strand breaks at defined sites, which are subsequently repaired through different cellular repair mechanisms. By designing guide RNAs (gRNAs) that lead the Cas9 enzyme to specific sRNA genes, scientists can precisely modify miRNAs, siRNAs, and other sRNA-related genes (Jinek et al., 2012). One prominent example of CRISPR/Cas targeting miRNAs is the work of Zhang et al. (2016a, b), who used CRISPR/Cas9 to edit miRNA genes in *Arabidopsis* to study their roles in developmental processes. They could observe changes in target gene expression and plant phenotypes by introducing specific mutations into miRNA genes. This strategy has also been applied to crops like rice, where CRISPR/Cas9 technology was employed to alter miRNA genes that play a role in regulating yield and responses to stress (Xu et al., 2017).

Likewise, CRISPR/Cas has been utilized to target siRNAs in plants. For example, research conducted by Wang et al. (2019) showcased the application of CRISPR/Cas9 to modify siRNA biogenesis pathways in maize. By editing the genes responsible for siRNA processing, the researchers were able to explore the function of siRNAs in silencing transposons and maintaining genome stability. This method underscores the adaptability of CRISPR/Cas in investigating and manipulating sRNA pathways across various plant species.

6.3.2 Modulating Small RNA Biogenesis Via CRISPR/Cas

The synthesis of sRNAs can be improved by targeting genes that are responsible for their biogenesis or by adding new regulatory elements. For instance, CRISPR/Cas can be utilized to activate genes that code for DCL proteins, essential for converting precursor RNAs into mature sRNAs. By stimulating these genes, the levels of sRNAs can be elevated, potentially boosting plant resistance to pathogens and enhancing tolerance to stress (Chen et al., 2019).

Conversely, CRISPR/Cas can suppress sRNA production by targeting genes involved in their synthesis or processing. For instance, CRISPR/Cas9 can create knockouts of genes encoding key components of the RNA silencing machinery, such as Argonaute (AGO) proteins. This suppression can be used to study the impact of reduced trim RNA levels on plant physiology and pathogen interactions (Kawasaki et al., 2016).

Modulating sRNA biogenesis using CRISPR/Cas can directly enhance pathogen resistance. Researchers can fine-tune the plant's defense mechanisms by increasing or decreasing sRNA levels by increasing or decreasing sRNA levels. For example, overexpressing sRNAs that target pathogen genes or key regulators of pathogen resistance pathways can improve resistance, while suppressing sRNA pathways may help to understand and mitigate the effects of pathogen-induced gene silencing (Li et al., 2018a, b).

6.3.3 Application in Pathogen-Resistant Crop Varieties

CRISPR/Cas-based modifications of sRNA pathways have been utilized to create crop varieties that are resistant to pathogens. For example, this approach has been employed to improve bacterial resistance in tomatoes. By focusing on sRNA pathways that are integral to the plant's immune system, researchers have developed tomato varieties with enhanced resistance to bacterial wilt and other diseases (Zhang et al., 2020a, b). CRISPR/Cas has been applied in rice to modify sRNA pathways to enhance resistance to viral infections. For instance, researchers utilized CRISPR/Cas9 to alter miRNAs and siRNAs that are crucial for antiviral responses, leading to rice varieties with improved resistance to the rice blast virus and other viral pathogens (Jiang et al., 2018). This method illustrates the capability of CRISPR/Cas to develop crops with customized sRNA profiles for better disease management. CRISPR/Cas-mediated editing of sRNA pathways has been used in wheat to enhance resistance to fungal pathogens. Researchers have developed wheat varieties with increased resistance to rust diseases and other fungal infections by targeting sRNA genes that regulate defense responses. This work highlights the utility of CRISPR/Cas in developing crops with robust pathogen resistance through precise manipulation of sRNA pathways (Liu et al., 2019).

6.4 Case Studies: Small RNA-Mediated Plant Protection Enhanced by CRISPR/Cas

The case studies of CRISPR/Cas application in crop plants are listed in Table 6.2.

6.4.1 Viral Pathogens

Integrating CRISPR/Cas technology with sRNA pathways has provided innovative approaches to combat viral pathogens in plants. One effective strategy involves using CRISPR/Cas to target and disrupt viral genomes directly. For example, researchers have used CRISPR/Cas9 to create transgenic plants expressing gRNAs that target viral RNA sequences. This method allows for precise cleavage of the viral genome, preventing replication and spread within the plant (Zhang et al., 2016a, b). A notable success story is using CRISPR/Cas technology in developing virus-resistant tobacco plants. In a study by Li et al. (2018a, b), CRISPR/Cas9 was used to target the RNA genome of Tobacco mosaic virus (TMV). The engineered plants exhibited significant resistance to TMV, demonstrating the potential of CRISPR/Cas-mediated sRNA pathways for viral pathogen control. Similarly, in rice, CRISPR/Cas9 was used to develop lines resistant to rice dwarf virus by targeting viral sequences with specific gRNAs, resulting in reduced viral load and improved plant health (Zhou et al., 2020). Ongoing research continues to explore and refine CRISPR/Cas-based strategies for viral pathogen management. Efforts are being made to improve the efficiency of gRNA delivery and enhance the specificity of CRISPR/Cas systems to minimize off-target effects. Advances in CRISPR/Cas technology, such as enhancing further development of more efficient Cas proteins and refined gRNA design, are expected further to enhance the effectiveness of sRNA-mediated viral resistance (Gao et al., 2021).

6.4.2 Bacterial and Fungal Pathogens

Modifying sRNA pathways through CRISPR/Cas has yielded encouraging results in the development of bacterial-resistant plant varieties. For instance, researchers have employed CRISPR/Cas9 to target and enhance sRNAs that control plant immune responses. A study conducted by Zhang et al. (2019) demonstrated the use of CRISPR/Cas9 to alter miRNAs that regulate defense genes in *Arabidopsis*. The modified plants showed increased resistance to bacterial pathogens like *Pseudomonas syringae*, highlighting the effectiveness of CRISPR/Cas in bolstering bacterial resistance via sRNA pathways.

Similar approaches have been applied to develop fungal-resistant crops. For instance, researchers have used CRISPR/Cas9 to target siRNAs that regulate genes associated with fungal resistance in wheat. The engineered wheat lines showed improved resistance to wheat rust pathogens, highlighting the effectiveness of CRISPR/Cas-mediated sRNA manipulation in combating fungal diseases (Liu et al., 2020). In maize, CRISPR/Cas9 was used to enhance sRNA pathways that regulate defense against *Aspergillus flavus*, a major fungal pathogen, resulting in reduced aflatoxin contamination and improved crop yield (Chen et al., 2021a, b).

Ongoing case studies and innovations reveal the potential of CRISPR/Cas technology in developing pathogen-resistant plants. For example, recent research has focused on optimizing the delivery of CRISPR/Cas components to plant tissues and improving the stability of transgenic lines. Advances in CRISPR/Cas technology, such as tissue-specific promoters and optimized gRNA design, are contributing to the development of crops with enhanced resistance to bacterial and fungal pathogens (Duan et al., 2022).

Table 6.2 Case studies of CRISPR/Cas-mediated small RNA pathway modifications in crops.

Study	Crop	Small RNA target	CRISPR/Cas application	Outcome/effect	References
Zhang et al. (2020)	Rice	miR156	CRISPR/Cas9-mediated knockout of miR156 precursor	Enhanced drought tolerance and improved root growth	Zhang et al. (2020)
Zhou et al. (2021)	Tomato	siRNA derived from transposons	CRISPR/Cas9-mediated modification of siRNA pathways to silence transposons	Increased resistance to tomato yellow leaf curl virus	Zhou et al. (2021)
Li et al. (2019)	Arabidopsis	miR399	CRISPR/Cas9-induced knockout of miR399 and its target genes	Improved phosphate uptake and enhanced disease resistance	Li et al. (2019)
Kim et al. (2022)	Maize	tasiRNA	CRISPR/Cas9 editing of tasiRNA pathway components	Enhanced resistance to maize stalk rot and increased yield	Kim et al. (2022)
Wang et al. (2023)	Wheat	miR166	CRISPR/Cas9-mediated knockout of miR166 precursor	Improved drought and heat tolerance	Wang et al. (2023)
Patel et al. (2024)	Soybean	endo-siRNA	CRISPR/Cas9 modification of endo-siRNA pathways	Enhanced resistance to soybean cyst nematode and improved growth	Patel et al. (2024)
Chen et al. (2021)	Barley	miR167	CRISPR/Cas9-targeted editing of miR167 precursor	Improved salt tolerance and enhanced yield under saline conditions	Chen et al. (2021)
Singh et al. (2020)	Cotton	miR168	CRISPR/Cas9-mediated knockout of miR168 and its target genes	Increased resistance to cotton leaf curl virus and improved fiber quality	Singh et al. (2020)
Jin et al. (2022)	Canola	siRNA from viral RNA	CRISPR/Cas9-induced modification of siRNA pathways to target viral RNAs	Enhanced resistance to canola blackleg disease	Jin et al. (2022)
Lee et al. (2023)	Potato	miR390	CRISPR/Cas9-mediated modification of miR390 pathway	Improved resistance to late blight and increased tuber yield	Lee et al. (2023)
Nguyen et al. (2024)	Pea	miR172	CRISPR/Cas9 editing of miR172 precursor	Enhanced resistance to powdery mildew and improved plant growth	Nguyen et al. (2024)
Tian et al. (2022)	Sorghum	siRNA derived from transposons	CRISPR/Cas9-targeted modification of siRNA pathways to silence transposons	Increased drought resistance and improved biomass	Tian et al. (2022)

6.4.3 Field Applications and Commercialization

The transition from laboratory research to field applications and commercialization of CRISPR/Cas-modified crops involves several critical steps. First, successful laboratory results must be validated through field trials to assess the performance and stability of the engineered traits under natural conditions. Field trials provide valuable data on the effectiveness of sRNA-mediated pathogen resistance and insights into potential agronomic benefits and challenges (Yang et al., 2019). Commercializing CRISPR/Cas-modified crops involves navigating regulatory requirements and addressing public concerns. Regulatory agencies assess the safety and efficacy of genetically modified crops before approval for commercial use. In addition, public acceptance of CRISPR/Cas technology and genetically modified organisms (GMOs) plays a significant role in the commercialization process. Effective communication of the benefits and safety of CRISPR/Cas-modified crops is essential for gaining public trust and facilitating market adoption (Graham et al., 2021). Despite challenges, successful examples of CRISPR/Cas-modified crops have entered the market. For example, CRISPR/Cas technology has facilitated the creation of disease-resistant crop varieties, including soybeans and tomatoes, which have been successfully commercialized for farmer use. These notable instances illustrate the potential of CRISPR/Cas technology to transform crop protection and enhance sustainability in agriculture (Kang et al., 2020).

6.5 Challenges and Limitations

6.5.1 Off-Target Effects and Unintended Consequences

A major concern associated with CRISPR/Cas technology is the risk of off-target effects, where unintended mutations may occur in areas of the genome that were not intended for editing. This issue becomes especially problematic when targeting sRNA genes, as the precision of gRNAs is essential to prevent interference with nontarget genes and pathways (Fu et al., 2013). Recent developments have aimed at enhancing the accuracy of CRISPR/Cas systems by employing high-fidelity variants of Cas9 and refining gRNA design to reduce off-target activity (Kleinstiver et al., 2016). Additionally, the intricate nature of sRNA pathways complicates the application of CRISPR/Cas technology. sRNAs participate in various regulatory mechanisms, and their alteration can lead to unintended consequences for gene expression and cellular activities. For example, modifying levels of miRNAs or siRNAs could inadvertently disrupt the regulation of multiple target genes, resulting in unforeseen impacts on plant growth, development, and stress responses (Bartel, 2018). Therefore, meticulous design and comprehensive validation of CRISPR/Cas modifications are crucial to achieving intended results while minimizing adverse effects.

6.5.2 Regulatory and Ethical Considerations

The commercialization of CRISPR/Cas-modified crops involves navigating complex regulatory approval processes. Regulatory agencies assess the safety and efficacy of GMOs before approving field trials and commercial release. In the case of CRISPR/Cas-modified crops, regulators must evaluate potential risks related to unintended effects, gene flow, and ecological impacts (Nordström et al., 2021). The regulatory framework varies by country, with some regions having more stringent requirements for GMOs than others. CRISPR/Cas technology in agriculture intersects with broader debates about GMOs. Concerns about safety, environmental impact, and ethical considerations influence public perception of genetically modified crops. Effective communication of the benefits and safety of

CRISPR/Cas-modified crops is crucial for gaining public acceptance. Addressing concerns transparently and providing evidence of the technology's safety and efficacy can help foster positive perceptions and facilitate the adoption of CRISPR/Cas-modified crops (Graham et al., 2021).

Ethical issues surrounding CRISPR/Cas technology encompass concerns regarding the long-term effects of genetic modifications on ecosystems and biodiversity. There is apprehension about the possible unintended outcomes of releasing modified crops into natural habitats as well as the ethical ramifications of altering the plant genome. It is crucial to involve stakeholders – such as scientists, policymakers, and the general public – in discussions to tackle these ethical dilemmas and promote the responsible application of CRISPR/Cas technology in agriculture (Schmidt et al., 2020).

6.5.3 Technical Challenges

A key technical challenge in CRISPR/Cas-mediated sRNA engineering is the effective delivery of CRISPR/Cas components into plant cells. The success of genome editing relies heavily on efficient and dependable delivery methods. Currently, several systems are used for this purpose, including *Agrobacterium*-mediated transformation, particle bombardment, and transient expression techniques. Each of these methods presents its own set of benefits and drawbacks, and researchers continue to explore ways to optimize these delivery approaches for various plant species and tissue types (Bortesi et al., 2016).

Ensuring the stability and expression of edited sRNAs is another challenge. Once CRISPR/Cas components are delivered into plant cells, maintaining stable expression of edited sRNAs is essential for achieving long-term effects. The stability of edited sRNAs can be influenced by degradation, epigenetic modifications, and interactions with cellular RNA processing machinery. Researchers are exploring strategies to enhance the stability and expression of edited sRNAs, including using tissue-specific promoters and regulatory elements (Kawasaki et al., 2019).

Monitoring and mitigating off-target effects remain critical challenges in CRISPR/Cas technology. While high-fidelity Cas9 variants and improved gRNA design can reduce off-target activity, comprehensive validation and monitoring are necessary to ensure the accuracy and safety of genome edits. Techniques such as whole-genome and targeted sequencing are used to identify and assess off-target effects. Ongoing research aims to develop more refined methods for monitoring and minimizing unintended consequences (Hsu et al., 2014).

6.6 Future Prospects

6.6.1 Integration with Other Genomic Technologies

Integrating CRISPR/Cas technology with RNAi presents exciting opportunities for enhancing plant immunity and stress tolerance. RNAi is a well-established method for silencing gene expression by targeting mRNA for degradation, while CRISPR/Cas provides precise genome-editing capabilities. Combining these technologies can enable multilayered approaches to gene regulation. For example, CRISPR/Cas can knockout or modify specific genes, while RNAi can target residual or compensatory gene expression. This synergistic approach can improve the efficiency and specificity of genetic modifications, leading to enhanced resistance against pathogens and environmental stresses (Kumagai & Barry, 2018).

TALENs represent another genome-editing technology that can work in conjunction with CRISPR/Cas. TALENs are known for their high specificity and can be employed alongside

CRISPR/Cas to target various genomic regions or make more precise modifications. The combination of TALENs and CRISPR/Cas enhances both the flexibility and accuracy of genome editing, enabling more intricate and customized alterations. Additionally, emerging technologies such as base editors and prime editors hold great potential for improving genome editing results. The integration of these tools with CRISPR/Cas may broaden the spectrum of applications and increase the efficiency of genetic engineering in plants (Rees & Liu, 2018; Anzalone et al., 2020).

6.6.2 Broadening the Applications of CRISPR/Cas in Enhancing Plant Immunity

Next-generation CRISPR technologies, such as CRISPR/Cas13, offer new avenues for enhancing plant immunity. Unlike CRISPR/Cas9, which creates double-strand breaks in the DNA, CRISPR/Cas13 targets RNA for degradation. This feature makes CRISPR/Cas13 particularly useful for targeting viral RNAs and other noncoding RNAs involved in pathogen defense. The application of CRISPR/Cas13 in plants could lead to new strategies for controlling RNA viruses and other pathogens by directly targeting their RNA genomes (Abudayyeh et al., 2016).

The development of improved CRISPR/Cas systems with enhanced specificity and efficiency continues to advance the field of plant genome editing. Innovations such as Cas9 variants with reduced off-target activity and optimized gRNA designs are improving the precision of genome edits. Additionally, the development of CRISPR/Cas systems with expanded protospacer adjacent motif (PAM) recognition sites and more efficient delivery methods are enhancing the versatility and effectiveness of CRISPR/Cas technology in plants (Hsu et al., 2014; Kleinstiver et al., 2016).

6.6.3 Sustainable Agriculture and Climate Resilience

CRISPR/Cas-mediated sRNA engineering presents significant opportunities for promoting sustainable agriculture and enhancing climate resilience. By creating crops that exhibit improved resistance to both biotic and abiotic stresses, CRISPR/Cas technology can play a vital role in developing climate-smart farming practices. For instance, crops engineered for drought resistance, enhanced nutrient utilization, and tolerance to extreme temperatures can help alleviate the adverse effects of climate change on agricultural yields (Zhang et al., 2018). Moreover, employing CRISPR/Cas technology to bolster crop resilience can advance sustainable farming methods. By diminishing the reliance on chemical inputs such as pesticides and fertilizers and by improving crop performance in challenging conditions, CRISPR/Cas-modified crops can foster more environmentally sustainable agricultural practices. Additionally, breeding crops with heightened resistance to pests and diseases can lessen the dependence on chemical interventions, further promoting sustainability in agriculture (Bortesi et al., 2016).

6.7 Conclusion

This chapter has examined the dynamic interplay between sRNA-mediated plant defense mechanisms and CRISPR/Cas genome-editing technology, underscoring their combined potential to bolster plant protection against a range of biotic stresses. It commenced with an introduction to sRNAs, outlining their functions in gene silencing and defense responses. Key sRNA classes, such as miRNAs and siRNAs, are essential in regulating plant immunity by targeting and degrading RNAs derived from pathogens and modulating the expression of plant genes involved in stress responses. The chapter discussed the fundamental principles of these sRNA pathways, highlighting their mechanisms of action, contributions to SAR, and their effectiveness in providing resistance against viral, bacterial, and fungal pathogens.

The chapter further explored the integration of CRISPR/Cas technology with sRNA pathways. Through CRISPR/Cas-mediated editing of sRNA genes, researchers can make precise modifications to these essential regulatory elements, thereby enhancing or inhibiting their functions to strengthen plant immunity. We examined case studies showcasing the successful use of CRISPR/Cas to target specific miRNAs or siRNAs in various crops, the modulation of sRNA biogenesis to enhance pathogen resistance, and the development of pathogen-resistant crop varieties via CRISPR/Cas modifications.

It also examined various case studies illustrating the practical applications of CRISPR/Cas-mediated sRNA engineering. These included strategies to combat viral pathogens, engineering resistance to bacterial and fungal diseases, and translating successful lab results into real-world agricultural practices. Despite these advancements, various challenges persist, such as off-target effects, regulatory obstacles, ethical implications, and technical difficulties associated with the delivery and expression of edited sRNAs.

References

Abudayyeh, O. O., Gootenberg, J. S., & Kellner, M. J. (2016). RNA targeting with CRISPR-Cas13. *Nature, 550*(7675), 280–284.

Agrios, G. N. (2005). *Plant Pathology*, 5th edition, Elsevier Academic Press.

Allen, E., Xie, Z., Gustafson, A. M., & Carrington, J. C. (2005). microRNA-directed phasing during trans-acting siRNA biogenesis in plants. *Cell, 121*(2), 207–221.

Anzalone, A. V., Randolph, P. B., & Davis, J. R. (2020). Search-and-replace genome editing without double-strand breaks or donor DNA. *Nature, 576*(7785), 149–157.

Axtell, M. J. (2013). Classification and comparison of small RNAs derived from transcriptional noise. *Plant Cell, 25*(9), 3114–3127.

Barrangou, R., Fremaux, C., Deveau, H., Richards, M., Boyaval, P., Moineau, S., et al. (2007). CRISPR provides acquired resistance against viruses in prokaryotes. *Science, 315*(5819), 1709–1712.

Bartel, D. P. (2004). MicroRNAs: genomics, biogenesis, mechanism, and function. *Cell, 116*(2), 281–297.

Bartel, D. P. (2018). Metazoan microRNAs. *Cell, 173*(1), 20–51.

Baulcombe, D. (2004). RNA silencing in plants. *Nature, 431*(7006), 356–363.

Bortesi, L., Fischer, R., & Schubert, D. (2016). The regulatory status of genome-edited crops. *Nature Biotechnology, 34*(8), 725–727.

Chen, X., Zhang, L., & Li, X. (2019). Engineering small RNA pathways with CRISPR/Cas9 for crop improvement. *Frontiers in Plant Science, 10*, 872.

Chen, X., Li, Z., & Zhang, L. (2021a). Improving maize resistance to *Aspergillus flavus* using CRISPR/Cas9-mediated small RNA pathways. *Plant Cell Reports, 40*(7), 975–988.

Chen, L., Wang, F., Liu, M., Zhang, Q., Li, H., & Zhang, H. (2021b). CRISPR/Cas9-targeted editing of miR167 enhances salt tolerance and yield in barley. *Plant Physiology, 185*(1), 230–244.

Cong, L., Ran, F. A., Cox, D., Lin, S., Barretto, R., Habib, N., et al. (2013). Multiplex genome engineering using CRISPR/Cas systems. *Science, 339*(6121), 819–823.

Ding, S. W., & Voinnet, O. (2007). Small RNAs direct antiviral immunity. *Cell, 130*(3), 413–426.

Duan, C., Liu, H., & Zhang, Z. (2022). Optimizing CRISPR/Cas delivery and stability in plants for enhanced pathogen resistance. *Frontiers in Plant Science, 13*, 832234.

Durrant, W. E., & Dong, X. (2004). Systemic acquired resistance. *Annual Review of Phytopathology, 42*, 185–209.

Fire, A., Xu, S., Montgomery, M. K., Kostas, S. A., Driver, S. E., & Mello, C. C. (1998). Potent and specific genetic interference by double-stranded RNA in *Caenorhabditis elegans*. *Nature, 391*(6669), 806–811.

Fu, Y., Foden, J. A., & Somerville, S. C. (2013). High-frequency off-target mutagenesis induced by CRISPR/Cas nucleases in human cells. *Nature Biotechnology, 31*(9), 822–828.

Gao, C., Zhang, Y., & Liu, J. (2021). Advances in CRISPR/Cas-mediated viral resistance in plants. *Annual Review of Phytopathology, 59*, 341–364.

Graham, S., Alexander, D., & Li, M. (2021). Regulatory and public acceptance challenges for CRISPR/Cas technology in agriculture. *Nature Sustainability, 4*(1), 18–32.

Hsu, P. D., Lander, E. S., & Zhang, F. (2014). Development and applications of CRISPR/Cas9 for genome engineering. *Cell, 157*(6), 1262–1278.

Jiang, W., Zhou, H., & Zhang, L. (2018). Engineering antiviral resistance in rice using CRISPR/Cas9. *Nature Biotechnology, 36*(11), 1150–1153.

Jin, Y., Huang, X., Yang, W., Zhang, C., Liu, Z., & Zhou, X. (2022). Targeting viral RNAs through CRISPR/Cas9-modified siRNA pathways enhances resistance to canola blackleg disease. *Plant Pathology Journal, 38*(2), 230–242.

Jinek, M., Chylinski, K., Fonfara, I., Hauer, M., Doudna, J. A., & Charpentier, E. (2012). A programmable dual-RNA-guided DNA endonuclease in adaptive bacterial immunity. *Science, 337*(6096), 816–821.

Jones-Rhoades, M. W., & Bartel, D. P. (2004). Computational identification of plant microRNAs and their targets, including a stress-induced miRNA. *Molecular Cell, 14*(6), 787–799.

Jones-Rhoades, M. W., Bartel, D. P., & Bartel, B. (2006). MicroRNAs and their regulatory roles in plants. *Annual Review of Plant Biology, 57*, 19–53.

Kang, H., Wang, X., & Zheng, H. (2020). Commercialization of CRISPR/Cas-modified crops: examples and perspectives. *Plant Biotechnology Journal, 18*(12), 2614–2624.

Kawasaki, K., Saito, K., & Hiraguri, A. (2016). CRISPR/Cas9-mediated suppression of small RNA production in plants. *Journal of Experimental Botany, 67*(4), 1021–1032.

Kawasaki, K., Saito, K., & Hiraguri, A. (2019). Advances in CRISPR/Cas9 technology for plant genome editing and its applications. *Journal of Plant Research, 132*(5), 691–706.

Kim, H., Lee, H., Park, J., Choi, Y., Kim, Y., & Kang, H. (2022). Editing of tasiRNA pathways by CRISPR/Cas9 enhances resistance to maize stalk rot and increases yield. *Journal of Integrative Plant Biology, 64*(3), 678–690.

Kleinstiver, B. P., Pattanayak, V., & Prewett, J. (2016). High-fidelity CRISPR/Cas9 nucleases with altered PAM specificities. *Nature, 523*(7561), 481–485.

Kumagai, M. H., & Barry, S. M. (2018). RNAi and CRISPR/Cas for plant immunity: a synergistic approach. *Plant Science, 270*, 116–126.

Lee, D., Lim, H., Kim, S., Choi, J., Lee, K., & Kim, J. (2023). CRISPR/Cas9-mediated modification of miR390 improves resistance to late blight and enhances tuber yield in potato. *Journal of Agricultural and Food Chemistry, 71*(11), 3456–3467.

Li, F., Wang, Y., & Xu, Z. (2015). The role of small RNAs in plant defense against fungal pathogens. *Molecular Plant-Microbe Interactions, 28*(4), 471–480.

Li, J., Zhang, Y., & Liu, X. (2018a). Modulation of small RNA pathways for pathogen resistance using CRISPR/Cas9. *Journal of Experimental Botany, 69*(18), 4377–4386.

Li, J., Zhang, Y., & Liu, X. (2018b). CRISPR/Cas9-mediated virus resistance in tobacco. *Journal of Experimental Botany*, *69*(10), 2577–2587.

Li, X., Yang, G., Zhang, C., Zhang, Y., Zhao, Y., & Huang, J. (2019). CRISPR/Cas9-induced knockout of miR399 enhances phosphate uptake and disease resistance in Arabidopsis. *Plant Cell Reports*, *38*(9), 1127–1138.

Liu, Y., Xie, Z., & Wang, L. (2012). Small RNAs: key regulators of plant immunity. *Plant Pathology Journal*, *28*(4), 355–362.

Liu, X., Wang, X., & Zhang, Y. (2019). Enhancing fungal pathogen resistance in wheat with CRISPR/Cas-mediated small RNA engineering. *Plant Biotechnology Journal*, *17*(5), 922–931.

Liu, X., Zhang, Y., & Wang, H. (2020). CRISPR/Cas9-mediated small RNA engineering for fungal resistance in wheat. *Journal of Experimental Botany*, *71*(12), 3657–3667.

Miao, C., Xiao, L., Hua, K., Zou, C., Zhao, Y., & Bressan, R. A. (2018). Mutations in a subfamily of ABA receptor genes promote rice growth and productivity. *Proceedings of the National Academy of Sciences*, *115*(23), 6058–6063.

Nguyen, T., Tran, H., Hoang, M., Pham, T., Nguyen, H., & Vu, L. (2024). Enhancing resistance to powdery mildew in pea by editing miR172 using CRISPR/Cas9. *Frontiers in Plant Science*, *15*, 634–646.

Nordström, K., Smit, M. A., & van Zanten, M. (2021). Regulation of CRISPR/Cas9-edited crops: current status and future perspectives. *Trends in Plant Science*, *26*(3), 304–317.

Patel, R., Sharma, P., Singh, A., Deshmukh, R., Mehta, P., & Kumar, V. (2024). Enhancing resistance to soybean cyst nematode through CRISPR/Cas9 modification of endo-siRNA pathways. *Theoretical and Applied Genetics*, *137*(4), 1049–1060.

Petersen, M., Brodersen, P., & Naested, H. (2007). Vascular small RNA: signals of systemic acquired resistance. *Plant Physiology*, *144*(3), 1400–1410.

Rees, H. A., & Liu, D. R. (2018). Base editing: precision chemistry on the genome and transcriptome of living cells. *Nature Reviews Genetics*, *19*(12), 770–788.

Schmidt, M., Mühlbach, H. P., & Klemm, J. (2020). Ethical and social implications of CRISPR/Cas9 technology in plant science. *Journal of Agricultural and Environmental Ethics*, *33*(6), 939–958.

Singh, S., Kumar, A., Gupta, N., Meena, S., Yadav, R., & Sharma, V. (2020). Knockout of miR168 using CRISPR/Cas9 improves resistance to cotton leaf curl virus and fiber quality. *Molecular Plant-Microbe Interactions*, *33*(9), 1294–1305.

Tian, Z., Zhang, H., Liu, X., Wang, X., Zhao, Q., & Li, J. (2022). Editing siRNA pathways using CRISPR/Cas9 increases drought resistance and biomass in sorghum. *Field Crops Research*, *275*, 108–119.

Vaucheret, H. (2006). Post-transcriptional small RNA pathways in plants: mechanisms and regulations. *Current Opinion in Plant Biology*, *9*(2), 122–128.

Voinnet, O. (2009). Origin, biogenesis, and activity of plant microRNAs. *Cell*, *136*(4), 669–687.

Wang, Y., Wu, X., & Li, X. (2019). CRISPR/Cas9-mediated modification of siRNA pathways in maize. *Molecular Plant*, *12*(1), 128–139.

Wang, J., Yang, S., Liu, H., Zhao, J., Li, Z., & Zhang, X. (2023). CRISPR/Cas9-mediated knockout of miR166 improves drought and heat tolerance in wheat. *Crop Science*, *63*(5), 1245–1256.

Xu, R., Xu, M., & Zhang, Y. (2017). Targeted modification of miRNAs in rice using CRISPR/Cas9 technology. *Plant Biotechnology Journal*, *15*(6), 725–732.

Yang, B., Chen, M., & Zhang, X. (2019). Field applications of CRISPR/Cas-modified crops: challenges and successes. *Field Crops Research*, *234*, 106–117.

Yoo, S. D., Cho, Y. H., & Tena, G. (2004). The role of small RNAs in systemic acquired resistance. *Journal of Experimental Botany*, *55*(408), 239–245.

Zaidi, S. S., Mansoor, S., & Mahfouz, M. M. (2017). Engineering plants for virus resistance using CRISPR/Cas9. *Current Opinion in Virology*, *26*, 9–15.

Zhang, Y., Zhang, F., & Liu, Y. (2016a). CRISPR/Cas9-mediated genome editing of miRNA genes in Arabidopsis. *Nature Communications*, *7*, 1321.

Zhang, Y., Zhang, F., & Liu, Y. (2016b). CRISPR/Cas9-mediated genome editing of viral genomes in plants. *Nature Communications*, *7*, 1321.

Zhang, Y., Zhang, F., & Liu, Y. (2018). CRISPR/Cas9-mediated genome editing for climate-resilient crops. *Frontiers in Plant Science*, *9*, 945.

Zhang, Q., Li, X., & Wang, Y. (2019). Enhancing bacterial resistance in Arabidopsis using CRISPR/Cas9 to modify small RNA pathways. *Molecular Plant-Microbe Interactions*, *32*(5), 482–493.

Zhang, Y., Chen, X., & Zhou, J. (2020a). CRISPR/Cas9-mediated enhancement of bacterial pathogen resistance in tomatoes. *Plant Cell Reports*, *39*(4), 541–553.

Zhang, X., Liu, Q., Yang, X., Wang, Y., Zhang, H., & Li, Y. (2020b). CRISPR/Cas9-mediated knockout of miR156 enhances drought tolerance and root growth in rice. *Journal of Experimental Botany*, *71*(12), 3678–3689.

Zhou, X., Zhang, Y., & Xu, Z. (2020). Development of rice lines resistant to rice dwarf virus using CRISPR/Cas9 technology. *Plant Biotechnology Journal*, *18*(3), 782–794.

Zhou, Y., Zhang, X., Liu, J., Wang, L., Zhang, Z., & Li, Y. (2021). Modification of siRNA pathways via CRISPR/Cas9 to target transposons improves resistance to tomato yellow leaf curl virus. *Plant Biotechnology Journal*, *19*(7), 1234–1246.

7

Plant Immunity Released by CRISPR/Cas Technology-Mediated Modifying Long Noncoding RNAs

Ananya Choudhury[1], Pritilagna Panigrahi[1], Swatismita Deo[1], Bhupati Nayak[1], Dhaarani V.[1], Swarnalata Tripathy[1], Jyoti Prakash Sahoo[1], and Swapan Kumar Tripathy[2]

[1] *Faculty of Agriculture and Allied Science, C. V. Raman Global University, Bhubaneswar, India*
[2] *College of Agriculture, Odisha University of Agriculture and Technology, Bhawanipatna, India*

7.1 Introduction

Plant immunity is a complex and multilayered defense system that protects plants from a wide range of pathogens, including bacteria, viruses, fungi, and nematodes. Traditional breeding methods and chemical treatments have been used to enhance plant resistance, but these approaches often come with limitations such as long development times, environmental impact, and potential for resistance development in pathogens. In recent years, the advent of clustered regularly interspaced palindromic repeats (CRISPR)/CRISPR-associated protein (Cas) technology has revolutionized genetic engineering, offering precise and efficient tools for modifying plant genomes to enhance immunity (Sattar et al., 2021). While it has been extensively used to study protein-coding genes, its application to noncoding RNAs (ncRNAs) such as long noncoding RNAs (lncRNAs) and microRNAs, which regulate gene expression, is still emerging (Vats et al., 2019). CRISPR/Cas9, originally discovered as part of the microbial adaptive immune system, was first identified in *Escherichia coli* and later in various bacterial and archaeal genomes (Golubov, 2021). It functions as a bacterial immune response mechanism against viral invasion. Recent research suggests that CRISPR/Cas-based editing of ncRNAs holds great potential for plant breeding and understanding the roles of lncRNAs in plants. This approach could overcome the limitations of current functional genomic techniques, such as partial knockdowns and off-target effects. The review highlights the progress and potential of CRISPR/Cas-mediated lncRNA editing in advancing plant science and agricultural improvement (Andolfo & Ercolano, 2015).

lncRNAs have emerged as critical regulators of gene expression, playing significant roles in various biological processes, including plant defense mechanisms (Sattar et al., 2021). Unlike protein-coding RNAs, lncRNAs do not translate into proteins but function through interactions with other RNAs, DNA, and proteins to modulate gene expression (Vats et al., 2019). The involvement of lncRNAs in plant immunity is a relatively new area of research, yet it holds immense potential for developing novel strategies to enhance disease resistance in crops (Andolfo & Ercolano, 2015). This chapter focuses on the innovative application of CRISPR/Cas technology to modify lncRNAs in plants, aiming to enhance their immune responses. By targeting lncRNAs, CRISPR/Cas can

Genome and Epigenome Editing for Stress-Tolerant Crops, First Edition. Edited by Jen-Tsung Chen.

precisely alter their sequences or expression levels, thereby modulating the plant's defense pathways. This approach offers a promising avenue for creating crops with improved resistance to various pathogens, contributing to sustainable agriculture and food security.

7.2 Plant Immunity

The immune system of plants is essential for preventing pathogen spread and maintaining a balanced microbial community (Noman et al., 2021). Effective plant responses require the integration of microbial and abiotic signals. Plants have evolved inducible immune systems with sensors like pattern recognition receptors (PRRs) and nucleotide-binding domain leucine-rich repeat-containing proteins (NLRs) that detect pathogen-associated molecular patterns (PAMPs) and effectors. Plant immunity includes two main types: pattern-triggered immunity (PTI) and effector-triggered immunity (ETI). PTI is activated by PAMPs, while ETI is triggered by pathogen effectors (Lee et al., 2017). An adaptable, low-energy immune system is beneficial for plant survival. Plants can mount robust immune responses, adjust them according to circumstances, and balance them with other physiological responses (Nykiel et al., 2022). The plant immune system involves complex networks of transcription factors, hormones, and receptors. While similar to mammalian innate immunity, plants lack adaptive immunity. Plant immunity is divided into cell-surface immunity, where receptors detect external pathogens, and intracellular immunity, where receptors detect internal pathogen effectors (Liang & Zhou, 2018). Intracellular immunity often involves localized cell death to restrict infection spread. Recent research highlights the cooperation between cell-surface and intracellular immune mechanisms. The intersection of lncRNA research and CRISPR/Cas technology offers new opportunities for enhancing plant immunity (Noman et al., 2021). Using CRISPR/Cas to edit lncRNAs can lead to the development of more resilient crops and a stable food supply. This introduction sets the stage for exploring the mechanisms, applications, and implications of CRISPR/Cas-mediated lncRNA modification in plant immunity (Altaf et al., 2023).

7.3 Long Noncoding RNAs and CRISPR/Cas Technology

The RNA world hypothesis suggests that RNA was the foundation of early life, initially handling information storage and catalytic functions. This role later transitioned to DNA for stability and proteins for adaptability, positioning RNA mainly as an intermediary in the central dogma: "DNA makes RNA makes protein." However, the discovery of new regulatory ncRNAs and the evidence that most of the genome in complex organisms is transcribed in a regulated manner challenge this view. These findings imply that RNA has continued to evolve and play significant roles alongside proteins and DNA (Mercer et al., 2009). lncRNAs are arbitrarily defined as ncRNA transcripts longer than 200 nucleotides. This cutoff excludes many well-known short RNAs like 5S rRNAs, tRNAs, snRNAs, and miRNAs. Other small RNAs, such as Y RNAs, and vault RNAs, are also excluded by this definition. lncRNAs can be further classified into three groups: small RNAs (less than 50 nt), RNA polymerase III transcripts (such as tRNAs, 5S rRNA, and Y RNAs), and small Pol II transcripts (~50–500 nt), which include most snRNAs and intron-derived snoRNAs. lncRNAs often undergo splicing and polyadenylation, making them "mRNA-like," but some are expressed from Pol I or Pol III promoters and are not polyadenylated (Mercer et al., 2009). lncRNAs can be intronic, antisense, or intergenic relative to protein-coding genes, and they can also originate from pseudogenes. Functional pseudogenes are present in both mouse and human genomes (Mercer et al., 2009).

lncRNAs include *trans*-acting regulatory RNAs and circular RNAs, which are generated by back-splicing and have functional roles. Efforts to name and classify lncRNAs, such as by the HUGO Gene Nomenclature Committee and the GENCODE collaboration, typically focus on their genomic orientation and location relative to protein-coding genes. Early studies emphasized long intergenic noncoding RNAs (lincRNAs) to distinguish their roles from proteins (Mattick et al., 2023). The discovery of lncRNAs has challenged miRNA-centered regulatory networks, providing new insights into gene regulation. Some lncRNAs act as sponges or target mimics for miRNAs, adding another layer to posttranscriptional regulation. lncRNAs can also form RNA duplexes with coding or antisense noncoding transcripts, influencing miRNA interactions with target mRNAs. Studies in non-small cell lung carcinoma (NSCLC) suggest that gene regulation networks are more complex than previously understood, with target mRNAs showing little variation despite changes in miRNA levels (Mattick et al., 2023).

The next generation of CRISPR-Cas technologies has significantly advanced our capabilities to manipulate, detect, image, and annotate specific DNA and RNA sequences across diverse species, revolutionizing genome editing from fundamental science to translational medicine (Pickar-Oliver & Gersbach, 2019). Building on initial successes, efforts have focused on discovering new systems such as Cas9, Cas12, Cascade, and Cas13 orthologs to enhance genome editing through mechanisms like nonhomologous end joining, homology-directed repair (HDR), and single-base editing enzymes (Tronick & Hunter, 2016). Nucleic acid, CRISPR-Cas-grounded RNA-directing appliances are being evolved for research, diagnostics, and medicine with nuclease-inert and RNA directing Cas proteins merging to effector proteins to modulate expression of gene, epigenetic moderation, and interactivity of chromatin, thereby propelling these tools toward clinical use in gene and cell therapies. These advancements significantly improv our understanding of biological processes and the potential for clinical applications (Zhang & Zhou, 2014). The development of these technologies marks a transformative era in genome editing, providing unprecedented precision and efficiency in genetic manipulation and offering promising applications in biology, medicine, and agriculture (Mojica & Montoliu, 2016).

7.4 Enhancing Plant Immunity Through lncRNA Modification

lncRNAs are molecules that do not code for proteins but play crucial roles in regulating gene expression in plants. They are involved in numerous biological processes such as development, stress responses, and immune responses (Li et al., 2018). In every part of their life, plants confront various strains (biotic and abiotic stresses) and to resist it, they evolve their immunity (Mattick et al., 2023). These systems include cell surface immune receptors and immune receptors in between the cells that pick up indications of microbial activities and switch on prior responses such as ROS and MAPK. These reciprocations lead to the transcriptional reprogramming of defense-related genes and the formation of late immune responses (Zhang et al., 2020). lncRNAs are defined as endogenous single-stranded nonprotein-coding transcripts longer than 200 nucleotides (Buscaill & Rivas, 2014). Despite their crucial roles, lncRNAs were initially overlooked because of the protein which is noncoding in nature, the species conservative which is inadequate, specific cell type, and scarcity of it (Rivas et al., 2017). However, recent advances have enhanced the knowledge of plant defense mechanisms and resistance toward the expression of genes, highlighting the importance of lncRNAs in plant immunity and stress responses (Amaral, 2023).

7.5 Key Roles of lncRNAs in Plant Immunity

lncRNAs play significant regulatory roles in gene expression despite their low expression levels compared to protein-coding genes and their cell- and tissue-specific expression patterns (Ravasi et al., 2006). They respond to environmental and developmental conditions similarly to protein-coding genes (Wiklund et al., 2006; Yang et al., 2013). The conservation of lncRNAs is a debated topic, covering aspects such as sequence, structure, function, and expression from syntenic loci (Meng et al., 2017). While many functional lncRNAs show multifaceted conservation (Goyal et al., 2017; Clark et al., 2012), some, like XIST, display little sequence conservation (Plath et al., 2002). Interestingly, the promoters of lncRNAs are often as conserved as those of protein-coding genes (Guttman et al., 2009).

This suggests that partial sequences or local spatial structures may suffice for their crucial functions. lncRNAs can modulate immune-related gene expression at transcriptional and posttranscriptional levels by acting as molecular scaffolds (Amaral, 2023). lncRNAs are present in almost all life forms, including plants, where numerous candidates have been identified through genome sequencing and bioinformatics. Examples include *Arabidopsis* (Liu et al., 2015a; Meng et al., 2019), rice (Zheng et al., 2019), maize (Eichten et al., 2014), cotton (Wang et al., 2015; Zhou et al., 2016), and Medicago (Wang et al., 2015). These lncRNAs are cataloged in various databases, including TAIR, Araport, PLncDB, GREENC, NONCODE, CANTATAdb, PNRD, and PlantNATsDB (Liu et al., 2015b; Wierzbicki et al., 2021). As the most abundant class of ncRNAs, lncRNAs regulate gene expression in various biological processes (Böhmdorfer & Wierzbicki, 2015). They are classified based on their genomic location relative to protein-coding genes: sense lncRNAs, antisense lncRNAs, intronic lncRNAs, bidirectional lncRNAs, and lincRNAs (Kapusta & Feschotte, 2014). Among these, lncRNAs and lincRNAs are the most well-characterized in plants.

7.6 Application of CRISPR/Cas Technology-Mediated Modifying Long Noncoding RNAs

7.6.1 Studies on Soybean lncRNAs

lncRNAs in *Glycine max* (soybean) play crucial roles in growth and flower production and its regulation, immune system, and reaction to salt stress (Ghose & Rai, 2021). Some studies identified about 69,000 lncRNA genes in *G. max* (Lin et al., 2020). After all these significant findings, the study of soybeans with CRISPR/Cas genome editing is in its initial phase to date. Niu et al. (2021) utilized a binary-sgRNA CRISPR/Cas9 appliance to create DNA deletions in lncRNA77580, a lncRNA associated with salinity stress resistance (Ghose & Rai, 2021). They observed dysregulation of neighboring protein-coding genes involved in salt stress responses in mutant soybean roots lacking lncRNA77580 (Niu et al., 2021). These findings highlight the potential of CRISPR/Cas systems for the functional characterization of lncRNAs in *G. max* and other plants (Ghose & Rai, 2021).

7.6.2 Studies on Rice lncRNAs

Despite limited reports on lncRNAs in rice, experimental and functional characterizations have confirmed their involvement in significant biological processes. lncRNAs in rice play crucial roles in tolerance to biotic and abiotic stresses (Xu et al., 2016) and regulate the expression of adjacent genes (Wang et al., 2015). They are also vital in growth and development, including germ cell development before meiosis (Komiya et al., 2014) and ovule development in female gametophytes (Liu et al., 2019). Additionally, lncRNAs are linked to male sterility (Ding et al., 2012), grain yield

(Wang et al., 2015), and regulation of leaf and seed morphology. Furthermore, lncRNAs respond to various abiotic stresses such as heat, drought, cold, salt, heavy metals, nutrient deficiency, frost, and radiation. Yuan et al. (2018) found that multiple downregulated polyadenylation (DPA) lncRNAs are involved in the activation of stress response genes related to abscisic acid (ABA) biosynthesis, transport, and metabolism, highlighting their role in drought stress response (Hirayama & Shinozaki, 2010).

7.6.3 Studies on Tomato lncRNAs

lncRNAs perform analytic executive parts in opposition to biotic stresses in plants (Zaynab et al., 2018). During plant–pathogen interactions, such as those involving *Pseudomonas syringae* pv. Tomato (Pst), significant changes in lncRNA expression occur. These changes can potentially be used as biosensors for the invasion of viruses (Prasad & Prasad, 2021). When Pest invades *Solanum lycopersicum* plants, it introduces effector proteins that disrupt PTI and induced ETI, leading to significant lncRNA responses. In tomato leaves, unequal expression of lncRNA has been picked out at various time points postinvasion, influencing photosynthesis and the expression of genes related to cell wall kinases and glutathione S-transferases (GST) during ETI and PTI (Rosli et al., 2021).

lncRNAs in tomatoes also play crucial roles in ethylene signaling and modulate fruit color, flavor, and texture. At least 3679 lncRNAs have been spotted in wild-type and ripening mutant tomato fruits, with 187 downregulated and 490 upregulated in ripening mutants (Zhu et al., 2015). Additionally, *S. lycopersicum* lncRNAs are vital in reaction toward viral and fungal infections (Sharma et al., 2022). The use of the CRISPR/Cas system to explore the parts of tomato lncRNAs is still in its initial phases. For example, genes related to ethylene biosynthesis and the critical fruit ripening gene DEMETER-LIKE 2 (SlDML2) were significantly downregulated in mutants. lncRNA1459 mutants showed expression alternation of ripening-related lncRNAs, such as lncRNA26, lncRNA28, and lncRNA48, among others (Li et al., 2018). These findings highlight the need for further exploration of lncRNA functions in fruit ripening and other processes (Table 7.1). CRISPR/Cas9 technology has been effectively utilized in various plants to enhance gene targeting and study gene function. For example, Miki et al. (2018) implemented a sequential transformation strategy in *Arabidopsis thaliana*, which improved the efficiency of CRISPR gene targeting. Tsutsui & Higashiyama (2017) explored the effectiveness of the pKAMA-ITACHI Red vector in *A. thaliana*, targeting genes such as PDS3, AG, and DUO1, demonstrating the potential of CRISPR/Cas9 in plant genetic research. These studies highlight the broad applications of CRISPR/Cas9 across different plant species, providing a valuable tool for genetic modifications and functional genomics.

7.7 Ethical Regulations, Bioethical Uses, and Future Prospects

CRISPR genome-editing technology presents several technical and ethical concerns. Technical challenges include mosaicism, where not all cells in an organism are edited uniformly, and the occurrence of HDR, which can introduce unintended changes (Champer et al., 2017, 2018; Carrami et al., 2018; Oberhofer et al., 2018; Pham et al., 2019). There is also a risk of reduced on-target editing efficiency and off-target impacts, where unintended genomic regions are altered (Hiranniramol et al., 2020). Ethical concerns revolve around the long-term implications of creating modified organisms and the potential unforeseen consequences on ecosystems and human health (Parsaeimehr et al., 2022). Moreover, some major technical limitations (Figure 7.1) for ethical concerns with CRISPR/Cas-mediated genome engineering are indexed in Figure 7.1a, and some technical limitations of CRISPR/Cas technology are indexed in Figure 7.1b.

Table 7.1 Application of CRISPR/Cas technology-mediated modifying long noncoding RNAs.

Plants	Targeted genes	Traits	Method	References
Malus pumila	*MdDIP4*	Disease resistance	Gene inactivation	Pompili et al. (2020)
Zea mays	*ZmPHYC1, ZmPHYC2*	Flowering time/plant height	Gene knockout and overexpression	Li et al. (2019)
Cucumis melo	*CmPDS*	Albinism	Gene knockout	Hooghvorst et al. (2019)
Elaeis guineensis	*EgIFR*	Disease resistance	Base editing	Budiani et al. (2018)
Brassica napus	*BnALS1*	Herbicides resistance	Base editing	Wu et al. (2020)
Oryza sativa	*Os8N3*	Disease resistance	Gene knockout	Kim et al. (2019)
Glycine max	*GmPRR37; GmFT2a/5a*	Flowering time and regional adaptability	Site-directed mutagenesis	Cai et al. (2022)
Nicotiana tabacum	*NtHL1*	Hybrid lethality	Frameshift mutation	Ma et al. (2020)
Citrullus lanatus	*CIPDS*	Albinism	Gene knockout	Tian et al. (2017)

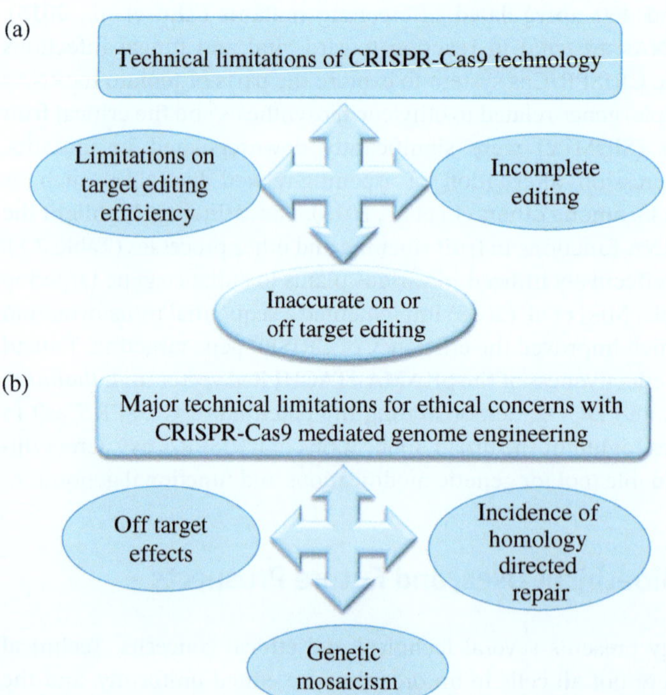

Figure 7.1 Major technical limitations; (a) ethical concerns with CRISPR/Cas-mediated genome engineering; (b) technical limitations of CRISPR/Cas technology.

CRISPR-Cas9 technology presents a range of risks and bioethical issues that warrant careful consideration. These include concerns such as gene drift, nontarget mutations, potential impacts on wild species leading to extinction, and ecological imbalances (Shinwari et al., 2017; Lawrenson et al., 2022;

Uddin et al., 2020; Piergentili et al., 2021). Ethical and legal dimensions include issues related to legislation, patenting of gene-edited organisms, germline disruption, and broader societal implications (Bellver Capella, 2016; Bier, 2022; Chennuri et al., 2022; Segers, 2023). Researchers emphasize the need for comprehensive ethical policies and regulations to guide the responsible use of CRISPR-Cas9 technology across different applications, including human embryos and crops (Feng et al., 2013; Zhang et al., 2020; Zhou et al., 2020). Moreover, Brokowski & Adli (2019) suggest that ethical considerations should evolve alongside technological advancements, advocating for evidence-based regulation by national and international bodies to ensure the ethical and safe application of CRISPR technologies. Despite these concerns, advancements in CRISPR-Cas technology are incorporating built-in safeguards to mitigate risks, potentially reducing regulatory barriers and accelerating its adoption in agriculture and other fields. This could lead to transformative improvements in crop traits, addressing global food security challenges more effectively (Brokowski & Adli, 2019).

CRISPR-based genome editing has emerged as a powerful tool in plant transcriptomics and functional genomics, showcasing its potential to enhance plant protection against pathogens through targeted manipulation of lncRNAs. This approach not only elucidates the complex regulatory networks involving lncRNAs in plant immunity but also offers a sustainable alternative to conventional chemical treatments, thereby bolstering crop resilience and food security (Iqbal et al., 2023). The integration of CRISPR/Cas-mediated alterations in lncRNAs holds promise for sustainable agricultural practices, higher crop yields, and improved stress tolerance, highlighting its pivotal role in addressing global agricultural challenges (Singh et al., 2023). Moreover, CRISPR/Cas technology has elucidated the dependence of regulatory plant ncRNAs on their secondary structure and genomic sequence, where mutations induced by CRISPR/Cas, such as deletions, alter ncRNA biogenesis processes and enable functional studies of ncRNAs in plants (Wierzbicki et al., 2021). Despite these advancements, further research is needed to fully comprehend the biological roles of ncRNAs and optimize CRISPR/Cas-mediated editing techniques for specific agronomic traits (Prasad & Prasad, 2021). Current applications of CRISPR/Cas in plant biology encompass gene targeting, knock-outs, gene expression modulation, chromosome tagging, and insertion of heterologous DNA sequences, with ongoing efforts to enhance multiplexing and tackle challenges in plastome and chondriome editing (Sharma et al., 2022).

7.8 Conclusion

The modification of lncRNAs using CRISPR/Cas technology represents a groundbreaking approach to enhancing plant immunity. This innovative strategy leverages the precision and efficiency of CRISPR/Cas systems to target and manipulate lncRNAs, which play crucial roles in regulating plant defense mechanisms. By fine-tuning the expression and function of these noncoding elements, it is possible to develop crops with heightened resistance to a wide array of pathogens, thereby reducing the reliance on chemical treatments and traditional breeding methods. This chapter underscores the potential of integrating advanced genomic tools with plant biotechnology to achieve sustainable agricultural practices and improve food security. Future research and practical applications of CRISPR/Cas-mediated lncRNA modifications hold promise for significant advancements in crop protection and resilience.

References

Altaf, M. A., Mandal, S., Behera, B., Mangal, V., Naz, S., Kumar, R., et al. (2023). Salinity stress tolerance in Solanaceous crops: current understanding and its prospects in genome editing. *Journal of Plant Growth Regulation*, *42*(7), 4020–4036. https://www.researchgate.net/publication/366633865_Salinity_Stress_Tolerance_in_Solanaceous_Crops_Current_Understanding_and_Its_Prospects_in_Genome_Editing

Amaral, P. S. (2023). The demographic transition and the asset supply channel. *European Economic Review*, *151*, 104317. https://www.sciencedirect.com/science/article/pii/S0014292122001970

Andolfo, G., & Ercolano, M. R. (2015). Plant innate immunity multicomponent model. *Frontiers in Plant Science*, *6*, 153141. https://doi.org/10.3389/fpls.2015.00987

Bellver Capella, V. (2016). Crispr-Cas9 gene editing revolution and its ethical and legal challenges. *Cuadernos de Bioética*, *27*(90), 223–239. https://pubmed.ncbi.nlm.nih.gov/27637196/

Bier, E. (2022). Gene drives gaining speed. *Nature Reviews Genetics*, *23*, 5–22. https://pubmed.ncbi.nlm.nih.gov/34363067/

Böhmdorfer, G., & Wierzbicki, A. T. (2015). Control of chromatin structure by long noncoding RNA. *Trends in Cell Biology*, *25*(10), 623–632. https://pubmed.ncbi.nlm.nih.gov/26410408/

Brokowski, C., & Adli, M. (2019). CRISPR ethics: moral considerations for applications of a powerful tool. *Journal of Molecular Biology*, *431*(1), 88–101. https://doi.org/10.1016/j.jmb.2018.05.044

Budiani, A., Putranto, R. A., Riyadi, I., Minarsih, H., & Faizah, R. (2018). Transformation of oil palm calli using CRISPR/Cas9 System: toward genome editing of oil palm. In: *IOP Conference Series: Earth and Environmental Science, Vol. 183*, No. 1, IOP Publishing, 012003.

Buscaill, P., & Rivas, S. (2014). Transcriptional control of plant defence responses. *Current Opinion in Plant Biology*, *20*, 35–46. https://pubmed.ncbi.nlm.nih.gov/24840291/

Cai, Y., Chen, L., & Hou, W. (2022). Genome Editing Technologies Contribute for Precision Breeding in Soybean. In: *Genome Editing Technologies for Crop Improvement*, Springer Nature Singapore, Singapore, 349–366.

Carrami, E. M., Eckermann, K. N., Ahmed, H. M. M., Sánchez C, H. M., Dippel, S., Marshall, J. M., et al. (2018). Consequences of resistance evolution in a Cas9-based sex conversion-suppression gene drive for insect pest management. *Proceedings of the National Academy of Sciences*, *115*, 6189–6194.https://doi.org/10.1073/pnas.1713825115. https://pubmed.ncbi.nlm.nih.gov/29844184/

Champer, J., Reeves, R., Oh, S. Y., Liu, C., Liu, J., Clark, A. G., et al. (2017). Novel CRISPR/Cas9 gene drive constructs reveal insights into mechanisms of resistance allele formation and drive efficiency in genetically diverse populations. *PLoS Genetics*, *13*, e1006796. https://doi.org/10.1371/journal.pgen.1006796. https://pubmed.ncbi.nlm.nih.gov/28727785/

Champer, J., Liu, J., Oh, S. Y., Reeves, R., Luthra, A., Oakes, N., et al. (2018). Reducing resistance allele formation in CRISPR gene drive. *Proceedings of the National Academy of Sciences*, *115*, 5522–5527. https://doi.org/10.1073/pnas.1720354115. https://www.pnas.org/doi/pdf/10.1073/pnas.1720354115

Chennuri, P. R., Adelman, Z. N., & Myles, K. M. (2022). Genetic approaches for controlling CRISPR-based autonomous homing gene drives. *Frontiers in Bioengineering and Biotechnology*, *10*, 1–9. https://www.frontiersin.org/journals/bioengineering-and-biotechnology/articles/10.3389/fbioe.2022.897231/full

Clark, A. J., Weiss, S. J., Kain, J. S., Jirak, I. L., Coniglio, M., Melick, C. J., et al. (2012). An overview of the 2010 hazardous weather testbed experimental forecast program spring experiment. *Bulletin of the American Meteorological Society*, *93*(1), 55–74.

Ding, J., Lu, Q., Ouyang, Y., Mao, H., Zhang, P., Yao, J., et al. (2012). A long noncoding RNA regulates photoperiod-sensitive male sterility, an essential component of hybrid rice. *Proceedings of the National Academy of Sciences*, *109*(7), 2654–2659. https://doi.org/10.1073/pnas.1121374109

Eichten, S. R., Schmitz, R. J., & Springer, N. M. (2014). Epigenetics: beyond chromatin modifications and complex genetic regulation. *Plant Physiology, 165*(3), 933–947. https://pubmed.ncbi.nlm.nih.gov/24872382/

Feng, Z., Zhang, B., Ding, W., Liu, X., Yang, D.-L., Wei, P., Cao, F., et al. (2013). Efficient genome editing in plants using a CRISPR/Cas system. *Cell Research, 23*, 1229. https://pubmed.ncbi.nlm.nih.gov/23958582/

Ghose, K., & Rai, K. M. (2021). Long noncoding RNAs and their implication in novel trait development in soybean. In: *Long Noncoding RNAs in Plants*, Academic Press, 133–150. https://doi.org/10.1016/B978-0-12-821452-7.00009-X

Golubov, A. (2021). CRISPR-bacterial immune system. In: *Genome Stability*, Academic Press, 91–105. https://pubmed.ncbi.nlm.nih.gov/33685600/

Goyal, A., Myacheva, K., Groß, M., Klingenberg, M., Duran Arqué, B., & Diederichs, S. (2017). Challenges of CRISPR/Cas9 applications for long non-coding RNA genes. *Nucleic Acids Research, 45*(3), e12-e12

Guttman, M., Amit, I., Garber, M., French, C., Lin, M. F., Feldser, D., et al. (2009). Chromatin signature reveals over a thousand highly conserved large non-coding RNAs in mammals. *Nature, 458*(7235), 223–227. https://pubmed.ncbi.nlm.nih.gov/19182780/

Hiranniramol, K., Chen, Y., Liu, W., & Wang, X. (2020). Generalizable sgRNA design for improved CRISPR/Cas9 editing efficiency. *Bioinformatics, 36*, 2684–2689. https://pubmed.ncbi.nlm.nih.gov/31971562/

Hirayama, T., & Shinozaki, K. (2010). Research on plant abiotic stress responses in the post-genome era: past, present and future. *The Plant Journal, 61*(6), 1041–1052. https://doi.org/10.1111/j.1365-313X.2010.04124.x

Hooghvorst, I., López-Cristoffanini, C., & Nogués, S. (2019). Efficient knockout of phytoene desaturase gene using CRISPR/Cas9 in melon. *Scientific Reports, 9*(1), 17077.

Iqbal, B., Li, G., Alabbosh, K. F., Hussain, H., Khan, I., Tariq, M., et al. (2023). Advancing environmental sustainability through microbial reprogramming in growth improvement, stress alleviation, and phytoremediation. *Plant Stress*, 100283. https://www.researchgate.net/publication/375453399_Advancing_Environmental_Sustainability_through_Microbial_Reprogramming_in_Growth_Improvement_Stress_Allevi-ation_and_Phytoremediatio

Kapusta, A., & Feschotte, C. (2014). Volatile evolution of long noncoding RNA repertoires: mechanisms and biological implications. *Trends in Genetics, 30*(10), 439–452. https://pubmed.ncbi.nlm.nih.gov/25218058/

Kim, S., Chen, J., Cheng, T., Gindulyte, A., He, J., He, S., et al. (2019). PubChem 2019 update: improved access to chemical data. *Nucleic Acids Research, 47*(D1), D1102–D1109. https://pubmed.ncbi.nlm.nih.gov/30371825/

Komiya, R., Ohyanagi, H., Niihama, M., Watanabe, T., Nakano, M., Kurata, N., & Nonomura, K. I. (2014). Rice germline-specific Argonaute MEL 1 protein binds to phasiRNAs generated from more than 700 linc RNAs. *The Plant Journal, 78*(3), 385–397. https://doi.org/10.1111/tpj.12483

Lawrenson, T., Chhetry, M., Clarke, M., Hundleby, P., & Harwood, W. (2022). Improved SpCas9 and LbCas12a genome editing systems in Brassica oleracea and Brassica napus. *bioRxiv*, 2022-05

Lee, H. A., Lee, H. Y., Seo, E., Lee, J., Kim, S. B., Oh, S., et al. (2017). Current understandings of plant nonhost resistance. *Molecular Plant-Microbe Interactions, 30*(1), 5–15. https://pubmed.ncbi.nlm.nih.gov/27925500/

Li, R., Fu, D., Zhu, B., Luo, Y., & Zhu, H. (2018). CRISPR/Cas9-mediated mutagenesis of lncRNA1459 alters tomato fruit ripening. *The Plant Journal, 94*(3), 513–524. https://doi.org/10.1111/tpj.13872

Li, J. R., Walker, S., Nie, J. B., & Zhang, X. Q. (2019). Experiments that led to the first gene-edited babies: the ethical failings and the urgent need for better governance. *Journal of Zhejiang University-SCIENCE B Biomedicine & Biotechnology, 20*, 32–38. https://pubmed.ncbi.nlm.nih.gov/30614228/

Liang, X., & Zhou, J. M. (2018). Receptor-like cytoplasmic kinases: central players in plant receptor kinase–mediated signalling. *Annual Review of Plant Biology*, *69*(1), 267–299. https://pubmed.ncbi.nlm.nih.gov/29719165/

Lin, X., Lin, W., Ku, Y. S., et al. (2020). Analysis of soybean long non coding RNAs reveals a subset of small peptide-coding transcripts. *Plant Physiology*, *182*, 1359–1374. https://doi.org/10.1104/pp.19.0132

Liu, X., Hao, L., Li, D., Zhu, L., & Hu, S. (2015a). Long non-coding RNAs and their biological roles in plants. *Genomics, Proteomics and Bioinformatics*, *13*(3), 137–147.

Liu, L., Oza, S., Hogan, D., Perin, J., Rudan, I., Lawn, J. E., et al. (2015b). Global, regional, and national causes of child mortality in 2000–13, with projections to inform post-2015 priorities: an updated systematic analysis. *The Lancet*, *385*(9966), 430–440.

Liu, H., Wang, R., Mao, B., Zhao, B., & Wang, J. (2019). Identification of lncRNAs involved in rice ovule development and female gametophyte abortion by genome-wide screening and functional analysis. *BMC Genomics*, *20*, 1–16. https://doi.org/10.1016/j.rsci.2020.03.003

Ma, J., Hancock, W. G., Nifong, J. M., Kernodle, S. P., & Lewis, R. S. (2020). Identification and editing of a hybrid lethality gene expands the range of interspecific hybridization potential in Nicotiana. *Theoretical and Applied Genetics*, *133*, 2915–2925.

Mattick, J. S., Amaral, P. P., Carninci, P., Carpenter, S., Chang, H. Y., Chen, L. L., et al. (2023). Long non-coding RNAs: definitions, functions, challenges and recommendations. *Nature Reviews Molecular Cell Biology*, *24*(6), 430–447. https://pubmed.ncbi.nlm.nih.gov

Meng, Q., Cai, K., Chen, Y., & Chen, L. (2017). Research progress on conducting polymer based supercapacitor electrode materials. *Nano Energy*, *36*, 268–285. https://www.infona.pl/resource/bwmeta1.element.elsevier-cef81d39-a858-328a-b022-87aabe

Meng, Y., Liu, X., Lu, S., Zhang, T., Jin, B., Wang, Q., et al. (2019). A review on occurrence and risk of polycyclic aromatic hydrocarbons (PAHs) in lakes of China. *Science of the Total Environment*, *651*, 2497–2506. https://pubmed.ncbi.nlm.nih.gov/30336439/

Mercer, T. R., Dinger, M. E., & Mattick, J. S. (2009). Long non-coding RNAs: insights into functions. *Nature Reviews Genetics*, *10*(3), 155–159. https://www.nature.com/articles/nrg2521

Miki, D., Zhang, W., Zeng, W., Feng, Z., & Zhu, J. K. (2018). CRISPR/Cas9-mediated gene targeting in Arabidopsis using sequential transformation. *Nature Communications*, *9*(1), 1967.

Mojica, F. J., & Montoliu, L. (2016). On the origin of CRISPR-Cas technology: from prokaryotes to mammals. *Trends in Microbiology*, *24*(10), 811–820. https://pubmed.ncbi.nlm.nih.gov/27401123/

Niu, F., Jiang, Q., Sun, X., et al. (2021). Large DNA fragment deletion in lncRNA77580 regulates neighbouring gene expression in soybean (*Glycine max*). *Functional Plant Biology*, *48*, 1139–1147. https://doi.org/10.1071/FP20400

Noman, M., Ahmed, T., Ijaz, U., Shahid, M., Azizullah, Li, D., et al. (2021). Plant–Microbiome crosstalk: dawning from composition and assembly of microbial community to improvement of disease resilience in plants. *International Journal of Molecular Sciences*, 22(13), 6852. https://doi.org/10.3390/ijms22136852

Nykiel, M., Gietler, M., Fidler, J., Prabucka, B., Rybarczyk-Płońska, A., Graska, J., et al. (2022). Signal transduction in cereal plants struggling with environmental stresses: from perception to response. *Plants*, *11*(8), 1009. https://doi.org/10.3390/plants11081009

Oberhofer, G., Ivy, T., & Hay, B. A. (2018). Behavior of homing endonuclease gene drives targeting genes required for viability or female fertility with multiplexed guide RNAs. *Proceedings of the National Academy of Sciences*, *115*, E9343–E9352. https://doi.org/10.1073/pnas.1805278115

Parsaeimehr, A., Ebirim, R., & Ozbay, G. (2022). CRISPR-cas technology a new era in genomic engineering. *Biotechnology Reports*, *12*, e00731, https://doi.org/10.1016/j.btre.2022.e00731. https://pubmed.ncbi.nlm.nih.gov/35686011/

Pham, T. B., Phong, C. H., Bennett, J. B., Hwang, K., Jasinskiene, N., Parker, K., et al. (2019). Experimental population modification of the malaria vector mosquito, *Anopheles stephensi*. *PLOS Genetics*, *15*, e1008440. https://doi.org/10.1371/journal.pgen.1008440

Pickar-Oliver, A., & Gersbach, C. A. (2019). The next generation of CRISPR-Cas technologies and applications. *Nature Reviews Molecular Cell Biology*, *20*(8), 490–507. https://www.nature.com/articles/s41580-019-0131-5

Piergentili, R., Del Rio, A., Signore, F., Umani Ronchi, F., Marinelli, E., & Zaami, S. (2021). CRISPR-Cas and its wide-ranging applications: from human genome editing to environmental implications, technical limitations. *Hazards Bioethical Issues Cells*, *10*(5), 969. https://www.ncbi.nlm.nih.gov/pmc/articles/PMC8143109/

Plath, K., Mlynarczyk-Evans, S., Nusinow, D. A., & Panning, B. (2002). Xist RNA and the mechanism of X chromosome inactivation. *Annual Review of Genetics*, *36*(1), 233–278. https://pubmed.ncbi.nlm.nih.gov/12429693/

Pompili, V., Dalla Costa, L., Piazza, S., Pindo, M., & Malnoy, M. (2020). Reduced fire blight susceptibility in apple cultivars using a high-efficiency CRISPR/Cas9-FLP/FRT-based gene editing system. *Plant Biotechnology Journal*, *18*(3), 845–858.

Prasad, A., & Prasad, M. (2021). Host-virus interactions mediated by long non-coding RNAs. *Virus Research*, *298*, 198402. https://doi.org/10.1016/j.virusres.2021.198402

Ravasi, T., Suzuki, H., Pang, K. C., Katayama, S., Furuno, M., Okunishi, R., et al. (2006). Experimental validation of the regulated expression of large numbers of non-coding RNAs from the mouse genome. *Genome Research*, *16*(1), 11–19. https://www.ncbi.nlm.nih.gov/pmc/articles/PMC1356124/

Rivas, E., Clements, J., & Eddy, S. (2017). A statistical test for conserved RNA structure shows lack of evidence for structure in lncRNAs. *Nature Methods*, *14*, 45–48.

Rosli, H. G., Sirvent, E., Bekier, F. N., Ramos, R. N., & Pombo, M. A. (2021). Genome-wide analysis uncovers tomato leaf lncRNAs transcriptionally active upon *Pseudomonas syringae* pv. tomato challenge. *Scientific Reports*, *11*(1), 24523. https://doi.org/10.1038/s41598-021-04005-0

Sattar, M. N., Iqbal, Z., Al-Khayri, J. M., & Jain, S. M. (2021). Induced genetic variations in fruit trees using new breeding tools: food security and climate resilience. *Plants*, *10*(7), 1347. https://doi.org/10.3390/plants10071347

Segers, S. (2023 Aug). Heritable genome editing: ethical aspects of a developing domain. *Human Reproduction*, *15*, dead167. https://doi.org/10.1093/humrep/dead167

Sharma, Y., Sharma, A., Madhu, S., Singh, K., & Upadhyay, S. K. (2022). Long non-coding RNAs as emerging regulators of pathogen response in plants. *Non-coding RNA*, *8*(1), 4. https://www.mdpi.com/journal/ncrna

Shinwari, S. H. E. H. L. A., Ahmad, M., Zhang, G., Jahan, S., & Sultana, S. (2017). Medicinal plant diversity use for gynecological disorders among the rural communities of Northern Pakistan. *Pakistan Journal of Botany*, *49*(5), 1787–1799.

Singh, A., Mazahar, S., Chapadgaonkar, S. S., Giri, P., & Shourie, A. (2023). Phyto-microbiome to mitigate abiotic stress in crop plants. *Frontiers in Microbiology*, *14*, 1210890. https://doi.org/10.3389/fmicb.2023.1210890

Tian, S., Jiang, L., Gao, Q., Zhang, J., Zong, M., Zhang, H., et al. (2017). Efficient CRISPR/Cas9-based gene knockout in watermelon. *Plant Cell Reports*, *36*, 399–406.

Tronick, E., & Hunter, R. G. (2016). Waddington, dynamic systems, and epigenetics. *Frontiers in Behavioural Neuroscience*, *10*, 107. https://www.frontiersin.org/journals/behavioral-neuroscience/articles/10.3389/fnbeh.2016.001

Tsutsui, H., & Higashiyama, T. (2017). pKAMA-ITACHI vectors for highly efficient CRISPR/Cas9-mediated gene knockout in Arabidopsis thaliana. *Plant and Cell Physiology*, *58*(1), 46–56.

Uddin, F., Rudin, C. M., & Sen, T. (2020). CRISPR gene therapy: applications, limitations, and implications for the future. *Frontiers in Oncology*, *10*, 1387. https://www.frontiersin.org/journals/oncology/articles/10.3389/fonc.2020.01387/full

Vats, S., Kumawat, S., Kumar, V., Patil, G. B., Joshi, T., Sonah, H., et al. (2019). Genome editing in plants: exploration of technological advancements and challenges. *Cells*, *8*(11), 1386. https://doi.org/10.3390/cells811138. https://www.researchgate.net/publication/337014113_Genome_Editing_in_Plants_Exploration_of_Technological_Advancements_and_Challenges

Wang, H., Niu, Q. W., Wu, H. W., Liu, J., Ye, J., Yu, N., & Chua, N. H. (2015). Analysis of non-coding transcriptomes in rice and maize uncovers roles of conserved lncRNA associated with agriculture traits. *The Plant Journal*, *84*(2), 404–416. https://doi.org/10.1111/tpj.1301

Wierzbicki, A. T., Blevins, T., & Swiezewski, S. (2021). Long noncoding RNAs in plants. *Annual Review of Plant Biology*, *72*(1), 245–271. https://pubmed.ncbi.nlm.nih.gov/33752440/

Wiklund, I., Carlsson, J., & Vakil, N. (2006). Gastroesophageal reflux symptoms and well-being in a random sample of the general population of a Swedish community. *Official Journal of the American College of Gastroenterology| ACG*, *101*(1), 18–28.

Wu, J., Chen, C., Xian, G., Liu, D., Lin, L., Yin, S., et al. (2020). Engineering herbicide-resistant oilseed rape by CRISPR/Cas9-mediated cytosine base-editing. *Plant Biotechnology Journal*, *18*(9), 1857.

Xu, X. W., Zhou, X. H., Wang, R. R., Peng, W. L., An, Y., & Chen, L. L. (2016). Functional analysis of long intergenic non-coding RNAs in phosphate-starved rice using competing endogenous RNA networks. *Scientific Reports*, *6*(1), 20715. https://doi.org/10.1016/j.rsci.2020.03.003

Yang, H., Wang, H., Shivalila, C. S., Cheng, A. W., Shi, L., & Jaenisch, R. (2013). One-step generation of mice carrying reporter and conditional alleles by CRISPR/Cas-mediated genome engineering. *Cell*, *154*(6), 1370–1379.

Yuan, J., Li, J., Yang, Y., Tan, C., Zhu, Y., Hu, L., et al. (2018). Stress-responsive regulation of long non-coding RNA polyadenylation in *Oryza sativa*. *The Plant Journal*, *93*(5), 814–827. https://doi.org/10.1111/tpj.13804

Zaynab, M., Fatima, M., Abbas, S., Umair, M., Sharif, Y., & Raza, M. A. (2018). Long non-coding RNAs as molecular players in plant defense against pathogens. *Microbial Pathogenesis*, *121*, 277–282. https://doi.org/10.1111/j.1365-313X.2010.04124.x

Zhang, L., & Zhou, Q. (2014). CRISPR/Cas technology: a revolutionary approach for genome engineering. *Science China Life Sciences*, *57*, 639–640.

Zhang, D., Hussain, A., Manghwar, H., Xie, K., Xie, S., Zhao, S., et al. (2020). Genome editing with the CRISPR-Cas system: an art, ethics and global regulatory perspective. *Plant Biotechnology Journal*, *18*(8), 1651–1669. https://doi.org/10.1111/pbi.13383

Zheng, B., Chevallier, F., Yin, Y., Ciais, P., Fortems-Cheiney, A., Deeter, M. N., et al. (2019). Global atmospheric carbon monoxide budget 2000–2017 inferred from multi-species atmospheric inversions. *Earth System Science Data*, *11*(3), 1411–1436.

Zhou, M., Wang, H., Zhu, J., Chen, W., Wang, L., Liu, S., et al. (2016). Cause-specific mortality for 240 causes in China during 1990–2013: a systematic subnational analysis for the Global Burden of Disease Study 2013. *The Lancet*, *387*(10015), 251–272. https://pubmed.ncbi.nlm.nih.gov/26510778/

Zhou, Z., Ren, L., Zhang, L., Zhong, J., Xiao, Y., Jia, Z., et al. (2020). Heightened innate immune responses in the respiratory tract of COVID-19 patients. *Cell Host & Microbe*, *27*(6), 883–890.

Zhu, B., Yang, Y., Li, R., Fu, D., Wen, L., Luo, Y., & Zhu, H. (2015). RNA sequencing and functional analysis implicate the regulatory role of long non-coding RNAs in tomato fruit ripening. *Journal of Experimental Botany*, *66*(15), 4483–4495. https://doi.org/10.1093/jxb/erv203

8

Plant Stress Memory Rewrite by CRISPR/Cas-Mediated Approaches

Srutirekha Mishra[1], Subhashree Priyadarshinee Sahoo[1], Subhalaxmi Swain[1], Laxmipriya Pati[1], Dhaarani V.[1], Swarnalata Tripathy[1], Jyoti Prakash Sahoo[1], and Swapan Kumar Tripathy[2]

[1] *Faculty of Agriculture and Allied Science, C. V. Raman Global University, Bhubaneswar, India*
[2] *College of Agriculture, Odisha University of Agriculture and Technology, Bhawanipatna, India*

8.1 Introduction

Environmental factors such as droughts, heat waves, and disease outbreaks can severely affect plant growth and productivity (Nadeem et al., 2018a, b; Liu et al., 2022). Over time, plants have evolved the ability to "remember" these stressors and respond more robustly upon exposure, a phenomenon known as "plant stress memory," which has significant potential for enhancing crop resilience (Nadeem et al., 2021). However, harnessing this process has been challenging. Recent advances in clustered regularly interspaced palindromic repeats (CRISPRs) gene-editing technology provide new opportunities to improve and modify plant stress memory, leading to the development of more resilient crops (Ray et al., 2016). The complexity of stress responses, involving multiple interactions and communication channels, makes it difficult to devise a universal strategy for rewriting stress memory (Suzuki et al., 2014). Traditional plant breeding techniques, chemical methods like ethyl methane sulfonate (EMS), physical techniques like gamma radiation, and biological methods like T-DNA and transposon insertion have historically altered plant genomes, causing point mutations, deletions, and gene duplications. Scientists aim to develop gene-editing tools that can modify genomes without introducing new transgenes (Suzuki et al., 2014).

Genome editing, initiated by transcription activator-like effector nucleases (TALENs) and zinc finger nucleases (ZFNs), is now further advanced by CRISPR/Cas technology (Shabala, 2013). One significant challenge in agriculture is the limited genetic diversity in elite breeding stocks. Since the 1950s, random mutagenesis using genotoxic agents has generated new alleles, albeit through a laborious and inefficient process that introduced multiple mutations simultaneously (Pacher & Puchta, 2017). Environmental stresses are a major issue in the twenty-first century, posing an unexpected barrier to ensuring a sustainable food supply for the growing global population. Enhancing crops' ability to adapt to diverse environmental conditions is crucial. Plants respond to abiotic stress by altering their morphology, biology, physiology, and molecular frameworks (Wang et al., 2003a, b). To feed the projected 9.7 billion people by 2050, agricultural production must increase by at least 85% (Alexandratos & Bruinsma, 2012). Researchers have found that various

Genome and Epigenome Editing for Stress-Tolerant Crops, First Edition. Edited by Jen-Tsung Chen.

physical and chemical stressors significantly reduce plant productivity, hindering their growth potential and survival (Shao et al., 2015).

The unpredictability of environmental stress exacerbates the challenge of feeding the expanding population (Nadeem et al., 2018a, b). Therefore, developing cultivars with higher adaptability to diverse environmental conditions is essential for CRISPR/Cas9-based multiplexing, which targets multiple genes in a single plant and has proven effective in various crops, including rice, wheat, maize (Char et al., 2017), and cotton (Gao et al., 2017). CRISPR-mediated gene activation holds great promise for creating genome-edited crops with enhanced stress tolerance by overexpressing tolerance genes (Zafar et al., 2020). This chapter discusses the intriguing world of plant stress memory and explores how the revolutionary CRISPR-Cas gene editing tool can be harnessed to manipulate this phenomenon for enhanced plant resilience. Moreover, this chapter will examine the molecular mechanisms underlying stress memory, discuss the potential of CRISPR-Cas to modify these mechanisms, and explore the implications for developing crops better equipped to withstand environmental stress.

8.2 Understanding Plant Stress Response

Plants serve as important model systems for studying Earth's ecological relationships, often facing various abiotic and biotic stresses in their environments (Hashem et al., 2017). Biotic stress refers to the disruption of normal plant metabolism by living organisms such as bacteria, fungi, viruses, nematodes, insects, and weeds. These stressors negatively affect plant vigor, growth, and productivity (Hashem et al., 2017). Research has shown that endophytic fungi can combat numerous plant diseases. According to Hartley & Gange (2009), these fungi help plants manage biotic stress by activating their defense systems. Initially discovered that the endophyte *Acremonium loliae* protected perennial ryegrass (*Lolium perenne* L.) from the Argentine stem weevil (*Listronotus bonariensis Kuschel*), an insect pest. Further studies have demonstrated that endophytic fungi can shield host plants from a range of biotic stressors. For instance, Waqas et al. (2021) found that the endophytic fungi *Penicillium citrinum* and *Aspergillus terreus* significantly reduced fungal infections caused by *Alternaria alternata* in sunflowers (*Helianthus annuus* L.), thereby promoting plant growth. Plants inoculated with endophytes develop defense mechanisms against biotic stress through induced systemic resistance (ISR), functional amino acids, antioxidants, hormone regulation, and other compounds (Waqas et al., 2021).

Abiotic stress, since they are sedentary and unable to escape, plants are always under stress from their environment. Plants protect themselves from such obstacles by controlling their physiological, metabolic, and developmental responses to outside environmental stimuli. While certain environmental stresses – like heat, drought, salt, and cold – occur frequently during plant life, others are repeated. One of the main environmental factors limiting agricultural productivity worldwide is abiotic stressors (Bartels & Sunkar, 2005). These pressures cause different morphological, physiological, biochemical, and metabolic changes in plants through different processes. Increased salinization of arable land is predicted to have catastrophic worldwide repercussions, causing up to 50% land loss by the year 2050 and 30% land loss over the following 25 years (Wang et al., 2003a, b). Plants have evolved several tactics to surmount these obstacles, such as embracing adaptation mechanisms that enable them to endure unfavorable circumstances or adopting certain development patterns that steer clear of stressors. Additionally, a large number of genes and the products of those genes react at the transcriptional and translational levels to abiotic stressors (Bartels et al., 2007; Cramer et al., 2011).

8.3 Plant Stress Memory and Its Affecting Factors

Recent findings reveal that plants possess the ability to acquire a stress memory, which can enhance their resilience to future stress events. This phenomenon, known as stress memory, can be mediated through mechanisms such as priming and epigenetic modifications. However, there are still significant gaps in our understanding of why stress memory is not always retained and the mechanisms behind memory dissipation or "forgetfulness" (Xu et al., 2024). Plants have developed highly sophisticated and adaptive survival strategies in response to frequent environmental stresses, which include the ability to learn from past experiences, prime themselves for future challenges, and modify their phenotypic responses (Zafar et al., 2020). Plants are highly adept at rapidly adjusting to stressful situations, but they also have the remarkable ability to reinforce their defenses by recalling past stressful experiences. This stress memory enables plants to mount more effective or accelerated responses in the event of future stress, including natural disasters (Xu et al., 2024). One significant and innovative regulatory route in this complex balancing process is RNA metabolism, which plays a crucial role in managing stress memory and recovery. Abiotic stress factors, such as prolonged dryness, extreme temperatures, and elevated levels of salt and heavy metals, can significantly impact plant health and productivity. These conditions cause various physiological and biochemical changes that challenge plant survival. For example, salt stress leads to necrosis, accelerated leaf death, and severe ion imbalances within plant cells, which are major limitations to crop productivity in coastal and semi-arid regions (Bita & Gerats, 2013). Similarly, drought stress affects plants by disrupting water balance and hindering growth. However, plants can develop a drought stress memory that enhances their ability to adapt and recover from future drought conditions (Jacques et al., 2023).

A key component in the plant stress response is the ethylene response factor (ERF), a member of the ERF/EREB family, which is essential for mediating responses to various abiotic and biotic stress stimuli (Tiwari et al., 2015). Additionally, plants undergo extensive chromatin remodeling in response to physiological and environmental changes, including temperature fluctuations, light intensity variations, microbial infections, and hormonal signals for cell differentiation (Liu & Weigel, 2015). Ethylene, a critical phytohormone, is involved in several physiological pathways that regulate the abiotic stress response, including reactions to heat and drought. The formation of stress memory involves a range of physiological processes, including photosynthesis, cellular defense mechanisms, osmotic adjustment, and interactions with soil microbes. Drought avoidance, a key strategy for managing water scarcity, is facilitated by morpho-physiological adjustments such as stomatal regulation. Abscisic acid (ABA) and ABA-responsive element (ABRE)-binding proteins orchestrate the stomatal opening and closing mechanisms through a complex network of transcription factors (Ilyas et al., 2021). Given the challenges posed by abiotic stresses, one of the primary goals for plant scientists is to develop crop varieties that can adapt to stress conditions and thrive in changing environments. Advances in understanding plant stress memory and the associated molecular mechanisms offer promising avenues for improving crop resilience and productivity (Henry et al., 2020). It may be beneficial to use mechanisms for memory reset in the event that the memory produces an unfavorable outcome. Resetting or amnesia is the term used to describe this final process, in which memory is not consolidated (Crisp et al., 2016).

The stress memory of plants to abiotic factors stimuli. An organism's somatic stress memory is limited to one generation of organisms. Consequently, plants have evolved a defense mechanism known as cold acclimation in response to low temperatures, which enables them to prepare for approaching freezing conditions. Memory impairment brought on by salt salinity stress, which affects plant development by reducing photosynthesis, transpiration productivity, and closure of stomatal pores, is one of the major dangers to agricultural productivity. As stated by Rasool et al. (2013) and Sudhir & Murthy (2004), the integrity of plasma membranes, crucialation homeostasis, and metabolic activity are all impacted by the ionic stress brought on by salt intake that is over optimal levels (Miao et al., 2014).

8.4 Molecular and Epigenetic Mechanism of Plant Stress Memory

Plants possess various biochemical processes to manage and remember stress, which is crucial for their growth, development, and resistance to environmental challenges. One notable method of stress memory is through the persistent modification of transcription factors (Conrath et al., 2006). These modifications allow plants to adapt to stress conditions, demonstrating a form of memory that can influence future stress responses. While plants can exhibit short-term stress memory through immediate acclimation responses, the duration of this memory is often limited by the half-lives of stress-induced proteins, RNAs, and metabolites (Conrath, 2011). To sustain stress memory over longer periods, plants may reprogram their phenology and morphology (Bruce et al., 2009). This reprogramming enables plants to endure future stress more effectively by modifying their developmental processes. Another crucial mechanism for long-term stress memory involves epigenetic modifications (Crisp et al., 2016).

These include permanent or heritable changes such as DNA methylation (DM) and alterations to histone structure. These epigenetic processes can affect gene expression and help plants adapt to environmental stresses such as salinity, temperature fluctuations, and nutrient deficiencies. Manipulating these epigenetic changes holds significant potential for enhancing crop growth, yield, quality, and productivity (Suelves et al., 2016). Epigenetic mechanisms play a crucial role in promoting sustainable agriculture by regulating key agronomic traits in crops, such as growth, seed germination, and fruit development. These traits are influenced by gene expression changes driven by DM, histone modifications (HMs), and small RNAs. These epigenetic pathways have a direct impact on agricultural yields, quality, and productivity (Ashapkin et al., 2020).

Creating strategies to enhance agricultural output under harsh environmental conditions requires a thorough understanding of plant stress responses and the role of epigenetics (Lee et al., 2023). While there is substantial evidence linking epigenetic changes to various stress responses, there remains a gap in hard data connecting these changes to physiological adaptability in plants across different stress conditions and environments. One important area of research involves enzymes such as DNA methyltransferase DRM2, which is known to mediate stress-induced epigenetic modifications (Alves de Freitas et al., 2019). DRM2 plays a significant role in the stress response through modifications that influence gene expression and plant adaptability. Additionally, histone acetyltransferases and histone deacetylases (HDACs) are involved in stress-responsive HMs, further contributing to the plant's ability to manage stress. To develop plants capable of withstanding environmental stresses and improving agricultural outcomes, it is essential to investigate these underlying epigenetic processes in detail. This involves exploring how generational inheritance affects plant stress responses and examining recent research on the mechanisms of plant stress-induced epigenetic regulation (Lee et al., 2023).

8.4.1 Mechanisms of DNA Methylation in Plant Development

In the study of plant epigenetics, DM is a thoroughly investigated process. During DM, a methyl group is added to cytosine residues, particularly at CpG dinucleotides, which can significantly impact gene expression and phenotypic plasticity (Kovalchuk, 2024). Recent research has highlighted that DM is a dynamic and responsive process that adapts to various environmental stresses. For example, environmental stress can induce changes in DM patterns, which in turn affect chromatin structure and gene expression (Kawakatsu et al., 2021).

Stress-induced DM changes can lead to the recruitment of chromatin-modifying proteins, which may inhibit the binding of certain transcription factors to DNA (Sadakierska-Chudy & Filip, 2015). This process can create patterns of HM and facilitate the recruitment of repressor complexes containing methyl-CpG-binding domain (MBD) proteins, DNMTs, and HDACs. Unlike genetic mutations, which are fixed and stable, DM can create complex and varied epigenetic states in crossbreeding populations. These states can lead to distinct and potentially advantageous phenotypes that genetic variation alone may not achieve (Tonosaki et al., 2022). Moreover, altered DM patterns can affect *cis*-regulatory elements, which are regions of the DNA that regulate gene expression and can result in new phenotypes and altered gene expression (Lee et al., 2023). The intricate process involving numerous enzymes and cofactors begins with a DNMT enzyme identifying a CpG dinucleotide (Tonosaki et al., 2022).

This modification affects chromatin structure and decreases gene transcription, as DM can regulate gene expression during stress response and plant growth (Wang et al., 2003a, b). DM regulates important plant characteristics such as leaf shape, disease resistance, and tolerance to environmental stress. In *Arabidopsis thaliana*, DM's effective inhibition of certain genes necessary for floral development leads to delayed flowering. This suppression is caused by the methylation of specific CpG islands in the promoter regions of these genes (Surdonja et al., 2017). Research indicates that DM may trigger the activation of hormone and defense pathways in response to the bacterial pathogen *Pseudomonas syringae* pv. tomato (Pst). *Arabidopsis* deploys a defensive mechanism to increase its resistance to *Hyaloperonospora arabidopsidis* (Hpa), the fungus causing downy mildew disease. Furthermore, in *Arabidopsis*, DNA hypermethylation is essential for enhancing the plant's defenses against two fungal pathogens: *Alternaria brassicicola* and *Plectosphaerella cucumerina* (López Sánchez et al., 2016).

8.5 Exploring CRISPR/Cas Approaches to Write Plant Stress Memory

CRISPR technology began to be explored by microbiology and bioinformatics labs around 2000, leading to the discovery of the transcription of CRISPR loci and the Cas gene coding for proteins with belicane and nucleus domains (Tang et al., 2002). The viral origin of CRISPR spacers was identified in 2005, which, along with the earlier development of ZFNs and TALENs, laid the foundation for genome modification. The advent of CRISPR/Cas technology has revolutionized plant genome editing, enabling significant advancements beyond traditional breeding methods (Bolotin et al., 2005). CRISPR/Cas9's significance in genome engineering cannot be overstated, as it involves Cas9 and guide RNA (gRNA) to enable precise genome modifications. The CRISPR system, an adaptive immune mechanism in prokaryotes, was first described in 1987, with *Escherichia coli* studies revealing crucial CRISPR components (Pourcel et al., 2005). The technology's core includes the Cas9 protein and gRNA (CRISPR RNA [crRNA] and *trans*-activating crRNA [tracrRNA]), with CRISPR activator (CRISPRa) and interference (CRISPRi) techniques utilizing deactivated Cas9 (dCas9) for gene regulation (Mali et al., 2013).

8.6 Genome-Editing Technique Based on CRISPR/Cas9

The CRISPR/Cas9 system is composed of one gRNA, or sgRNA, and CRISPR-associated protease 9 (Cas9). It is based on an adaptive immune system seen in the genomes of bacteria and archaea that guards against the entry of foreign plasmids or viral 67 DNA (Marraffini & Sontheimer, 2010).

The tracrRNA and the protospacer-matching crRNA, both required for CRISPR activity, are combined to form the synthetic sgRNA (Marraffini & Sontheimer, 2010).

As a component of the sgRNA/Car complex, the 20 nucleotides at a sgRNA's S 5 terminus attach to the target genomic region (Gilbert et al., 2013). Immediately upstream of the Staphylococcus protospacer-associated domain (PAM, NGG for SpCas9). This particular target site needs to be found in this particular pyogenes. Le is a short-conserved DNA sequence that lies downstream of the cleavage site and varies in length depending on the type of bacterium. It typically has between two and five base pairs (Larson et al., 2013). Target DNA in the genome is cleaved by the large (1368 amino acid) multidomain DNA endonuclease SpCas9, resulting in a blunt ending double-strand break (DSB). Ultimately, the host's biological mechanism heals the DSB (Mei et al., 2016). Two techniques are employed to fix the DSBs brought on by the Cas9 protein 22: nonhomologous end joining (NHEJ) and homologous direct recombination (HDR) procedures (Mojica et al., 2000). HDR uses a homologous DNA template and is very accurate. A significant number of dancing DNA templates with a sequence of target DNA are required since HDR is mostly active during the late S and G2 stages of the cell cycle (Jansen et al., 2002).

A donor DNA template with sequence homology at the anticipated DSB spot is introduced to achieve the desired gene insertion or replacement (Lin et al., 2018; Yang et al., 2022). Many technical developments in recent years have made it easier to develop and research crop plants using CRISPR editing. Plants' poor intrinsic HDR efficiency is a significant technological barrier to the often-advised applications of CRISPR, including large-scale chromosomal deletions and gene swapping. Even with the development of numerous methods to improve HDR, there are still some drawbacks. To produce VirD2 relaxes and precise and targeted DNA DSBs, Ali et al. (2020) combined Cas and VirD12.·To support HDR, this enables the repair template to be in close proximity to the DSBs. Increased herbicide resistance in sugarcane was attained by modifying many alleles of the ACETOLACTATE SYNTHASE (ALS) gene using the nuclease CRISPR/Cas9. These CRISPR-edited sugarcane plants have herbicide resistance because the target ALS gene was specifically replaced by HDR of DSB using a design template (Bosley et al., 2015).

8.7 The CRISPR-Cas Genome Modification Mechanism

The CRISPR-Cas9 type II defensive mechanism in bacteria comprises three stages: adaptation (spacer acquisition), crRNA processing/biogenesis, and interference. Recognition of foreign DNA requires a conserved protospacer adjacent motif (PAM) between 2 and 5 base pairs. The crRNA maturation process involves the breakdown of pre-crRNA by Cas proteins (Cas1, Cas2, Cas9, and Cas4), leading to the formation of a crRNA–tracrRNA duplex (Doudna & Charpentier, 2014). This duplex, along with Cas proteins, forms an effector complex that binds to the target DNA. During the interference stage, crRNA guides Cas proteins to the target sequence, resulting in DSBs in the DNA. CRISPR technology enables functional genomics, epigenetic modulation, and genome engineering to improve crop varieties. Epigenetics involves nongenetic heredity mechanisms like DM that influence gene expression (Hsu et al., 2014). Gain-of-function techniques help identify gene functions without disrupting gene families. Advanced genome engineering methods like CRISPR/Cas, TALENs, and ZFNs are now used to enhance crops for sustainable agriculture, replacing traditional breeding and genetically modified organism (GMO) techniques (Jinke et al., 2012).

The intricate process involving numerous enzymes and cofactors begins with a DNMT enzyme identifying a CpG dinucleotide. This modification affects chromatin structure and decreases gene transcription, as DM can regulate gene expression during stress response and plant growth

(Wang et al., 2003a, b). DM regulates important plant characteristics such as leaf shape, disease resistance, and tolerance to environmental stress. In *A. thaliana*, DM's effective inhibition of certain genes necessary for floral development leads to delayed flowering. This suppression is caused by the methylation of specific CpG islands in the promoter regions of these genes (Surdonja et al., 2017). Research indicates that DM may trigger the activation of hormone and defense pathways in response to the bacterial pathogen *P. syringae* pv. tomato (Pst). *Arabidopsis* deploys a defensive mechanism to increase its resistance to Hpa, the fungus causing downy mildew disease. Furthermore, in *Arabidopsis*, DNA hypermethylation is essential for enhancing the plant's defenses against two fungal pathogens: *Alternaria brassicicola* and *P. cucumerina* (López Sánchez et al., 2016). However, Table 8.1 summarizes information on various stressors, their first encounter and recurrence stages, the affected crops, and the targeted genes. Moreover, Figure 8.1 describes a concept flowchart of plant stress memory rewrite by CRISPR-Cas.

Table 8.1 Various stressors, their first encounter and recurrence stages, the affected crops, and the targeted genes.

Stressor	First encounter	Recurrent	Crops	Targeted gene	References
Salinity stress	Reproductive stage	Reproductive stage	Wheat	*Gene Q*	Jacob et al. (2010)
Abiotic stress	Reproductive stage	Reproductive stage	Maize	*Tb1*	Zhang et al. (2022a)
Biotic stress	Seedling stage	Seedling stage	Sorghum	*Csif and csih*	Khan et al. (2021)

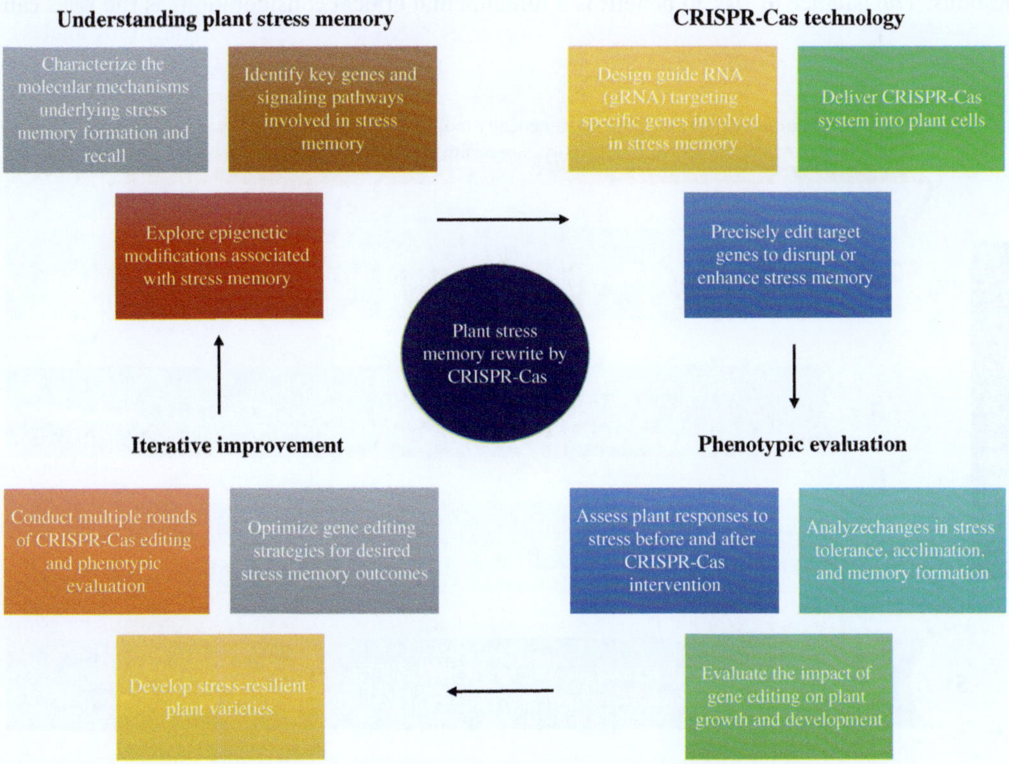

Figure 8.1 Concept flowchart of plant stress memory rewrite by CRISPR-Cas.

8.8 Ethical Issues

Upon discovering DNA, scientists have strived to identify site-specific differences in the DNA sequence. Since the 1990s, advancements in genome engineering technology have led to the development of the CRISPR/Cas9 method of genome editing, which is highly complex yet straightforward, cost-effective, and cutting-edge (Mekler et al., 2020). Earlier genome-editing platforms like ZFNs and TALENs were initially used but eventually fell out of favor due to their complexity, expense, and time-consuming nature (Carroll & Charo, 2015). The rapidly evolving CRISPR/Cas9 technology is revolutionizing molecular biology by enabling researchers to make precise changes to the DNA of various organisms. Since its introduction in 2012, CRISPR/Cas9 has been quickly adopted due to its simplicity and ease of use. Its applications range from therapeutic drugs to agriculture, including potential alterations to the human germline to address genetic diseases (Doudna & Charpentier, 2014). However, this technology has raised significant ethical concerns, particularly regarding off-target mutations, which can have harmful effects on individuals and the environment (Wang et al. 2003a, b).

Research indicates that humans, compared to zebrafish, have a higher number of off-target mutations, which can lead to cellular death and abnormal morphology. In addition to ethical concerns, regulatory issues are critical as the ease of generating genetic alterations makes it challenging to identify transformed organisms outside the laboratory (Hsu et al., 2014). Regulations surrounding CRISPR/Cas9 are necessary to manage the growing market for GMOs (Sontheimer & Barrangou, 2015). The technology's potential for genetic improvement, such as modifying human embryos or altering somatic cells to introduce desired phenotypic traits, also raises ethical questions. The balance of risk to benefit is a fundamental ethical consideration, as the risks can

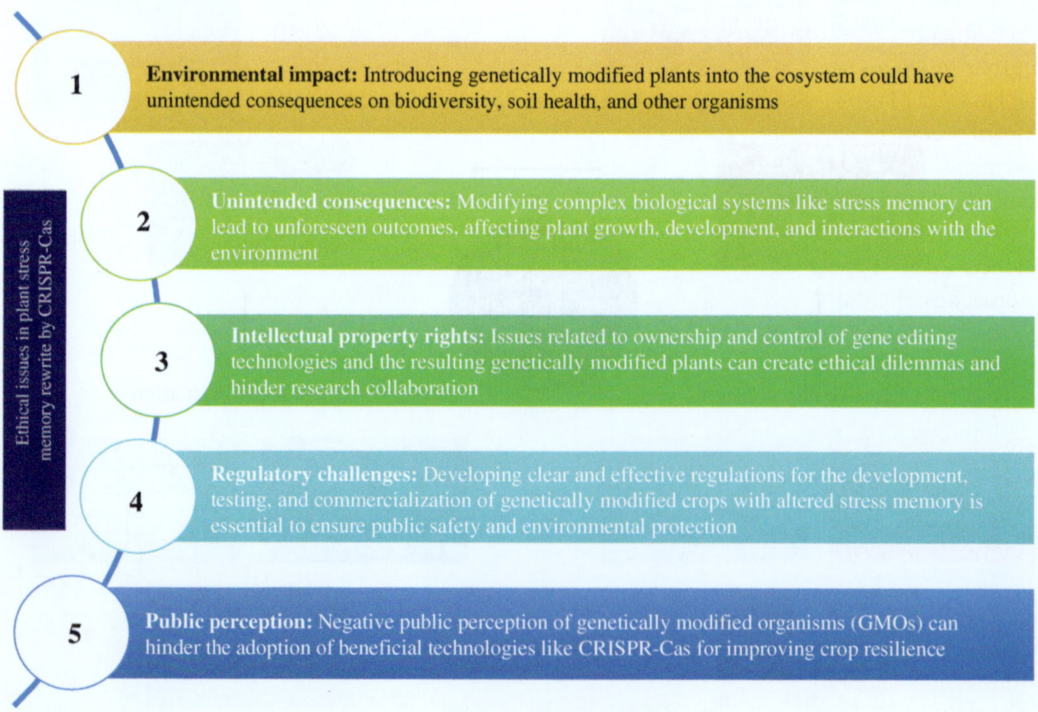

Figure 8.2 Ethical issues in plant stress memory rewrite by CRISPR-Cas.

endanger people or the environment (Sontheimer & Barrangou, 2015). The global impact of CRISPR/Cas9 extends beyond the lab, influencing biotech and pharmaceutical industries and leading to industry collaboration agreements, such as those between Caribou Biosciences and Novartis (Hsu et al., 2014). However, the five most important ethical issues in plant stress memory rewrite by CRISPR-Cas are indexed in Figure 8.2.

8.9 Future Prospects

CRISPR/Cas-mediated approaches hold significant promise for advancing plant stress memory research and application (Wang et al., 2003a, b). By precisely editing plant genomes, scientists can enhance plants' ability to "remember" and respond more effectively to recurring stressors such as salinity, drought, and biotic attacks (Zhang et al., 2022b). This genetic modification can target key regulatory genes involved in stress responses, enabling plants to activate their defense mechanisms more rapidly and robustly upon subsequent stress exposures (Mekler et al., 2020). Prospects include developing crops with improved resilience and yield stability under fluctuating environmental conditions, reducing the need for chemical inputs, and contributing to sustainable agriculture (Wang et al., 2003a, b). These advancements could revolutionize crop management and productivity, especially in the face of climate change and increasing global food demands (Wang et al., 2003a, b).

8.10 Conclusion

Stress memory in plants enables them to learn from past experiences to become more resilient to future challenges such as biotic, abiotic, salinity, or drought stress. This memory is regulated by nucleosome occupancy, microRNAs, and changes in chromatin and histones. The most well-known memory mechanisms occur at the shoot level, but understanding the relationship between soil microbial legacy and root system memory is also crucial. Insights into plant stress memory can significantly impact food security in the face of climate change. The decline of plant species involves multiple factors, and the concept of plant memory introduces a new dimension to plant science. Plant stress priming supports the idea that plants can retain memories. Comprehensive approaches, including precise experimental designs and data from various fields, are necessary to understand plant memory. The recovery period after stress is vital for forming or losing memories, with RNA turnover playing a key role. Epigenomic, transcriptomic, and metabolomic approaches help unravel how plants respond to drought stress, and integrating these with structural analyses can improve understanding. Future research should focus on memory and integrated approaches to better control water in plant production.

References

Alexandratos, N., & Bruinsma, J. (2012). World Agriculture Towards 2030/2050: The 2012 Revision. https://doi.org/10.22004/ag.econ.288998

Ali, S. T., Wang, L., Lau, E. H., Xu, X. K., Du, Z., Wu, Y., et al. (2020). Serial interval of SARS-CoV-2 was shortened over time by nonpharmaceutical interventions. *Science*, *369*(6507), 1106–1109.

Alves de Freitas Guedes, F., Menezes-Silva, P. E., DaMatta, F. M., & Alves-Ferreira, M. (2019). Using transcriptomics to assess plant stress memory. *Theoretical and Experimental Plant Physiology*, *31*, 47–58.

Ashapkin, V. V., Kutueva, L. I., Aleksandrushkina, N. I., & Vanyushin, B. F. (2020). Epigenetic mechanisms of plant adaptation to biotic and abiotic stresses. *International Journal of Molecular Sciences*, *21*(20), 7457.

Bartels, D., & Sunkar, R. (2005). Drought and salt tolerance in plants. *Critical Reviews in Plant Sciences*, *24*(1), 23–58.

Bartels, D., Phillips, J., & Chandler, J. (2007). Desiccation tolerance: gene expression, pathways, and regulation of gene expression. *Plant Desiccation Tolerance*, *35*, 115–137.

Bita, C. E., & Gerats, T. (2013). Plant tolerance to high temperature in a changing environment: scientific fundamentals and production of heat stress-tolerant crops. *Frontiers in Plant Science*, *4*, 273.

Bolotin, A., Quinquis, B., Sorokin, A., & Ehrlich, S. D. (2005). Clustered regularly interspaced short palindrome repeats (CRISPRs) have spacers of extrachromosomal origin. *Microbiology*, *151*(8), 2551–2561.

Bosley, K. S., Botchan, M., Bredenoord, A. L., Carroll, D., Charo, R. A., Charpentier, E., et al. (2015). CRISPR germline engineering – the community speaks. *Nature Biotechnology*, *33*(5), 478–486. https://doi.org/10.1038/nbt.3227

Bruce, B., Fries, J. F., Ambrosini, D., Lingala, B., Gandek, B., Rose, M., & Ware, J. E. (2009). Better assessment of physical function: item improvement is neglected but essential. *Arthritis Research & Therapy*, *11*, 1–11.

Carroll, D., & Charo, R. A. (2015). The societal opportunities and challenges of genome editing. *Genome Biology*, *16*, 1–9.

Char, S. N., Neelakandan, A. K., Nahampun, H., Frame, B., Main, M., Spalding, M. H., et al. (2017). An Agrobacterium-delivered CRISPR/Cas9 system for high-frequency targeted mutagenesis in maize. *Plant Biotechnology Journal*, *15*(2), 257–268.

Conrath, U. (2011). Molecular aspects of defence priming. *Trends in Plant Science*, *16*(10), 524–531.

Conrath, U., Beckers, G. J., Flors, V., García-Agustín, P., Jakab, G., Mauch, F., et al. (2006). Priming: getting ready for battle. *Molecular Plant-Microbe Interactions*, *19*(10), 1062–1071.

Cramer, G. R., Urano, K., Delrot, S., Pezzotti, M., & Shinozaki, K. (2011). Effects of abiotic stress on plants: a systems biology perspective. *BMC Plant Biology*, *11*, 1–14.

Doudna, J. A., & Charpentier, E. (2014). The new frontier of genome engineering with CRISPR-Cas9. *Science*, *346*(6213), 1258096.

Gao, Z. K., Cai, Q., Yang, Y. X., Dong, N., & Zhang, S. S. (2017). Visibility graph from adaptive optimal kernel time-frequency representation for classification of epileptiform EEG. *International Journal of Neural Systems*, *27*(04), 1750005.

Gilbert, L. A., Larson, M. H., Morsut, L., Liu, Z., Brar, G. A., Torres, S. E., et al. (2013). CRISPR-mediated modular RNA-guided regulation of transcription in eukaryotes. *Cell*, *154*(2), 442–451. https://doi.org/10.1016/j.cell.2013.06.044

Hartley, S. E., & Gange, A. C. (2009). Impacts of plant symbiotic fungi on insect herbivores: mutualism in a multitrophic context. *Annual Review of Entomology*, *54*(1), 323–342.

Hashem, S. I., Murphy, A. N., Divakaruni, A. S., Klos, M. L., Nelson, B. C., Gault, E. C., et al. (2017). Impaired mitophagy facilitates mitochondrial damage in Danon disease. *Journal of Molecular and Cellular Cardiology*, *108*, 86–94.

Henry, B. M., Lippi, G., & Plebani, M. (2020). Laboratory abnormalities in children with novel coronavirus disease 2019. *Clinical Chemistry and Laboratory Medicine (CCLM)*, *58*(7), 1135–1138.

Hsu, S. D., Tseng, Y. T., Shrestha, S., Lin, Y. L., Khaleel, A., Chou, C. H., et al. (2014). miRTarBase update 2014: an information resource for experimentally validated miRNA-target interactions. *Nucleic Acids Research*, *42*(D1), D78–D85. https://doi.org/10.1093/nar/gkt1266

Ilyas, R. A., Sapuan, S. M., Atikah, M. S. N., Asyraf, M. R. M., Rafiqah, S. A., Aisyah, H. A., et al. (2021). Effect of hydrolysis time on the morphological, physical, chemical, and thermal behavior of sugar palm nanocrystalline cellulose (*Arenga pinnata* (Wurmb.) Merr). *Textile Research Journal*, *91*(1–2), 152–167.

Jacob, Y., Stroud, H., LeBlanc, C., Feng, S., Zhuo, L., Caro, E., et al. (2010). Regulation of heterochromatic DNA replication by histone H3 lysine 27 methyltransferases. *Nature*, *466*(7309), 987–991.

Jacques, O., Arpin, E., Ammi, M., & Noël, A. (2023). The political and fiscal determinants of public health and curative care expenditures: evidence from the Canadian provinces, 1980–2018. *Canadian Journal of Public Health*, *114*(4), 584–592.

Jansen, R., Embden, J. D. V., Gaastra, W., & Schouls, L. M. (2002). Identification of genes that are associated with DNA repeats in prokaryotes. *Molecular Microbiology*, *43*(6), 1565–1575. https://doi.org/10.1046/j.1365-2958.2002.02839.x

Jinke, G., Chan, T., & Gang, W. (2012). Numerical simulation on distribution characteristics of particle distribution uniformity in a radial style diesel particulate filter. *Energy Procedia*, *17*, 404–413.

Kawakatsu, T., Teramoto, S., Takayasu, S., Maruyama, N., Nishijima, R., Kitomi, Y., & Uga, Y. (2021). The transcriptomic landscapes of rice cultivars with diverse root system architectures grown in upland field conditions. *The Plant Journal*, *106*(4), 1177–1190.

Khan, I., Khan, S., Zhang, Y., Zhou, J., Akhoundian, M., & Jan, S. A. (2021). CRISPR-Cas technology based genome editing for modification of salinity stress tolerance responses in rice (*Oryza sativa* L.). *Molecular Biology Reports*, *48*(4), 3605–3615.

Kovalchuk, N. (2024). Trends in the use of information and communication technologies in the teaching of foreign languages and translator training. Teaching languages at higher educational establishments at the present stage. *Intersubject Relations*, *44*, 77–94.

Larson, M. H., Gilbert, L. A., Wang, X., Lim, W. A., Weissman, J. S., & Qi, L. S. (2013). CRISPR interference (CRISPRi) for sequence-specific control of gene expression. *Nature Protocols*, *8*(11), 2180–2196. https://doi.org/10.1038/nprot.2013.132

Lee, S., Choi, J., Park, J., Hong, C. P., Choi, D., Han, S., et al. (2023). DDM1-mediated gene body DNA methylation is associated with inducible activation of defense-related genes in Arabidopsis. *Genome Biology*, *24*(1), 106.

Lin, C. S., Hsu, C. T., Yang, L. H., Lee, L. Y., Fu, J. Y., Cheng, Q. W., et al. (2018). Application of protoplast technology to CRISPR/Cas9 mutagenesis: from single-cell mutation detection to mutant plant regeneration. *Plant Biotechnology Journal*, *16*(7), 1295–1310. https://doi.org/10.1111/pbi.12870

Liu, C., & Weigel, D. (2015). Chromatin in 3D: progress and prospects for plants. *Genome Biology*, *16*, 1–6.

Liu, H., Able, A. J., & Able, J. A. (2022). Priming crops for the future: rewiring stress memory. *Trends in Plant Science*, *27*(7), 699–716.

López Sánchez, A., Stassen, J. H., Furci, L., Smith, L. M., & Ton, J. (2016). The role of DNA (de) methylation in immune responsiveness of Arabidopsis. *The Plant Journal*, *88*(3), 361–374.

Mali, P., Yang, L., Esvelt, K. M., Aach, J., Guell, M., DiCarlo, J. E., et al. (2013). RNA-guided human genome engineering via Cas9. *Science*, *339*(6121), 823–826.

Marraffini, L. A., & Sontheimer, E. J. (2010). Self versus non-self discrimination during CRISPR RNA-directed immunity. *Nature*, *463*(7280), 568–571.

Mei, F., Lehmann-Horn, K., Shen, Y. A. A., Rankin, K. A., Stebbins, K. J., Lorrain, D. S., et al. (2016). Accelerated remyelination during inflammatory demyelination prevents axonal loss and improves functional recovery. *Elife*, *5*, e18246. https://doi.org/10.7554/eLife.18246.001

Mekler, V., Kuznedelov, K., & Severinov, K. (2020). Quantification of the affinities of CRISPR-Cas9 nucleases for cognate protospacer adjacent motif (PAM) sequences. *Journal of Biological Chemistry, 295*(19), 6509–6517. https://doi.org/10.1074/jbc.RA119.012239

Miao, C., Duan, Q., Sun, Q., Huang, Y., Kong, D., Yang, T., et al. (2014). Assessment of CMIP5 climate models and projected temperature changes over Northern Eurasia. *Environmental Research Letters, 9*(5), 055007.

Mojica, F. J., Díez-Villaseñor, C., Soria, E., & Juez, G. (2000). Biological significance of a family of regularly spaced repeats in the genomes of Archaea, bacteria and mitochondria. *Molecular Microbiology, 36*(1), 244–246. https://doi.org/10.1046/j.1365-2958.2000.01838.x

Nadeem, M. A., Nawaz, M. A., Shahid, M. Q., Doğan, Y., Comertpay, G., Yıldız, M., et al. (2018a). DNA molecular markers in plant breeding: current status and recent advancements in genomic selection and genome editing. *Biotechnology & Biotechnological Equipment, 32*(2), 261–285.

Nadeem, M., Li, J., Wang, M., Shah, L., Lu, S., Wang, X., & Ma, C. (2018b). Unraveling field crops sensitivity to heat stress: mechanisms, approaches, and future prospects. *Agronomy, 8*(7), 128. https://doi.org/10.3390/agronomy8070128

Nadeem, M. S., Kazmi, I., Ullah, I., Muhammad, K., & Anwar, F. (2021). Allicin, an antioxidant and neuroprotective agent, ameliorates cognitive impairment. *Antioxidants, 11*(1), 87. https://doi.org/10.3390/antiox11010087

Pacher, M., & Puchta, H. (2017). From classical mutagenesis to nuclease-based breeding-directing natural DNA repair for a natural end-product. *The Plant Journal, 90*(4), 819–833.

Pourcel, C., Salvignol, G., & Vergnaud, G. (2005). CRISPR elements in *Yersinia pestis* acquire new repeats by preferential uptake of bacteriophage DNA, and provide additional tools for evolutionary studies. *Microbiology, 151*(3), 653–663.

Rasool, R., Ganai, B. A., Kamili, A. N., Akbar, S., & Masood, A. (2013). Synergistic effect of auxins and cytokinins on propagation of *Artemisia amygdalina* (Asteraceae), a critically endangered plant of Kashmir. *Pakistan Journal of Botany, 45*(2), 629–634.

Ray, D., Ghosh, A., Mustafi, S. B., & Raha, S. (2016). Plant stress response: Hsp70 in the spotlight. *Heat Shock Proteins and Plants*, 123–147. https://doi.org/10.1007/978-3-319-46340-7_7

Sadakierska-Chudy, A., & Filip, M. (2015). A comprehensive view of the epigenetic landscape. Part II. Histone post-translational modification, nucleosome level, and chromatin regulation by ncRNAs. *Neurotoxicity Research, 27*, 172–197.

Shabala, S. (2013). Learning from halophytes: physiological basis and strategies to improve abiotic stress tolerance in crops. *Annals of Botany, 112*(7), 1209–1221.

Shao, L., Takeda, K., Kato, S., Mori, S., & Kodama, T. (2015). Communication between lymphatic and venous systems in mice. *Journal of Immunological Methods, 424*, 100–105.

Sontheimer, E. J., & Barrangou, R. (2015). The bacterial origins of the CRISPR genome-editing revolution. *Human gene therapy, 26*(7), 413–424.

Sudhir, P., & Murthy, S. D. S. (2004). Effects of salt stress on basic processes of photosynthesis. *Photosynthetica, 42*, 481–486.

Suelves, M., Carrió, E., Núñez-Álvarez, Y., & Peinado, M. A. (2016). DNA methylation dynamics in cellular commitment and differentiation. *Briefings in Functional Genomics, 15*(6), 443–453.

Surdonja, K., Eggert, K., Hajirezaei, M. R., Harshavardhan, V. T., Seiler, C., Von Wirén, N., et al. (2017). Increase of DNA methylation at the HvCKX2.1 promoter by terminal drought stress in barley. *Epigenomes, 1*(2), 9.

Suzuki, N., Rivero, R. M., Shulaev, V., Blumwald, E., & Mittler, R. (2014). Abiotic and biotic stress combinations. *New Phytologist, 203*(1), 32–43.

Tang, T. H., Bachellerie, J. P., Rozhdestvensky, T., Bortolin, M. L., Huber, H., Drungowski, M., Elge, T., Brosius, J., & Hüttenhofer, A. (2002). Identification of 86 candidates for small nonmessenger RNAs from the archaeon *Archaeoglobus fulgidus*. *Proceedings of the National Academy of Sciences*, *99*, 7536–7541. https://doi.org/10.1073/pnas.112047299

Tiwari, S., Hopke, P. K., Pipal, A. S., Srivastava, A. K., Bisht, D. S., Tiwari, S., et al. (2015). Intra-urban variability of particulate matter (PM2.5 and PM10) and its relationship with optical properties of aerosols over Delhi. *India. Atmospheric Research*, *166*, 223–232.

Tonosaki, K., Fujimoto, R., Dennis, E. S., Raboy, V., & Osabe, K. (2022). Will epigenetics be a key player in crop breeding? *Frontiers in Plant Science*, *13*, 958350.

Wang, W., Vinocur, B., & Altman, A. (2003a). Plant responses to drought, salinity and extreme temperatures: towards genetic engineering for stress tolerance. *Planta*, *218*, 1–14.

Wang, L., Wang, L., Tan, M., Wang, L., Zhao, W., You, J., et al. (2023b). The pattern of alternative splicing and DNA methylation alteration and their interaction in linseed (*Linum usitatissimum* L.) response to repeated drought stresses. *Biological Research*, *56*(1), 12.

Waqas, M., Yaning, C., Iqbal, H., Shareef, M., ur Rehman, H., & Bilal, H. M. (2021). Synergistic consequences of salinity and potassium deficiency in quinoa: linking with stomatal patterning, ionic relations and oxidative metabolism. *Plant Physiology and Biochemistry*, *159*, 17–27.

Xu, W. B., Cao, F., Liu, P., Yan, K., & Guo, Q. H. (2024). The multifaceted role of RNA-based regulation in plant stress memory. *Frontiers in Plant Science*, *15*, 1387575.

Yang, H., Li, P., Jin, G., Gui, D., Liu, L., & Zhang, C. (2022). Temporal regulation of alternative splicing events in rice memory under drought stress. *Plant Diversity*, *44*(1), 116–125.

Zafar, S. A., Zaidi, S. S. E. A., Gaba, Y., Singla-Pareek, S. L., Dhankher, O. P., Li, X., et al. (2020). Engineering abiotic stress tolerance via CRISPR/Cas-mediated genome editing. *Journal of Experimental Botany*, *71*(2), 470–479. (land arc, NE Pakistan by Hussain et al., 2021. Lithos, 398, 106298)

Zhang, D., Wang, G., Yu, X., Wei, T., Farbiak, L., Johnson, L. T., et al. (2022a). Enhancing CRISPR/Cas gene editing through modulating cellular mechanical properties for cancer therapy. *Nature Nanotechnology*, *17*(7), 777–787.

Zhang, H., Zhu, J., Gong, Z., & Zhu, J. K. (2022b). Abiotic stress responses in plants. *Nature Reviews Genetics*, *23*(2), 104–119.

9

Salinity-Tolerant Crop Breeding Through Reprogramming Noncoding RNAs Mediated by CRISPR/Cas9 Technology

Santanu Samanta[1] *and Aryadeep Roychoudhury*[2]

[1] *Department of Biotechnology, St. Xavier's College (Autonomous), Kolkata, West Bengal, India*
[2] *Discipline of Life Sciences, School of Sciences, Indira Gandhi National Open University, New Delhi, Delhi, India*

9.1 Introduction

The sessile lifestyle of plants subjects them to a multitude of environmental constraints, including water scarcity, soil salinity, extreme temperature, flooding, and radiation. These challenges collectively constitute various forms of abiotic stresses, which can significantly impede plant growth, development, biomass production, and yield, and in extreme cases, lead to plant mortality (Koyro et al., 2011). Among these stresses, salinity is unequivocally recognized as the most persistent and damaging issue in agriculture, resulting in a substantial reduction in cultivable land and crop productivity (Tuteja et al., 2012). Saline soil is generally defined in terms of electrical conductivity of the saturation extract exceeding 4 deciSiemens per meter ($dS\,m^{-1}$), which detrimentally affects diverse physiological processes in plants. Furthermore, salinity hinders soil performance, agricultural sustainability, and food safety. Data indicates that over the past few decades, annual global income losses due to salt-induced land degradation have been estimated at US$ 27.3 billion (Qadir et al., 2006). Salinity-affected soil encompasses approximately 1060.1 million hectares globally, including regions such as India, China, Pakistan, Egypt, Mexico, the United States, Turkey, Syria, and Australia. It is estimated that salinity has impacted over 45 million hectares of irrigated land, representing 20% of the total arable land worldwide (Zhu, 2001; Qadir et al., 2014). Additionally, around 2 million hectares of fertile soil are degraded annually due to salinity, accounting for approximately 1% of the global agricultural land (Etesami & Beattie, 2018). Projections suggest that by 2050, more than 50% of the arable land on Earth will succumb to salinity (Thiem et al., 2018).

Most modern cultivars of major crop species exhibit high susceptibility to salinity, leading to poor performance under high saline conditions. Therefore, it is imperative to identify an effective method for enhancing salt tolerance to improve global food production and accelerate sustainable agricultural development. In recent decades, various approaches have been employed to address salinity stress and develop salt-tolerant crop cultivars. Traditional breeding methods, while having achieved limited success in improving salt tolerance, are laborious and cannot meet the growing global food demand, particularly in cereals such as rice (Muthu et al., 2020) and wheat (Al-Ashkar et al., 2019). Hence, the exploration of innovative and advanced strategies is necessary to address these challenges. Recently, the introduction of clustered regularly interspaced short palindromic

Genome and Epigenome Editing for Stress-Tolerant Crops, First Edition. Edited by Jen-Tsung Chen.

repeat (CRISPR)-based genome engineering, facilitated by CRISPR-associated protein 9 (Cas9), has opened up new possibilities in plant breeding. This technology provides a precise and rapid tool for creating desirable traits in plants (Samanta et al., 2016; Schaeffer & Nakata, 2016; Soda et al., 2018). While CRISPR/Cas9 has been extensively utilized for studying protein-encoding regions in plants, progress in the functional analysis of nonprotein-coding ribonucleic acids (ncRNAs) has only just begun (Basak & Nithin, 2015; Bravo-Vázquez et al., 2024). This chapter aims to summarize the current status and progress of CRISPR/Cas9-mediated reprogramming of ncRNAs in plant molecular breeding, which holds potential for the rapid development of salt-resilient genotypes in agriculturally significant crop plants.

9.2 Negative Impact of Salinity Stress on Plants

The accumulation of electrolytic solutes in the soil at concentration exceeding 40 mM NaCl exerts significant stress on plant species, leading to substantial impacts on plant growth and subsequent yield variations among crops. Sodium (Na^+) and chloride (Cl^-) are recognized as the major ions contributing to ion toxicity (Foronda, 2022). The excessive presence of inorganic salt reduces the water potential of the soil matrix, ultimately resulting in hyperosmotic stress in plants (Lei et al., 2015). Salt-induced osmotic stress and reduced water conductivity primarily hinder plant growth in saline conditions (Munns, 2005; Abbasi et al., 2016). The responses of plants to salinity stress are depicted schematically in Figure 9.1. Most crop species can tolerate salt up to a certain threshold. Based on their adaptive evolution, plants can be categorized as (i) glycophytes and (ii) halophytes. Glycophytes are plants that cannot tolerate high concentration of salt (100–200 mM), inhibiting their growth and resulting in plant death (Jia et al., 2011; Kataria & Verma, 2018). The majority of terrestrial plants, including crop species, fall into the glycophyte category. However, significant variation in salt tolerance exists among cereal crops; for instance, barley and wheat exhibit relative tolerance, while corn and rice are at the susceptible end of the range (Mass, 1990).

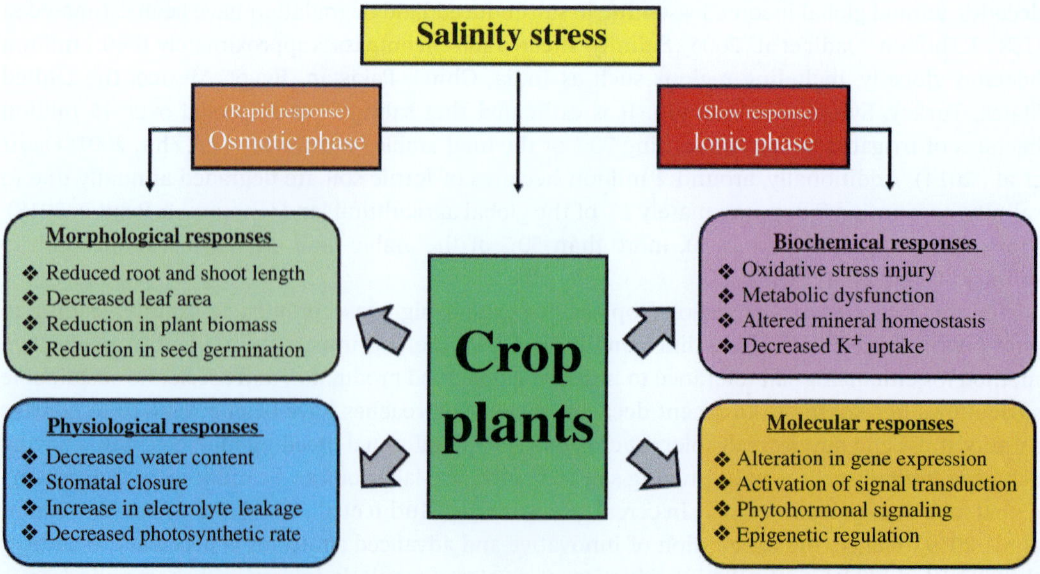

Figure 9.1 Different effects of salinity stress in plants.

Among cereals, rice is considered the most salt-susceptible (Qadir et al., 2014). In contrast, halophytes are plants that naturally inhabit saline environment and complete their life cycle in the presence of high salt concentration (300–500 mM), such as saltbush (*Atriplex* spp.), *Spartina*, and *Salicornia* (Parida & Das, 2005; Flowers & Colmer, 2015). Most crop plants are nitrophobic and thus do not thrive in soil with excessive salt. Plant responses to salinity occur in two distinct phases: (i) an immediate response largely due to the osmotic effect of increasing salt concentration outside the plant roots, known as the "osmotic phase," and (ii) a slower response when salt accumulates to toxic concentration, referred to as the "ionic phase." In this phase, growth reduction takes time to develop due to internal salt injury resulting from excessive Na^+ accumulation (Ismail et al., 2014). Salt-susceptible cultivars tend to accumulate ions more rapidly than salt-tolerant cultivars. A high Na^+ concentration displaces K^+, an essential element for growth and development, leading to reduced plant productivity (Arif et al., 2020). Uptake of excess Cl^- also disrupts photosynthetic activity due to impaired chlorophyll production (Agastian et al., 2000).

9.3 Why Developing Salinity-Tolerant Crops Is a Necessity?

The Earth is characterized by a significant presence of NaCl, with approximately 3.0% of its water composed of this compound. The utilization of this saline solution has had a significant impact on, and continues to influence, the agricultural land where crops are cultivated. The expansion of saline areas is anticipated due to the inundation of seawater during flooding or the excessive extraction of groundwater for intensive agricultural practices, particularly in arid and semiarid regions. Additionally, the extent and severity of salinity-affected agricultural land are predicted to increase as a result of the global climate change-induced rise in mean sea level resulting in flooding, especially in the coastal areas causing further salt accumulation (Dasgupta et al., 2015). With no prospect of expanding arable land due to rapid industrialization, and emerging water scarcity in highly populated developing countries of the world, the situation is further worsened. The burgeoning population is expanding at an alarming rate and it is expected to touch the mark of ~8.3 billion by 2030 and ~9.3 billion by 2050 (Obaid, 2001). Thus, providing food crops to the ever-increasing human population is crucial for agricultural researchers in the 21st century. The increase in global population is highly reliant on food crops, which play a major role in meeting future food demand (Godfray et al., 2010; Shiferaw et al., 2011). It is estimated that grain production will need to be enhanced at least 87% by 2050 to meet its nutritional demand. This poses a serious challenge for plant scientists and agricultural biotechnologists to ensure food supply for the growing human population (Kim et al., 2021). In terms of worldwide production, wheat, rice, and maize are the three most widely grown crops and are responsible for over 50% of daily caloric uptake by the human population. Since all these widely consumed crops are susceptible to salinity, novel mechanisms to induce salinity tolerance in crop plants are an urgent need for maximizing cereal crop production and removing the food gap.

9.4 Breeding for Salt-Tolerant Crops Against Salinity Stress

Breeding is the classical strategy for developing salinity-tolerant crops. To address salinity stress mitigation, proper mapping/identification of genes is the key factor in conventional breeding. Thus, this technique largely depends on well-characterized germplasms and involves screening genotypes for the improvement of crop plants through hybridization, recurrent selection, pedigree selection, polyploidy, backcrossing, and introgression events (Ashraf et al., 2008). The presence of variability of genetic resources is a prime consideration to expedite successful breeding efforts, particularly the gene pool of wild relatives, landraces, and traditional cultivars (Babu et al., 2017). Accurate phenotyping procedures are critical for identifying useful germplasm for crop improvement programs as well as for deciphering the genetic basis of the mechanisms associated with salt tolerance. Through breeding, it is easy to manipulate simply inherited or qualitative traits, which often are less sensitive to environmental fluctuations. However, it is difficult to manipulate quantitative traits like salinity tolerance, since these traits generally have low heritability and susceptibility to environmental fluctuations. Therefore, using conventional breeding, it is challenging for breeders to improve such kinds of attributes (Varshney et al., 2011). Additionally, genetic variability for salt stress is highly limited in the germplasm for most crop species, and it is further hindered by reproductive barriers. As salinity tolerance is a multigenic trait, such approaches have only limited success. As a procedure, this technique is quite slow, more time-consuming, and laborious (Chinnusamy et al., 2005). Multiple undesirable genes are often transferred along with the desirable traits, which makes this method inefficient and unreliable (Muehlbauer et al., 2006; Rahman et al., 2016). Great effort is, therefore, being diverted toward advanced techniques to minimize salinity-induced growing concern on crop productivity. This requires the investigation of useful genes or gene modulators that may assist the salinity-tolerant crop breeding programs and their successful incorporation into the crop plants.

9.5 An Overview of ncRNAs in Plants

Plants have evolved an array of complex mechanisms to recognize environmental constraints and respond by reprogramming the expression of multiple genes at various levels to maintain cellular homeostasis under harsh environmental conditions (Hackenberg et al., 2015; Yi et al., 2015). A thorough understanding of these mechanisms is crucial for utilizing biotechnological tools to enhance important agronomic traits. The discovery of ncRNAs has opened a new avenue to uncover and comprehend intricate regulatory systems in plants, particularly those involved in stress responses, despite their inability to undergo protein translation. Besides protein-coding genes, ncRNAs are equally important regulators of gene expressions in plants, ranging from regulating developmental processes to coping with abiotic stress (Wang et al., 2017; Waititu et al., 2020; Bhogireddy et al., 2021; Gill et al., 2024). Instead of playing biological roles as proteins, they function at the RNA transcript level (Palazzo & Lee, 2015). Various classes of ncRNAs including microRNAs (miRNAs) and long noncoding RNAs (lncRNAs) have been identified in plants. According to their sequence length, ncRNAs can be further classified into (i) small ncRNAs (20–40 nucleotides in length) and (ii) lncRNAs (typically $>$ 200 nucleotides). miRNAs consist of endogenously expressed short (21–24 nucleotides), single-stranded, long interfering ncRNAs that are capable of modulating gene expression in plants. They recognize and bind with their target transcripts based on near-perfect complementarity and repress their translation by degradation of the messenger RNA (mRNA) molecules, chromatin modifications, etc. (Contreras-Cubas et al., 2012; Choudhary et al., 2021). Unlike animal miRNAs, which specifically target 3′ untranslated regions (3′ UTRs) of the mRNA, plant miRNAs bind with their targets with near-perfect complementarity (Meng et al., 2011). The processing and biogenesis of miRNAs are closely regulated by environmental cues. lncRNAs represent a diverse and functionally important class of ncRNAs. Recent sequencing studies have shown that most of the lncRNAs transcribed by RNA polymerase II share features of mRNA, such as 5′-capping, poly-A tails at 3′ end, and undergo posttranscriptional modifications such as alternative splicing (Beaulieu et al., 2012; Srijyothi et al., 2018). Additionally, other RNA polymerases like RNA polymerase III, IV, and V are also reported to play a central role in the biogenesis of lncRNAs (Liu et al., 2012; Chekanova, 2015; Song & Zhang, 2017). lncRNAs transcribed by RNA polymerases IV and V via RNA-dependent DNA methylation serve as precursors for small interfering RNAs (siRNAs). Unlike RNA polymerase II transcripts, lncRNAs produced by RNA polymerases IV and V possess some structural differences such as lack of poly-A tails. Furthermore, lncRNAs transcribed by RNA polymerases IV and V exhibit extremely low expression and instability. These small ncRNAs are key regulatory elements, also known as "riboswitches" or "riboregulators" (Gong et al., 2021; Jain et al., 2021; Jiang et al., 2024). The strong regulatory role of ncRNAs and their involvement in fine-tuning salt stress in various crop plants are listed in Table 9.1. Activation of different regulatory ncRNAs by salinity leads to the modulation of a wide range of critical, interconnected, stress-responsive, gene regulatory networks by targeting a specific subset of transcripts either synergistically or antagonistically. This intricate regulation safeguards the genome from both external and internal factors (Zhang et al., 2022). An in-depth comprehension of the regulatory mechanisms governed by ncRNAs presents substantial potential for the development of sustainable strategies for enhancing crop productivity. Consequently, this area of scientific inquiry has assumed critical importance in ensuring global food security.

Table 9.1 Representative examples of ncRNAs involved in mediating salinity stress responses across various crop plants.

Types of ncRNA	Crop plants	Concentration of NaCl	Key role	References
miRNA	*Hordeum vulgare*	100 mmol L^{-1}	A total of 152 miRNAs were identified, with 44 miRNAs (39 conserved and 5 novel) being documented as responsive to salinity	Deng et al. (2015)
miRNA	*Triticum aestivum*	200 mM	*TaMIR159a*, *TaMIR160*, *TaMIR160*, *TaMIR167*, *TaMIR174*, *TaMIR399*, *TaMIR408*, *TaMIR11124*, and *TaMIR1133* were upregulated and found to be salt responsive	Lu et al. (2011)
miRNA	*Pennisetum glaucum*	250 mM	Analysis of gene expression revealed that 81 conserved miRNAs (30 triggered, 51 suppressed) and 14 novel miRNAs (3 triggered, 11 suppressed) exhibited differential regulation in response to salinity stress, indicating their responsiveness to salt-induced stimuli	Shinde et al. (2020)
miRNA	*Triticum turgidum* ssp. *dicoccoides*	50–150 mmol L^{-1}	A total of 88 conserved and 124 novel miRNAs were identified, with 50 being responsive to salinity. Among these, 32 exhibited increased expression, while 18 showed suppressed expression. Notably, miR172b, miR1120a, and miR393a displayed the most significant differential expression	Feng et al. (2017)

(Continued)

Table 9.1 (Continued)

Types of ncRNA	Crop plants	Concentration of NaCl	Key role	References
miRNA	*Psidium guajava*	200 mM	A total of 40 potential miRNAs were identified and characterized. Among them, *pgu-miR156f-5p, pgu-miR160c-5p, pgu-miR162-3p, pgu-miR164b-5p, pgu-miR166t, pgu-miR167a-5p*, and *pgu-miR390b-5p* displayed distinct expression patterns. Notably, *pgu-miR162-3p, pgu-miR164b-5p*, and *pgu-miR166t* exhibited the most significant alterations	Sharma et al. (2020)
miRNA	African rice (*Oryza glaberrima*)	EC of ~10 $dS\,m^{-1}$	A total of 148 known and 335 novel miRNAs were documented, with 246 representing the conserved ones. Notably, OglmiR399i, OglmiR396-5p, OglmiR159b, OglmiR398b, OglmiR169a, OglmiR10-3p, OglmiR11, OglmiR94, and OglmiR172b were observed to exhibit salt-responsive behavior	Mondal et al. (2018)
miRNA	*Raphanus sativus*	200 mM	A total of 136 known miRNAs and 68 potential novel miRNAs were recorded. Notably, miR172b, miR1120a, and miR393a exhibited the most significant differential expression under salt stress conditions	Sun et al. (2015)
miRNA	*Ipomoea batatas*	150 mM	475 known miRNAs and 175 novel miRNAs were reported; among them, 51 and 13 were significantly triggered and 76 and 9 were suppressed in leaf and root tissues, respectively	Yang et al. (2020)
miRNA	*Vicia faba*	150 mM	A total of 527 and 693 miRNAs were reported to be differentially expressed in salinity-susceptible and salinity-resistant cultivars, respectively. Moreover, 284 and 298 miRNAs were enhanced and 243 and 395 miRNAs were suppressed in salt-susceptible and salt-tolerant cultivars, respectively	Alzahrani et al. (2019)
miRNA, lncRNA	*Oryza sativa*	EC of 4 $dS\,m^{-1}$	214 lncRNAs and 79 miRNAs were reported to be differentially expressed	Jiang et al. (2024)
lncRNA	*Chenopodium quinoa*	300 mM	4460 differently expressed lncRNAs were identified; among them, only 54 were differentially expressed at all the stress time points	Luo et al. (2022)
miRNA, lncRNA	*Gossypium hirsutum*	250 mM	1117 unique lncRNAs were reported in response to salinity stress. Six miRNAs were also identified to be salt-responsive and lnc_973 and lnc_253 were found to regulate the transcript level of ghr-miR399 and ghr-156e as a target mimic during exposure to salt stress	Deng et al. (2018)
lncRNA	*Solanum lycopersicum* and *Solanum pennellii*	200 mM	1044 putative lncRNAs were recognized, of which 154 and 137 lncRNAs were differentially expressed	Li et al. (2022)

Table 9.1 (Continued)

Types of ncRNA	Crop plants	Concentration of NaCl	Key role	References
lncRNA	*Oryza sativa*	110 mM	In root and leaf tissues, 1626 and 2208 transcripts were reported as putative high-confidence lncRNAs, respectively, from salt-tolerant and salt-susceptible cultivars. Among the identified candidate lncRNAs, 1529 and 2103 were categorized as novel lncRNAs, in root and leaf tissues, respectively	Jain et al. (2021)
lncRNA	*Cicer arietinum*	150 mM	A total of 3450 lncRNAs were identified, of which 3373 lncRNAs were linked to regulate their target genes in a *cis*-regulating manner and 80 unique lncRNAs were found to interact with 136 different miRNAs	Kumar et al. (2021)
lncRNA	*Sorghum bicolor*	150 mM	In the study, 133 and 126 differentially transcribed lncRNAs were identified in Roma (a salinity-susceptible cultivar) and M-81E (a salinity-resistant cultivar), respectively. Among these, three new lncRNAs (lncRNA13472, lncRNA11310, and lncRNA2846) exhibited enhanced expression, while two new lncRNAs (lncRNA26929 and lncRNA14798) showed inhibited expression levels during exposure to salt	Sun et al. (2020)
lncRNA	*Medicago truncatula*	150 mM	A total of 23,324 putative lncRNAs underwent screening, resulting in the identification of 7361 lncRNAs in salt-affected leaf samples and 7874 lncRNAs in salt-affected root samples that demonstrated responsiveness to salinity	Wang et al. (2015)
lncRNA	*Sesamum indicum*	150 mM	When subjected to salt stress, a total of 2482 lncRNAs were documented. Analysis of their expression patterns revealed that 700 lncRNAs exhibited differential expression in response to salinity	Gong et al. (2021)

9.6 Mechanistic Insights of CRISPR/Cas9 Tool

New advancements in plant breeding techniques are imperative to surmount the limitations of conventional methods. Over the past decade, the CRISPR/Cas9 approach has arisen as a potent and versatile technology for precise gene/genome engineering across various systems (Jinek et al., 2012). Often described as a "gift from mother nature," the CRISPR/Cas9 machinery is a prokaryotic (bacterial and archaeal) RNA-mediated adaptive immune system that has evolved to resist entry of bacteriophages and foreign nucleic acids (Ran et al., 2013; Ma et al., 2014). The technique gained prominence in 2012 when its potential for genome editing was postulated. CRISPR mediates gene knock-in, knockout, and replacement, along with editing of single to multiple genes by binding a specific sequence at the genomic as well as epigenetic levels (Bortesi & Fischer, 2015; Paul & Qi, 2016). Compared to first-generation genome engineering methods (zinc finger nucleases

[ZFNs]) and second-generation genome engineering methods (transcriptional activator-like effector nucleases [TALENs]), third-generation genome engineering tools such as CRISPR/Cas9 are more efficient and can edit multiplexed gene simultaneously (Nekrasov et al., 2013; Lowder et al., 2015; Zhao et al., 2015). Moreover, this technique is easier to design, is cost-effective and robust, and allows simultaneous editing at multiple sites across the genome (Li et al., 2021; Zhang et al., 2021; Gan & Ling, 2022). This efficient system demands only two components: noncoding crRNAs, specifically a combination of precursor CRISPR RNA (pre-crRNA) and *trans*-activating crRNA (tracrRNA), and the Cas9 protein. The tracrRNA contains a sequence that complements the repeat sequence in the crRNA, forming a unique dual RNA hybrid structure, referred to as single-guide RNA (sgRNA). A CRISPR locus is composed of a CRISPR array, comprising short direct repeats separated by short variable DNA sequences (often termed as "spacer"), which is flanked by numerous *Cas* genes. The CRISPR array is flanked by a series of genes encoding the Cas proteins, which facilitate the three phases of immunity: adaptation, expression, and interference (Barrangou, 2013; Mei et al., 2016; Ahmar et al., 2023). During the adaptation phase, resistance is acquired by the insertion of a new spacer sequence in a CRISPR locus. In the next phase, memory is retrieved when the CRISPR array is transcribed to yield mature CRISPR RNAs (crRNAs). In the third phase, crRNAs guide the respective complexes of Cas proteins to specifically target and cleave the targeted nucleic acids (Janik et al., 2019; Tyagi et al., 2020; Zhou et al., 2020). The precise targeting ability of CRISPR/Cas9 ensures the elimination of genes responsible for undesirable traits and at the same time insertion of desired genes in plant genomes.

There are two types of CRISPR/Cas9 methods available: Class 1 and Class 2, wherein both classes are further characterized into three types. Class 1 systems, which encompass types I and III as well as the putative type IV. Class 2 systems, which include types II, V, and VI. In case of types I, II, and V, the target is DNA, type VI targets RNA, and type III can cleave both RNA and single stranded DNA (Makarova et al., 2015; Hille et al., 2018). Of all these, most research is focused on the principles and applications of Class 2 (type II) CRISPR/Cas systems and type V CRISPR/Cas12a (Bortesi & Fischer, 2015). The Cas9 protein isolated from *Streptococcus thermophilus* or *Streptococcus pyogenes* can be guided by crRNAs with a distinct HNH (McrA-like fold) and RuvC-like (RNaseH fold) nuclease domain, which cleaves the target DNA to generate a double-strand break (Gasiunas et al., 2012; Jinek et al., 2012). The HNH domain is responsible for cleaving the complementary strand, while the RuvC-like domain specifically cleaves the noncomplementary strand, resulting in a blunt cut in the target DNA (Jinek et al., 2012). The other alternative enzyme, Cas12a from the *S. pyogenes* and *Francisellano vicida* is a second Class 2 effector protein, which possesses a RuvC domain, but lacks an HNH domain. Instead of canonical G-rich (5′-NGG-3′) protospacer adjacent motif (PAM) sequence required by Cas9 located at the 3′ target site, Cas12a recognizes T-rich (5′-TTTN-3′) PAM sequences juxtaposed at the 5′ position of the target site (Zetsche et al., 2015). Moreover, Cas12a creates sticky ends with 4–5 nucleotide overhangs distal from the PAM sequence. CRISPR/Cas9 exhibits high efficiency in precisely cleaving DNA, ensuring the elimination of genes responsible for undesirable or harmful traits and enabling the insertion of desired genes into plant genomes. Consequently, CRISPR/Cas9 can be harnessed to enhance cultivars that demonstrate robust performance in adverse environment.

9.7 Targeting/Reprogramming ncRNAs Using CRISPR/Cas9 Tools

The CRISPR/Cas9 system has been successfully implemented in diverse organisms, including bacteria, animals, and plants. Its capability for precise gene editing makes it particularly well suited for use in crop improvement programs, thereby presenting numerous opportunities. While CRISPR/Cas9 technology has been extensively employed in zebrafish, mouse, and human cell lines to engineer miRNA or lncRNA genes, its application in plant systems remains relatively unexplored. CRISPR/Cas9 can be harnessed to reprogram ncRNA modification for the upregulation or downregulation of target genes, facilitating the dissection of complex regulatory pathways. In the realm of protein-coding genes, it is crucial to note that the deletion of nucleotides adjacent to the PAM sequence induces a nucleotide frameshift, resulting in gene silencing or editing. However, the application of the CRISPR/Cas9 tool for knock-out or knock-down of ncRNA genes presents a complex challenge, making the use of CRISPR/Cas9 a demanding task. It is important to recognize that the mere deletion of nucleotides may not efficiently silence miRNA genes. This limitation primarily arises from the inadequate space available for targeting noncoding genes without impacting proximal genes. This concern is particularly pertinent in the context of silencing miRNA, as a significant portion of these genes is encoded within the introns of protein-coding genes. Only a few studies have reported successful implementation of the CRISPR/Cas9 approach for reprogramming miRNA (*MIR*) genes. The initial CRISPR/Cas9 approach was employed to reprogram the expression of miR159A, miR159B, and miR319 in *Arabidopsis* (Lowder et al., 2015). CRISPR was utilized to target two *MIR* genes, miR1514 and miR1509, in soybean using biolistic delivery to achieve transient expression of the CRISPR/Cas9 vector containing Cas9 nuclease and guide RNAs (gRNAs) (Jacobs et al., 2015). Subsequently, Zhao et al. (2016) outlined an alternative strategy for leveraging the CRISPR/Cas9 approach for targeted gene replacement in *Arabidopsis*. They devised a combinatory dual-sgRNA/Cas9 vector to bring about the successful deletion of miRNA gene regions (*MIR169a* and *MIR827a*), enabling targeted deletion and donor repair for gene replacement. In a recent study, the CRISPR/Cas9 tool was effectively employed to induce a specific mutation in the miR156 recognition sites of the *ideal plant architecture 1* (*ipa1*) gene in rice (Li et al., 2016). Additionally, Zhou et al. (2017) demonstrated the potency of the CRISPR/Cas9 system in generating miRNA knockout mutants in plants. They established stable miRNA mutants in rice and validated that single-strand conformation polymorphism is a more dependable approach for screening mutant plants derived from the CRISPR/Cas9 tool. Their findings revealed that a single base pair addition or deletion often leads to the formation of redundant miRNAs. However, a larger deletion in close proximity to the mature miRNA or complementary miRNA* results in the loss of function for target miRNAs. *OsmiR528* unequivocally plays a crucial role in the positive response to salt stress in rice. In addition, the manipulation of mutants of OsmiR408 and OsmiR528 has undeniably led to the significant alteration of transcript levels of many seemingly unrelated miRNAs. A robust study utilizing CRISPR/Cas9-facilitated gene knockout has unequivocally confirmed OsmiR535 as a prime target for salt resistance in rice (Yue et al., 2020). These irrefutable experiments strongly suggest the utilization of CRISPR/Cas9-based genome editing tools in various plant species for the modification of miRNA genes. Efforts are relentlessly being made to further enhance CRISPR/Cas9 technology to ensure its applicability in editing miRNA genes in future. The newly developed Cas12a enzyme has conclusively demonstrated powerful and efficient gene-editing capabilities with significantly reduced off-target effects (Begemann et al., 2017; Zaidi et al., 2017; Zhang et al., 2017). Tang et al. (2017) decisively employed CRISPR/Cas12a to systematically attenuate the transcription of miRNA159b by utilizing deactivated Cas12a fused to the SRDX transcriptional repressor domain. Furthermore, CRISPR/Cas13a has conclusively demonstrated its potential to be employed to target miRNA and other ncRNA transcripts by directly cleaving ncRNA in vivo (Aman et al., 2018).

9.8 Concluding Remarks

Salinity poses a widespread problem for crop plants, significantly reducing their yield potential. Recent years have seen a surge in ncRNA-based research, garnering unprecedented attention from plant scientists. The successful application of CRISPR/Cas9 in editing protein-coding

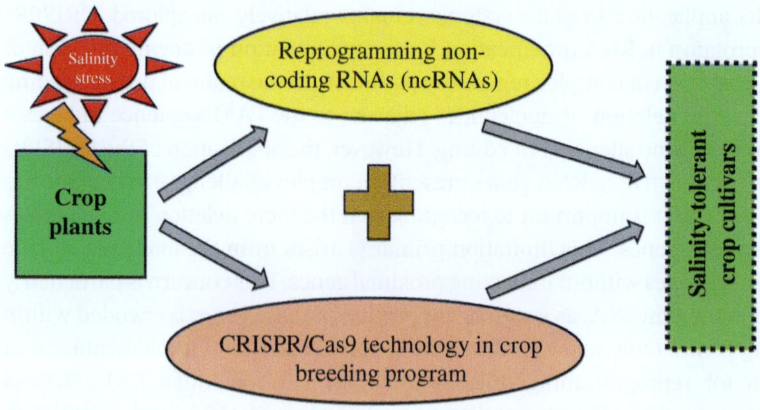

Figure 9.2 Outline of CRISPR/Cas9 technology to develop salt-tolerant crops.

genes has been notable, yet its exploration of nonprotein-coding genes or ncRNAs is still in its early stages. The advent of emerging gene editing technologies, particularly CRISPR/Cas9, is poised to accelerate the functional characterization and manipulation of novel ncRNAs. The amalgamation of CRISPR/Cas9 technology and ncRNAs holds immense potential for generating precise novel alleles that can affect crucial agronomic traits. This chapter presents recent endeavors to edit various ncRNAs using the CRISPR/Cas9 toolbox for crop improvement programs. The insights provided in this chapter will empower plant breeders and agricultural scientists to deepen their comprehension of diverse ncRNAs, the mechanism of CRISPR/Cas9 technology, and CRISPR/Cas9-mediated ncRNA regulation in various crop species. Despite prevailing challenges, strategic utilization of this promising technique is poised to furnish a sustainable, transgene-free solution for controlling diverse agronomic traits and developing salinity-tolerant germplasm and elite crop varieties (Figure 9.2) to tackle global food security challenges.

References

Abbasi, H., Jamil, M., Haq, A., Ali, S., Ahmad, R., Malik, Z., & Parveen, Z. (2016). Salt stress manifestation on plants, mechanism of salt tolerance and potassium role in alleviating it: a review. *Zemdirbyste-Agriculture*, *103*, 229–238.

Agastian, P., Kingsley, S. J., & Vivekanandan, M. (2000). Effect of salinity on photosynthesis and biochemical characteristics in mulberry genotypes. *Photosynthetica*, *38*, 287–290.

Ahmar, S., Hensel, G., & Gruszka, D. (2023). CRISPR/Cas9-mediated genome editing techniques and new breeding strategies in cereals – current status, improvements, and perspectives. *Biotechnology Advances*, *69*, 108248.

Al-Ashkar, I., Alderfasi, A., El-Hendawy, S., Al-Suhaibani, N., El-Kafafi, S., & Seleiman, M. F. (2019). Detecting salt tolerance in double haploid wheat lines. *Agronomy*, *9*, 211.

Alzahrani, S. M., Alaraidh, I. A., Khan, M. A., Migdadi, H. M., Alghamdi, S. S., & Alsahli, A. A. (2019). Identification and characterization of salt-responsive microRNAs in *Vicia faba* by high-throughput sequencing. *Genes*, *10*, 303.

Aman, R., Ali, Z., Butt, H., Mahas, A., Aljedaani, F., Khan, M. Z., Ding, S., & Mahfouz, M. (2018). RNA virus interference via CRISPR/Cas13a system in plants. *Genome Biology*, *19*, 1.

Arif, Y., Singh, P., Siddiqui, H., Bajguz, A., & Hayat, S. (2020). Salinity induced physiological and biochemical changes in plants: an omic approach towards salt stress tolerance. *Plant Physiology and Biochemistry*, *156*, 64–77.

Ashraf, M., Athar, H. R., Harris, P. J. C., & Kwon, T. R. (2008). Some prospective strategies for improving crop salt tolerance. *Advances in Agronomy*, *97*, 45–110.

Babu, N. N., Krishnan, S. G., Vinod, K. K., Krishnamurthy, S. L., Singh, V. K., Singh, M. P., Singh, R., Ellur, R. K., Rai, V., Bollinedi, H., Bhowmick, P. K., Yadav, A. K., Nagarajan, M., Singh, N. K., Prabhu, K. V., & Singh, A. K. (2017). Marker aided incorporation of *Saltol*, a major QTL associated with seedling stage salt tolerance, into *Oryza sativa* 'pusa basmati 1121'. *Frontiers in Plant Science*, *8*, 1–14.

Barrangou, R. (2013). CRISPR–Cas systems and RNA-guided interference. *Wiley Interdisciplinary Reviews: RNA*, *4*, 267–278.

Basak, J., & Nithin, C. (2015). Targeting non-coding RNAs in plants with the CRISPR-Cas technology is a challenge yet worth accepting. *Frontiers in Plant Science*, *6*, 1001.

Beaulieu, Y. B., Kleinman, C. L., Landry-Voyer, A. M., Majewski, J., & Bachand, F. (2012). Polyadenylation-dependent control of long noncoding RNA expression by the poly (a)-binding protein nuclear 1. *PLoS Genetics*, *8*, e1003078.

Begemann, M. B., Gray, B. N., January, E., Gordon, G. C., He, Y., Liu, H., Wu, X., Brutnell, T. P., Mockler, T. C., & Oufattole, M. (2017). Precise insertion and guided editing of higher plant genomes using Cpf1 CRISPR nucleases. *Scientific Reports*, *7*, 11606.

Bhogireddy, S., Mangrauthia, S. K., Kumar, R., Pandey, A. K., Singh, S., Jain, A., Budak, H., Varshney, R. K., & Kudapa, H. (2021). Regulatory non-coding RNAs: a new frontier in regulation of plant biology. *Functional & Integrative Genomics*, *21*, 313–330.

Bortesi, L., & Fischer, R. (2015). The CRISPR/Cas9 system for plant genome editing and beyond. *Biotechnology Advances*, *33*, 41–52.

Bravo-Vázquez, L. A., Méndez-Garcia, A., Chamu-García, V., Rodríguez, A. L., Bandyopadhyay, A., & Paul, S. (2024). The applications of CRISPR/Cas-mediated microRNA and lncRNA editing in plant biology: shaping the future of plant non-coding RNA research. *Planta*, *259*, 32.

Chekanova, J. A. (2015). Long non-coding RNAs and their functions in plants. *Current Opinion in Plant Biology*, *27*, 207–216.

Chinnusamy, V., Jagendorf, A., & Zhu, J. K. (2005). Understanding and improving salt tolerance in plants. *Crop Science*, *45*, 437–448.

Choudhary, A., Kumar, A., Kaur, H., & Kaur, N. (2021). MiRNA: the taskmaster of plant world. *Biologia*, *76*, 1551–1567.

Contreras-Cubas, C., Palomar, M., Arteaga-Vázquez, M., Reyes, J. L., & Covarrubias, A. A. (2012). Non-coding RNAs in the plant response to abiotic stress. *Planta*, *236*, 943–958.

Dasgupta, S., Hossain, M. M., Huq, M., & Wheeler, D. (2015). Climate change and soil salinity: the case of coastal Bangladesh. *Ambio*, *44*, 815–826.

Deng, P., Wang, L., Cui, L., Feng, K., Liu, F., Du, X., Tong, W., Nie, X., Ji, W., Weining, S., & Zhang, B. (2015). Global identification of microRNAs and their targets in barley under salinity stress. *PLoS One*, *10*, e0137990.

Deng, F., Zhang, X., Wang, W., Yuan, R., & Shen, F. (2018). Identification of *Gossypium hirsutum* long non-coding RNAs (lncRNAs) under salt stress. *BMC Plant Biology*, *18*, 23.

Etesami, H., & Beattie, G. A. (2018). Mining halophytes for plant growth-promoting halotolerant bacteria to enhance the salinity tolerance of non-halophytic crops. *Frontiers in Microbiology*, *9*, 148.

Feng, K., Nie, X., Cui, L., Deng, P., Wang, M., & Song, W. (2017). Genome-wide identification and characterization of salinity stress-responsive miRNAs in wild emmer wheat (*Triticum turgidum* ssp. *dicoccoides*). *Genes*, *8*, 156.

Flowers, T. J., & Colmer, T. D. (2015). Plant salt tolerance: adaptations in halophytes. *Annals of Botany, 115*, 327–331.

Foronda, D. A. (2022). Reclamation of a saline-sodic soil with organic amendments and leaching. *Environmental Sciences Proceedings, 16*, 56.

Gan, W. C., & Ling, A. P. K. (2022). CRISPR/Cas9 in plant biotechnology: applications and challenges. *Biotechnologia, 103*, 81–93.

Gasiunas, G., Barrangou, R., Horvath, P., & Siksnys, V. (2012). Cas9–crRNA ribonucleoprotein complex mediates specific DNA cleavage for adaptive immunity in bacteria. *Proceedings of the National Academy of Sciences, 109*, E2579–E2586.

Gill, S. S., Khan, N. A., Agarwala, N., Singh, K., Sunkar, R., & Tuteja, N. (2024). ncRNAs in plant development and stress responses. *Plant Physiology and Biochemistry, 214*, 108950.

Godfray, H. C. J., Beddington, J. R., Crute, I. R., Haddad, L., Lawrence, D., Muir, J. F., Pretty, J., Robinson, S., Thomas, S. M., & Toulmin, C. (2010). Food security: the challenge of feeding 9 billion people. *Science, 327*, 812–818.

Gong, H., You, J., Zhang, X., Liu, Y., Zhao, F., Cui, X., & Zhang, Y. (2021). Genome-wide identification and functional analysis of long non-coding RNAs in sesame response to salt stress. *Journal of Plant Biology, 64*, 555–565.

Hackenberg, M., Gustafson, P., Langridge, P., & Shi, B. J. (2015). Differential expression of microRNAs and other small RNAs in barley between water and drought conditions. *Plant Biotechnology Journal, 13*, 2–13.

Hille, F., Richter, H., Wong, S. P., Bratovič, M., Ressel, S., & Charpentier, E. (2018). The biology of CRISPR-Cas: backward and forward. *Cell, 172*, 1239–1259.

Ismail, A., Takeda, S., & Nick, P. (2014). Life and death under salt stress: same players, different timing? *Journal of Experimental Botany, 65*, 2963–2979.

Jacobs, T. B., LaFayette, P. R., Schmitz, R. J., & Parrott, W. A. (2015). Targeted genome modifications in soybean with CRISPR/Cas9. *BMC Biotechnology, 15*, 16.

Jain, P., Hussian, S., Nishad, J., Dubey, H., Bisht, D. S., Sharma, T. R., & Mondal, T. K. (2021). Identification and functional prediction of long non-coding RNAs of rice (*Oryza sativa* L.) at reproductive stage under salinity stress. *Molecular Biology Reports, 48*, 2261–2271.

Janik, E., Niemcewicz, M., Ceremuga, M., Krzowski, L., & Bijak, M. (2019). Various aspects of a gene editing system—CRISPR–Cas9. *International Journal of Molecular Sciences, 21*, 9604.

Jia, Z., Luo, W., Xie, J., Pan, Y., Chen, Y., Tang, S., & Liu, W. (2011). Salinity dynamics of wetland ditches receiving drainage from irrigated agricultural land in arid and semi-arid regions. *Agricultural Water Management, 100*, 9–17.

Jiang, C., Wang, Y., He, Y., Peng, Y., Xie, L., Li, Y., Sun, W., Zhou, J., Zheng, C., & Xie, X. (2024). Identification and characterization of miRNAs and lncRNAs associated with salinity stress in rice panicles. *International Journal of Molecular Sciences, 25*, 8247.

Jinek, M., Chylinski, K., Fonfara, I., Hauer, M., Doudna, J. A., & Charpentier, E. (2012). A programmable dual-RNA-guided DNA endonuclease in adaptive bacterial immunity. *Science, 337*, 816–821.

Kataria, S., & Verma, S. K. (2018). Salinity stress responses and adaptive mechanisms in major glycophytic crops: the story so far. In: V. Kumar, S. Wani, P. Suprasanna, L. S. Tran (Eds) *Salinity Responses and Tolerance in Plants, Vol. Vol. 1*, Springer, Cham, 1–39.

Kim, N., Jeon, J., Elbert, J., Kim, C., & Su, X. (2021). Redox-mediated electrochemical desalination for waste valorization in dairy production. *Chemical Engineering Journal, 428*, 131082.

Koyro, H. W., Ahmad, P., & Geissler, N. (2011). Abiotic stress responses in plants: an overview. In: P. Ahmad, M. Prasad (Eds) *Environmental Adaptations and Stress Tolerance of Plants in the Era of Climate Change*, Springer, New York, 1–28.

Kumar, N., Bharadwaj, C., Sahu, S., Shiv, A., Shrivastava, A. K., Reddy, S. P. P., Soren, K. R., Patil, B. S., Pal, M., Soni, A., Roorkiwal, M., & Varshney, R. K. (2021). Genome-wide identification and functional prediction of salt-stress related long non-coding RNAs (lncRNAs) in chickpea (*Cicer arietinum* L.). *Physiology and Molecular Biology of Plants, 27*, 2605–2619.

Lei, X. G., Zhu, J. H., Cheng, W. H., Bao, Y., Ho, Y. S., Reddi, A. R., Holmgren, A., & Arnér, E. S. (2015). Paradoxical roles of antioxidant enzymes: basic mechanisms and health implications. *Physiological Reviews, 96*, 307–364.

Li, M., Li, X., Zhou, Z., Wu, P., Fang, M., Pan, X., Lin, Q., Luo, W., Wu, G., & Li, H. (2016). Reassessment of the four yield-related genes *Gn1a, DEP1, GS3,* and *IPA1* in rice using a CRISPR/Cas9 system. *Frontiers in Plant Science, 7*, 377.

Li, Y., Li, W., & Li, J. (2021). The CRISPR/Cas9 revolution continues: from base editing to prime editing in plant science. *Journal of Genetics and Genomics, 48*, 661–670.

Li, N., Wang, Z., Wang, B., Wang, J., Xu, R., Yang, T., Huang, S., Wang, H., & Yu, Q. (2022). Identification and characterization of long non-coding RNA in tomato roots under salt stress. *Frontiers in Plant Science, 13*, 834027.

Liu, J., Jung, C., Xu, J., Wang, H., Deng, S., Bernad, L., Arenas-Huertero, C., & Chua, N. H. (2012). Genome-wide analysis uncovers regulation of long intergenic noncoding RNAs in Arabidopsis. *Plant Cell, 24*, 4333–4345.

Lowder, L. G., Paul, J. W., Baltes, N. J., Voytas, D. F., Zhang, Y., Zhang, D., Tang, X., Zheng, X., Hsieh, T. F., & Qi, Y. (2015). A CRISPR/Cas9 toolbox for multiplexed plant genome editing and transcriptional regulation. *Plant Physiology, 169*, 971–985.

Luo, C., He, B., Shi, P., Xi, J., Gui, H., Pang, B., Cheng, J., Hu, F., Chen, X., & Lv, Y. (2022). Transcriptome dynamics uncovers long non-coding RNAs response to salinity stress in *Chenopodium quinoa*. *Frontiers in Plant Science, 13*, 988845.

Ma, Y., Zhang, L., & Huang, X. (2014). Genome modification by CRISPR/Cas9. *The FEBS Journal, 281*, 5186–5193.

Makarova, K. S., Wolf, Y. I., Alkhnbashi, O. S., Costa, F., Shah, S. A., Saunders, S. J., Barrangou, R., Brouns, S. J. J., Charpentier, E., Haft, D. H., Horvath, P., Moineau, S., Mojica, F. J. M., Terns, R. M., Terns, M. P., White, M. F., Yakunin, A. F., Garrett, R. A., van der Oost, J., Backofen, R., & Koonin, E. V. (2015). An updated evolutionary classification of CRISPR–Cas systems. *Nature Reviews. Microbiology, 13*, 722–736.

Mass, E. V. (1990). Crop salt tolerance. Chapter 13. In: K. K. Tanji (Ed) *Agricultural Salinity Assessment and Management* ASCE Manuals and Reports on Engineering No. 71, American Society of Civil Engineers, New York, 262–304.

Mei, Y., Wang, Y., Chen, H., Sun, Z. S., & Ju, X. (2016). Recent progress in CRISPR/Cas9 technology. *Journal of Genetics and Genomics, 43*, 63–75.

Meng, Y., Shao, C., Wang, H., & Chen, M. (2011). The regulatory activities of plant microRNAs: a more dynamic perspective. *Plant Physiology, 157*, 1583–1595.

Mondal, T. K., Panda, A. K., Rawal, H. C., & Sharma, T. R. (2018). Discovery of microRNA-target modules of African rice (*Oryza glaberrima*) under salinity stress. *Scientific Reports, 8*, 570.

Muehlbauer, F. J., Cho, S., Sarker, A., McPhee, K. E., Coyne, C. J., Rajesh, P. N., & Ford, R. (2006). Application of biotechnology in breeding lentil for resistance to biotic and abiotic stress. *Euphytica, 147*, 149–165.

Munns, R. (2005). Genes and salt tolerance: bringing them together. *The New Phytologist, 167*, 645–663.

Muthu, V., Abbai, R., Nallathambi, J., Rahman, H., Ramasamy, S., Kambale, R., Thulasinathan, T., Ayyenar, B., Muthurajan, R., & Struik, P. C. (2020). Pyramiding QTLs controlling tolerance against drought, salinity, and submergence in rice through marker assisted breeding. *PLoS One*, *15*, e0227421.

Nekrasov, V., Staskawicz, B., Weigel, D., Jones, J. D., & Kamoun, S. (2013). Targeted mutagenesis in the model plant *Nicotiana benthamiana* using Cas9 RNA-guided endonuclease. *Nature Biotechnology*, *31*, 691.

Obaid, T. A. (2001). Footprints and milestones: population and environmental change. In: *The State of World Population*. https://www.unfpa.org/sites/default/files/pub-pdf/swp2001_eng.pdf

Palazzo, A. F., & Lee, E. S. (2015). Non-coding RNA: what is functional and what is junk? *Frontiers in Genetics*, *6*, 2.

Parida, A. K., & Das, A. B. (2005). Salt tolerance and salinity effects on plants: a review. *Ecotoxicology and Environmental Safety*, *60*, 324–349.

Paul, J. W., & Qi, Y. (2016). CRISPR/Cas9 for plant genome editing: accomplishments, problems and prospects. *Plant Cell Reports*, *35*, 1417–1427.

Qadir, M., Noble, A. D., Schibert, S., Thomas, R. J., & Arslan, A. (2006). Sodicity-induced land degradation and its sustainable management: problems and prospects. *Land Degradation and Development*, *17*, 661–676.

Qadir, M., Quillérou, E., Nangia, V., Murtaza, G., Singh, M., Thomas, R. J., Drechsel, P., & Noble, A. D. (2014). Economics of salt-induced land degradation and restoration. *Natural Resources Forum*, *38*, 282–295.

Rahman, M. A., Thomson, M. J., Shah-E-Alam, M., De Ocampo, M., Egdane, J., & Ismail, A. M. (2016). Exploring novel genetic sources of salinity tolerance in rice through molecular and physiological characterization. *Annals of Botany*, *117*, 1083–1097.

Ran, F. A., Hsu, P. D., Wright, J., Agarwala, V., Scott, D. A., & Feng, Z. (2013). Genome engineering using the CRISPR-Cas9 system. *Nature Protocols*, *8*, 2281–2308.

Samanta, M. K., Dey, A., & Gayen, S. (2016). CRISPR/Cas9: an advanced tool for editing plant genomes. *Transgenic Research*, *25*, 561–573.

Schaeffer, S. M., & Nakata, P. A. (2016). The expanding footprint of CRISPR/Cas9 in the plant sciences. *Plant Cell Reports*, *35*, 1451–1468.

Sharma, A., Ruiz-Manriquez, L. M., Serrano-Cano, F. I., Reyes-Pérez, P. R., Alfaro, C. K. T., Andrade, Y. E. B., Aros, A. K. H., Srivastava, A., & Paul, S. (2020). Identification of microRNAs and their expression in leaf tissues of guava (*Psidium guajava* L.) under salinity stress. *Agronomy*, *10*, 1920.

Shiferaw, B., Prasanna, B. M., Hellin, J., & Bänziger, M. (2011). Crops that feed the world 6 past successes and future challenges to the role played by maize in global food security. *Food Security*, *3*, 307–327.

Shinde, H., Dudhate, A., Anand, L., Tsugama, D., Gupta, S. K., Liu, S., & Takano, T. (2020). Small RNA sequencing reveals the role of pearl millet miRNAs and their targets in salinity stress responses. *South African Journal of Botany*, *132*, 395–402.

Soda, N., Verma, L., & Giri, J. (2018). CRISPR-Cas9 based plant genome editing: significance, opportunities and recent advances. *Plant Physiology and Biochemistry*, *131*, 2–11.

Song, Y., & Zhang, D. (2017). The role of long noncoding RNAs in plant stress tolerance. *Methods in Molecular Biology*, *1631*, 41–68.

Srijyothi, L., Saravanaraman, P., Talukdar, P., Cheemala, A., & Sudhakar, B. (2018). Roles of non-coding RNAs in transcriptional regulation. In: K. Ghedira (Ed) *Transcriptional and Post-Transcriptional Regulation*, IntechOpen.

Sun, X., Xu, L., Wang, Y., Yu, R., Zhu, X., Luo, X., Gong, Y., Wang, R., Limera, C., Zhang, K., & Liu, L. (2015). Identification of novel and salt-responsive miRNAs to explore miRNA-mediated regulatory network of salt stress response in radish (*Raphanus sativus* L.). *BMC Genomics*, *16*, 197.

Sun, X., Zheng, H., Li, J., Liu, L., Zhang, X., & Na, S. (2020). Comparative transcriptome analysis reveals new lncRNAs responding to salt stress in sweet sorghum. *Frontiers in Bioengineering and Biotechnology*, *8*, 331.

Tang, X., Lowder, L. G., Zhang, T., Malzahn, A. A., Zheng, X., Voytas, D. F., Zhong, Z., Chen, Y., Ren, Q., Li, Q., Kirkland, E. R., Zhang, Y., & Qi, Y. (2017). A CRISPR–Cpf1 system for efficient genome editing and transcriptional repression in plants. *Nature Plants*, *3*, 17018.

Thiem, D., Gołębiewski, M., Hulisz, P., Piernik, A., & Hrynkiewicz, K. (2018). How does salinity shape bacterial and fungal microbiomes of *Alnus glutinosa* roots? *Frontiers in Microbiology*, *9*, 651.

Tuteja, N., Gill, S. S., Tiburcio, A. F., & Tuteja, R. (2012). Salinity stress: a major constraint in crop production. In: N. Tuteja, S. S. Gill, A. F. Tiburcio, R. Tuteja (Eds) *Improving Crop Resistance to Abiotic Stress*, 1st edition, Wiley-VCH Verlag GmbH, New York, 71–96.

Tyagi, S., Kumar, R., Das, A., Won, S. Y., & Shukla, P. (2020). CRISPR-Cas9 system: a genome-editing tool with endless possibilities. *Journal of Biotechnology*, *319*, 36–53.

Varshney, R. K., Bansal, K. C., Aggarwal, P. K., Datta, S. K., & Craufurd, P. Q. (2011). Agricultural biotechnology for crop improvement in a variable climate: hope or hype. *Trends in Plant Science*, *16*, 363–371.

Waititu, J. K., Zhang, C., Liu, J., & Wang, H. (2020). Plant non-coding RNAs: origin, biogenesis, mode of action and their roles in abiotic stress. *International Journal of Molecular Sciences*, *21*, 8401.

Wang, T. Z., Liu, M., Zhao, M. G., Chen, R., & Zhang, W. H. (2015). Identification and characterization of long non-coding RNAs involved in osmotic and salt stress in *Medicago truncatula* using genome-wide high-throughput sequencing. *BMC Plant Biology*, *15*, 131.

Wang, J., Meng, X., Dobrovolskaya, O. B., Orlov, Y. L., & Chen, M. (2017). Non-coding RNAs and their roles in stress response in plants. *Genomics, Proteomics & Bioinformatics*, *15*, 301–312.

Yang, Z., Zhu, P., Kang, H., Liu, L., Cao, Q., Sun, J., Dong, T., Zhu, M., Li, Z., & Xu, T. (2020). High-throughput deep sequencing reveals the important role that microRNAs play in the salt response in sweet potato (*Ipomoea batatas* L.). *BMC Genomics*, *21*, 164.

Yi, F., Chen, J., & Yu, J. (2015). Global analysis of uncapped mRNA changes under drought stress and microRNA-dependent endonucleolytic cleavages in foxtail millet. *BMC Plant Biology*, *15*, 1–15.

Yue, E., Cao, H., & Liu, B. (2020). *OsmiR535*, a potential genome editing target for drought and salinity stress tolerance in *Oryza sativa*. *Plants*, *9*, 1337.

Zaidi, S. S., Mahfouz, M. M., & Mansoor, S. (2017). CRISPR-Cpf1: a new tool for plant genome editing. *Trends in Plant Science*, *22*, 550–553.

Zetsche, B., Gootenberg, J. S., & Abudayyeh, O. O. (2015). Cpf1 is a single RNA-guided endonuclease of a class 2 CRISPR-Cas system. *Cell*, *163*, 759–771.

Zhang, X., Wang, J., Cheng, Q., Zheng, X., Zhao, G., & Wang, J. (2017). Multiplex gene regulation by CRISPRddCpf1. *Cell Discovery*, *3*, 17018.

Zhang, D., Zhang, Z., Unver, T., & Zhang, B. (2021). CRISPR/Cas: a powerful tool for gene function study and crop improvement. *Journal of Advanced Research*, *29*, 207–221.

Zhang, Y., Zhou, Y., Zhu, W., Liu, J., & Cheng, F. (2022). Non-coding RNAs fine-tune the balance between plant growth and abiotic stress tolerance. *Frontiers in Plant Science*, *13*, 965745.

Zhao, H. W., Lv, X., & Yin, W. (2015). The CRISPR/Cas9 system: a novel strategy for targeted genome engineering. *Journal of Pathogen Biology*, *10*, 281–284.

Zhao, Y., Zhang, C., Liu, W., Gao, W., Liu, C., Song, G., Li, W. X., Mao, L., Chen, B., Xu, Y., Li, X., & Xie, C. (2016). An alternative strategy for targeted gene replacement in plants using a dual-sgRNA/Cas9 design. *Scientific Reports*, *6*, 23890.

Zhou, J., Deng, K., Cheng, Y., Zhong, Z., Tian, L., Tang, X., Tang, A., Zheng, X., Zhang, T., Qi, Y., & Zhang, Y. (2017). CRISPR-Cas9 based genome editing reveals new insights into microRNA function and regulation in rice. *Frontiers in Plant Science*, *8*, 1598.

Zhou, J., Li, D., Wang, G., Wang, F., Kunjal, M., Joldersma, D., & Liu, Z. (2020). Application and future perspective of CRISPR/Cas9 genome editing in fruit crops. *Journal of Integrative Plant Biology*, *62*, 269–286.

Zhu, J. K. (2001). Plant salt tolerance. *Trends in Plant Science*, *6*, 66–71.

10

Genetic Engineering of *Cis*-Regulatory Elements by CRISPR/Cas Technology for Crop Improvement

Afreen Anis, Soumyashree Dash, Soumita Karmakar, and Jyoti Prakash Sahoo

Faculty of Agriculture and Allied Sciences, C.V. Raman Global University, Bhubaneswar, India

10.1 Introduction

Clustered regularly interspaced palindromic repeat (*CRISPR*)-associated (*Cas*) systems, found in many microorganisms and archaea, protect against phages, viruses, and other genetic invaders. This system consists of *CRISPR*-associated genes and *CRISPR* repeat-spacer arrays (Acharya et al., 2014). The arrays include spacers derived from past invaders' DNA, interspersed with repeating sequences (Bai et al., 2017). Transcription of these arrays produces *trans-activating CRISPR RNA* (tracrRNA) and *CRISPR RNA* (crRNA), which guide *Cas* proteins in cleaving foreign DNA (Butardo et al., 2017). When a new infection occurs, the crRNA recognizes the invader, allowing *Cas* proteins to cut the DNA, preventing infection (Hsu et al., 2014). The *CRISPR-Cas9* system, comprising an RNA-guided *Cas*9 endonuclease and sgRNA, targets specific DNA sequences via a protospacer adjacent motif (PAM) sequence, causing double-strand breaks (DSBs) (Chen et al., 2019). These DSBs are repaired by nonhomologous end joining (NHEJ) or homology-directed repair (HDR) (Hsu et al., 2014). NHEJ is efficient but error-prone, while HDR allows precise modifications (Cong et al., 2013). Originally, a bacterial defense, *CRISPR/Cas* is now a powerful tool for editing *cis*-regulatory elements (CREs) like promoters and enhancers, enhancing crop traits such as yield, stress resistance, and disease resistance (Knott & Doudna 2018). The chapter highlights *CRISPR/Cas* mechanisms and their application in controlling gene expression for desired phenotypes.

10.2 *CRISPR/Cas* System Advancements

CRISPR-Cas systems, widely used for genetic alterations since 2012, have achieved significant milestones like resolving interspecies organ transplantation issues, repairing disease-causing mutations, and identifying key genes for cancer immunotherapy (Deb et al., 2018). However, challenges remain, including unintended genome modifications and limited target ranges due to specific PAM sequences (Dey et al., 2015). Researchers are addressing these by engineering more accurate *Cas* proteins, developing advanced guide RNA (gRNA), discovering novel *Cas* proteins for a broader range of PAM sequences, optimizing delivery methods, enhancing DNA repair

Genome and Epigenome Editing for Stress-Tolerant Crops, First Edition. Edited by Jen-Tsung Chen.

mechanisms, and refining component designs (Dong et al., 2020). These efforts aim to improve the precision and reliability of *CRISPR-Cas* systems for genetic and medical applications (Wright et al., 2016).

10.2.1 Dead-Cas9 System

The nuclease-dead *Cas*9 (d*Cas*9) variant retains the ability to bind specific DNA sequences but cannot cut DNA due to specific mutations (Du et al., 2020). This allows scientists to use d*Cas*9 in gene activation (*CRISPR*a) or suppression (*CRISPR*i) by attaching transcriptional activators or repressors. d*Cas*9 can also be fused with various effector domains, enabling precise recruitment of proteins for genome imaging or epigenetic modifications (Gaj et al., 2013). This system's efficiency and simplicity allow for the concurrent targeting and alteration of multiple genes within a single cell, making it a versatile and user-friendly tool for gene regulation, imaging, and studying complex genetic interactions (Wright et al., 2016). Moreover, the process of altering plant genes using *CRISPR/Cas9* is indexed in Figure 10.1.

10.2.2 System Cas9 Variant

CRISPR/Cas9 technology relies on the NGG PAM sequence for gene targeting, but other less efficient PAM (Protospacer Adjacent Motifs) sequences like NGA and NAG cover only about 1/6th of the human genome, limiting targetable regions. To address this, researchers developed *Cas*9 variants such as x*Cas*9, Sp*Cas*9-NG, and Sp*Cas*9 targeting NRRH, NRCH, and NRTH PAMs, broadening the range of targetable PAMs (Hille et al., 2018; Jia et al., 2022; Joung & Sander 2013;

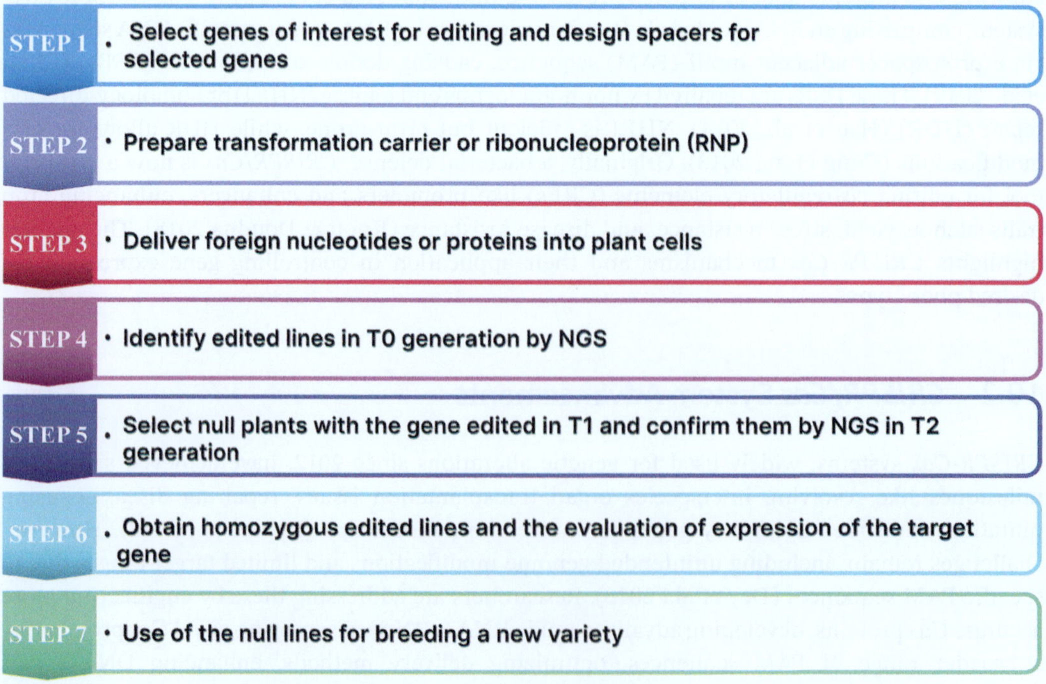

Figure 10.1 The process of altering plant genes using CRISPR/Cas9.

Karkute et al., 2017; Koltun et al., 2018). Scientists created SpG and SpRY, which can target a wider array of PAMs, including near-PAMless versions. Despite these advancements, the variants still face challenges like low cleavage activity and efficiency, necessitating further molecular engineering improvements for effective disease-relevant genome editing (GE) (Ran et al., 2013).

10.2.3 RNA Editing System

The Type VI *CRISPR-Cas*13 system can edit RNA by targeting and cleaving single-stranded RNA (ssRNA). It includes four *Cas13* proteins: *Cas13a*, *Cas13b*, *Cas13c*, and *Cas13d* (*C2c2*) (Minkenberg et al., 2017). These proteins are used in viral detection, splicing regulation, transcript tagging, and RNA knockdown. Scientists have developed two RNA editing frameworks: REPAIR, for A-to-I (G) conversion, and RESCUE, for C-to-U conversion, using catalytically inactive *Cas*13 fused with the ADAR2 deaminase domain. RNA editing offers high efficiency, specificity, and temporary, reversible changes, avoiding the ethical concerns of permanent DNA edits. RNA editing is now widely used in preclinical trials for various disorders (Minkenberg et al., 2017).

10.2.4 Primer Editing System

Prime editing is a recent GE method that does not require donor DNA templates or DSBs. This technique uses a modified *Cas*9 enzyme, a prime editing guide RNA (pegRNA), and a reverse transcriptase to precisely insert, remove, or substitute bases in DNA (Komor et al., 2017; Kumar 2019; Kumar et al., 2015; Langner et al., 2018; Lee et al., 2019). The pegRNA directs the system to the target site and encodes the intended genetic change, while the reverse transcriptase uses the pegRNA as a template for incorporating new genetic information (Li et al., 2014, 2019; Lino et al., 2018; Lu et al., 2017; Ma et al., 2016, 2021; Mao et al., 2022). Prime editing offers improved precision and efficiency, potentially correcting up to 89% of disease-causing gene mutations. Despite its promise, further research is needed to fully understand and optimize this technique. Its impact on genetic research and medical treatments is expected to grow, providing new opportunities for addressing genetic disorders (Zhang et al., 2014).

10.3 *CRISPR/Cas9* Application for Improving Crop Quality

Advancements in agricultural technologies have shifted focus from merely increasing crop yields to improving crop quality, nutritional value, and stress resistance. Traditional methods like crossbreeding and mutagenesis are labor-intensive, whereas GE technologies, especially *CRISPR/Cas9*, offer precise and heritable genetic modifications with minimal off-target effects. GE allows a range of DNA changes, from single-nucleotide polymorphisms (SNPs) to large-scale insertions and deletions, enhancing crop quality efficiently. *Site-directed nucleases (SDNs)* such as *zinc-finger nucleases (ZFNs)*, *transcription activator-like effector nucleases (TALENs)*, and *CRISPR/Cas* systems target DNA to create DSBs, repaired by the plant's mechanisms through NHEJ or homologous-directed recombination (HDR) (Parry et al., 2009). *CRISPR*, favored for its simplicity and efficiency, uses gRNA for specificity, eliminating the need for complex protein engineering. Discovered in *Escherichia coli* in 1987, *CRISPR/Cas9* systems, particularly Sp*Cas*9 from *Streptococcus pyogenes*, have been extensively adopted for gene functional analysis and introducing desirable traits in crops (Rodríguez-Leal et al. 2017). The system employs guide RNA (gRNA or sgRNA) and *Cas*9 protein to induce DSBs, repaired by NHEJ for gene knockout or HDR for precise modifications (Miller et al., 2007; Potenza et al., 2004; Qian et al., 2017; Ricroch

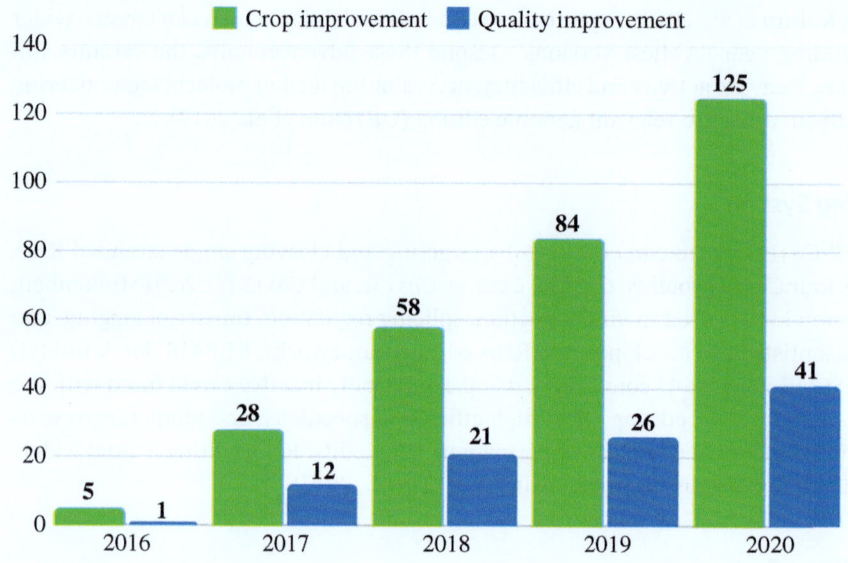

Figure 10.2 Publication data of CRISPR/Cas from 2016 to 2020.

et al., 2017; Rushton et al., 2002; Sadhukhan et al., 2017; Shi et al., 2015; Swinnen et al., 2016). Since its introduction to plant gene editing in 2013, *CRISPR/Cas9* has enabled gene knockout, knock-in, knockdown, activation, and simultaneous multigene editing. Publications on *CRISPR/Cas9* in agriculture have surged, focusing on enhancing crop quality and stress resilience and addressing detrimental regulatory issues (Figure 10.2).

10.4 *Cis*-regulatory Components in Introns, Promoters, and Intergenic Areas

Gene expression in eukaryotes is tightly controlled during growth and development, with transcription being a key component. Transcription starts near genes and involves transcription factors (TFs) and regulatory proteins interacting with specific DNA regions called CREs, found in promoters, enhancers, and other regulatory sequences (Tang et al., 2019; Vats et al., 2019; Venter 2007; Wang et al., 2016, 2020; Yang et al., 2018). These interactions ensure precise and timely gene expression, activating the transcription of neighboring genes as needed. TFs and regulatory proteins binding to CREs initiate and regulate transcription, ensuring accurate gene expression throughout the organism's development (Lambert et al., 2018). Moreover, the structure of plant genes is indexed in Figure 10.3.

10.4.1 The Promoter Region's *Cis*-Regulatory Components

Promoters are vital for plant genetic engineering, categorized into constitutive, inducible, and tissue-specific types. They consist of basal elements and CREs, which include transcription factor binding sites (TFBSs) essential for gene regulation (Wittkopp & Kalay, 2011). The regulation of *AtMFT* by the *RY-motif* and *AtABI3* in *Arabidopsis*, and the similarity to *OsLFL1* in rice, indicates a conserved mechanism for gene expression modulation in response to environmental cues, supported by experimental data linking *OsLFL1* to the *RY-motif* in the *OsMFT1* promoter (Lambert et al., 2018).

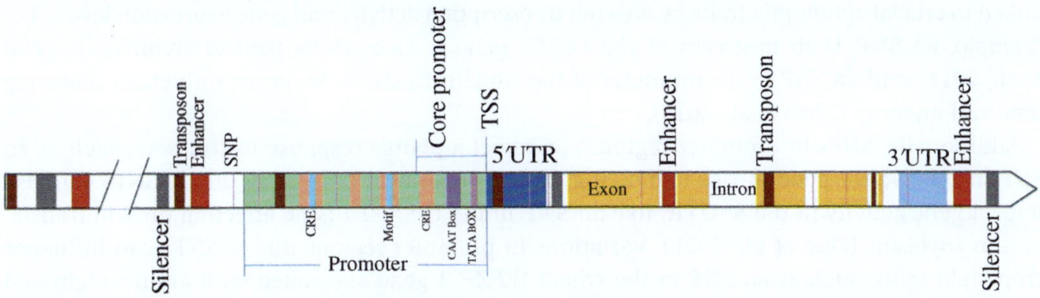

Figure 10.3 The structure of plant genes, distinguishing between coding (exons and introns) and noncoding regions (UTRs, transposons, silencers, enhancers, and promoters). Promoters contain transcription factor binding sites (TFBSs) and RNA polymerase-binding sites near the transcription start site (TSS). Basal components like TATA-box and CAAT-box form the core promoter, while gene components are represented by different colors.

10.4.2 Intergenic and Intron Regions Contain *Cis*-Regulatory Elements

Introns and intergenic regions in plants harbor CREs like enhancers and silencers, which regulate gene expression by recruiting TFs and regulatory proteins. These CREs, influenced by chromatin structure and DNA accessibility, modulate gene expression spatially and temporally (Lambert et al., 2018). Early studies, such as those on the AB80 gene in peas and intergenic enhancers in Arabidopsis, highlighted their regulatory significance. Examples in maize and yeast emphasize their role across species, and advancements in biotechnology enable their use to optimize crop traits for sustainable agriculture (Lambert et al., 2018).

10.5 Natural Variation Affecting *Cis*-Regulatory Elements in Crops by InDels, SNPs, and Transposable Elements

Natural variation in main crops, driven by environmental stresses, genetic recombination, and spontaneous mutations, is crucial for evolutionary adaptation and breeding programs aimed at improving crop resilience, productivity, and quality. At the genomic level, this variation includes SNPs, insertions or deletions (indels), and structural variants like inversions, duplications, and large deletions (Zhu et al., 2020). These genetic changes can impact gene function, expression, and regulation, influencing traits like yield, disease resistance, and environmental adaptation. Advances in genomic technologies, such as high-throughput sequencing and genome-wide association studies (GWAS), have accelerated the identification and utilization of beneficial genetic variants, enhancing crop improvement efforts to meet agricultural challenges (Doni Jayavelu et al., 2020).

10.5.1 SNPs Located in the Area of *Cis*-Regulation

SNPs are DNA sequence changes that can occur as transitions or transversions and are found throughout the genome (Ye et al., 2019; Yuan et al., 2020; Zafar et al., 2020; Zhan et al., 2019; Zhang et al., 2004). They are categorized based on their location relative to genes: coding SNPs in coding regions and noncoding SNPs in noncoding regions, also known as regulatory SNPs. Regulatory SNPs, located in *cis*-regulatory regions or introns, can significantly impact gene expression by affecting the binding of regulatory proteins like TFs. SNPs in *cis*-regulatory regions are linked to crucial agronomic traits by altering transcription activity and gene expression levels. For example, an SNP 11 kb upstream of the *OsLG1* gene in rice affects panicle architecture (Zhu et al., 2013), and an SNP in the promoter of the GmGBP1 gene in soybeans influences flowering time and maturity (Zhao et al., 2018).

Additionally, SNPs in promoter regions can affect a plant's response to stressors, such as an SNP in the *DROT1* gene in rice enhancing drought tolerance (Sun et al. 2020). SNPs can also impact gene activity in the 5′ UTR, like an SNP in the *GmPHF1* gene affecting protein translation in soybeans (Guo et al., 2021). Variations in promoter regions due to SNPs can influence crop yield traits, such as an SNP in the wheat *WFZP-A* gene associated with grain weight and spikelet number (Guo et al., 2021). Thus, SNPs in regulatory regions play a critical role in determining gene expression and crop productivity by modifying protein translation and gene expression patterns.

10.5.2 InDels in *Cis*-Regulatory Regions

Genetic insertions or deletions (InDels) can significantly affect gene function and expression by altering the binding of TFs to CREs in promoter regions. For example, a 21-base pair InDel in the promoter of the *STP1* gene in tomatoes affects soluble solid levels in fruit, with the insertion linked to higher levels (Wang et al., 2023). Similarly, a 12-base pair InDel in the TFBS of the rice *OsNOG1* gene promoter influences *OsNOG1* expression (Liu et al., 2024), and a 3-base pair InDel in the promoter of *Sl-ALMT9* in tomatoes affects fruit malate content by disrupting a WRKY repressor binding site (Ye et al., 2017). Identifying such genetic variations is crucial for crop improvement efforts, as they directly impact traits like nutritional content, yield, and disease resistance (Wang et al., 2023).

10.5.3 Components that are Transposable in *Cis*-Regulatory Areas

Transposable elements (TEs) are dynamic DNA sequences that can move within the genome using "cut and paste" or "copy and paste" mechanisms. These movements can alter genome size, rearrange chromosomes, cause mutations, and influence gene expression by inserting into or near critical regulatory regions such as promoters, enhancers, or introns (Hayward & Gilbert, 2022). For instance, in rice, the auxin response element within the *INDITTO2 TE* in the promoter of the *DEEPER ROOTING 1* (*OsDRO1*) gene facilitates deeper root growth and enhances drought tolerance by regulating neighboring genes in response to auxin (Zhao et al., 2021). Conversely, in wheat, TEs in the promoter and 3′ UTR of the powdery mildew resistance gene *Pm41* suppress the gene, increasing disease susceptibility; reactivating *Pm41* restores resistance (Liu et al., 2024).

TEs can also create new splicing sites, affecting gene expression and leading to alternative splicing, as seen in lettuce where a CACTA transposon insertion resulted in pale-green leaves (Hayward & Gilbert, 2022). Numerous TEs have been identified in plants like *Brassica rapa*, *Arabidopsis*, maize, and wheat, highlighting their role in genome evolution and plant traits. For example, TE insertions in *B. rapa* are linked to changes in gene expression levels and structures (Hayward & Gilbert, 2022). TEs expand TFBSs, influencing regulatory changes and genome-wide transcriptional regulation (Zhang et al., 2021). TEs are integral to species evolution, affecting genome stability, transcriptional regulation, and genetic diversity, and their natural diversity can be harnessed for molecular breeding to develop crops with improved traits (Zhang et al., 2021).

10.6 Ways to Use the *CRISPR/Cas*-Editing System for Transgene-Free Genetic Engineering

The *CRISPR/Cas* system leverages an organism's DNA repair mechanisms to enable precise gene changes but faces challenges such as difficulties in introducing the *Cas*9–sgRNA complex, off-target mutations, and stringent regulatory scrutiny for transgenic plants. Developing nontransgenic genetic modification techniques is crucial for crop commercialization. One method involves using preassembled *Cas*9 protein–gRNA ribonucleoproteins (RNPs) in plant cells, effective across various species without needing codon optimization (Liang et al., 2017). This method achieved targeted mutagenesis in *Arabidopsis*, tobacco, lettuce, and rice (Liang et al., 2017), though it needs refinement for grass crops. Another method involves temporary DNA production in wheat, creating transgene-free mutants without herbicide selection (Sun et al., 2020). Virus-induced, transgene-free editing produced *Cas*9-free wheat mutants by crossing *Cas*9 transgenic and wild-type wheat, followed by removing the *Cas*9 transgene (Li et al., 2023). A *Cas*9/*CRISPR* vector with a fluorescent tag was developed to identify transgene-free plants easily (He et al., 2018+). Another technique used tRNA-like sequence motifs for heritable gene editing, proven in *B. rapa* and *Arabidopsis thaliana*, eliminating the need for transgene removal or viral vectors (Yang et al., 2023). Transgene-free GE reduces off-target effects and toxicity, offering a new way to create genome-edited plants immune to transgenic regulation and preventing reediting in genetic research (Yang et al., 2023).

10.7 The Genome-Wide Association Study's (GWAS) Identification of CREs that are Vital to Agriculture

GWAS have become crucial for studying complex agronomic traits influenced by multiple genetic regions and environmental factors, especially with advancements in high-density genotyping arrays and sequencing technology. High-throughput phenotyping has further enhanced GWAS, as shown in various studies (Ersoz et al., 2007; Varshney et al., 2012). GWAS has identified candidate regulatory elements (CREs) linked to agronomic traits and stress resilience. GR can target these CREs to induce gradual trait changes. For example, promoter changes in rice genes like *GW7* and *GSE5* affect grain characteristics (Wang et al., 2015; Ruan et al., 2020). GWAS has confirmed genes such as *CYP734A4* and *OsNPF6.1HapB* associated with yield and nitrogen use efficiency and identified CREs related to fruit and grain quality in rice and tomatoes (Yang et al., 2019). Additionally, GWAS has identified genes related to biotic and abiotic stress tolerance such as *Xa22*, *Osbsr-d1*, and *OsSTL1* in rice and maize (Wang et al., 2015) and CREs associated with microelement accumulation and heavy metal tolerance like AtMATE in *Arabidopsis* and zinc transporter genes in maize (Yang et al., 2023). These findings highlight the potential of CRE editing via *CRISPR/Cas* to enhance agronomic traits, emphasizing GWAS's critical role in discovering significant CREs and its anticipated impact on agriculture as genome-editing technologies advance (Yang et al., 2023).

10.8 Customized CRE Alterations

Gene expression is largely controlled by elements of DNA known as *cis*-regulatory ones (CREs), which are found close to or inside genes. Both geographically and chronologically expression-specific genetic regulation is carried out by various components, including insulators, silencers, enhancers, and promoters. Conventional methods for modifying CREs have been fundamental in understanding gene regulation and manipulating gene expression for various applications in genetics, biotechnology, and medicine (Mao et al., 2022).

10.8.1 Mutagenesis Targeted at a Location

Site-directed mutagenesis is a well-researched technique for altering CREs, enabling the introduction of specific mutations at predetermined DNA points. By using oligonucleotide primers with the desired mutation, researchers can amplify and introduce these changes into the DNA (Zhao et al., 2020; Zheng et al., 2011,). This method helps dissect the functional regions of CREs by altering individual nucleotides to determine their specific roles in gene regulation (Zhu et al., 2020; Zilberman et al., 2007).

10.8.2 Reporter Gene Assays

Reporter gene experiments are a popular method for examining CRE function. By linking a regulatory element to a reporter gene like luciferase or GFP, researchers can quantify its activity. Changes in reporter gene expression reveal how modifications to the CRE affect gene regulation. This method is especially useful for high-throughput screening of CRE activity under various conditions or stimuli (Ruan et al., 2020).

10.8.3 Deletion and Insertion Mutations

Creating deletion and insertion mutations within CREs helps understand their function. Deletion mutations remove specific DNA segments to identify critical regulatory regions, while insertion mutations add sequences to see how they affect gene expression. These mutations can be introduced using restriction enzyme digestion followed by ligation or modern methods like *CRISPR/Cas9*-mediated GE (Ruan et al., 2020).

10.8.4 Electrophoretic Mobility Shift Assays (EMSAs)

EMSAs, or gel shift assays, study protein–DNA interactions essential to CRE function. Researchers incubate labeled DNA fragments with nuclear extracts or purified TFs to observe protein–DNA complex formation. The mobility of these complexes during gel electrophoresis indicates binding affinity and specificity. Mutations within the CRE can be analyzed to determine their impact on protein binding (Ruan et al., 2020).

10.8.5 Chromatin Immunoprecipitation (ChIP)

ChIP experiments investigate TFs' in vivo binding to CREs within chromatin. Using specific antibodies, researchers precipitate DNA–protein complexes from cross-linked chromatin (Ruan et al., 2020). ChIP-seq or ChIP-PCR analysis of the precipitated DNA identifies binding locations, elucidating the regulatory landscape of CREs. Mutations in CREs can be analyzed to assess their impact on gene expression resulting from TF binding (Ruan et al., 2020).

10.8.6 Functional Characterization Through Transgenic Models and Comparative Genomics

Transgenic animal models, especially mice, are used to study the in vivo function of CREs. By introducing modified CREs linked to a reporter or endogenous gene into the genome, researchers can observe physiological effects. This approach allows for studying CRE function in a whole organism, providing insights into complex regulatory networks beyond what in vitro methods can offer. Comparative genomics aims to identify conserved CREs across species, which are likely crucial for regulatory function. By analyzing sequence conservation, researchers can predict functional regions and design experiments to test these predictions. This involves sequence alignment, identifying conserved motifs, and performing functional assays to validate their regulatory roles.

10.9 CRE Editing with *CRISPR/Cas*

CRISPR/Cas9 accelerates crop improvement by creating DSBs repaired by NHEJ or HDR (Lin et al., 2020). Moreover, Mushtaq et al. (2021) showed precise editing without DSBs using *CRISPR/Cas* base editors. Prime editing with reverse transcriptase and *Cas*9 nickase enables accurate GE in crops (Tang et al., 2020). CRE GE allows for studying functional genes and achieving transgene-free offspring (Tang et al., 2020). Combining *CRISPR/Cas* and CRE editing can modify gene expression through promoter engineering, altering gene regulatory profiles to enhance agronomic traits (Pandiarajan et al., 2018).

10.9.1 Enhancement of Agronomic Features Using CRE Editing

To alter the boost of agricultural characteristics in crop plants, like resilience to disease and abiotic tolerance, productivity, and excellence parameters, CREs of promoters have undergone many CRE edits utilizing the *CRISPR/Cas* system (Tang et al., 2020).

10.9.1.1 Disease Resistance

Transcription activator-like (TAL) effectors from *Xanthomonas oryzae* pv. oryzae (Xoo) target regulatory effector binding elements (EBEs) in *OsSWEET* rice gene promoters, crucial for disease resistance. Altering the *OsSWEET11* promoter sequence near the *TATA box* prevents *TAL effector PthXo1 binding*, conferring resistance (Chu et al., 2006; Yang et al., 2006). However, *OsSWEET11* mutants had issues like pollen formation deficiency. Editing the *PthXo1* EBE in the *OsSWEET11* promoter produced fertile, resistant rice (Singh et al., 2021). *CRISPR/Cas9* editing of the pathogen-induced element in the *Xa13 (OsSWEET11)* promoter created *Xoo*-tolerant rice (Singh et al., 2021). Other TAL effectors like *PthXo3*, *AvrXa7*, *TalC*, and *TalF* activate *OsSWEET14*, while *PthXo2* induces *OsSWEET13*. Natural resistance alleles with EBE alterations in *OsSWEET13* and *OsSWEET14* led to broad-spectrum resistant rice without compromising function (Xu et al., 2019). *CRISPR/Cas9* editing of EBEs in *OsSWEET11, OsSWEET13,* and *OsSWEET14* conferred broad-spectrum resistance against all tested *Xoo* strains. Similar strategies in *Citrus sinensis* and Duncan grapefruit improved resistance to citrus canker (Peng et al., 2017). However, deleting *OsSWEET14* or *CsLOB1* may enhance resistance but impact growth negatively (Antony et al., 2010; Jia et al., 2017). Editing the *OsSULTR3;6 promote*r EBE in rice created immunity to *Xoc* strains *Tal5d* and *Tal2g*, demonstrating that altering EBEs can interfere with TAL effector-mediated susceptibility genes (Hummel et al., 2012). Broad-spectrum resistance was achieved by activating R genes with TAL effector nucleases targeting EBEs (Hummel et al., 2012). Thus, *CRISPR/Cas9* and CRE editing of EBEs offer promising strategies for enhancing disease resistance, though trade-offs in growth and development must be considered.

10.9.1.2 Tolerance for Abiotic Stress

Abiotic stresses significantly affect plant development and production. Plants adapt to these stresses by regulating stress-responsive genes, which can be enhanced through CRE editing of key genes. For instance, in rice, the *GT-1* element in the *OsRAV2* promoter is crucial for salt stress response. *CRISPR/Cas9*-mediated editing of this element confirmed its role, with mutants showing impaired responses to salinity while maintaining normal growth under nonstress conditions (Duan et al. 2017). Similarly, in *Arabidopsis*, knockdown of the *ANAC069* gene, which encodes an *NAC* protein that binds to CREs of stress-responsive genes, improves resistance to osmotic and salt stress (Shi et al., 2017). In maize, modifying the *ARGOS8* gene promoter through *CRISPR/Cas9* enhanced drought tolerance and increased grain yield under stress (Shi et al., 2017). Moreover, editing the promoter of the *bZIP68* gene, a negative regulator of cold tolerance in maize, could develop cold-tolerant varieties (Shi et al., 2017). These findings highlight the potential of promoter editing in improving crop resistance to various abiotic stresses.

10.9.1.3 Productivity and Attributes of Excellence

Increasing crop production often involves selecting rare spontaneous mutations in gene regulatory regions. *Cis*-Regulatory alleles generated by *CRISPR/Cas*-mediated CRE editing are used to enhance quantitative traits in breeding. In tomatoes, *CRISPR/Cas9* mutagenesis of the *CLAVATA3* promoter using eight sgRNAs produced a range of fruit sizes and locule numbers, creating new alleles with diverse phenotypic traits (Rodríguez-Leal et al., 2017; Yuste-Lisbona et al., 2020). Additionally, *CRISPR/Cas9* editing of the *SlWUS* gene's *CArG* element in tomatoes resulted in higher locules and larger fruits (Yuste-Lisbona et al., 2020). The insertion of the *CaMV 35S* promoter upstream of the *ANT1* gene increased anthocyanin levels, causing purple plant tissues and fruits (Čermák et al., 2015). In rice, mutations in the Wx gene promoter, containing CREs like the endosperm-box, A-box, and CAAT-box, led to variations in amylose content (Chen et al., 2020). Inserting novel CREs near the TATA-box enhanced amylose content and grain quality (Huang et al., 2020). In barley, *CRISPR/Cas9* mutations in the HvPAPhya gene's CREs (*GCN4*, *Skn1*, and *RY* elements) significantly reduced grain phytase activity (Holmes et al., 2017). These examples illustrate the potential of CRE editing in improving crop traits.

10.9.2 Epigenetic Qualities

Enhancing crop plants through epigenetic modifications, such as histone alterations and DNA methylation, is a promising strategy. Promoters, which regulate gene expression, often harbor DNA methylation and demethylation sites. Modified *CRISPR*/d*Cas9* systems can target these epigenetic regions, facilitating the recruitment of enzymes that induce epigenetic changes (Noman et al., 2016). For instance, a *CRISPR*/d*Cas9* system was developed to target DNA methylation in *Nicotiana tabacum*, silencing the FWA promoter and causing early flowering in *Arabidopsis* (Papikian et al., 2019). Similarly, *CRISPR*/d*Cas9* coupled with AtHAT1 improved drought resistance in *Arabidopsis* by enhancing *AREB1 promoter* activity (Paixão et al. 2019). Additionally, the *H3K27* demethylase *REF6,* which binds to specific DNA motifs, can be targeted using *CRISPR/Cas9* to remove CRE motifs, thereby preventing *REF6* from accessing its target genes (Li et al., 2018). Understanding the mechanisms of epigenetic regulation and stability is crucial for integrating epigenetics into crop improvement (Li et al., 2018). Advanced *CRISPR*/d*Cas9* tools and precise epigenetic profiling techniques can significantly boost agricultural productivity by enabling targeted epigenome editing (Tang et al., 2020).

10.9.3 CRE Editing's Obstacles and Opportunities

Crop biotechnology has advanced through genetic transformation, with *CRISPR/Cas* enabling precise gene edits. Promoter rewriting faces technical challenges but editing CREs can control gene expression, impacting traits like tomato fruit diameter and rice amylose levels (Gao et al., 2021;

Chen et al., 2020). Editing CREs enhances stress tolerance in *Arabidopsis*, rice, and cotton (Cheng et al., 2013). Combining GWAS with *CRISPR/Cas* can accelerate breeding, while multiomics tools help evaluate CRE-edited crops. This approach promises significant advancements in crop productivity, quality, and stress resistance (Yang et al., 2019).

10.10 Methods for Improving Horticultural Crops Through *CRISPR/Cas*-Mediated *Cis*-Engineering Applications

10.10.1 De Novo CRE Identification

The use of *cis*-engineering for crop enhancement benefits from prior knowledge of CREs within the cis-regulatory areas. Transcripts can be submitted to multiple databases, including TRANSFAC, PLACE, Plant CARE, JASPAR Core PLANTAE, Plant TFDB, and Plant Regulomics, to locate a number of previously discovered CREs, including plant promoters containing TFBSs. These regions are able to undergo many validation processes when TFBSs are anticipated (Lin et al., 2020; Tang et al., 2020). The following are some examples of in vitro techniques according to DNA–protein interactions.

The DNA EMSA detects protein–DNA interactions by observing mobility shifts when DNA binds to proteins during electrophoresis. The DNA pull-down assay isolates proteins that bind to specific DNA sequences, while the yeast one-hybrid assay identifies protein–DNA interactions in yeast cells. ChIP methods, such as ChIP with DNA microarray (ChIP-chip) and ChIP sequencing (ChIP-seq), map protein–DNA interactions genome-wide by combining ChIP with microarray technology or high-throughput sequencing. These techniques support *cis*-engineering for crop enhancement by validating TFBSs (Lin et al., 2020).

The discovery of new CREs is crucial, and bioinformatics and experimental approaches, using genomic and transcriptome data, play a vital role (Lin et al., 2020). Sequence conservation among coexpressed, gene families, and orthologous genes aids in CRE identification. Tools like matrix motifs and MEME, along with comparative methods like phylogenetic footprinting and mVISTA, help find common CREs (Lin et al., 2020). For instance, analyzing CLV3 promoters in Solanaceae species has revealed potential CREs (Tang et al., 2020). Experimental methods further characterize tissue-specific and inducible CREs, complementing bioinformatics strategies (Tang et al., 2020).

10.10.2 Selecting the Best Method for *Cis*-Engineering Mediated by CRISPR/Cas

CRISPR/Cas-based technologies provide versatile methods for altering *cis*-regulatory regions to achieve specific objectives. Techniques include creating reduced-function alleles, replacing promoters, inducing indel mutations via NHEJ, and introducing point mutations or CRE exchanges. These methods can modulate gene expression patterns and levels effectively. Examples such as modifying the G-box element in soybeans and introducing the R6 motif in apples and kiwifruit illustrate how these approaches can enhance gene expression related to anthocyanin production and other desirable traits across different horticultural species (Tang et al., 2020).

10.10.3 Editing the Genome Without Transgenes

The preferred method for enhancing crops through *cis*-engineering while ensuring market acceptance is transgene-free GE. This approach avoids integrating *CRISPR/Cas* DNA constructs into the host genome, addressing biosecurity and off-target modification concerns

(Tang et al., 2020). Strategies include herbicide-dependent isolation, programmed self-elimination, and transient expression systems. Transient expression is particularly effective for polyploid and vegetatively propagated crops, minimizing off-target effects and public concerns. Preassembled *Cas*9 protein–gRNA complexes enhance specificity and are used in protoplast transformation across various horticultural crops. Planta transformation methods offer promise by utilizing natural biological processes to deliver edited DNA, exemplified by recent successes in wheat and rice (Tang et al., 2020).

10.11 Opportunities and Challenges

10.11.1 The Intricacy of Horticulture Crops' Genomes

The paragraph discusses the challenges and advancements in using *CRISPR*/*Cas* technologies to edit the genomes of horticultural crops, which vary greatly in genome size and often are polyploid. Editing polyploid genomes requires high efficiency to target multiple alleles simultaneously. Despite these challenges, advancements such as highly active *Cas* nucleases and effective expression frameworks have made *CRISPR*/*Cas* useful in polyploid crops (Tang et al., 2020). An example is the successful editing of all five alleles of the *FaTM6* gene in the octoploid and highly heterozygous cultivated strawberry, Fragaria × ananassa cv. Camarosa, using a dual single-guide RNA *CRISPR/Cas9* method. This was achieved using the reference genome of the diploid wild strawberry *F. vesca* due to the lack of a sequenced genome for Fragaria × ananassa. The paragraph suggests a *CRISPR*/*Cas*-mediated mutagenesis method for use in horticultural crops with high heterozygosity or ploidy or without genome sequence data (Tang et al., 2020).

10.11.2 De Novo High-Throughput Identification of CREs Inside Their Natural Environments

In vitro methods for validating CREs have traditionally been inaccurate and low throughput. Recent innovations like ChIP-seq, DNase-seq, and ATAC-seq have enhanced our understanding of transcriptional evidence but cannot assess CREs in their natural context. To address this, *CRISPR*/*Cas*-based tiling screens were developed to identify functional CREs in mammalian cells by densely mapping gRNAs across a *cis*-regulatory region. Although Rodríguez-Leal et al. demonstrated this method's potential in tomatoes, it has not yet been widely adopted for identifying CREs in plants (Tang et al., 2020).

10.11.3 Accurate and Effective Genome Editing

The paragraph discusses the role of nucleotide-level *cis*-engineering in precise GE, highlighting the use of cytidine and adenine base editors (CBEs and ABEs) for targeted base modifications without causing DSBs. In horticultural crops like tomatoes, potatoes, and watermelons, only CBEs have been effectively used for gene function research. While base editors offer an alternative to HDR-mediated point mutations, introducing or replacing desired CREs in plants remains challenging. Efforts to improve HDR efficiency involve donor configuration and repair pathway manipulation. A recent technique in moss (*Physcomitrella patens*) uses *CRISPR/Cas9* and oligonucleotide templates for efficient and accurate multiplexed GE, which could be beneficial for horticultural crops (Xu et al., 2019).

10.11.4 Modification of the Epigenome

Phenotypic variety in horticulture crops has been limited by the small number of naturally occurring epialleles. Epigenome editing with nuclease-dead *Cas9* has the potential to increase this variety. Some epialleles are transient, while others are heritable over several generations (Brand et al., 1985). However, it is uncertain if these editing-induced modifications can be reliably inherited, and continuous expression of *CRISPR* components might be necessary to maintain traits in progeny, which restricts the technology's use in crop development. Improving this approach involves developing better *CRISPR* tools and identifying beneficial epialleles in horticulture crops (Xu et al., 2019). However, Table 10.1 details various genes and their functions in plant development, which could be targeted for improving crops through *CRISPR/Cas*-mediated *cis*-engineering applications.

Table 10.1 List of targeted genes for improving crops through CRISPR/Cas-mediated *cis*-engineering application.

Genes	Function	Development
Deeper Rooting 1 (DRO1)	It enhances the deeper root systems that improve drought tolerance and nutrient absorption	Enhance drought tolerance in rice by promoting deeper root systems
Xa13	The *Xa13* gene in rice encodes a plasma membrane protein	The *Xa13* gene was identified through genetic mapping and positional cloning
CsLOB1	The *CsLOB1* gene in rice encodes a protein similar to cellulose synthase that is essential to plant growth and cell wall production	It plays a role in the biosynthesis of cell walls and its impact on plant development
Anthocyanin mutant 1 (AN1)	It regulates the biosynthesis of anthocyanin pigments, influencing the coloration of plant tissues such as flowers, fruits, and leaves	It was identified through genetic studies focused on pigmentation mutants, revealing its role in regulating anthocyanin biosynthesis in plants
CLAVATA 3 (CLV3)	It regulates stem cell maintenance and meristem activity in plants	Maintain the stem cell populations and regulate growth patterns in *Arabidopsis thaliana*
Arabidopsis Histone Acetyl Transferase 1 (AtHAT1)	It adapts the histone proteins to regulate chromatin and gene expression dynamics within *A. thaliana*	It regulates chromatin and gene expression structure within *A. thaliana*
OsERF922	In rice, it regulates stress responses and enhances disease resistance	Increase disease resistance
WRKY17	In cotton, it controls reactions to both biotic and abiotic stimuli, enhancing the plant's defense mechanisms and stress tolerance	It contributes to stress response regulation
Cas9	It encodes the *Cas9* protein, an endonuclease that can cut DNA at specific locations	*CRISPR/Cas9* development has transformed genetic engineering by making it possible for accurate and effective genome editing across various organisms
CRISPR RNA (crRNA)	It encodes RNA sequences that guide the *Cas9* protein to specific DNA targets for precise cleavage	The crRNA gene involves harnessing bacterial adaptive immunity from *CRISPR* arrays to guide *Cas* proteins for targeted genome editing

(*Continued*)

Table 10.1 (Continued)

Genes	Function	Development
Trans-activating CRISPR RNA (tracrRNA)	It forms a guide RNA by binding with the crRNA complex, enabling *Cas*9 to precisely target and cleave specific DNA sequences in genome editing using the *CRISPR/Cas9* system	The development of tracrRNA participants understands its role in enhancing *CRISPR/Cas9* functionality by facilitating the formation of guide RNA complexes.
Alliinase	Present in garlic and it encodes an enzyme that converts alliin to allicin, providing garlic's characteristic aroma and antimicrobial properties	In garlic development, the *alliinase* gene facilitates the formation of sulfur-containing compounds that contribute to the plant's defense mechanisms and flavor profile
MdMYB1	Present in apples and it regulates the production of anthocyanin pigments, contributing to red coloration in apple skins and potentially enhancing fruit health	In apples, *MdMYB10* gene expression influences anthocyanin accumulation, impacting fruit coloration
VRN1	It regulates flowering time in plants, particularly in response to vernalization (cold exposure), influencing the transition from vegetative to reproductive growth	*VRN1* gene expression influences flowering time in plants, crucial for determining the timing of reproductive development in response to cold temperatures
Flowering Locus T1 (FT1)	The *FT1* gene promotes flowering in plants by encoding a protein that starts the process of changing from vegetative to reproductive growth by moving from leaves to the apical meristem of the shoot	*FT1* gene expression facilitates the precise timing of flowering in response to environmental cues, crucial for reproductive success in plants
Gibberellin 2-oxidase (GA2ox)	It regulates plant growth by controlling the levels of active gibberellin hormones, influencing processes such as stem elongation and seed germination	*GA2ox* genes regulate plant development by modulating gibberellin levels, affecting growth processes
The CO1	It regulates flowering time in plants by promoting the expression of flowering activators in response to photoperiodic cues	The *CO1* gene coordinates flowering time in plants by sensing day length and activating genes that initiate the change in growth from vegetative to flowering
AcLFY	Inside onions, it regulates floral development by promoting the change in growth from vegetative to floral	The *AcLFY* gene inside onions orchestrates the timing and formation of flowers, crucial for reproductive success and seed production
ASMT	Present in garlic and it catalyzes the formation of allicin and other sulfur compounds, which impart garlic's distinct flavor and contribute to its medicinal properties	The *ASMT* gene in garlic contributes to the biosynthesis of sulfur compounds, crucial for flavor development and potential health benefits during bulb maturation

Table 10.1 (Continued)

Genes	Function	Development
ARGOS8	It regulates the seed size and weight in plants, including soybeans, by modulating cell expansion and growth	It involves its expression in developing seeds, where it modulates cell expansion and growth processes to influence seed size and weight in plants
Maize bZIP68	It is known as a transcription factor implicated in response regulation to environmental stresses and developmental processes in maize plants	Throughout various stages of maize growth, it regulates stress responses and developmental pathways through its expression and activity via transcriptional regulation
ANAC069	As a transcription factor in plants, it regulates stress responses, senescence, and developmental pathways by controlling its downstream target genes	Throughout plant growth stages, its expression and activity contribute to stress responses and regulate senescence
Xa27	It is a resistance gene in rice, providing resistance against bacterial blight caused by *Xanthomonas oryzae* pv. oryzae	Involves its expression and activation inside rice plants, leading to the recognition and defense against *X. oryzae* pv. oryzae
OsSWEET14	It is a sugar transporter in rice, facilitating the efflux of sugars from plant cells to support growth, development, and responses to environmental stresses	It involves its expression and activity across various stages of rice growth
CsLOB1	As a transcription factor in Citrus species, it regulates fruit size and shape by controlling cell division and expansion during fruit development	Its expression and activity during citrus fruit growth modulate cell division and expansion, determining fruit size and shape
TaMYBL1	It is a *MYB* transcription factor in wheat, responsible for controlling several facets of plant development, growth, and stress responses	It involves its expression and activity throughout different growth stages of wheat

10.12 Conclusion

The genetic engineering of CREs using *CRISPR/Cas* technology holds significant promise for enhancing various traits in organisms. This approach allows precise modifications in gene expression, enabling the fine-tuning of biological processes without altering the coding sequences. As a result, improvements in characteristics, for example, disease tolerance, stress tolerance, and productivity, can be achieved with a great degree of specificity and efficiency. *CRISPR/Cas* is a potent implement for developing genetic engineering because of its accuracy and adaptability, offering new possibilities for research and applications in agriculture, medicine, and biotechnology. However, ongoing research is essential to fully understand the implications and to guarantee this technology is used in a morally and safely manner.

References

Acharya, S., Ranjan, R., Pattanaik, S., Maiti, I. B., & Dey, N. (2014). Efficient chimeric plant promoters derived from plant infecting viral promoter sequences. *Planta, 239*, 381–396.

Antony, G., Zhou, J., Huang, S., Li, T., Liu, B., White, F., & Yang, B. (2010). Rice xa13 recessive resistance to bacterial blight is defeated by induction of the disease susceptibility gene Os-11N3. *Plant Cell, 22*, 3864–3876.

Bai, X., Huang, Y., Hu, Y., Liu, H., Zhang, B., Smaczniak, C., Hu, G., Han, Z., & Xing, Y. (2017). Duplication of an upstream silencer of FZP increases grain yield in rice. *Nature Plants, 3*, 885–893.

Brand, A. H., Breeden, L., Abraham, J., Sternglanz, R., & Nasmyth, K. (1985). Characterization of a 'silencer' in yeast: a DNA sequence with properties opposite to those of a transcriptional enhancer. *Cell, 41*, 41–48.

Butardo, V. M. Jr., Anacleto, R., Parween, S., Samson, I., de Guzman, K., Alhambra, C. M., Misra, G., & Sreenivasulu, N. (2017). Systems genetics identifies a novel regulatory domain of amylose synthesis. *Plant Physiology, 173*, 887–906.

Čermák, T., Baltes, N. J., Čegan, R., Zhang, Y., & Voytas, D. F. (2015). High-frequency, precise modification of the tomato genome. *Genome Biology, 16*, 1–15.

Chen, K., Wang, Y., Zhang, R., Zhang, H., & Gao, C. (2019). *CRISPR/Cas* genome editing and precision plant breeding in agriculture. *Annual Review of Plant Biology, 70*, 667–697.

Chen, H., Zeng, Y., Yang, Y., Huang, L., Tang, B., Zhang, H., et al. (2020). Allele-aware chromosome-level genome assembly and efficient transgene-free genome editing for the autotetraploid cultivated alfalfa. *Nature Communications, 11*(1), 2494.

Cheng, M.-C., Liao, P.-M., Kuo, W.-W., & Lin, T.-P. (2013). The Arabidopsis ETHYLENE RESPONSE FACTOR1 regulates abiotic stress-responsive gene expression by binding to different cis-acting elements in response to different stress signals. *Plant Physiology, 162*, 1566–1582.

Chu, Z., Yuan, M., Yao, J., Ge, X., Yuan, B., Xu, C., Li, X., Fu, B., Li, Z., & Bennetzen, J. L. (2006). Promoter mutations of an essential gene for pollen development result in disease resistance in rice. *Genes & Development, 20*, 1250–1255.

Cong, L., Ran, F. A., Cox, D., Lin, S., Barretto, R., Habib, N., Hsu, P. D., Wu, X., Jiang, W., Marraffini, L. A., et al. (2013). Multiplex genome engineering using *CRISPR/Cas* systems. *Science, 339*, 819–823. https://doi.org/10.1126/science.1231143

Deb, D., Shrestha, A., Maiti, I. B., & Dey, N. (2018). Recombinant promoter (MUASCsV8CP) driven totiviral killer protein 4 (KP4) imparts resistance against fungal pathogens in transgenic tobacco. *Frontiers in Plant Science, 9*, 278.

Dey, N., Sarkar, S., Acharya, S., & Maiti, I. B. (2015). Synthetic promoters in planta. *Planta, 242*, 1077–1094.

Doni Jayavelu, N., Jajodia, A., Mishra, A., & Hawkins, R. D. (2020). Candidate silencer elements for the human and mouse genomes. *Nature communications*, 11, 1061

Dong, O. X., Yu, S., Jain, R., et al. (2020). Marker-free carotenoid-enriched rice generated through targeted gene insertion using *CRISPR-Cas9*. *Nature Communications, 11*, 1178.

Du, Y., Liu, L., Peng, Y., Li, M., Li, Y., Liu, D., Li, X., & Zhang, Z. (2020). UNBRANCHED3 expression and inflorescence development is mediated by UNBRANCHED2 and the distal enhancer, KRN4, in maize. *PLoS Genetics, 16*, e1008764.

Duan, P., Xu, J., Zeng, D., Zhang, B., Geng, M., Zhang, G., Huang, K., Huang, L., Xu, R., & Ge, S. (2017). Natural variation in the promoter of GSE5 contributes to grain size diversity in rice. *Molecular Plant, 10*, 685–694.

Ersoz, E. S., Yu, J., & Buckler, E. S. (2007). Applications of linkage disequilibrium and association mapping in crop plants. In Genomics-Assisted Crop Improvement: Vol. 1: Genomics Approaches and Platforms (pp. 97–119). Dordrecht, Springer Netherlands.

Gaj, T., Gersbach, C. A., Barbas, C. F., & III. (2013). ZFN, TALEN, and *CRISPR/Cas*-based methods for genome engineering. *Trends in Biotechnology, 31*, 397–405. https://doi.org/10.1016/j.tibtech.2013.04.004

Guo, Z., Yang, Q., Huang, F., Zheng, H., Sang, Z., Xu, Y., et al. (2021). Development of high-resolution multiple-SNP arrays for genetic analyses and molecular breeding through genotyping by target sequencing and liquid chip. *Plant Communications*, *2*(6). https://doi.org/10.1016/j.xplc.2021.100230

Hayward, A., & Gilbert, C. (2022). Transposable elements. *Current Biology, 32*(17), R904–R909.

He, Y., Zhu, M., Wang, L., Wu, J., Wang, Q., Wang, R., & Zhao, Y. (2018). Programmed self-elimination of the CRISPR/Cas9 construct greatly accelerates the isolation of edited and transgene-free rice plants. *Molecular Plant, 11*(9), 1210–1213.

Hille, F., Richter, H., Wong, S. P., Bratovič, M., & Ressel, S. (2018). The biology of *CRISPR-Cas*: backward and forward. *Cell, 172*, 1239–1259. [PubMed] [Google Scholar]

Holmes, I. B., Wendt, T., Gil-Humanes, J., Deleuran, L. C., Starker, C. G., Voytas, D. F., & Brinch-Pedersen, H. (2017). Evaluation of the mature grain phytase candidate HvPAPhy_a gene in barley (Hordeum vulgare L.) using CRISPR/Cas9 and TALENs. *Plant Molecular Biology, 95*, 111–121.

Hsu, P. D., Lander, E. S., & Zhang, F. (2014). Development and applications of *CRISPR-Cas*9 for genome engineering. *Cell, 157*, 1262–1278. https://doi.org/10.1016/j.cell.2014.05.010

Huang, H. Y., Lin, Y. C. D., Li, J., Huang, K. Y., Shrestha, S., Hong, H. C., et al. (2020). miRTarBase 2020: updates to the experimentally validated microRNA–target interaction database. *Nucleic Acids Research, 48*(D1), D148–D154.

Hummel, A. W., Doyle, E. L., & Bogdanove, A. J. (2012). Addition of transcription activator-like effector binding sites to a pathogen strain-specific rice bacterial blight resistance gene makes it effective against additional strains and against bacterial leaf streak. *New Phytologist, 195*, 883–893.

Jia, H., Zhang, Y., Orbovíc, V., Xu, J., White, F. F., Jones, J. B., & Wang, N. (2017). Genome editing of the disease susceptibility gene CsLOB1 in citrus confers resistance to citrus canker. *Plant Biotechnology Journal, 15*, 817–823.

Jia, H., Omar, A. A., Orbovíc, V., & Wang, N. (2022). Biallelic editing of the LOB1 promoter via *CRISPR/Cas9* creates canker-resistant 'Duncan' grapefruit. *Phytopathology, 112*, 308–314.

Joung, J. K., & Sander, J. D. (2013). TALENs: a widely applicable technology for targeted genome editing. *Nature Reviews Molecular Cell Biology, 14*, 49–55. https://doi.org/10.1038/nrm3486 [PMC free article] [PubMed] [CrossRef] [Google Scholar]

Karkute, S. G., Singh, A. K., Gupta, O. P., Singh, P. M., & Singh, B. (2017). *CRISPR/Cas9* mediated genome engineering for improvement of horticultural crops. *Frontiers in Plant Science, 8*, 1635.

Knott, G. J., & Doudna, J. A. (2018). *CRISPR-Cas* guides the future of genetic engineering. *Science, 361*, 866–869. [PMC free article] [PubMed] [Google Scholar]

Koltun, A., Corte, L. E.-D., Mertz-Henning, L. M., & Gonçalves, L. S. (2018). Genetic improvement of horticultural crops mediated by *CRISPR/Cas*: a new horizon of possibilities. *Horticultura Brasileira, 36*, 290–298.

Komor, A. C., Badran, A. H., & Liu, D. R. (2017). *CRISPR*-based technologies for the manipulation of eukaryotic genomes. *Cell, 169*, 559.

Kumar, S. (2019). Epigenomics for crop improvement: current status and future perspectives. *Journal of Genetics and Cell Biology, 3*, 128–134.

Kumar, S., AlAbed, D., Whitteck, J. T., Chen, W., Bennett, S., Asberry, A., Wang, X., DeSloover, D., Rangasamy, M., & Wright, T. R. (2015). A combinatorial bidirectional and bicistronic approach for coordinated multi-gene expression in corn. *Plant Molecular Biology*, *87*, 341–353.

Lambert, S. A., Jolma, A., Campitelli, L. F., Das, P. K., Yin, Y., Albu, M., et al. (2018). The human transcription factors. *Cell*, *172*(4), 650–665.

Langner, T., Kamoun, S., & Belhaj, K. (2018). *CRISPR* crops: plant genome editing toward disease resistance. *Annual Review of Phytopathology*, *56*, 479–512. [PubMed] [Google Scholar]

Lee, J. E., Neumann, M., Duro, D. I., & Schmid, M. (2019). *CRISPR*-based tools for targeted transcriptional and epigenetic regulation in plants. *PLoS One*, *14*, e0222778.

Li, Y., Fan, C., Xing, Y., Yun, P., Luo, L., Yan, B., Peng, B., Xie, W., Wang, G., & Li, X. (2014). Chalk5 encodes a vacuolar H^+-translocating pyrophosphatase influencing grain chalkiness in rice. *Nature Genetics*, *46*, 398–404.

Li, C., Chen, C., Chen, H., Wang, S., Chen, X., & Cui, Y. (2018). Verification of DNA motifs in Arabidopsis using *CRISPR/Cas9*-mediated mutagenesis. *Plant Biotechnology Journal*, *16*, 1446–1451.

Li, C., et al. (2019). A new rice breeding method: *CRISPR/Cas9* system editing of the Xa13 promoter to cultivate transgene-free bacterial blight-resistant rice. *Plant Biotechnology Journal*, *18*, 313–315.

Li, T., Yang, Y., Qi, H., Cui, W., Zhang, L., Fu, X., et al. (2023). CRISPR/Cas9 therapeutics: progress and prospects. *Signal Transduction and Targeted Therapy*, *8*(1), 36.

Liang, Z., Chen, K., Li, T., Zhang, Y., Wang, Y., Zhao, Q., et al. (2017). Efficient DNA-free genome editing of bread wheat using CRISPR/Cas9 ribonucleoprotein complexes. *Nature Communications*, *8*(1), 14261.

Lin, Q., Zong, Y., Xue, C., Wang, S., Jin, S., Zhu, Z., Wang, Y., Anzalone, A. V., Raguram, A., & Doman, J. L. (2020). Prime genome editing in rice and wheat. *Nature Biotechnology*, *38*, 582–585.

Lino, C. A., Harper, J. C., Carney, J. P., & Timlin, J. A. (2018). Delivering *CRISPR*: a review of the challenges and approaches. *Drug Delivery*, *25*, 1234–1257. https://doi.org/10.1080/10717544.2018.1474964 [PMC free article] [PubMed] [CrossRef] [Google Scholar]

Liu, X., Zheng, S., Tian, S., Si, Y., Ma, S., Ling, H. Q., & Niu, J. (2024). Natural variant of Rht27, a dwarfing gene, enhances yield potential in wheat. *Theoretical and Applied Genetics*, *137*(6), 1–13.

Lu, Z., Hofmeister, B. T., Vollmers, C., DuBois, R. M., & Schmitz, R. J. (2017). Combining ATAC-seq with nuclei sorting for discovery of cis-regulatory regions in plant genomes. *Nucleic Acids Research*, *45*, e41.

Ma, X., Feng, F., Wei, H., Mei, H., Xu, K., Chen, S., Li, T., Liang, X., Liu, H., & Luo, L. (2016). Genome-wide association study for plant height and grain yield in rice under contrasting moisture regimes. *Frontiers in Plant Science*, *7*, 1801.

Ma, L., Qing, C., Zhang, M., Zou, C., Pan, G., & Shen, Y. (2021). GWAS with a PCA uncovers candidate genes for accumulations of microelements in maize seedlings. *Physiologia Plantarum*, *172*, 2170–2180.

Mao, H., Li, S., Chen, B., Jian, C., Mei, F., Zhang, Y., Li, F., Chen, N., Li, T., & Du, L. (2022). Variation in cis-regulation of a NAC transcription factor contributes to drought tolerance in wheat. *Molecular Plant*, *15*, 276–292.

Miller, J. C., Holmes, M. C., Wang, J., Guschin, D. Y., & Lee, Y. L. (2007). An improved zinc-finger nuclease architecture for highly specific genome editing. *Nature Biotechnology*, *25*, 778–785. [PubMed] [Google Scholar]

Minkenberg, B., Wheatley, M., & Yang, Y. (2017). *CRISPR/Cas9*-enabled multiplex genome editing and its application. *Progress in Molecular Biology and Translational Science*, *149*, 111–132. [PubMed] [Google Scholar]

Mushtaq, M., Pir, H. A., Zaid, A., & Wani, S. H. (2021). CRISPR/Cas system: A powerful approach for enhanced resistance against rice blast. In: *Biocontrol Agents and Secondary Metabolites*, Woodhead Publishing, 649–658.

Noman, A., Aqeel, M., & He, S. (2016). CRISPR-Cas9: tool for qualitative and quantitative plant genome editing. *Frontiers in Plant Science*, *7*, 1740.

Paixão, J. F. R., Gillet, F.-X., Ribeiro, T. P., Bournaud, C., Lourenço-Tessutti, I. T., Noriega, D. D., Melo, B. P., de Almeida-Engler, J., & Grossi-de-Sa, M. F. (2019). Improved drought stress tolerance in Arabidopsis by *CRISPR*/d*Cas*9 fusion with a histone AcetylTransferase. *Scientific Reports*, *9*, 1–9.

Pandiarajan, A., Kamaraj, R., Vasudevan, S., & Vasudevan, S. (2018). OPAC (orange peel activated carbon) derived from waste orange peel for the adsorption of chlorophenoxyacetic acid herbicides from water: adsorption isotherm, kinetic modelling and thermodynamic studies. *Bioresource Technology*, *261*, 329–341.

Papikian, A., Liu, W., Gallego-Bartolomé, J., & Jacobsen, S. E. (2019). Site-specific manipulation of Arabidopsis loci using CRISPR-Cas9 SunTag systems. *Nature Communications*, *10*(1), 729.

Parry, M. A., Madgwick, P. J., Bayon, C., Tearall, K., Hernandez-Lopez, A., Baudo, M., Rakszegi, M., Hamada, W., Al-Yassin, A., Ouabbou, H., et al. (2009). Mutation discovery for crop improvement. *Journal of Experimental Botany*, *60*, 2817–2825. https://doi.org/10.1093/jxb/erp189

Peng, A., Chen, S., Lei, T., Xu, L., He, Y., Wu, L., Yao, L., & Zou, X. (2017). Engineering canker-resistant plants through *CRISPR/Cas9*-targeted editing of the susceptibility gene CsLOB1 promoter in citrus. *Plant Biotechnology Journal*, *15*, 1509–1519.

Potenza, C., Aleman, L., & Sengupta-Gopalan, C. (2004). Targeting transgene expression in research, agricultural, and environmental applications: promoters used in plant transformation. *In Vitro Cellular and Developmental Biology*, *40*, 1–22.

Qian, W., Wu, C., Fu, Y., Hu, G., He, Z., & Liu, W. (2017). Novel rice mutants overexpressing the brassinosteroid catabolic gene CYP734A4. *Plant Molecular Biology*, *93*, 197–208.

Ran, F. A., Hsu, P. D., Wright, J., Agarwala, V., & Scott, D. A. (2013). Genome engineering using the *CRISPR-Cas*9 system. *Nature Protocols*, *8*, 2281–2308. [PMC free article] [PubMed] [Google Scholar]

Ricroch, A., Clairand, P., & Harwood, W. (2017). Use of *CRISPR* systems in plant genome editing: toward new opportunities in agriculture. *Emerging Topics in Life Sciences*, *1*, 169–182. [PMC free article] [PubMed] [Google Scholar]

Rodríguez-Leal, D., Lemmon, Z. H., Man, J., Bartlett, M. E., & Lippman, Z. B. (2017). Engineering quantitative trait variation for crop improvement by genome editing. *Cell*, *171*, 470–480. https://doi.org/10.1016/j.cell.2017.08.030 [PubMed] [CrossRef] [Google Scholar]

Ruan, B., Shang, L., Zhang, B., Hu, J., Wang, Y., Lin, H., Zhang, A., Liu, C., Peng, Y., & Zhu, L. (2020). Natural variation in the promoter of TGW2 determines grain width and weight in rice. *New Phytologist*, *227*, 629–640.

Rushton, P. J., Reinstadler, A., Lipka, V., Lippok, B., & Somssich, I. E. (2002). Synthetic plant promoters containing defined regulatory elements provide novel insights into pathogen-and wound-induced signaling. *Plant Cell*, *14*, 749–762.

Sadhukhan, A., Kobayashi, Y., Nakano, Y., Iuchi, S., Kobayashi, M., Sahoo, L., & Koyama, H. (2017). Genome-wide association study reveals that the aquaporin NIP1; 1 contributes to variation in hydrogen peroxide sensitivity in *Arabidopsis thaliana*. *Molecular Plant*, *10*, 1082–1094.

Shi, J., Habben, J. E., Archibald, R. L., Drummond, B. J., Chamberlin, M. A., Williams, R. W., Lafitte, H. R., & Weers, B. P. (2015). Overexpression of ARGOS genes modifies plant sensitivity to ethylene, leading to improved drought tolerance in both Arabidopsis and maize. *Plant Physiology*, *169*, 266–282.

Shi, J., Gao, H., Wang, H., Lafitte, H. R., Archibald, R. L., Yang, M., Hakimi, S. M., Mo, H., & Habben, J. E. (2017). ARGOS8 variants generated by *CRISPR-Cas*9 improve maize grain yield under field drought stress conditions. *Plant Biotechnology Journal*, *15*, 207–216.

Singh, J., James, D., Murali Achary, V. M., Patel, M. K., Thakur, J. K., Reddy, M. K., & Tripathy, B. C. (2021). Coordinated overexpression of OsSUT1, OsSWEET11 and OsSWEET14 in rice impairs carbohydrate metabolism that has implications in plant growth, yield and susceptibility to Xanthomonas oryzae pv oryzae (Xoo). bioRxiv, 2021-01.

Sun, Y., Shi, Y., Liu, G., Yao, F., Zhang, Y., Yang, C., Guo, H., Liu, X., Jin, C., & Luo, J. (2020). Natural variation in the OsbZIP18 promoter contributes to branched-chain amino acid levels in rice. *New Phytologist, 228*, 1548–1558.

Swinnen, G., Goossens, A., & Pauwels, L. (2016). Lessons from domestication: targeting cis-regulatory elements for crop improvement. *Trends in Plant Science, 21*, 506–515.

Tang, W., Ye, J., Yao, X., Zhao, P., Xuan, W., Tian, Y., Zhang, Y., Xu, S., An, H., & Chen, G. (2019). Genome-wide associated study identifies NAC42-activated nitrate transporter conferring high nitrogen use efficiency in rice. *Nature Communications, 10*, 1–11.

Tang, X., Sretenovic, S., Ren, Q., Jia, X., Li, M., Fan, T., Yin, D., Xiang, S., Guo, Y., & Liu, L. (2020). Plant prime editors enable precise gene editing in rice cells. *Molecular Plant, 13*, 667–670.

Varshney, R. K., Ribaut, J.-M., Buckler, E. S., Tuberosa, R., Rafalski, J. A., & Langridge, P. (2012). Can genomics boost productivity of orphan crops? *Nature Biotechnology, 30*, 1172–1176.

Vats, S., Kumawat, S., Kumar, V., Patil, G. B., Joshi, T., Sonah, H., Sharma, T. R., & Deshmukh, R. (2019). Genome editing in plants: exploration of technological advancements and challenges. *Cells, 8*, 1386.

Venter, M. (2007). Synthetic promoters: genetic control through cis engineering. *Trends in Plant Science, 12*, 118–124.

Wang, S., Li, S., Liu, Q., Wu, K., Zhang, J., Wang, S., Wang, Y., Chen, X., Zhang, Y., & Gao, C. (2015). The OsSPL16-GW7 regulatory module determines grain shape and simultaneously improves rice yield and grain quality. *Nature Genetics, 47*, 949–954.

Wang, H., La Russa, M., & Qi, L. S. (2016). *CRISPR/Cas9* in genome editing and beyond. *Annual Review of Biochemistry, 85*, 227–264. [PubMed] [Google Scholar]

Wang, D., Yu, K., Jin, D., Sun, L., Chu, J., Wu, W., Xin, P., Gregová, E., Li, X., & Sun, J. (2020). Natural variations in the promoter of Awn Length Inhibitor 1 (ALI-1) are associated with awn elongation and grain length in common wheat. *The Plant Journal, 101*, 1075–1090.

Wang, Y., Shi, C., Ge, P., Li, F., Zhu, L., Wang, Y., et al. (2023). A 21-bp InDel in the promoter of STP1 selected during tomato improvement accounts for soluble solid content in fruits. *Horticulture Research, 10*(3), uhad009.

Wittkopp, P. J., & Kalay, G. (2011). Cis-regulatory elements: molecular mechanisms and evolutionary processes underlying divergence. *Nature Reviews Genetics, 13*, 59–69.

Wright, A. V., Nuñez, J. K., & Doudna, J. A. (2016). Biology and applications of *CRISPR* systems: harnessing nature's toolbox for genome engineering. *Cell, 164*, 29–44. [PubMed] [Google Scholar]

Xu, Z., Xu, X., Gong, Q., Li, Z., Li, Y., Wang, S., Yang, Y., Ma, W., Liu, L., & Zhu, B. (2019). Engineering broad-spectrum bacterial blight resistance by simultaneously disrupting variable TALE-binding elements of multiple susceptibility genes in rice. *Molecular Plant, 12*, 1434–1446.

Yang, B., Sugio, A., & White, F. F. (2006). Os8N3 is a host disease-susceptibility gene for bacterial blight of rice. *Proceedings of the National Academy of Sciences, 103*, 10503–10508.

Yang, M., Lu, K., Zhao, F.-J., Xie, W., Ramakrishna, P., Wang, G., Du, Q., Liang, L., Sun, C., & Zhao, H. (2018). Genome-wide association studies reveal the genetic basis of ionomic variation in rice. *Plant Cell, 30*, 2720–2740.

Yang, Y., Guo, M., Sun, S., Zou, Y., Yin, S., Liu, Y., Tang, S., Gu, M., Yang, Z., & Yan, C. (2019). Natural variation of OsGluA2 is involved in grain protein content regulation in rice. *Nature Communications, 10*, 1–12.

Yang, Y., Kong, Q., Tee, W. T., Li, Y., Low, P. M., Patra, B., et al. (2023). Transcription factor bZIP52 modulates Arabidopsis seed oil biosynthesis through interaction with WRINKLED1. *Plant Physiology, 192*(4), 2628–2639.

Ye, J., Wang, X., Hu, T., Zhang, F., Wang, B., Li, C., Yang, T., Li, H., Lu, Y., & Giovannoni, J. J. (2017). An InDel in the promoter of Al-ACTIVATED MALATE TRANSPORTER9 selected during tomato domestication determines fruit malate contents and aluminum tolerance. *Plant Cell*, *29*, 2249–2268.

Ye, J., Li, W., Ai, G., Li, C., Liu, G., Chen, W., Wang, B., Wang, W., Lu, Y., & Zhang, J. (2019). Genome-wide association analysis identifies a natural variation in basic helix-loop-helix transcription factor regulating ascorbate biosynthesis via d-mannose/l-galactose pathway in tomato. *PLoS Genetics*, *15*, e1008149.

Yuan, J., Wang, X., Zhao, Y., Khan, N. U., Zhao, Z., Zhang, Y., Wen, X., Tang, F., Wang, F., & Li, Z. (2020). Genetic basis and identification of candidate genes for salt tolerance in rice by GWAS. *Scientific Reports*, *10*, 1–9.

Yuste-Lisbona, F. J., Fernandez-Lozano, A., Pineda, B., Bretones, S., Ortíz-Atienza, A., García-Sogo, B., Müller, N. A., Angosto, T., Capel, J., & Moreno, V. (2020). ENO regulates tomato fruit size through the floral meristem development network. *Proceedings of the National Academy of Sciences*, *117*, 8187–8195.

Zafar, K., Sedeek, K. E., Rao, G. S., Khan, M. Z., Amin, I., Kamel, R., Mukhtar, Z., Zafar, M., Mansoor, S., & Mahfouz, M. M. (2020). Genome editing technologies for rice improvement: progress, prospects, and safety concerns. *Frontiers in Genome Editing*, *2*, 5.

Zhan, C., Hu, J., Pang, Q., Yang, B., Cheng, Y., Xu, E., Zhu, P., Li, Y., Zhang, H., & Cheng, J. (2019). Genome-wide association analysis of panicle exsertion and uppermost internode in rice (*Oryza sativa* L.). *Rice*, *12*, 1–13.

Zhang, J. Z., Creelman, R. A., & Zhu, J.-K. (2004). From laboratory to field. Using information from Arabidopsis to engineer salt, cold, and drought tolerance in crops. *Plant Physiology*, *135*, 615–621.

Zhang, F., Wen, Y., & Guo, X. (2014). *CRISPR/Cas9* for genome editing: progress, implications and challenges. *Human Molecular Genetics*, *23*, R40–R46. [PubMed] [Google Scholar]

Zhang, F., Hu, Z., Wu, Z., Lu, J., Shi, Y., Xu, J., Wang, X., Wang, J., Zhang, F., & Wang, M. (2021). Reciprocal adaptation of rice and *Xanthomonas oryzae* pv. oryzae: cross-species 2D GWAS reveals the underlying genetics. *Plant Cell*, *33*, 2538–2561.

Zhao, Q., Feng, Q., Lu, H., Li, Y., Wang, A., Tian, Q., Zhan, Q., Lu, Y., Zhang, L., & Huang, T. (2018). Pan-genome analysis highlights the extent of genomic variation in cultivated and wild rice. *Nature Genetics*, *50*, 278–284.

Zhao, H., Zhang, W., Zhang, T., Lin, Y., Hu, Y., Fang, C., & Jiang, J. (2020). Genome-wide MNase hypersensitivity assay unveils distinct classes of open chromatin associated with H3K27me3 and DNA methylation in *Arabidopsis thaliana*. *Genome Biology*, *21*, 1–17.

Zhao, Y., Wu, L., Fu, Q., et al. (2021). INDITTO2 transposon conveys auxin-mediated DRO1 transcription for rice drought avoidance. *Plant, Cell & Environment*, *44*, 1846–1857.

Zheng, H., Lei, Y., Lin, S., Zhang, Q., & Zhang, Z. (2011). Bidirectionalization of a methyl jasmonate-inducible plant promoter. *Biotechnology Letters*, *33*, 387–393.

Zhu, M., Zhu, B., Wang, Y. H., Wu, Y., Xu, L., Guo, L. P., et al. (2013). Linkage disequilibrium estimation of Chinese beef Simmental cattle using high-density SNP panels. *Asian-Australasian Journal of Animal Sciences*, *26*(6), 772.

Zhu, Z., Yin, J., Chern, M., Zhu, X., Yang, C., He, K., Liu, Y., He, M., Wang, J., & Song, L. (2020). New insights into bsr-d1-mediated broad-spectrum resistance to rice blast. *Molecular Plant Pathology*, *21*, 951–960.

Zilberman, D., Gehring, M., Tran, R. K., Ballinger, T., & Henikoff, S. (2007). Genome-wide analysis of *Arabidopsis thaliana* DNA methylation uncovers an interdependence between methylation and transcription. *Nature Genetics*, *39*, 61–69.

11

Harnessing CRISPR Technology to Edit Noncoding RNAs for Enhanced Disease Resistance in Crops

Abhik Sarkar[1,2]*, Subham Ghosh*[1,2]*, Dipro Sinha*[1]*, Parinita Das*[1,2]*, Sneha Murmu*[1]*, and Himanshushekhar Chaurasia*[3]

[1] *Division of Agricultural Bioinformatics, ICAR-Indian Agricultural Statistics Research Institute, New Delhi, Delhi, India*
[2] *Discipline of Bioinformatics, The Graduate School, ICAR-Indian Agricultural Research Institute, New Delhi, Delhi, India*
[3] *Mechanical Processing Division, ICAR-Central Institute for Research on Cotton Technology, Mumbai, Maharashtra, India*

11.1 Introduction

Genome editing involves modifying the genomic DNA at a specific target spot across a range of cell types and creatures. This might involve the insertion, deletion, and replacement of DNA, which can lead to the acquisition of novel genetic features, the silencing of target genes, and the repair of dangerous gene mutations (Xu & Li, 2020). Recent developments in programmable nuclease-based genome-editing technologies have significantly increased our capacity to make precise modifications to eukaryotic cells' genomes (Cox et al., 2015; Wang et al., 2016b). Currently, there are three widely used methods for editing genomes: RNA-guided CRISPRs (clustered regularly interspaced short palindromic repeats)-Cas (CRISPR-associated) systems' nucleases, transcription activator-like effector nucleases (TALENs), and zinc finger nucleases (ZFNs) (Miller et al., 2007; Cox et al., 2015; Wang et al., 2016b; Figures 11.1a–c). CRISPR-Cas systems (the most popular genome-editing tool in molecular biology labs worldwide) are utilized because of their straightforward design, low cost, high efficiency, good repeatability, and quick cycle (Komor et al., 2017).

An RNA-mediated adaptive immune system called CRISPR-Cas defends against mobile genomes (Charpentier & Marraffini, 2014). Most bacteria and archaea have an adaptive defense system called CRISPR-Cas, which keeps them safe from phage, viruses, and other foreign genetic materials (Deveau et al., 2010). Two fundamental elements of the CRISPR/Cas9 gene-editing technique are a guide RNA that matches a desired target gene and CRISPR-associated protein 9 (Cas9), an endonuclease that breaks double strands of DNA to allow alterations to the genome (Redman et al., 2016a; Figure 11.2).

The term CRISPRs describes bacterial genomic sequences. They provide defense against invasive viruses when paired with a set of proteins called CRISPR-associated (Cas). One of the related proteins, Cas9, is an endonuclease that splits DNA into its two strands. A segment of RNA directs Cas9 to its target. This is synthesized as a synthetic single-guide RNA (sgRNA), a single strand of RNA with 18–20 nucleotides in the part that binds to the genomic DNA. The protospacer adjacent motif (PAM) is a specific DNA sequence that must be at the 3′ end of the guide RNA in order for it to cut. The exact length, which ranges from 2 to 5 nucleotides, depends on the bacteria that generates the

Genome and Epigenome Editing for Stress-Tolerant Crops, First Edition. Edited by Jen-Tsung Chen.

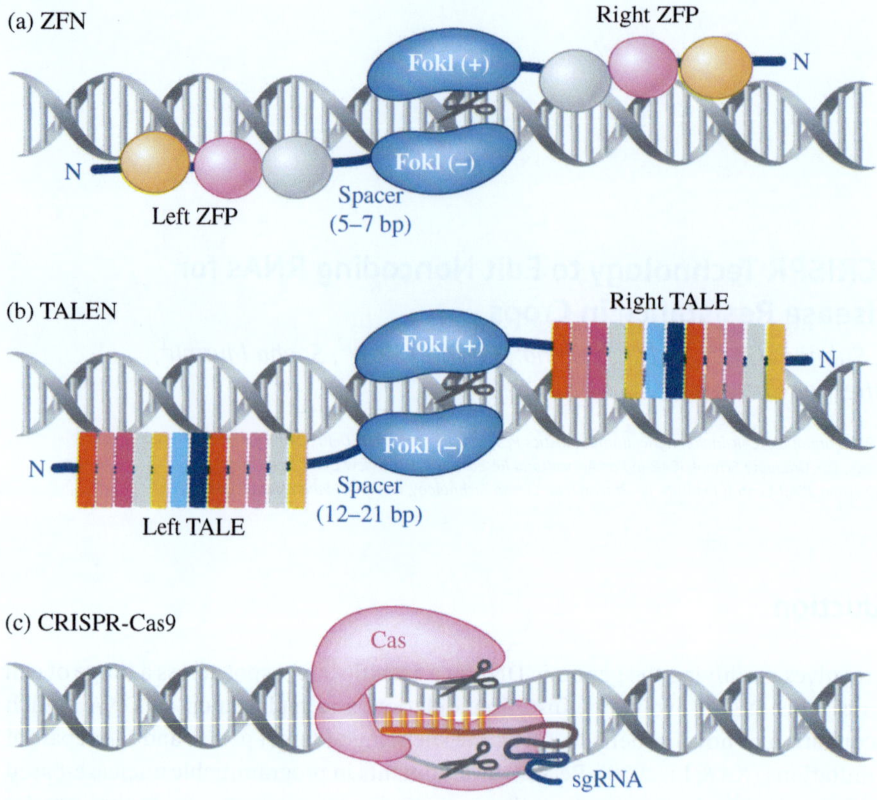

Figure 11.1 Three classes of programmable nucleases. (a) ZFN, (b) TALEN, and (c) CRISPR-Cas9. *Source:* Langner et al. (2018) / with permission of Annual Reviews.

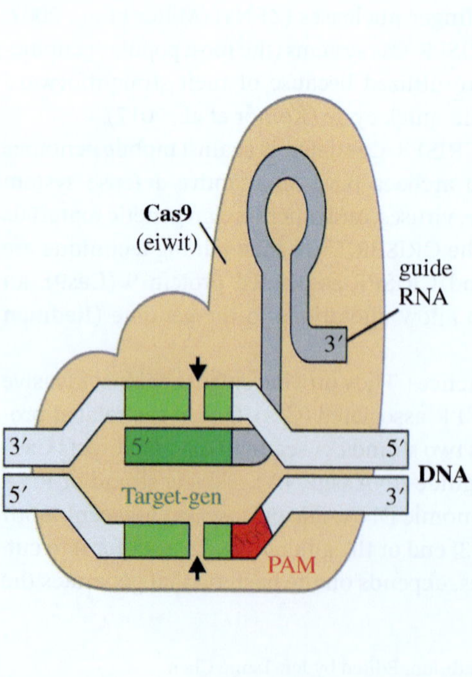

Figure 11.2 The CRISPR/Cas9 system. *Source:* Redman et al. (2016b) / CC BY 4.0 / Walter de Gruyter GmbH.

Cas9 enzyme. After a DNA break, two main repair mechanisms can occur: homology-directed repair, which relies on a homologous DNA fragment as a repair template, or nonhomologous end joining, which often leads to random insertions or deletions. The latter enables precise genome editing. By using the Cas9 nuclease and single guide RNA (sgRNA), a homologous DNA template containing the desired sequence change can be introduced, allowing for precise modifications, even down to a single base pair.

Noncoding RNAs (ncRNAs) are RNA molecules that are transcribed from DNA but do not encode proteins. Instead, they play key roles in regulating gene expression and various cellular processes. Besides CRISPR-Cas, ncRNAs also play a crucial role in gene regulation by controlling gene expression at multiple levels, despite not being translated into proteins. Key types of ncRNAs include the following:

- ✓ **MicroRNAs (miRNAs)**: These small ncRNAs (about 22 nucleotides) bind to messenger RNAs (mRNAs) to either degrade them or block their translation, thereby reducing gene expression.
- ✓ **Long Noncoding RNAs (lncRNAs)**: These are longer transcripts that regulate gene expression through various mechanisms, such as chromatin remodeling, transcriptional interference, and posttranscriptional regulation.
- ✓ **Small Interfering RNAs (siRNAs)**: siRNAs are involved in the RNA interference (RNAi) pathway, where they guide the degradation of specific mRNA targets, silencing gene expression.
- ✓ **Piwi-Interacting RNAs (piRNAs)**: Primarily found in germ cells, piRNAs protect the genome from transposable elements (TEs) by silencing these elements through DNA methylation and chromatin modification.
- ✓ **Circular RNAs (circRNAs)**: These are stable, covalently closed RNA molecules that can act as "sponges" for miRNAs, thereby preventing miRNAs from repressing their target mRNAs.

Overall, ncRNAs contribute to various biological processes, including development, cell differentiation, and stress responses, by fine-tuning the regulation of genes at the transcriptional and posttranscriptional levels.

11.2 A Brief Overview of CRISPR/Cas Technology

11.2.1 History and Development

DNA sequences from *Escherichia coli* bacteria contained CRISPRs, which were initially identified and published in 1987 (Ishino et al., 1987). In 1995, they were also discovered in archaea. Scientists first proposed using the information encoded in CRISPR loci for genotyping different strains of bacteria in medical research, first on *Mycobacterium tuberculosis* (Groenen et al., 1993) and then on *Streptococcus pyogenes* (Hoe et al., 1999), even though the biological function of the CRISPR system had not yet been fully understood. It found that CRISPR loci in distinct strains of the same pathogenic bacterium exhibited a significant degree of polymorphism, making it possible to identify bacterial strains in clinical settings. Francisco Mojica of the University of Alicante (Spain) discovered comparable structures in the archaeal genome of *Haloferax mediterranei* in 1995, which marked a significant advancement in our understanding of the biological role of CRISPR loci (Mojica et al., 1993).

The study of two French food scientists, Rodolphe Barrangou and Philippe Horvath, provided the first experimental understanding of the functioning of the CRISPR system in 2007. They were employed by the Danish corporation Danisco and were researching *Streptococcus thermophilus*, a microbe used in yogurt cultures (Barrangou et al., 2007). Due to the company's extensive collection of bacterial strains dating back to the 1980s, researchers were able to track how bacteria acquired new spacers at the CRISPR locus in response to bacteriophage attacks. In *S. thermophilus*, the addition of these new spacers resulted in immunity to the corresponding phages. This discovery led to one of the first patents in the CRISPR field and, in 2005, Danisco began using CRISPR-based technology to "vaccinate" bacterial cultures against phage infections (De Leo, 2022).

The present nomenclature for CRISPR was proposed by Mojica in correspondence between the two cooperating research organizations and was first published in 2002 (Jansen et al., 2002). The existence of genes linked to CRISPRs (dubbed by the authors cas1–4, CRISPR-associated genes) was initially reported. Researchers discovered genes located near the CRISPR loci in various prokaryotes, two of which had motifs typical of helicases and nucleases (Gostimskaya, 2022). This supported the hypothesis that the Cas genes were not randomly associated with the CRISPR locus and played a role in DNA metabolism. In 2002, a team led by Eugene Koonin at NCBI (Bethesda, United States) also identified these neighboring genes but did not link them to CRISPR arrays at the time. Since the initial discovery, a vast diversity of CRISPR-associated genes, including helicases, nucleases, polymerases, and others, has been found in prokaryotic cells. These proteins are classified into two main groups: the adaptive module, responsible for acquiring immunity (with Cas1 and Cas2 as key players), and the effector module, which recognizes and cleaves mobile genetic elements. Additional regulatory proteins are also associated with the system.

CRISPR-Cas systems are currently classified into two classes and six types, with numerous subtypes. As of the review (Makarova et al., 2019), over 30 subtypes had been identified (Figure 11.2). The key difference between the two classes lies in their effector modules: Class 1 systems use a complex of multiple proteins, while Class 2 systems rely on a single multidomain protein, such as Cas9, Cas12, or Cas13 (Makarova et al., 2013; Figure 11.3).

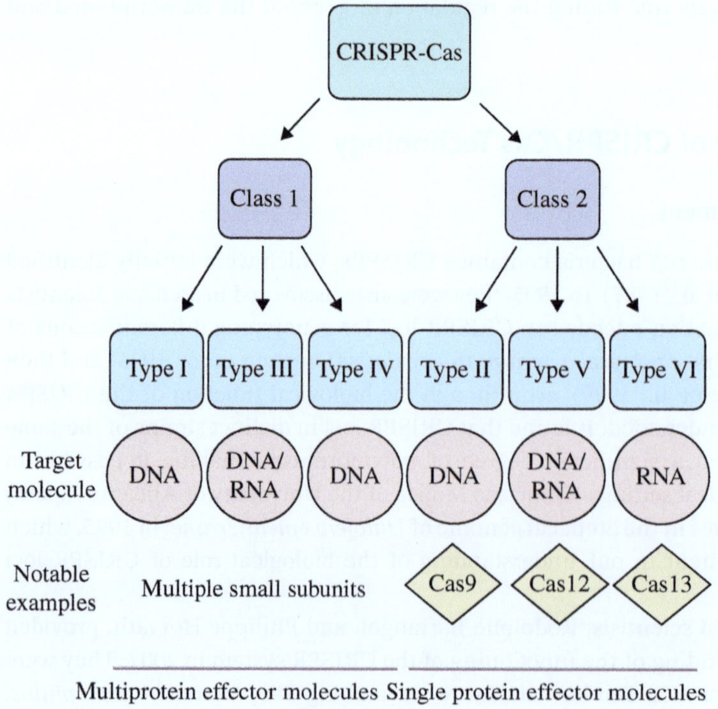

Figure 11.3 Classification of known CRISPR-Cas systems. *Source:* Gostimskaya (2022) / with permission of Springer Nature.

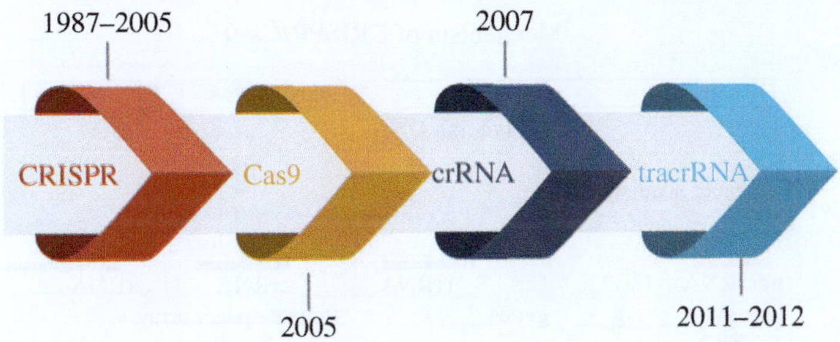

Figure 11.4 A historical chronology of the CRISPR-Cas9 system's component discoveries (Deltcheva et al., 2011).

The Cas9 nuclease is one of the proteins that make up the so-called genetic scissors, a system of directed cutting of foreign DNA (and, as was later discovered, RNA) that is the subject of extensive research among all known Cas proteins. This protein was first described in connection with its association with CRISPRs (Bolotin et al., 2005) where it was originally named Cas5 (other alternative names are Csn1 and Csx12).

A key component of the CRISPR/Cas9 system is a short RNA molecule, called CRISPR RNA (crRNA), which guides the immune proteins of prokaryotes to target foreign genetic material. This crRNA is derived from the transcription of the CRISPR locus, initially produced as a precursor molecule (pre-crRNA) containing multiple spacers and repeats, which is later cleaved into individual crRNA fragments. John van der Oost's team (Wageningen University, Netherlands) first described crRNA, while Virginijus Siksnys' group (Vilnius University, Lithuania) demonstrated that a 20-base pair sequence within the crRNA, complementary to the target DNA, is necessary and sufficient for Cas9 nuclease activity. This study was one of two independent investigations that explained how Cas9 uses crRNA to target foreign DNA.

The last component needed to put together a functional CRISPR/Cas9 system in vitro was found to be a different short RNA molecule that Emmanuelle Charpentier's group identified in 2011 (Deltcheva et al., 2011). This molecule is involved in crRNA processing (Figure 11.4). This molecule was called *trans*-activating CRISPR RNA (tracrRNA), and it was necessary for nuclease action.

11.2.2 Mechanisms of Action of CRISPR-Cas

The CRISPR-Cas system functions as a prokaryotic immune mechanism, defending bacteria and archaea against foreign genetic elements like viruses (bacteriophages). The mechanism of action (Figure 11.5) is stated below.

11.2.2.1 Acquisition (Adaptation)

When a bacterium is invaded by a virus, a portion of the virus's DNA is captured and inserted into the bacterium's genome at a specific location known as the CRISPR locus. This viral DNA segment, known as a "spacer," is stored between repetitive sequences and acts as a genetic memory of past infections.

11.2.2.2 Expression

The CRISPR locus, containing the stored spacers, is transcribed into a long RNA molecule called pre-crRNA. This molecule is then processed into shorter crRNAs, each containing a spacer sequence. Alongside, another RNA called tracrRNA (in some systems) is expressed, which helps in crRNA maturation and guiding the Cas protein.

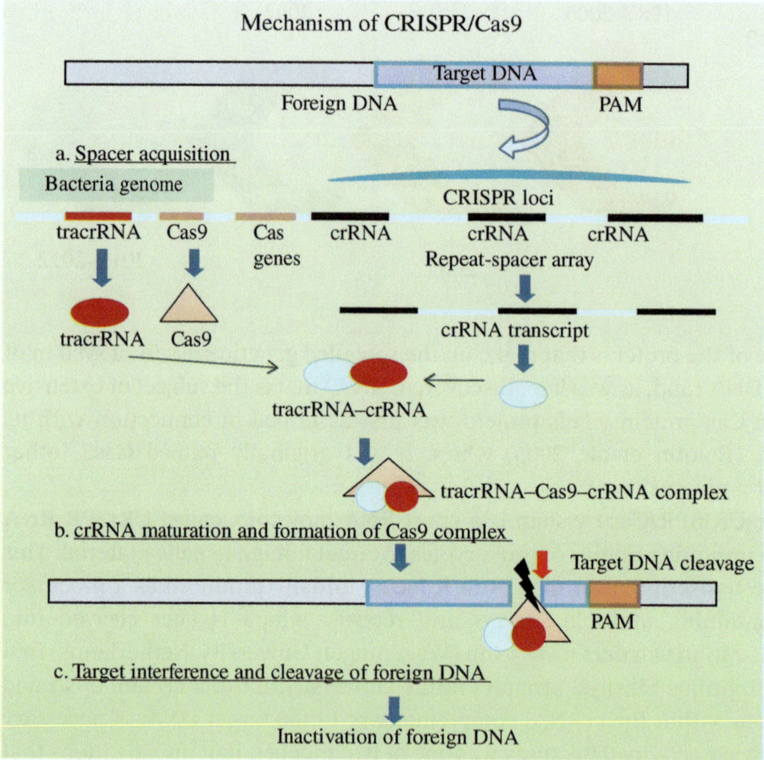

Figure 11.5 Mechanism of action of CRISPR-Cas9. *Source:* Rani et al., (2016) / with permission of Springer Nature.

11.2.2.3 Interference

The crRNA, which contains a sequence complementary to the target virus's DNA, guides a Cas protein (like Cas9, Cas12, or Cas13) to the invading DNA. Cas proteins are nucleases that cut DNA. Once the crRNA finds its matching sequence in the viral DNA, the Cas protein cleaves the DNA, effectively neutralizing the virus.

- ✓ In Class 1 systems, multiple Cas proteins form a complex that works together to target and destroy the foreign DNA.
- ✓ In Class 2 systems (e.g. CRISPR-Cas9), a single Cas protein, such as Cas9, performs the recognition and cleavage, simplifying the process.

The precise targeting ability of CRISPR-Cas9, guided by the crRNA, allows for targeted genome editing in various applications by cutting DNA at specific locations and enabling subsequent DNA repair or modifications.

11.2.3 Applications of CRISPR-Cas in Agriculture

CRISPR-Cas technology has become a transformative tool in agriculture, offering precise genome-editing capabilities. Its applications in crop improvement include enhancing yield, improving resistance to diseases and pests, and increasing tolerance to environmental stresses such as drought and salinity. In agricultural breeding, CRISPR is used to accelerate the domestication of wild species by introducing desirable traits, such as better nutritional content and longer shelf life. Additionally, CRISPR-based biotechnologies, such as gene-expression modulation and high-throughput gene editing, hold great potential for advancing agricultural productivity and sustainability.

11.2.3.1 Improvement of Yield

One strategy to boost cereal production is by modifying cytokinin homeostasis. In rice, altering the C-terminus of *Oryza sativa* LOGL5, which activates cytokines, increased yield under various conditions. Similarly, knocking down the gene for cytokinin oxidase/dehydrogenase (CKX), which breaks down cytokinin, led to higher yields in wheat. Deleting the gene for amino acid permease 3 in rice enhanced yield and tiller numbers without affecting grain quality. CRISPR-Cas editing of genes like *O. sativa* PIN5b, GS3, GW2, and GW5, which regulate grain size and weight, also improved crop yields.

11.2.3.2 Improving Quality

A crop's other qualities, aside from yield, are as important for agricultural output. Lower amylose content grains are better for cooking and eating, and they are used extensively in the adhesives and textile industries. Granule-bound starch synthase 1 (GBSS1) is essential for the biosynthesis of amylose. Therefore, using CRISPR/Cas9 to disrupt GBSS1, 12 elite inbred lines produced waxy maize varieties. Similarly, using cytosine base editing (CBE) to change the amino acid sequence of GBSS1, rice lines with a range of low amylose contents were produced (Xu et al., 2021).

11.2.3.3 Disease Resistance

Plant protection against biotic stress may be achieved more successfully by employing CRISPR-Cas disruption of host susceptibility factors, as opposed to the introduction of dominant resistance genes, which tends to encourage the reciprocal evolution of resistance in pathogens. The destructive disease is known as *Xanthomonas oryzae* pv. *oryzae*, which causes bacterial blight and poses a serious danger to the world's rice crop. A set of bacterial agents has the ability to trigger the transcription of the SWEET genes, whose products are required for the vulnerability to illness.

Example:

Researchers have created rice lines with broad-spectrum resistance to *X. oryzae* pv. *oryzae* by utilizing CRISPR-Cas to alter the promoter region of *O. sativa* SWEET11, *O. sativa SWEET13*, and *O. sativa SWEET14*. Similar to this, in citrus, *Xanthomonas citri* subsp. *citri* can be resistant when Citrus × sinensis LOB1's promoter region is targeted (Oliva et al., 2019).

11.2.3.4 Herbicide Resistance

The development of herbicide-resistant germplasms becomes a financially viable strategy to sustain high agricultural yield and stop soil degradation as weed issues spread around the world. When weighed against traditional transgenic techniques that incorporate foreign herbicide-resistant genes, like bar, which encodes phosphinothricin *N*-acetyltransferase into crops, CRISPR-Cas editing of herbicide-targeted genes to confer endogenous resistance is appealing due to its transgene-free, quick, and flexible nature.

Example:

A crucial enzyme in the manufacture of branched-chain amino acids, acetolactate synthase (ALS), is the target of herbicides like imidazolinone and sulfonylurea. Herbicide resistance to rice was provided by the insertion of specific base transitions into *O. sativa* ALS with the use of CBEs while maintaining ALS activity.

11.2.3.5 Applications in Breeding Technologies

Even though CRISPR-Cas has demonstrated a great deal of potential for crop improvement, agricultural productivity might be further enhanced by combining it with traditional breeding techniques (Figure 11.6).

i) **Haploid induction**:
In two generations, doubled haploid technology may stabilize the genetic background of hybrid lines, while traditional methods need six to eight generations of selfing to achieve homozygosity.

Example:
Haploid induction lines of wheat and rice have also been created by CRISPR-mediated mutagenesis of MATRILINEAL (MTL) (Liu et al., 2020).

ii) **Generation of male sterile lines**:
Hybrid vigor has been extensively utilized in agriculture to enhance yields and quality. To commercially produce hybrid seeds, preventing self-pollination in the female parent is necessary to avoid producing homozygous seeds. Among various solutions, inducing male sterility in the female parent has proven to be the most efficient and practical method. Male sterility in transformable lines can be quickly established by CRISPR-Cas gene editing.

Example:
A suspected strictosidine synthase gene was mutated to produce a male-sterile tomato line as well (Du et al., 2020).

iii) **Hybrid vigor fixation**:
Despite their established nature, male-sterile line systems for producing hybrid seeds are nonetheless expensive and time-consuming for certain crops. Another option to correct elite hybrid backgrounds would be to induce apomixis, a naturally occurring asexual reproduction mechanism.

Example:
The disruption of MTL (which results in the deletion of the paternal genome after conception) and MiMe genes produced clonal diploid embryos in rice (Wang et al., 2019).

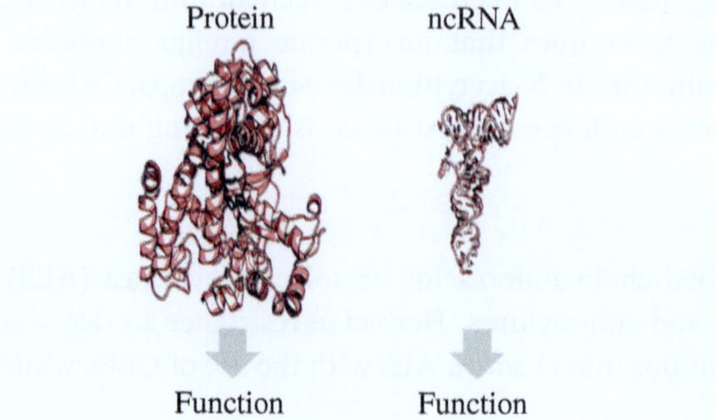

Figure 11.6 Applications of CRISPR-Cas9 in breeding technologies. *Source:* Zhu et al. (2020) / CC BY 4.0 / Thomas Shafee.

iv) **Self-incompatibility manipulation**:
Because some crops, like potatoes, are inherently self-incompatible, the lack of inbred lines has hampered genetic improvement. Self-compatible potato lines have been produced by mutating S-RNase using CRISPR-Cas, a codominant gene that causes gametophytic self-incompatibility in the Solanaceae.

Example:
A function for CRISPR-Cas and other gene mutations like farnesyl pyrophosphate synthase 2 in regaining self-incompatibility (Qin et al., 2018).

11.3 Overview of Noncoding RNAs

A functioning RNA molecule that is not translated into a protein is known as ncRNA (Mattick & Makunin, 2006). The precursor of generating functional ncRNA is termed an RNA gene. RNA molecules known as ncRNAs do not encode proteins, in contrast to mRNA. Rather, they function in the cell in various structural and regulatory capacities. These RNAs' vital roles in a wide range of biological activities have drawn a lot of attention to them in recent years. In the past, RNA was thought to be just a messenger that helps ribosomes synthesize proteins by carrying genetic information from DNA. But the discovery of ncRNAs (Mattick et al., 2023) has completely changed our knowledge of the functions of RNA, demonstrating that RNA molecules may control the expression of genes, preserve the integrity of genomes, and take part in a wide range of biological processes. Beyond protein coding, ncRNAs perform vital roles in a complex and varied landscape that is revealed by this enlarged perspective on RNA biology.

It is unknown how many ncRNAs are present in the human genome, however, new transcriptomic and bioinformatic research indicates that thousands of noncoding transcripts may exist. If any of the recently discovered ncRNAs have any roles at all, they are unknown (Jarroux et al., 2017). The amount of noncoding transcription that is functional is unknown; some people think that most noncoding transcripts are spurious transcriptions or nonfunctional "junk RNA," while others anticipate that many noncoding transcripts will be found to have roles.

11.3.1 Classification of Noncoding RNAs

ncRNAs (Ma et al., 2013) can be broadly classified into two main categories based on their size, which are small ncRNAs and lncRNAs. Additionally, some ncRNAs are categorized based on their specific functions.

11.3.1.1 Small Noncoding RNAs

Small ncRNAs are defined as ncRNAs that are small in size. Different types of these small ncRNAs are mentioned as follows:

- **miRNAs**: These are small ncRNAs, typically 20–25 nucleotides long, that regulate gene expression post transcriptionally. miRNAs (Hammond, 2015) bind to complementary sequences in the 3′ untranslated regions (UTRs) of target mRNAs, leading to either degradation of the mRNA or inhibition of its translation. This form of regulation is vital in controlling various biological processes such as cell proliferation, differentiation, apoptosis, and development. Dysregulation of miRNAs is often linked to diseases, particularly cancer, where specific miRNAs can function as oncogenes or tumor suppressors.

- **siRNAs**: siRNAs (Lares et al., 2010) are typically 20–24 nucleotides in length and play a role in the RNAi pathway. siRNAs arise from double-stranded RNA (dsRNA) precursors and guide the RNA-induced silencing complex (RISC) to complementary mRNA molecules, leading to their degradation. siRNA-mediated RNAi is a mechanism used by cells to defend against viral infections and TEs, and it is also a valuable tool in molecular biology for gene silencing.
- **piRNAs**: These are slightly longer, around 24–31 nucleotides, and they interact with Piwi proteins. piRNAs (Ozata et al., 2018) are primarily found in the germline and are involved in maintaining genome integrity by silencing TEs. piRNAs also play a role in regulating gene expression during spermatogenesis and other developmental processes.
- **Small Nucleolar RNAs (snoRNAs)**: snoRNAs (Maxwell & Fournier, 1995) are involved in the chemical modification of other RNA molecules, particularly rRNA, through processes such as 2′-O-methylation and pseudouridylation. By modifying rRNA, snoRNAs help ensure proper ribosome function and protein synthesis.

11.3.1.2 Long Noncoding RNAs (lncRNAs)

lncRNAs are defined as transcripts longer than 200 nucleotides that do not code for proteins. lncRNAs are incredibly diverse in both structure and function. Some of the key roles they play include regulating gene expression, chromatin remodeling, and serving as molecular scaffolds.

11.3.2 Functional Roles of Noncoding RNAs

ncRNAs play numerous functional roles in cells, and their importance spans across various biological processes.

11.3.2.1 Gene Silencing (RNA Interference)

siRNAs guide the RISC (Deng et al., 2014) to target mRNAs, leading to mRNA degradation and reduced gene expression.

11.3.2.2 Posttranscriptional Gene Regulation

miRNAs bind to the 3′ UTRs of target mRNAs, inhibiting translation or inducing mRNA degradation.

11.3.2.3 Chromatin Remodeling

lncRNAs can recruit chromatin-modifying complexes (like histone methyltransferases) to specific genomic loci, leading to changes in chromatin structure that either promote or inhibit gene transcription.

11.3.2.4 Epigenetic Regulation

ncRNAs such as lncRNAs and snoRNAs can influence DNA methylation or histone modification, thereby altering gene expression without changing the DNA sequence (Olmedo-Suárez et al., 2022).

11.3.2.5 Transcriptional Activation and Repression

Certain lncRNAs act as activators by facilitating the assembly of transcriptional machinery on promoters or enhancers, increasing gene expression. lncRNAs can recruit repressor complexes, such as Polycomb repressive complexes, to genes, resulting in transcriptional silencing (Mulvey, 2014).

11.3.2.6 Transcriptional Interference

Some ncRNAs overlap with promoter regions of genes and physically block the transcription machinery, inhibiting the initiation of transcription.

11.3.2.7 Alternative Splicing

lncRNAs can influence the splicing of pre-mRNAs, guiding the splicing machinery to generate alternative isoforms of a gene, and diversifying protein expression (Liu et al., 2021).

11.3.2.8 RNA Decoys

lncRNAs can function as decoys by binding to transcription factors, RNA-binding proteins, or miRNAs, preventing them from interacting with their target mRNAs.

11.3.2.9 Regulation of mRNA Stability

lncRNAs and miRNAs can bind to mRNAs, influencing their stability and determining whether they are degraded or translated into proteins.

11.3.2.10 X-chromosome Inactivation

The lncRNA Xist plays a crucial role in X-chromosome inactivation in female mammals by coating one X chromosome and recruiting silencing complexes to shut down its expression (Siniscalchi et al., 2022).

11.3.2.11 Cellular Stress Response

ncRNAs (Valadkhan & Valencia-Hipólito, 2015), particularly miRNAs, are involved in regulating genes responsible for the cellular stress response, allowing cells to adapt to oxidative stress, DNA damage, or viral infections by modifying gene expression patterns.

11.3.3 Importance in Plant Biology

ncRNAs play a crucial role in plant biology by regulating gene expression and helping plants adapt to environmental conditions. Here are some points highlighting the importance of ncRNAs in plants:

11.3.3.1 Developmental Processes

ncRNAs are vital for plant growth and development. They regulate key developmental processes, such as leaf morphology, root growth, and flowering time. These processes are critical for the proper formation of plant organs and the timing of reproductive events.

11.3.3.2 Stress Response and Adaptation

Plants are exposed to various abiotic (drought, salinity, and temperature) and biotic (pathogen attacks) stresses. ncRNAs help plants adapt by regulating stress-responsive genes.

11.3.3.3 Defense Against Pathogens

Plants use ncRNAs as part of their immune system to fend off pathogens. For example, siRNAs are involved in RNAi, which silences viral genes, protecting the plant from viral infections.

11.3.3.4 Hormone Signaling Regulation

ncRNAs modulate hormone signaling pathways, which are essential for plant growth, development, and stress responses. For instance, miRNAs are involved in regulating auxin signaling, affecting processes like cell division and elongation.

11.3.3.5 Nutrient Homeostasis

ncRNAs regulate genes involved in nutrient uptake and distribution. For instance, certain miRNAs regulate phosphorus and nitrogen uptake, ensuring that plants can efficiently use these essential nutrients under fluctuating environmental conditions.

11.3.3.6 Response to Environmental Cues

Plants rely on ncRNAs to respond to environmental signals like light and temperature. This regulation affects key developmental transitions, such as seed germination and flowering time, helping plants synchronize their life cycle with changing environmental conditions.

ncRNAs are essential regulators in plant biology, influencing nearly every aspect of plant growth, development, and environmental adaptation. They act as key modulators of gene expression, ensuring that plants can efficiently respond to environmental stresses, defend against pathogens, and carry out their life cycle. Understanding ncRNA functions in plants can lead to advances in agricultural biotechnology, such as improving crop resilience and productivity.

11.4 CRISPR/Cas-Edited Crops for Disease Resistance and Limitations

The prokaryotic CRISPR-Cas9 system is an RNA-mediated, sequence-specific adaptive immune system that defends against viruses and bacteriophages. This concept is widely used for the protection of plants from various diseases.

11.4.1 Enhancing Fungal Resistance by Modifying Host Susceptibility Genes

Fungal plant pathogens pose a major threat to agriculture due to their complex lifestyles, genetic flexibility, and ability to invade hosts, causing diseases like rot, smut, and mildew, leading to significant yield losses. Traditional methods of introducing resistance (R) genes into plants have been ineffective long term due to pathogen-specific resistance and the fungi's ability to mutate and overcome these defenses. An alternative strategy to improve fungal resistance involves modifying host susceptibility (S) genes using CRISPR-Cas9. This method is effective without affecting overall plant health or development, offering a more durable solution against fungal pathogens.

Example:
Magnaporthe grisea is the causal organism for the fungal blast in *O. sativa*. The target gene for fungal blasts is OsMPK5. It helps in pathogen infection. By the knockout method, resistance to this disease is obtained (Xie & Yang, 2013).

11.4.2 Enhancing Bacterial Resistance Through Modification of Host Susceptibility Genes

Bacteria, both beneficial and harmful, pose a significant threat to agriculture, causing diseases like spots and rots, which are difficult to control and result in major yield losses. Phytopathogenic bacteria can be crop specific, polyphagous (affecting a broad range of hosts), or kingdom crossers, capable of infecting different kingdoms, including plants and animals. Their evolving nature and ability to transfer genes horizontally help bacteria develop resistance against antibacterial agents.

Table 11.1 Exploitation of CRISPR-Cas9 to counteract crop bacterial disease.

Plant	Targeted area in gene	Disease	References
Rice	Mutagenesis of the ERF transcription factor gene *OsERF922*	Blast	Wang et al. (2016a)
Wanjinchen oranges	(CsLOB1G and CsLOB1−) alleles	Citrus bacterial canker (CBC)	Peng et al. (2017)

Bacteria invade plants through natural openings or mechanical wounds, and the plant's defense system responds accordingly. However, certain host genes, including susceptibility (S) genes, can assist pathogens in overcoming immunity and initiating infection. Modifying these S genes through genome editing, particularly with CRISPR-Cas9, offers the potential for more durable bacterial resistance (Table 11.1). However, progress in developing bacterial-resistant plant varieties using this approach has been limited, partly due to the complexity of plant genetics and bacterial evolution.

Example:
Erwinia amylovora is a causal bacterium for the Fire Blight disease in *Golden delicious.*

The target genes are DIPM-1, DIPM-2, and DIPM-4. These genes interact with pathogen effectors. By the knockout method of CRISPR-Cas, resistance to this bacterial disease is achieved significantly (Malnoy et al., 2016).

11.4.3 Viral Resistance by Editing Host Susceptible Genes

Plant viruses, with their obligate parasitic nature, rapidly evolving genomes, and high infectivity, pose a significant threat to crops, leading to substantial yield losses. Viruses are classified into DNA and RNA types, with DNA viruses being further divided into double-stranded DNA (dsDNA) and single-stranded DNA (ssDNA) forms, while RNA viruses have four types: dsRNA, negative-strand single-stranded DNA (−ssRNA), positive-strand ssRNA (+ssRNA), and retroviruses. Among these, DNA viruses, particularly the ssDNA viruses from the Geminiviridae family, cause significant damage to crops like cucumbers, tomatoes, and wheat by hijacking plant cellular machinery to replicate through the rolling circle method. Since viruses lack their own translational machinery, they rely on the plant's transcription and translation systems. Therefore, resistance to both DNA and RNA viruses can be developed by either targeting viral genes or modifying host susceptibility genes to prevent viral replication (Table 11.2).

Example:
Turnip mosaic virus (*TuMV*) is the causal organism for mosaic diseases in *Arabidopsis thaliana.* The target gene is eIF4E, which helps to replicate the virus. By the knockout method, mosaic disease can be suppressed (Pyott et al., 2016).

11.4.4 Limitation in Developing Resistance in Plants

In the past decade, numerous studies have highlighted the potential of the CRISPR system for editing the genomes of a wide variety of plant species, including both monocots and dicots, to enhance resistance against various plant pathogens. However, there are two significant limitations to consider.

Table 11.2 List of viral resistance in plants introduced via CRISPR-Cas9 against DNA and RNA viruses.

Virus	Type of nucleic acid	Involved protein	Plant under attack	References
Beet severe curly top virus	DNA	Cas9	Capsicum	Ji et al. (2015)
Bean yellow dwarf virus	DNA	Cas9	Oat	Baltes et al. (2015)
Turnip mosaic virus	RNA	Cas13	Cruciferous plants, Chinese cabbage, turnip, mustard, radish	Aman et al. (2018)
Tomato yellow leaf curl virus	DNA	Cas9	Invading a number of seeds, including tomato	Ali et al. (2015)
Yellowing virus	RNA	Cas13	Cucumber	Chandrasekaran et al. (2016)
Zucchini yellow mosaic virus	RNA	Cas13	Cucumber	Jia et al. (2017)

The first limitation arises from the constantly evolving nature of pathogens, which can alter their genomes to overcome existing resistance mechanisms in plants. This dynamic interaction means that even engineered resistance can become ineffective over time, as pathogens adapt and find ways to bypass these defenses. Additionally, there is often a trade-off involved when editing plant genes, particularly susceptibility (S) genes. Modifying these genes to enhance resistance may inadvertently affect the overall fitness and growth of the plant, potentially leading to reduced yield or vigor.

The second major limitation is related to the CRISPR-Cas9 technology itself, specifically the issue of "off-target mutations." This refers to unintended edits in the genome that occur when the CRISPR system mistakenly targets similar sequences, leading to potential disruptions in other important genes. Such off-target effects can have unpredictable consequences, complicating the development of stable and safe genetically modified plants.

Besides these two major limitations, other limitations or challenges are as follows:

- ✓ **Genetic Complexity:** The intricate genetic makeup of plants, involving multiple genes and interactions, makes it challenging to engineer effective resistance.
- ✓ **Pathogen Evolution:** Rapidly evolving pathogens can quickly overcome plant defenses, rendering resistance ineffective over time.
- ✓ **Horizontal Gene Transfer:** Some pathogens can acquire resistance genes from other organisms, complicating control efforts.
- ✓ **Variable Response:** Different plant species and varieties may respond differently to the same resistance strategy, making it difficult to achieve universal resistance.
- ✓ **Trade-Offs:** Enhancing resistance may come at the cost of other desirable traits, such as growth or yield, leading to reduced overall performance.
- ✓ **Environmental Factors:** External conditions, such as climate and soil health, can influence the effectiveness of resistance strategies.
- ✓ **Regulatory Challenges**: Genetic modification and gene editing face regulatory hurdles, which can delay the development and deployment of resistant varieties.

11.5 Modulation of Noncoding RNAs in CRISPR/Cas-Edited Crops

ncRNAs are known for their role in gene regulation and other cellular processes, and their modifications can affect their stability, function, and interactions.

11.5.1 Mechanisms of Noncoding RNA Modification

Some of the key mechanisms of ncRNA modifications (Eddy, 2001; Thakur et al., 2020) include the following:

11.5.1.1 Methylation

N6-methyladenosine (m6A) is one of the most common modifications in mRNA and ncRNAs, including lncRNAs and miRNAs. It influences RNA stability, splicing, and translation efficiency. Apart from this, 5-methylcytosine (m5C) can be found in tRNAs, rRNAs, and ncRNAs, and m5C helps with RNA stability and function.

11.5.1.2 Pseudouridylation

Pseudouridylation is the conversion of uridine to pseudouridine; this modification is widespread in rRNA and tRNA. It enhances the stability of RNA structure and can affect the translation of proteins.

11.5.1.3 ADAR-Mediated Deamination

Adenosine-to-inosine (A-to-I) editing is catalyzed by ADAR enzymes; this editing can alter the coding potential and secondary structure of ncRNAs like miRNAs.

11.5.1.4 Polyadenylation

While typically associated with mRNA stability and export, polyadenylation can also affect certain ncRNAs, particularly lncRNAs.

11.5.1.5 Phosphorylation

This modification can regulate the function and activity of certain ncRNAs by modifying their interaction with proteins.

11.5.1.6 Base Conversion or Oxidation

Certain ncRNAs may undergo oxidative modifications, which could affect their role in the cell, especially under conditions of oxidative stress.

11.5.2 Impact of CRISPR/Cas Editing on Noncoding RNA Expression

CRISPR/Cas-editing technology has had a profound impact on the study and manipulation of ncRNA expression, shedding light on their diverse biological roles. Here is an overview of how CRISPR/Cas editing (Yang et al., 2018) affects ncRNA expression:

11.5.2.1 Disruption of ncRNA Loci

CRISPR/Cas9 can be used to directly disrupt the genomic loci of ncRNAs, including lncRNAs (Zibitt et al., 2021) and miRNAs. By creating indels (insertions or deletions) at the target site, the transcription or function of these ncRNAs can be abolished, helping researchers study their role in gene regulation, cell differentiation, and disease processes. CRISPR/Cas9 can be used to disrupt the promoters or enhancers of ncRNAs, preventing their transcription. This helps in understanding the transcriptional regulation of ncRNAs.

11.5.2.2 CRISPR Interference (CRISPRi) for ncRNA Regulation

CRISPRi (Larson et al., 2013), which uses a catalytically inactive Cas9 (dCas9) fused to a transcriptional repressor, allows for the silencing of ncRNAs without permanently altering the genome. This is especially useful for ncRNAs with overlapping coding regions or essential regulatory elements.

11.5.2.3 CRISPR Activation (CRISPRa) for ncRNA Upregulation

CRISPRa (Ding et al., 2022), which uses dCas9 fused to a transcriptional activator, allows researchers to upregulate the expression of ncRNAs. This tool is valuable for investigating the roles of ncRNAs in contexts where overexpression or increased activity mimics physiological or pathological states, such as cancer or stress responses.

11.5.2.4 Editing ncRNA Processing and Maturation

CRISPR/Cas9 can target specific sequences within precursor forms of ncRNAs, such as pre-miRNAs, to study their maturation and processing. By introducing mutations or deletions at critical sites, researchers can alter the way ncRNAs are processed by enzymes like Drosha and Dicer, providing insights into how these processes are regulated.

11.5.2.5 Epigenetic Regulation of ncRNA Expression

CRISPR-based epigenetic editing tools (e.g. dCas9 fused with histone-modifying enzymes) enable the manipulation of chromatin states at ncRNA loci. This is useful for studying how ncRNAs are regulated epigenetically in different cell types or disease states (Luo, 2019; Fadul et al., 2023).

11.5.2.6 Targeting ncRNAs in Disease Models

CRISPR/Cas editing has been applied to study the role of oncogenic or tumor-suppressive ncRNAs (e.g. certain lncRNAs or miRNAs) in cancer development. By knocking out or activating ncRNAs in cancer cell lines or animal models, researchers can assess their contribution to tumorigenesis, metastasis, and drug resistance.

11.5.2.7 CRISPR-Linked Single-Cell Technologies for ncRNA Analysis

The integration of CRISPR/Cas systems with single-cell sequencing technologies allows researchers to study ncRNA expression at the single-cell level. This helps in uncovering cell type-specific functions of ncRNAs in heterogeneous tissues, such as the brain or the tumor (Kadhim & Al-Kelaby, 2022) microenvironment.

11.5.2.8 CRISPR Screens for Functional ncRNAs

Genome-wide CRISPR screens have been developed to identify functional ncRNAs. By targeting thousands of ncRNA loci in a high-throughput manner, researchers can systematically investigate which ncRNAs play crucial roles in specific biological pathways, such as cell cycle regulation, immune responses, or metabolism.

11.5.3 Functional Implications for Disease Resistance

The use of CRISPR/Cas editing to manipulate ncRNA expression has significant functional implications for disease resistance. ncRNAs, including lncRNAs, miRNAs, and other small ncRNAs, are involved in regulating immune responses, cellular stress pathways, and gene expression networks that contribute to disease resistance. Here is how CRISPR/Cas editing of ncRNAs can enhance our understanding and application of ncRNAs in disease resistance:

ncRNAs such as miRNAs and lncRNAs are known to regulate genes involved in innate and adaptive immune responses. For example, specific miRNAs target mRNAs that encode immune signaling proteins like cytokines, transcription factors, and receptors.

By upregulating ncRNAs that boost immune responses, CRISPRa can enhance the expression of ncRNAs that activate pro-inflammatory cytokines or enhance pathogen recognition, thus increasing resistance to infections.

Conversely, CRISPRi can silence ncRNAs that inhibit immune responses, promoting stronger defenses against pathogens. In viral infections like HIV or hepatitis, certain ncRNAs have been found to suppress immune genes that recognize and destroy infected cells. CRISPR/Cas editing could upregulate these immune-activating ncRNAs to bolster the body's natural defenses against viral replication.

Many autoimmune diseases, such as rheumatoid arthritis or multiple sclerosis, involve dysregulation of inflammatory pathways. ncRNAs play a critical role in these processes by either promoting or suppressing inflammatory gene expression.

CRISPR/Cas can be used to target ncRNAs involved in chronic inflammation. For example, knocking out or downregulating miRNAs that drive excessive inflammation could help reduce tissue damage in autoimmune conditions.

CRISPR-mediated editing of ncRNAs involved in the TNF-alpha signaling pathways could help restore immune balance, reducing autoimmune disease symptoms while maintaining overall immune function. Many ncRNAs function as oncogenes or tumor suppressors. miRNAs, for example, can regulate the expression of genes involved in cell cycle control, apoptosis, and DNA repair.

Using CRISPR/Cas9 to knockout oncogenic ncRNAs can block their ability to promote tumor growth and metastasis. By eliminating ncRNAs that silence tumor suppressor genes, cancer cells may lose their ability to proliferate uncontrollably Conversely, CRISPRa can be used to upregulate tumor-suppressive ncRNAs that help the body resist cancer development.

Targeting host miRNAs that viruses depend on for replication can provide a novel approach to viral resistance. For example, the miRNA, miR-122 is essential for HCV replication. By knocking out miR-122 with CRISPR, researchers have been able to reduce HCV viral loads in infected cells, making this approach a potential therapeutic strategy.

ncRNAs play roles in bacterial resistance to antibiotics by regulating stress responses and virulence factor production. CRISPR/Cas editing could target ncRNAs that contribute to the bacterial survival mechanisms. By disabling bacterial ncRNAs that mediate antibiotic resistance or virulence, CRISPR technology could improve the effectiveness of existing antibiotics. This approach could be crucial in combating multidrug-resistant bacteria.

Many ncRNAs are involved in the regulation of genes associated with neuroinflammation and neuronal survival, which play a role in neurodegenerative diseases like Alzheimer's disease and Parkinson's disease.

CRISPR/Cas9 can be used to edit ncRNAs that regulate toxic protein accumulation or neuroinflammatory pathways, potentially enhancing resistance to these conditions. For instance, targeting lncRNAs that promote beta-amyloid plaque accumulation in Alzheimer's could slow disease progression.

In summary, CRISPR/Cas editing of ncRNAs holds enormous potential for improving disease resistance across a wide range of conditions, from infectious diseases to cancer and autoimmune disorders. By precisely modulating ncRNA expression, researchers and clinicians can harness the regulatory power of ncRNAs to bolster natural defenses and design targeted therapies for enhancing disease resistance.

11.6 Example of CRISPR-Cas-Edited Crops with Modified Noncoding RNAs

The CRISPR-Cas system has revolutionized agricultural biotechnology by providing a precise and efficient method for genome editing in crops. This technology allows for targeted modifications of specific genes, leading to enhanced traits such as disease resistance, improved yield, and better nutritional profiles. Several crop species have been successfully edited using CRISPR-Cas, showcasing the potential of this technology to address pressing agricultural challenges.

The integration of CRISPR/Cas technology in agriculture has opened new avenues for enhancing crop resilience, particularly through the modification of ncRNAs (ncRNAs). These RNA molecules play crucial roles in regulating gene expression, including those involved in plant defenses against pathogens. By precisely editing ncRNAs, researchers can improve the expression of resistance genes and enhance stress response mechanisms. This approach not only addresses the challenges posed by various plant diseases but also allows for the preservation of desirable traits in crops. Here, we explore several notable examples of CRISPR/Cas-edited crops where modifications to ncRNAs have led to improved disease resistance and overall plant health.

Here some examples are stated below:

- **Rice**: Researchers have used CRISPR/Cas9 to modify ncRNAs that regulate stress-response pathways, enhancing drought tolerance and resistance to bacterial blight. By targeting specific miRNAs, they were able to improve the expression of stress-related genes.
- **Wheat**: In wheat, CRISPR has been applied to edit lncRNAs involved in disease resistance pathways. Modifying these ncRNAs has resulted in enhanced resistance to fungal pathogens like *Fusarium graminearum*, which causes wheat head blight.
- **Tomato**: Scientists have utilized CRISPR to alter miRNAs that control fruit development and pathogen resistance. These modifications have led to improved resistance against *Tomato yellow leaf curl virus* while maintaining desirable fruit traits.
- **Soybean**: In soybean, CRISPR/Cas9 has been employed to target ncRNAs involved in the regulation of key defense genes, resulting in enhanced resistance to soybean rust and other diseases.
- **Barley**: Edited miR156 to enhance resistance to powdery mildew and improve growth.
- **Potato**: Targeted lncRNAs to increase resistance to late blight and boost yield.
- **Pea**: Modified miR167 to improve resistance against aphids and other pests.
- **Cotton**: Altered lncRNAs for enhanced fiber quality and resistance to bollworms.
- **Sorghum**: Edited miR397 to increase drought tolerance and improve nutrient use efficiency.
- **Cabbage**: Targeted lncRNAs to enhance resistance to black rot and other diseases.

These examples illustrate the potential of CRISPR/Cas technology to modify ncRNAs for improved disease resistance in various crops, paving the way for more resilient agricultural systems.

11.7 Future Perspectives and Challenges

11.7.1 Future Perspectives

The potential of CRISPR-Cas technologies, particularly in the context of ncRNA editing, presents exciting opportunities for addressing genetic diseases and enhancing crop traits. As research continues to evolve, several future perspectives emerge to optimize the efficacy and safety of these approaches.

- The integration of RNP delivery systems for CRISPR tools will enhance specificity and reduce off-target effects, paving the way for safer clinical applications of ncRNA editing.
- Ongoing research to optimize electroporation conditions and alternative ex vivo delivery methods will aim to minimize cell stress and maintain high viability, improving outcomes for ncRNA interventions.
- Developing efficient in vivo editing techniques will be crucial for addressing genetic diseases, allowing for direct administration of genome-editing tools without the complications of ex vivo manipulation.
- Enhancing delivery specificity to ensure that ncRNA editing occurs only in target cells will be essential to prevent potential malignancies associated with off-target edits.
- Achieving high editing efficiencies in target populations will be a priority, with research focused on determining optimal thresholds for therapeutic effectiveness in various conditions.
- Continued exploration of biocompatible and mechanically stable delivery systems will help mitigate immunogenic responses to CRISPR components, facilitating more effective ncRNA therapies.
- Strategies to address immune responses to delivery agents, such as the co-administration of immunosuppressants, will be important for improving the efficacy of in vivo CRISPR applications targeting ncRNAs.
- Ongoing development of various nonviral and viral delivery systems will support the advancement of CRISPR technologies for editing ncRNAs, although overcoming challenges related to low editing frequencies and immunogenicity will remain key focus areas.

11.7.2 Challenges

The implementation of CRISPR-Cas technologies, particularly in the editing of ncRNAs, presents various challenges that must be overcome to realize their full potential in clinical and agricultural applications. These challenges encompass technical, biological, ethical, and regulatory dimensions, each impacting the effectiveness, safety, and acceptance of these innovative genome-editing tools.

- Off-target effects can lead to unintended edits at nontarget sites, causing functional disruptions and potential carcinogenic outcomes.
- Delivery systems often struggle to achieve high specificity and efficiency in vivo while avoiding immune responses, making it difficult to target desired tissues or cell types.
- Immune responses to CRISPR components, especially from viral vectors, can reduce treatment efficacy and lead to inflammatory complications due to preexisting immunity.
- Low editing efficiency in target cells can limit therapeutic effectiveness, particularly in genetic disorders where higher levels of editing are required.
- Maintaining the stability and functionality of edited cells postdelivery is crucial, as stress or contamination during handling can compromise viability.

- Ethical concerns regarding long-term effects, consent, and potential germline modifications necessitate the adaptation of regulatory frameworks for responsible use.
- The complex roles of ncRNAs in gene regulation introduce challenges in predicting editing outcomes, requiring thorough characterization before interventions.
- Current CRISPR-Cas systems may not adequately address all types of mutations, particularly those needing precise modifications rather than simple disruptions, limiting the range of treatable diseases.

11.8 Conclusion

Harnessing CRISPR technology to edit ncRNAs offers a novel strategy for enhancing disease resistance in crops. By precisely targeting ncRNAs that regulate plant defense pathways, researchers can optimize gene expression to bolster resistance against pathogens. This approach minimizes the risk of disrupting essential coding genes while enabling fine-tuning of plant immunity. Despite challenges such as off-target effects and delivery system limitations, advancements in CRISPR methodologies and our growing understanding of ncRNA roles provide significant opportunities for improving crop resilience. Successful application of these techniques could lead to sustainable agricultural practices and enhanced food security in response to emerging global challenges.

References

Ali, Z., Abulfaraj, A., Idris, A., Ali, S., Tashkandi, M., & Mahfouz, M. M. (2015). CRISPR/Cas9-mediated viral interference in plants. *Genome Biology*, *16*(1), 1–11. https://doi.org/10.1186/S13059-015-0799-6/FIGURES/4

Aman, R., Ali, Z., Butt, H., Mahas, A., Aljedaani, F., Khan, M. Z., Ding, S., & Mahfouz, M. (2018). RNA virus interference via CRISPR/Cas13a system in plants. *Genome Biology*, *19*(1), 1–9. https://doi.org/10.1186/S13059-017-1381-1/FIGURES/3

Baltes, N. J., Hummel, A. W., Konecna, E., Cegan, R., Bruns, A. N., Bisaro, D. M., & Voytas, D. F. (2015). Conferring resistance to geminiviruses with the CRISPR–Cas prokaryotic immune system. *Nature Plants*, *1*(10), 1–4. https://doi.org/10.1038/nplants.2015.145

Barrangou, R., Fremaux, C., Deveau, H., Richards, M., Boyaval, P., Moineau, S., Romero, D. A., & Horvath, P. (2007). CRISPR provides acquired resistance against viruses in prokaryotes. *Science*, *315*(5819), 1709–1712. https://doi.org/10.1126/SCIENCE.1138140/SUPPL_FILE/BARRANGOU.SOM.PDF

Bolotin, A., Quinquis, B., Sorokin, A., & Dusko Ehrlich, S. (2005). Clustered regularly interspaced short palindrome repeats (CRISPRs) have spacers of extrachromosomal origin. *Microbiology*, *151*(8), 2551–2561. https://doi.org/10.1099/MIC.0.28048-0/CITE/REFWORKS

Chandrasekaran, J., Brumin, M., Wolf, D., Leibman, D., Klap, C., Pearlsman, M., Sherman, A., Arazi, T., & Gal-On, A. (2016). Development of broad virus resistance in non-transgenic cucumber using CRISPR/Cas9 technology. *Molecular Plant Pathology*, *17*(7), 1140–1153. https://doi.org/10.1111/mpp.12375

Charpentier, E., & Marraffini, L. A. (2014). Harnessing CRISPR-Cas9 immunity for genetic engineering. *Current Opinion in Microbiology*, *19*, 114–119. https://doi.org/10.1016/j.mib.2014.07.001

Cox, D. B. T., Platt, R. J., & Zhang, F. (2015). Therapeutic genome editing: prospects and challenges. *Nature Medicine*, *21*(2), 121–131. https://doi.org/10.1038/nm.3793

De Leo, A. N. (2022). The code breaker: Jennifer Doudna, gene editing, and the future of the human race. *Practical Radiation Oncology*, *12*(4), e251–e252. https://doi.org/10.1016/J.PRRO.2022.02.014

Deltcheva, E., Chylinski, K., Sharma, C. M., Gonzales, K., Chao, Y., Pirzada, Z. A., Eckert, M. R., Vogel, J., & Charpentier, E. (2011). CRISPR RNA maturation by trans-encoded small RNA and host factor RNase III. *Nature*, *471*(7340), 602–607. https://doi.org/10.1038/nature09886

Deng, Y., Wang, C. C., Choy, K. W., Du, Q., Chen, J., Wang, Q., Li, L., Chung, T. K. H., & Tang, T. (2014). Therapeutic potentials of gene silencing by RNA interference: principles, challenges, and new strategies. *Gene*, *538*(2), 217–227. https://doi.org/10.1016/J.GENE.2013.12.019

Deveau, H., Garneau, J. E., & Moineau, S. (2010). CRISPR/Cas system and its role in phage-bacteria interactions. *Annual Review of Microbiology*, *64*, 475–493. https://doi.org/10.1146/ANNUREV.MICRO.112408.134123/CITE/REFWORKS

Ding, X., Yu, L., Chen, L., Li, Y., Zhang, J., Sheng, H., Ren, Z., Li, Y., Yu, X., Jin, S., & Cao, J. (2022). Recent progress and future prospect of CRISPR/Cas-derived transcription activation (CRISPRa) system in plants. *Cells*, *11*(19), 3045. https://doi.org/10.3390/CELLS11193045

Du, M., Zhou, K., Liu, Y., Deng, L., Zhang, X., Lin, L., Zhou, M., Zhao, W., Wen, C., Xing, J., Li, C., & Li, C. (2020). A biotechnology-based male-sterility system for hybrid seed production in tomato. *The Plant Journal*, *102*(5), 1090–1100. https://doi.org/10.1111/tpj.14678

Eddy, S. R. (2001). Non-coding RNA genes and the modern RNA world. *Nature Reviews Genetics*, *2*(12), 919–929. https://doi.org/10.1038/35103511

Fadul, S. M., Arshad, A., & Mehmood, R. (2023). CRISPR-based epigenome editing: mechanisms and applications. *Epigenomics*, *15*(21), 1137–1155. https://doi.org/10.2217/EPI-2023-0281

Gostimskaya, I. (2022). CRISPR–Cas9: a history of its discovery and ethical considerations of its use in genome editing. *Biochemistry (Moscow)*, *87*(8), 777–788. https://doi.org/10.1134/S0006297922080090

Groenen, P. M. A., Bunschoten, A. E., van Soolingen, D., & van Errtbden, J. D. A. (1993). Nature of DNA polymorphism in the direct repeat cluster of Mycobacterium tuberculosis; application for strain differentiation by a novel typing method. *Molecular Microbiology*, *10*(5), 1057–1065. https://doi.org/10.1111/J.1365-2958.1993.TB00976.X

Hammond, S. M. (2015). An overview of microRNAs. *Advanced Drug Delivery Reviews*, *87*, 3–14. https://doi.org/10.1016/J.ADDR.2015.05.001

Hoe, N., Nakashima, K., Grigsby, D., Pan, X., Dou, S. J., Naidich, S., Garcia, M., Kahn, E., Bergmire-Sweat, D., & Musser, J. M. (1999). Rapid molecular genetic subtyping of serotype M1 group A Streptococcus strains. *Emerging Infectious Diseases*, *5*(2), 254. https://doi.org/10.3201/EID0502.990210

Ishino, Y., Shinagawa, H., Makino, K., Amemura, M., & Nakata, A. (1987). Nucleotide sequence of the iap gene, responsible for alkaline phosphatase isozyme conversion in Escherichia coli, and identification of the gene product. *Journal of Bacteriology*, *169*(12), 5429–5433. https://doi.org/10.1128/jb.169.12.5429-5433.1987

Jansen, R., van Embden, J. D. A., Gaastra, W., & Schouls, L. M. (2002). Identification of genes that are associated with DNA repeats in prokaryotes. *Molecular Microbiology*, *43*(6), 1565–1575. https://doi.org/10.1046/j.1365-2958.2002.02839.x

Jarroux, J., Morillon, A., & Pinskaya, M. (2017). History, discovery, and classification of lncRNAs. *Advances in Experimental Medicine and Biology*, *1008*, 1–46. https://doi.org/10.1007/978-981-10-5203-3_1

Ji, X., Zhang, H., Zhang, Y., Wang, Y., & Gao, C. (2015). Establishing a CRISPR–Cas-like immune system conferring DNA virus resistance in plants. *Nature Plants*, *1*(10), 1–4. https://doi.org/10.1038/nplants.2015.144

Jia, H., Zhang, Y., Orbović, V., Xu, J., White, F. F., Jones, J. B., & Wang, N. (2017). Genome editing of the disease susceptibility gene *Cs LOB 1* in citrus confers resistance to citrus canker. *Plant Biotechnology Journal*, *15*(7), 817–823. https://doi.org/10.1111/pbi.12677

Kadhim, K., & Al-Kelaby, A. (2022). A review on CRISPR/Cas9 as a novel technique for cancer therapy. *Cellular Therapy and Transplantation*, *11*(3–4), 10–24. https://doi.org/10.18620/ctt-1866-8836-2022-11-3-4-10-24

Komor, A. C., Badran, A. H., & Liu, D. R. (2017). CRISPR-based technologies for the manipulation of eukaryotic genomes. *Cell*, *168*(1–2), 20–36. https://doi.org/10.1016/J.CELL.2016.10.044/ASSET/851B3364-E850-4775-8B05-0FD6ECBF84FC/MAIN.ASSETS/GR3.JPG

Langner, T., Kamoun, S., & Belhaj, K. (2018). CRISPR crops: plant genome editing toward disease resistance. *Annual Review of Phytopathology*, *56*(1), 479–512. https://doi.org/10.1146/annurev-phyto-080417-050158

Lares, M. R., Rossi, J. J., & Ouellet, D. L. (2010). RNAi and small interfering RNAs in human disease therapeutic applications. *Trends in Biotechnology*, *28*(11), 570–579.

Larson, M. H., Gilbert, L. A., Wang, X., Lim, W. A., Weissman, J. S., & Qi, L. S. (2013). CRISPR interference (CRISPRi) for sequence-specific control of gene expression. *Nature Protocols*, *8*(11), 2180–2196. https://doi.org/10.1038/nprot.2013.132

Liu, C., Zhong, Y., Qi, X., Chen, M., Liu, Z., Chen, C., Tian, X., Li, J., Jiao, Y., Wang, D., Wang, Y., Li, M., Xin, M., Liu, W., Jin, W., & Chen, S. (2020). Extension of the *in vivo* haploid induction system from diploid maize to hexaploid wheat. *Plant Biotechnology Journal*, *18*(2), 316–318. https://doi.org/10.1111/pbi.13218

Liu, Y., Liu, X., Lin, C., Jia, X., Zhu, H., Song, J., & Zhang, Y. (2021). Noncoding RNAs regulate alternative splicing in cancer. *Journal of Experimental and Clinical Cancer Research*, *40*(1), 1–14. https://doi.org/10.1186/S13046-020-01798-2

Luo, Y. (2019). Refining CRISPR-based genome and epigenome editing off-targets. *Cell Biology and Toxicology*, *35*(4), 281–283. https://doi.org/10.1007/S10565-019-09482-8/METRICS

Ma, L., Bajic, V. B., & Zhang, Z. (2013). On the classification of long non-coding RNAs. *RNA Biology*, *10*(6), 924–933. https://doi.org/10.4161/RNA.24604

Makarova, K. S., Wolf, Y. I., & Koonin, E. V. (2013). The basic building blocks and evolution of CRISPR–Cas systems. *Biochemical Society Transactions*, *41*(6), 1392–1400. https://doi.org/10.1042/BST20130038

Makarova, K. S., Wolf, Y. I., Iranzo, J., Shmakov, S. A., Alkhnbashi, O. S., Brouns, S. J. J., Charpentier, E., Cheng, D., Haft, D. H., Horvath, P., Moineau, S., Mojica, F. J. M., Scott, D., Shah, S. A., Siksnys, V., Terns, M. P., Venclovas, Č., White, M. F., Yakunin, A. F., Yan, W., Zhang, F., Garrett, R. A., Backofen, R., van der Oost, J., Barrangou, R., & Koonin, E. V. (2019). Evolutionary classification of CRISPR–Cas systems: a burst of class 2 and derived variants. *Nature Reviews Microbiology*, *18*(2), 67–83. https://doi.org/10.1038/s41579-019-0299-x

Malnoy, M., Viola, R., Jung, M. H., Koo, O. J., Kim, S., Kim, J. S., Velasco, R., & Kanchiswamy, C. N. (2016). DNA-free genetically edited grapevine and apple protoplast using CRISPR/Cas9 ribonucleoproteins. *Frontiers in Plant Science*, *7*(December), 230280. https://doi.org/10.3389/FPLS.2016.01904/BIBTEX

Mattick, J. S., & Makunin, I. V. (2006). Non-coding RNA. *Human Molecular Genetics*, *15*(suppl_1), R17–R29. https://doi.org/10.1093/HMG/DDL046

Mattick, J. S., Amaral, P. P., Carninci, P., Carpenter, S., Chang, H. Y., Chen, L. L., Chen, R., Dean, C., Dinger, M. E., Fitzgerald, K. A., Gingeras, T. R., Guttman, M., Hirose, T., Huarte, M., Johnson, R., Kanduri, C., Kapranov, P., Lawrence, J. B., Lee, J. T., Mendell, J. T., Mercer, T. R., Moore, K. J., Nakagawa, S., Rinn, J. L., Spector, D. L., Ulitsky, I., Wan, Y., Wilusz, J. E., & Wu, M. (2023). Long non-coding RNAs: definitions, functions, challenges and recommendations. *Nature Reviews Molecular Cell Biology*, *24*(6), 430–447. https://doi.org/10.1038/S41580-022-00566-8

Maxwell, E. S., & Fournier, M. J. (1995). The small nucleolar RNAs. *Annual Review of Biochemistry*, *64*, 897–934. https://doi.org/10.1146/ANNUREV.BI.64.070195.004341

Miller, J. C., Holmes, M. C., Wang, J., Guschin, D. Y., Lee, Y. L., Rupniewski, I., Beausejour, C. M., Waite, A. J., Wang, N. S., Kim, K. A., Gregory, P. D., Pabo, C. O., & Rebar, E. J. (2007). An improved zinc-finger nuclease architecture for highly specific genome editing. *Nature Biotechnology*, *25*(7), 778–785. https://doi.org/10.1038/nbt1319

Mojica, F. J. M., Juez, G., & Rodriguez-Valera, F. (1993). Transcription at different salinities of Haloferax mediterranei sequences adjacent to partially modified PstI sites. *Molecular Microbiology*, *9*(3), 613–621. https://doi.org/10.1111/J.1365-2958.1993.TB01721.X

Mulvey, B. B. (2014). Transcription Regulation; from Long Non-coding RNA Networks to Elongation [Doctoral Dissertation, Department of Biomedical Sciences, Florida State University].

Oliva, R., Ji, C., Atienza-Grande, G., Huguet-Tapia, J. C., Perez-Quintero, A., Li, T., Eom, J.-S., Li, C., Nguyen, H., Liu, B., Auguy, F., Sciallano, C., Luu, V. T., Dossa, G. S., Cunnac, S., Schmidt, S. M., Slamet-Loedin, I. H., Vera Cruz, C., Szurek, B., Frommer, W. B., White, F. F., & Yang, B. (2019). Broad-spectrum resistance to bacterial blight in rice using genome editing. *Nature Biotechnology*, *37*(11), 1344–1350. https://doi.org/10.1038/s41587-019-0267-z

Olmedo-Suárez, M. Á., Ramírez-Díaz, I., Pérez-González, A., Molina-Herrera, A., Coral-García, M. Á., Lobato, S., Sarvari, P., Barreto, G., & Rubio, K. (2022). Epigenetic regulation in exposome-induced tumorigenesis: emerging roles of ncRNAs. *Biomolecules*, *12*(4), 513. https://doi.org/10.3390/BIOM12040513/S1

Ozata, D. M., Gainetdinov, I., Zoch, A., O'Carroll, D., & Zamore, P. D. (2018). PIWI-interacting RNAs: small RNAs with big functions. *Nature Reviews Genetics*, *20*(2), 89–108. https://doi.org/10.1038/s41576-018-0073-3

Peng, A., Chen, S., Lei, T., Xu, L., He, Y., Wu, L., Yao, L., & Zou, X. (2017). Engineering canker-resistant plants through CRISPR/Cas9-targeted editing of the susceptibility gene CsLOB1 promoter in citrus. *Plant Biotechnology Journal*, *15*(12), 1509–1519. https://doi.org/10.1111/PBI.12733

Pyott, D. E., Sheehan, E., & Molnar, A. (2016). Engineering of CRISPR/Cas9-mediated potyvirus resistance in transgene-free *Arabidopsis* plants. *Molecular Plant Pathology*, *17*(8), 1276–1288. https://doi.org/10.1111/mpp.12417

Qin, X., Li, W., Liu, Y., Tan, M., Ganal, M., & Chetelat, R. T. (2018). A farnesyl pyrophosphate synthase gene expressed in pollen functions in S-RNase-independent unilateral incompatibility. *The Plant Journal*, *93*(3), 417–430. https://doi.org/10.1111/tpj.13796

Rani, R., Yadav, P., Barbadikar, K. M., Baliyan, N., Malhotra, E. V., Singh, B. K., Kumar, A., & Singh, D. (2016). CRISPR/Cas9: a promising way to exploit genetic variation in plants. *Biotechnology Letters*, *38*(12), 1991–2006. https://doi.org/10.1007/s10529-016-2195-z

Redman, M., King, A., Watson, C., & King, D. (2016a). What is CRISPR/Cas9? Archives of disease in childhood. *Education and Practice Edition*, *101*(4), 213. https://doi.org/10.1136/ARCHDISCHILD-2016-310459

Redman, M., King, A., Watson, C., & King, D. (2016b). What is CRISPR/Cas9? *Archives of Disease in Childhood. Education and Practice Edition*, *101*(4), 213–215. https://doi.org/10.1136/archdischild-2016-310459

Siniscalchi, C., Di Palo, A., Russo, A., & Potenza, N. (2022). The lncRNAs at X chromosome inactivation center: not just a matter of sex dosage compensation. *International Journal of Molecular Sciences*, *23*(2), 611. https://doi.org/10.3390/IJMS23020611

Thakur, P., Estevez, M., Lobue, P. A., Limbach, P. A., & Addepalli, B. (2020). Improved RNA modification mapping of cellular non-coding RNAs using C- and U-specific RNases. *Analyst*, *145*(3), 816–827. https://doi.org/10.1039/C9AN02111F

Valadkhan, S., & Valencia-Hipólito, A. (2015). lncRNAs in stress response. *Current Topics in Microbiology and Immunology*, *394*, 203–236. https://doi.org/10.1007/82_2015_489

Wang, F., Wang, C., Liu, P., Lei, C., Hao, W., Gao, Y., Liu, Y. G., & Zhao, K. (2016a). Enhanced rice blast resistance by CRISPR/Cas9-targeted mutagenesis of the ERF transcription factor gene OsERF922. *PLoS One, 11*(4), e0154027. https://doi.org/10.1371/JOURNAL.PONE.0154027

Wang, H., La Russa, M., & Qi, L. S. (2016b). CRISPR/Cas9 in genome editing and beyond. *Annual Review of Biochemistry, 85*, 227–264. https://doi.org/10.1146/ANNUREV-BIOCHEM-060815-014607/CITE/REFWORKS

Wang, C., Liu, Q., Shen, Y., Hua, Y., Wang, J., Lin, J., Wu, M., Sun, T., Cheng, Z., Mercier, R., & Wang, K. (2019). Clonal seeds from hybrid rice by simultaneous genome engineering of meiosis and fertilization genes. *Nature Biotechnology, 37*(3), 283–286. https://doi.org/10.1038/s41587-018-0003-0

Xie, K., & Yang, Y. (2013). RNA-guided genome editing in plants using a CRISPR-Cas system. *Molecular Plant, 6*(6), 1975–1983. https://doi.org/10.1093/mp/sst119

Xu, Y., & Li, Z. (2020). CRISPR-Cas systems: overview, innovations and applications in human disease research and gene therapy. *Computational and Structural Biotechnology Journal, 18*, 2401–2415. https://doi.org/10.1016/j.csbj.2020.08.031

Xu, Y., Lin, Q., Li, X., Wang, F., Chen, Z., Wang, J., Li, W., Fan, F., Tao, Y., Jiang, Y., Wei, X., Zhang, R., Zhu, Q., Bu, Q., Yang, J., & Gao, C. (2021). Fine-tuning the amylose content of rice by precise base editing of the *Wx* gene. *Plant Biotechnology Journal, 19*(1), 11–13. https://doi.org/10.1111/pbi.13433

Yang, J., Meng, X., Pan, J., Jiang, N., Zhou, C., Wu, Z., & Gong, Z. (2018). CRISPR/Cas9-mediated noncoding RNA editing in human cancers. *RNA Biology, 15*(1), 35–43. https://doi.org/10.1080/15476286.2017.1391443

Zhu, H., Li, C., & Gao, C. (2020). Applications of CRISPR–Cas in agriculture and plant biotechnology. *Nature Reviews Molecular Cell Biology, 21*(11), 661–677. https://doi.org/10.1038/s41580-020-00288-9

Zibitt, M. S., Hartford, C. C. R., & Lal, A. (2021). Interrogating lncRNA functions via CRISPR/Cas systems. *RNA Biology, 18*(12), 2097–2106. https://doi.org/10.1080/15476286.2021.1899500

12

Temperature-Smart Crops Through Edited Noncoding RNAs Using CRISPR/Cas Technology

Sharia Yesmin Labonno, Marium Khatun, Sumi Sarkar, and Abul Kalam Mohammad Aminul Islam

Department of Genetics and Plant Breeding, Faculty of Agriculture, Bangabandhu Sheikh Mujibur Rahman Agricultural University, Gazipur, Bangladesh

12.1 Introduction

Global agriculture faces immense challenges due to rising global temperatures and increasingly erratic weather patterns driven by climate change. Unpredictable temperature fluctuations result in both heat and cold stresses, severely impacting key physiological processes in plants, such as seed germination, growth, and reproduction (Zhao et al., 2021). Plant sensitivity to temperature stress is well established; heat stress leads to protein denaturation, membrane instability, and increased oxidative stress, while cold stress disrupts cell membrane fluidity and enzymatic activities, impairing growth and development (Mittler et al., 2022). As climate change continues to intensify, the urgency to develop temperature-smart crops has become paramount for ensuring global food security. One promising strategy includes editing noncoding RNAs (ncRNAs) using clustered regularly interspaced short palindromic repeat (CRISPR)/Cas technology to enhance crop resilience to extreme temperatures, including both heat and cold stresses.

Zhang et al. (2022) and Liu et al. (2023) reported that ncRNAs (microRNAs [miRNAs] and long noncoding RNAs [lncRNAs]) play vital roles in the regulation of gene expression in stress-response pathways and metabolic processes in plants. In the beginning, CRISPR/Cas genome-editing techniques have revolutionized plant improvement, providing precise and competent methods for the modification of targeted genes. CRISPR/Cas technology gives tremendous opportunities for crop improvement, far surpassing the capabilities of traditional breeding techniques. For instance, recent studies have successfully used CRISPR/Cas9 to target genes associated with temperature stress response, improving resilience in wheat and rice (Wang et al., 2023a; Li et al., 2024).

This approach holds promise for addressing the increasing pressures on agricultural systems due to climate change. For instance, CRISPR/Cas9 is employed to enhance biotic (disease resistance) and abiotic (drought and cold tolerance) stresses (Chen et al., 2024). However, concerns about off-target effects persist, driving research toward improving the accuracy of the CRISPR system. Recent advancements, such as the use of shorter single-guide RNAs (sgRNAs) and dual Cas9 nickases, have been explored to mitigate these risks (Nguyen et al., 2024). Additionally, new delivery methods, including RNA- and protein-based systems, offer potential pathways for CRISPR

Genome and Epigenome Editing for Stress-Tolerant Crops, First Edition. Edited by Jen-Tsung Chen.

applications without introducing foreign DNA into the plant genome, thereby addressing concerns over genetically modified organisms (GMOs) (Garcia et al., 2024).

The plant breeders confidently hybridize (crossing or backcrossing) between selected parents to generate new genotypes through combinations of desired traits from both parents. This breeding based on heterosis commenced at the end of the nineteenth century. Conventional breeding has been instrumentalized with techniques like crossbreeding, heterosis breeding, and mutation breeding (Wallace et al., 2018). These breeding techniques have struggled to keep pace with the rapid environmental changes, highlighting the need for innovative strategies to bolster crop resilience (Tang et al., 2024). The discovery of ncRNAs has ushered in a new era of molecular breeding as an important regulator in gene expression. ncRNAs, particularly miRNAs and lncRNAs, regulate complex gene networks in response to abiotic stresses like temperature fluctuations (Sunkar et al., 2024). These ncRNAs have now emerged as critical targets for genome editing via CRISPR/Cas9 technology, a revolutionary technique of precision genome editing. The integration of CRISPR/Cas technology with ncRNA research is setting the stage for the development of temperature-smart crops capable of withstanding both heat and cold stresses, ensuring agricultural productivity even under harsh environmental conditions (Borrelli et al., 2023).

12.2 The Function of ncRNAs in Response to Temperature Stress in Plants

ncRNAs play vital roles in the regulation of different physiological developments in plants, particularly in response to environmental stresses, including extreme temperatures. Unlike coding RNAs, which are translated into proteins, ncRNAs do not encode proteins but instead function in the regulation of important gene expression, serving as actors in the signaling pathways and response during stress. In the context of temperature stress, both miRNAs and lncRNAs have appeared as critical components in mediating plant adaptations to heat and cold conditions (Bartel, 2004; Figure 12.1).

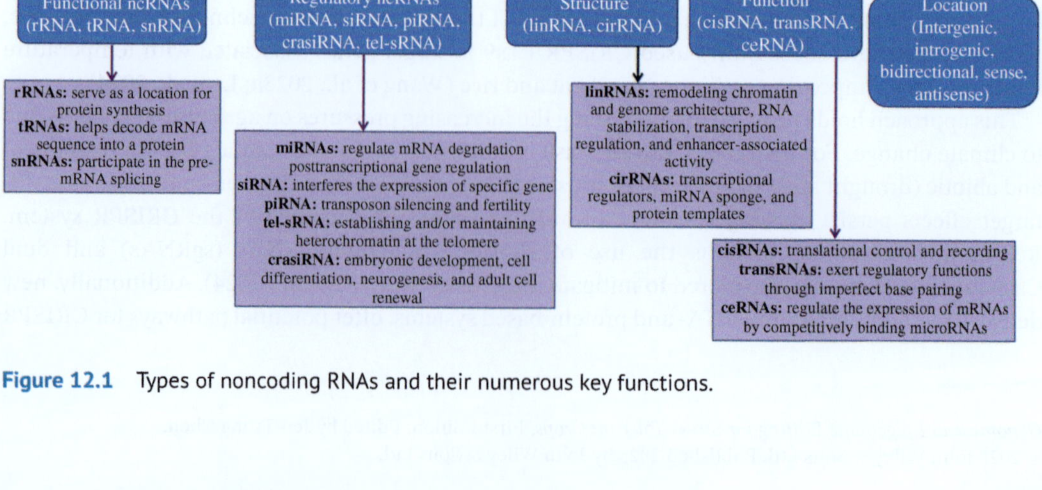

Figure 12.1 Types of noncoding RNAs and their numerous key functions.

12.2.1 MicroRNAs (miRNAs) in Temperature Stress Responses

miRNAs are small, approximately 20–24 nucleotide long ncRNAs that regulate gene expression at the posttranscriptional level by pointing definite miRNAs for translational repression or degradation. They display significant contribution in temperature stress responses in the following ways:

- **Heat Stress Regulation**: miRNAs like **miR398** have been implicated in the regulation of heat shock proteins (HSPs) and other stress-related genes. Under high-temperature conditions, miR398 levels are altered, which in turn affects target gene expression involved in the response against heat stress and hence enhances plant survival (Zhang et al., 2020).
- **Cold Stress Regulation**: Specific miRNAs, such as **miR156**, have been associated with cold stress responses. The miR156 is involved in the developmental transition and tolerance to stress by targeting the transcriptional factors, thereby modulating the expression of genes in response to low temperatures (Liu et al., 2020).
- **Signal Transduction**: miRNAs are involved in signaling pathways that mediate the perception of temperature changes. They can regulate various transcription factors and signaling molecules, thereby integrating environmental signals and activating stress response pathways (Khan et al., 2019).

12.2.2 lncRNAs (Long Noncoding RNAs) and Temperature Stress Responses

The lncRNAs are typically greater than 200 nucleotide spans and are known to have regulatory functions in gene expression. Their roles in temperature stress responses include:

- **Regulation of Gene Expression**: lncRNAs can control the gene expression associated with temperature stress by acting as scaffolds or guides in chromatin remodeling and transcriptional regulation. For example, certain lncRNAs revealed enhanced expression of genes responsive to cold stress, improving the cold tolerance of plants (Sunkar et al., 2007; Ding et al., 2020).
- **Interaction with miRNAs**: lncRNAs may serve as competitive endogenic RNAs (ceRNAs), sequester miRNAs, and hence prevent them from their mRNA targets. This mechanism can modulate the availability of miRNAs in temperature stress responses, allowing for fine-tuning of the expression of genes under the stress environments (Zhao et al., 2017).
- **Stress Memory**: lncRNAs can play a vital role in stressful environments, allowing the plants to "remember" their past stress conditions and help plants to response in future stresses more effectively. This adaptation is particularly beneficial for plants facing fluctuating temperatures (Rodriguez-Moreno et al., 2020).

12.2.3 Integration of miRNAs and lncRNAs in Temperature Stress Responses

The interaction between miRNAs and lncRNAs forms a complex regulatory network that enables plants to respond adaptively to temperature changes. This integration permits plants for a synchronized response toward stress, involving multiple layers of gene regulation:

- **Regulatory Feedback Loops**: miRNAs and lncRNAs can participate in feedback loops where the expression of one influences the other, leading to enhanced resilience against temperature stress.
- **Pathway Crosstalk**: The regulatory roles of miRNAs and lncRNAs extend to crosstalk with other signaling pathways, such as those involving hormones (e.g. abscisic acid and auxins), further amplifying the plant's ability to cope with temperature fluctuations (Xie et al., 2020; Table 12.1).

Table 12.1 Examples of noncoding RNAs (ncRNAs) involved in temperature stress response.

ncRNA type	Target genes	Stress type	Function in temperature stress	References
miR156	SPL (SQUAMOSA PROMOTER BINDING PROTEIN-LIKE)	Heat stress	Increases heat shock protein production	Wang et al. (2023b)
miR397	LACCASE	Cold and heat stress	Modulates antioxidant activity during temperature stress	Liu et al. (2021a)
lncRNA COLDAIR	CBF pathway genes	Cold stress	Enhances cold tolerance by activating cold-responsive genes	Zhao et al. (2020)
ZmNCED lncRNA	Heat shock protein genes	Heat stress	Promotes heat stress resilience through increased protein expression	Chen et al. (2021)

12.3 CRISPR/Cas Technology for Editing Noncoding RNAs

CRISPR/Cas technology, a precise revolutionary tool for genome editing, was developed originally from the bacterial immune system. In recent years, this tool has been effectively used to edit both coding and noncoding regions of the plant genome. The ncRNAs (miRNAs) and lncRNAs are the important controlling molecules that regulate the expression of genes in response to environmental stress, especially temperature. CRISPR/Cas technology provides a precise mechanism for manipulating these ncRNAs to enhance tolerance against stresses in crops, making it a powerful method for developing temperature-smart crops (Basak & Nithin, 2015; Table 12.2).

12.3.1 Overview of CRISPR/Cas Technology

Major components that consist of CRISPR/Cas9 are as follows:

- **Cas9 Nuclease**: A protein that induces double-stranded breaks (DSBs) in specific locations on DNA.
- **Guide RNA (gRNA)**: It is a synthetic RNA that is designed to guide the Cas9 enzyme for a specific sequence of DNA.

Table 12.2 CRISPR/Cas systems used in ncRNA editing for temperature stress tolerance.

CRISPR/Cas system	Edited ncRNA	Target crop	Outcome	References
CRISPR/Cas9	miR156	Tomato	Enhanced heat tolerance	Wang et al. (2023b)
CRISPR/Cas12b	miR397	Barley	Improved cold and heat tolerance	Liu et al. (2021a)
CRISPR/Cas9	lncRNA LAIR	Rice	Increased cold tolerance	Liu et al. (2022)
CRISPR/Cas12	lncRNA COLDAIR	Wheat	Enhanced cold tolerance	Zhao et al. (2020)

When Cas9 is guided to target a site by the gRNA, it cuts the DNA, leading to either a gene knockout or precise editing through the introduction of mutations, insertions, or deletions during DNA repair. This allows researchers to target and modify specific ncRNA sequences to control gene expression and regulatory networks encompass response against temperature stress (Jaganathan et al., 2018).

12.3.2 Editing of miRNAs (MicroRNAs)

The miRNAs are the small ncRNAs that regulate the expression of genes at the posttranscriptional level by inhibiting or degrading mRNA translation. Using CRISPR/Cas technology, specific miRNAs can be edited to alter their regulatory functions:

- **Targeting miR398 for Heat Tolerance**: The miR398 regulates the expression of the HSPs, which protect plants under heat stress. By editing miR398 using CRISPR/Cas, researchers can reduce its activity, allowing for the overexpression of HSPs, thereby improving heat tolerance (Zhang et al., 2020).
- **Manipulating miR156 for Cold Tolerance**: The miR156 targets transcription factors that regulate developmental processes and stress responses. To knockout or modify miR156 for enhancing cold tolerance by altering the expression of genes responsible for cold acclimation, CRISPR/Cas9 can be used successfully (Liu et al., 2020).

12.3.3 Editing of lncRNAs (Long Noncoding RNAs)

The lncRNAs are responsible for numerous biological processes including remodeling of chromatin, regulation of transcription, and signal transduction. CRISPR/Cas9 technology can be applied to edit lncRNAs to improve temperature stress responses in plants:

- **Modifying COLD1 lncRNA for Cold Tolerance**: COLD1 is a cold-responsive lncRNA in plants that regulates cold stress pathways. Genome editing of the COLD1 locus mediated by CRISPR/Cas9 can increase the cold stress response in plants by increasing the expression of cold-responsive genes (Ding et al., 2020).
- **lncRNA-Mediated Heat Stress Responses**: Certain lncRNAs are involved in regulating heat-responsive genes. Editing lncRNAs using CRISPR/Cas can alter the gene expression by maintaining cellular homeostasis under heat stress, leading to improved thermotolerance (Liu et al., 2020).

12.3.4 CRISPR/Cas Variants for ncRNA Editing

For precise and diverse editing of ncRNAs, various CRISPR/Cas methods have been developed as follows:

- **CRISPR/Cas9**: The most widely used version, ideal for knocking out ncRNAs or inducing small insertions/deletions.
- **CRISPR/dCas9**: A modified version of Cas9 that lacks nuclease activity, allowing for transcriptional regulation without cutting the DNA. This version can be used to modulate ncRNA expression by recruiting transcriptional activators or repressors (Qi et al., 2013).
- **CRISPR/Cas13**: Specifically targets RNA rather than DNA. Cas13 can be used to directly target and degrade ncRNAs, providing an alternative approach to gene regulation (Abudayyeh et al., 2017).

12.4 CRISPR/Cas9 Editing of Noncoding RNAs (ncRNAs) for Cold and Heat Tolerance

The ncRNAs are crucial regulatory components that perform a significant role in numerous biological events in plants, particularly in their responses to abiotic stresses such as temperature fluctuations. lncRNAs and small interfering RNAs (siRNAs) among different ncRNAs are particularly important to regulate gene expression and facilitate stress tolerance. The CRISPR/Cas9 gene-editing technology provides an innovative approach for modifying ncRNAs, enhancing the ability of crops to withstand cold and heat stress. This section explores the roles of ncRNAs in temperature tolerance, CRISPR/Cas9 for editing ncRNAs, and its implications for agricultural resilience.

12.4.1 Role of ncRNAs in Temperature Stress Responses

ncRNAs are involved in various regulatory mechanisms that help plants adapt to temperature stresses:

- **Regulation of Gene Expression**: ncRNAs can modulate gene expression response to stress by interacting with the transcription factors, chromatin modifiers, and other regulatory proteins. For example, some long ncRNAs can play a role as molecular scaffolds to bring together regulatory proteins, influencing the transcriptional activities of neighboring genes (Wang et al., 2019).
- **Stress Signaling Pathways**: ncRNAs are integral to stress signaling pathways. They can facilitate the crosstalk between different signaling molecules, enabling plants to mount appropriate responses to temperature changes (Ding et al., 2020).
- **Posttranscriptional Regulation**: Noncoding small RNAs, for example, miRNAs and siRNAs are involved in the posttranscriptional regulation of target mRNAs, affecting their stability and translation. This allows for rapid responses to environmental changes (Liu et al., 2020).

12.4.2 CRISPR/Cas9 Mechanism for Editing ncRNAs

CRISPR/Cas9 technology offers a powerful platform for precisely editing ncRNAs in plants, facilitating the enhancement of temperature resilience.

12.4.2.1 Design of Guide RNAs (gRNAs)

- The first step involves designing a specific gRNA that targets the genomic sequences encoding the ncRNAs. This gRNA must be carefully designed to ensure high specificity and minimal off-target effects.

12.4.2.2 CRISPR/Cas9 Components Delivery

- The gRNA and Cas9 complex can be introduced into the plant cells through various methods, including *Agrobacterium*-mediated transformation or biolistics. Efficient delivery is critical for successful editing.

12.4.2.3 Target Recognition and Cleavage

- The gRNA guides the Cas9 nuclease into the target ncRNA gene once it enters the cell. Cas9 then induces DSB at the specific position of the DNA.

12.4.3 DNA Repair Mechanisms

After DSBs, the plant's DNA repair machinery is activated. The primary repair mechanisms include the following:

- **Nonhomologous End Joining (NHEJ)**: This mechanism can result in the addition or deletion (Indels) at the break position, leading to loss of function mutation in the ncRNAs.
- **Homology Directed Repair (HDR)**: HDR can be used to introduce specific mutations if the donor template is provided and that enhances the function of the ncRNA (Chen et al., 2021).

12.5 Applications in Cold and Heat Tolerance

Editing ncRNAs using CRISPR/Cas9 has significant potential for developing crops that can better withstand cold and heat stress.

12.5.1 Cold Tolerance

- ncRNAs that regulate the cold-responsive gene expression and can be edited to enhance cold tolerance. For example, targeting lncRNAs responsible for cold stress can lead to increased expression of C-repeat binding factor (CBF) genes, which show a critical role in the acclimation of cold (Ding et al., 2020; Figure 12.2).

12.5.2 Heat Tolerance

- ncRNAs associated with heat stress responses can be modified to boost the HSPs expression. By editing lncRNAs that negatively regulate HSP expression, researchers can improve the ability of the plant to cope with high temperatures, promoting better plant growth and yield in heat stress environments (Mittler et al., 2018; Liu et al., 2020; Figure 12.2).

12.6 Case Studies: Temperature-Smart Crops Developed Through ncRNA Editing

Recent advancements in CRISPR/Cas-mediated editing of ncRNAs have opened new possibilities for enhancing crop resilience to extreme temperatures. This section provides an overview of specific case studies where ncRNA editing has been successfully employed to develop temperature-smart crops, focusing on improving cold and heat tolerance. The examples highlight the diverse strategies used and the outcomes observed in different crops (Table 12.3).

12.6.1 Heat-Resilient Maize

Maize (*Zea mays*) is a staple food crop for many parts of the world, but it is highly susceptible during the reproductive stage to heat stress. Heat stress at the flowering stage and during grain filling may lead to significant yield losses in maize. One key mechanism by which maize adapts to heat stress involves the regulation of stress-related genes through ncRNAs, specifically miRNAs like miR156.

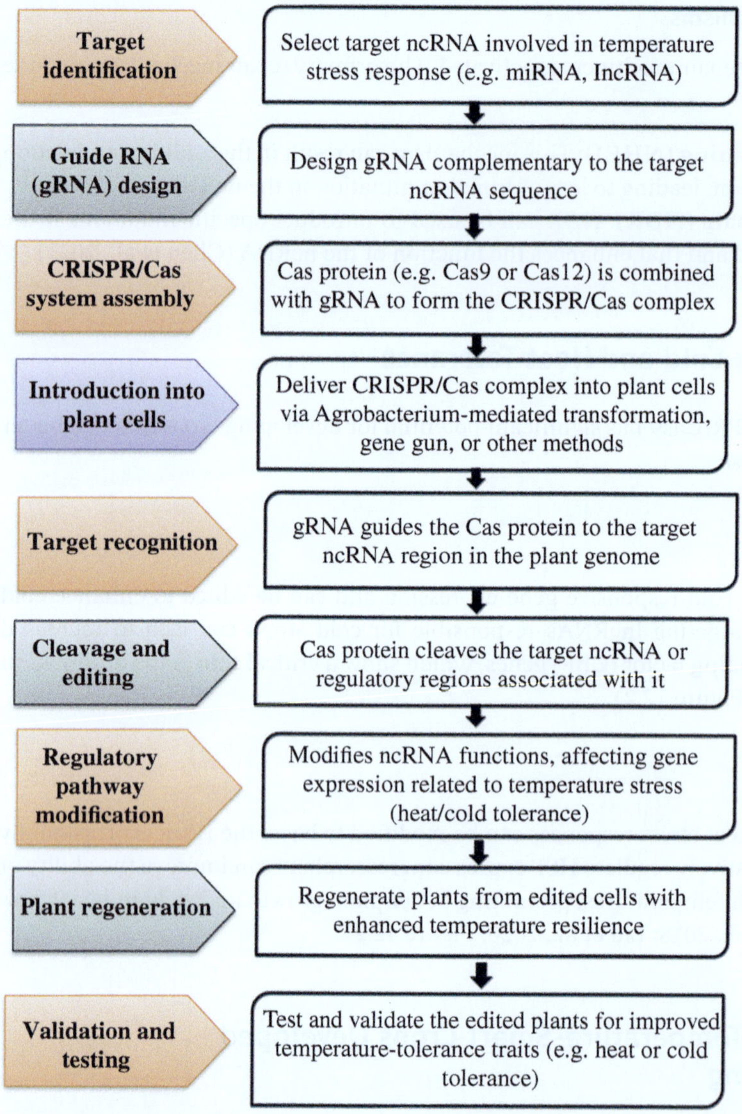

Figure 12.2 CRISPR/Cas mechanism on ncRNAs for temperature-smart crops.

Table 12.3 Potential applications of edited ncRNAs in temperature-smart crops.

Edited ncRNA	Crop	Improved trait	Mechanism	References
miR156	Tomato	Heat tolerance	Upregulation of heat shock proteins	Wang et al. (2023b)
miR397	Barley	Dual cold and heat tolerance	Regulation of antioxidant activity	Liu et al. (2021a)
lncRNA LAIR	Rice	Cold tolerance	Activation of CBF pathway genes	Liu et al. (2022)
lncRNA COLDAIR	Wheat	Cold tolerance	Activation of cold-responsive genes	Zhao et al. (2020)

Using CRISPR/Cas9 technology, researchers have edited miR156 to upregulate SQUAMOSA promoter-like (SPL) transcription factor expression, which enhances stress responses and promotes the growth of the plant. Zhang et al. (2020) demonstrated that CRISPR-mediated editing of miR156 in maize resulted in increased tolerance to high temperatures, improved photosynthetic efficiency, and delayed senescence, leading to higher grain yields under heat stress conditions (Zhang et al., 2020). The upregulation of SPL genes helped modulate processes such as flowering time and leaf development, which is critical for heat adaptation.

In addition to miR156, other miRNAs such as miR167 and miR164 have been implicated in heat stress regulation, making them potential candidates for future CRISPR/Cas-based editing to create even more heat-resilient maize varieties.

12.6.2 Cold-Tolerant Rice

Another staple food crop, rice (*Oryza sativa* L.), is highly sensitive to cold stress, especially during seedling and reproductive stage. Cold stress can severely reduce plant growth, delay flowering, and reduce grain yield. The ncRNA-mediated cold response pathways in rice have been a major focus of research aimed at improving cold tolerance.

A key target for cold tolerance in rice is the lncRNA COLD1, which plays a significant role in cold sensing and response. To enhance cold tolerance in rice, CRISPR/Cas9 was successfully applied to manipulate the expression of the COLD1 gene. The gene COLD1 was edited to improve cold sensing in rice seedlings, leading to enhanced survival rates and growth under cold stress conditions (Liu et al., 2020). Edited rice plants demonstrated improved cold tolerance at temperatures as low as 4 °C, with higher rates of photosynthesis, better root development, and less oxidative damage compared to nonedited controls.

Furthermore, miR319, another important miRNA responsible for cold stress, was targeted to regulate the expression of cold-responsive genes by applying CRISPR/Cas9 through improving both the early development and grain yield of rice under cold conditions (Wang et al., 2020).

12.6.3 Heat-Resilient Wheat

Wheat (*Triticum aestivum*) is another most widely cultivated crop in the world, but the productivity of this crop is significantly affected due to heat stress. Similar to maize and rice, wheat heat tolerance is regulated by complex gene networks, many of which are controlled by ncRNAs.

Research has focused on editing miRNAs that regulate heat-responsive genes in wheat. miR159, for example, regulates MYB transcription factor expression responsible for hormone signaling and response to stress. In wheat, CRISPR/Cas9 has been successfully used to knockout, leading to increased expression of MYB genes, which improved heat tolerance by enhancing ABA (abscisic acid) signaling pathways, which are crucial for stress adaptation (Khan et al., 2019).

The lncRNAs also showed a significant role in heat stress responses in wheat. For instance, the lncRNA HSFA1a was recently identified as a regulator of heat tolerance in wheat. Researchers enhanced the expression of HSPs by manipulating these lncRNAs, which are critical for protecting cellular structures during heat stress (Yu et al., 2021).

12.6.4 Cold-Tolerant Barley

Barley (*Hordeum vulgare*) is often grown in cooler climates but can still suffer from cold stress, particularly during early development. The manipulation of cold-responsive ncRNAs using CRISPR/Cas technology has shown promise in improving barley's resilience to low temperatures.

One successful example is the editing of miR172, which regulates the expression of APETALA 2 (AP2) transcription factors responsible for flowering and cold tolerance. CRISPR/Cas9-mediated suppression of miR172 in barley led to increased AP2 expression, enhancing cold tolerance and ensuring proper flowering under cold conditions (Zhao et al., 2017). The edited plants exhibited higher photosynthetic rates, reduced damage from frost, and improved grain quality. Similarly, the lncRNA HVRNX1 was targeted for editing, resulting in increased cold tolerance by modulating the vernalization process in barley (Franco-Zorrilla & Solano, 2017).

12.7 Challenges and Future Directions

12.7.1 Challenges in ncRNA Editing

While CRISPR/Cas9 technology has proven highly effective for DNA editing, editing ncRNAs presents unique challenges due to the structural complexity and multifunctionality of these molecules. Some specific challenges are as follows:

- **Off-Target Effects**: CRISPR/Cas9 has the potential for off-target mutations. In the context of ncRNA editing, off-target effects could disrupt essential regulatory networks, leading to unintended phenotypic consequences (Gao, 2016). Rigorous testing and validation are required to ensure that off-target edits do not impair other vital processes.
- **Pleiotropic Effects**: Many ncRNAs, especially lncRNAs, are involved in multiple regulatory pathways. Editing them may produce pleiotropic effects that could negatively impact plant development or other stress responses (Feng et al., 2020). For instance, editing a lncRNA involved in cold tolerance could inadvertently affect flowering time or other developmental processes.
- **Regulatory Barriers**: Despite significant advancements in CRISPR/Cas technology, the regulatory frameworks governing GMOs vary across different regions. In some countries, CRISPR-edited crops are classified as GMOs and are subject to strict regulations, while in others, they may be exempt if they do not contain foreign DNA (Jaganathan et al., 2018).

12.7.2 Growing Potentials of CRISPR/Cas as a Genome-Editing Tool

Base editing and prime editing have emerged as new CRISPR-based tools and offer more accurate and efficient tools for ncRNA editing. These technologies allow for single nucleotide changes without introducing double-strand breaks, reducing the risks of off-target effects and enabling finer control over ncRNA functions (Komor et al., 2016; Lin et al., 2020).

- **Base Editing**: This technique enables the conversion of specific base pairs (e.g. T to C or G to A) without causing a DSB. This approach is ideal for ncRNA editing, where even a single nucleotide change can drastically affect function.
- **Prime Editing**: Prime editing allows for more precise insertions, deletions, or base pair changes and could be particularly useful for modifying lncRNAs without introducing large disruptions to the surrounding genomic region.

These newer tools, combined with CRISPR/Cas, hold immense potential for future applications in developing climate-resilient crops. Researchers are also exploring the use of CRISPR interference (CRISPRi) and CRISPR activation (CRISPRa) to regulate ncRNA expression dynamically, allowing for temporary or conditional expression of gene response changes to environmental signals (Qi et al., 2013).

12.7.3 Field Implementation and Large-Scale Adoption

While CRISPR/Cas9-based editing of ncRNAs has shown promising results in the lab, field trials are essential to evaluate the practicality and long-term benefits of temperature-smart crops. In controlled laboratory settings, environmental conditions are stable, and it is easier to observe specific stress responses. However, field environments are variable, with a complex interplay of biotic stress as well as abiotic stress, which can influence the growth of the plant and its development.

Key challenges for field implementation are as follows:

- **Environmental Variability**: Heat and cold stresses in the field are often accompanied by other stressors such as drought, salinity, or pest pressures. Edited crops must be tested under a range of environmental conditions to ensure robustness.
- **Stability Across Generations**: Edited plants need to maintain the desired traits across multiple generations. The stability of the introduced traits under varying field conditions and across different crop cycles is critical.
- **Economic Viability**: Large-scale adoption will require economic analysis to ensure that edited crops provide a return on investment for farmers. Higher yields under stress conditions, reduced input costs, and resilience to climate variability are key factors that influence adoption.

Field trials for temperature-smart crops edited through ncRNAs will be necessary to ensure that these crops perform well not only under experimental conditions but also in real-world agricultural settings. Such trials are already underway in crops like maize, rice, and wheat, but scaling up these efforts across a wider range of crops and geographical regions is essential for widespread adoption.

12.8 Future Prospects for Temperature-Smart Crops

As the impacts of climate change intensify, the demand for temperature-smart crops will increase globally. The use of CRISPR/Cas technology to edit ncRNAs holds immense potential for addressing the challenges posed by fluctuating temperatures in agriculture. Looking forward, several exciting avenues for research and development are emerging.

12.8.1 Developing Multistress-Resilient Crops

One key area of future research is the development of crops that can tolerate multiple abiotic stresses simultaneously. In the real world, crops are not just exposed to extreme temperatures but also face drought, salinity, and flooding. By targeting ncRNAs that regulate multiple stress-response pathways, researchers can engineer crops that are resilient to a broader range of environmental challenges. For example, miR159 and miR398 have been implicated in both heat and drought tolerance, making them ideal targets for multistress-resilience engineering (Zhang et al., 2019; Tang et al., 2020).

Future applications may involve the simultaneous editing of multiple ncRNAs using advanced CRISPR techniques, such as multiplexed CRISPR systems, to create crops that can thrive in complex and unpredictable environments (Xie et al., 2020).

12.8.2 CRISPR/Cas Systems Beyond Cas9

Cas9 is the most well-known protein associated with CRISPR-based genome editing, and the identification of different Cas proteins expands the range and versatility of the CRISPR toolkit (Makarova et al., 2020; Van Vu et al., 2022). These systems offer various advantages, such as different target recognition capabilities, enhanced specificity, and the capability of editing both RNA and DNA.

12.8.2.1 Cas12 (Cpf1)

Cas12, also known as Cpf1, offers several benefits over Cas9, such as producing staggered double-strand breaks, which can facilitate more precise edits. It is smaller in size than Cas9, making it easier to deliver via vectors like viral systems. Unlike Cas9, which requires a specific protospacer adjacent motif (PAM) sequence (NGG), Cas12 recognizes T-rich PAMs, expanding the targetable regions of the genome (Zetsche et al., 2015; Smith et al., 2024).

12.8.2.2 Cas13

Cas13 is unique in that it specifically targets RNA instead of DNA, making it ideal for RNA editing applications. This could be used for controlling gene expression without altering the genome itself. Cas13 has been employed for diagnostics, especially in viral detection, because of its ability to cleave RNA sequences after recognition (Zhou et al., 2024).

12.8.2.3 Cas12b

Cas12b is another variant of the Cas12 family that has been shown to work efficiently in both prokaryotic and eukaryotic systems. It offers high accuracy in gene editing and can function at a broader temperature range, making it suitable for editing in different environmental conditions (Li et al., 2024).

12.8.2.4 CasX

CasX is a smaller protein compared to both Cas9 and Cas12, offering a more compact and efficient tool for genome editing, especially in scenarios where size constraints are critical, such as viral delivery. It has been found to be highly specific and is still in the early stages of development, with potential in both research and therapeutic applications (Nguyen et al., 2024).

12.8.2.5 Cas14

Cas14 is another recently discovered system, notable for being one of the smallest Cas proteins. It has a unique ability to target single-stranded DNA and is particularly useful in settings requiring high specificity without significant off-target effects (Garcia et al., 2024).

12.8.2.6 Applications and Future Directions

These alternative CRISPR systems open up new avenues in genome editing, from plant breeding and disease resistance to precise RNA modulation and diagnostics (Borrelli et al., 2018). With ongoing research and refinement, Cas proteins beyond Cas9 hold significant promise for transforming biological research and agriculture.

12.8.3 Synthetic Biology and ncRNA Engineering

The intersection of CRISPR technology with synthetic biology holds great promise for creating new ncRNA molecules that do not exist in nature but can be engineered to perform specific functions in plant stress-response pathways (Guan et al., 2013). By designing synthetic ncRNAs, researchers could enhance the precision of gene regulation, enabling crops to switch on or off specific pathways in response to temperature changes. Such advances could lead to crops that are even more finely tuned to their environment, improving their ability to withstand extreme temperatures.

For example, synthetic miRNAs could be designed to target key genes responsible for heat stress, while synthetic lncRNAs could be engineered to modulate chromatin structure in response to cold stress. This approach could lead to the creation of crops with entirely new regulatory networks that are optimized for climate resilience (Rodriguez-Moreno et al., 2020).

12.8.4 Gene-Drive Systems for Rapid Trait Dissemination

Gene-drive technology, which promotes the rapid spread of specific genetic traits through populations, could accelerate the adoption of temperature-smart crops. By integrating CRISPR-based gene drives with ncRNA editing, researchers could potentially spread temperature-resilience traits through entire crop populations more quickly than traditional breeding or genetic modification methods (Esvelt et al., 2014).

For example, in crops like wheat or rice, gene-drive systems could be used to ensure that ncRNA modifications conferring heat or cold tolerance are passed on to future generations at a higher rate than would occur naturally. This could be particularly useful in regions facing rapid climate change, where traditional breeding methods might not keep pace with environmental shifts.

12.8.5 Precision Agriculture and ncRNA-Based Sensors

The development of **biosensors** based on ncRNAs could further enhance temperature-smart crops by allowing real-time monitoring of stress responses. These ncRNA-based sensors could be integrated into crops to detect changes in temperature or other stressors at the molecular level, triggering specific protective responses when thresholds are crossed (Kushwaha et al., 2021).

By combining CRISPR/Cas9-based ncRNA editing with precision agriculture tools, farmers could monitor crop stress levels in real time and apply targeted interventions such as irrigation or nutrient application only when needed. This approach would improve resource efficiency while ensuring optimal crop performance under temperature stress.

12.9 Conclusions

As global temperatures continue to rise, the development of temperature-smart crops will become increasingly vital for ensuring sustainable agriculture and protecting food systems from climate-induced disruptions. The availability of arable land is decreasing due to urbanization, increasing population growth and climate change adversely affects the sustainable production of food and faces many obstacles. To face these challenges, plant breeders need revolutionary breeding techniques, like CRISPR/Cas, to generate new traits and varieties that can increase crop yield and tolerate adverse environmental conditions. Through the precise editing of ncRNAs, CRISPR/Cas technology offers a powerful and adaptable tool for safeguarding the future of farming in an increasingly unpredictable world. By utilizing CRISPR/Cas9 tools, many researchers have successfully modified and improved many traits of interest related to yield and quality of the crops. Additionally, several genome-edited crops are released for commercial cultivation such as CRISPR-wx1 maize, TALEN-ppo potato, and TALEN-fad2 soybean. The future of temperature-smart crops lies in overcoming current challenges such as off-target effects and regulatory hurdles while expanding the toolkit with next-generation CRISPR systems, synthetic biology approaches, and multistress-resilience strategies. Field trials and large-scale implementation will play a crucial role in bringing these innovations from the lab to the farm, where they will make a tangible impact on agricultural productivity.

References

Abudayyeh, O. O., Gootenberg, J. S., Essletzbichler, P., Han, S., Joung, J., Belanto, J. J., & Zhang, F. (2017). RNA targeting with CRISPR–Cas13. *Nature*, *550*(7675), 280–284.

Bartel, D. P. (2004). MicroRNAs: genomics, biogenesis, mechanism, and function. *Cell*, *116*(2), 281–297.

Basak, J., & Nithin, C. (2015). Targeting non-coding RNAs in plants with the CRISPR-Cas technology is a challenge yet worth accepting. *Frontiers in Plant Science*, *6*, 1001.

Borrelli, V. M. G., Brambilla, V., Rogowsky, P., Marocco, A., & Lanubile, A. (2018). The enhancement of plant disease resistance using CRISPR/Cas9 technology. *Frontiers in Plant Science*, *9*, 1245.

Borrelli, V. M., De Luca, M., & Rocco, M. (2023). Exploring CRISPR/Cas9-mediated editing of non-coding RNAs for enhancing stress tolerance in crops. *Frontiers in Plant Science*, *14*, 983017.

Chen, L., Liu, X., Zhang, Y., & Wang, S. (2021). Targeted editing of lncRNA ZmNCED using CRISPR/Cas9 enhances heat stress tolerance in maize by increasing heat shock protein expression. *Plant Physiology*, *187*(3), 1534–1548.

Chen, Z., Wu, X., & Zhou, Q. (2024). CRISPR/Cas9 for enhancing crop resilience to abiotic stress: a review. *Plant Science*, *333*, 111442.

Ding, Y., Fang, Y., Guo, Z., Wu, J., Zhang, L., Chen, Y., et al. (2020). Changes in long non-coding RNA expression under drought and salt stresses in rice. *Scientific Reports*, *10*(1), 1–13.

Esvelt, K. M., Smidler, A. L., Catteruccia, F., & Church, G. M. (2014). Concerning RNA-guided gene drives for the alteration of wild populations. *eLife*, *3*, e03401.

Feng, J., Yang, G., & Wang, L. (2020). Long non-coding RNA regulation of gene expression during stress responses in plants. *Current Opinion in Plant Biology*, *54*, 23–29.

Franco-Zorrilla, J. M., & Solano, R. (2017). Long noncoding RNAs controlling transcriptional regulatory networks in plants. *Plant Cell Physiology*, *58*(7), 1206–1216.

Gao, C. (2016). The future of CRISPR technologies in agriculture. *Nature Reviews Molecular Cell Biology*, *17*(5), 343–350.

Garcia, D., Liu, Y., & Wang, X. (2024). Harnessing CRISPR/Cas14 for compact and efficient genome editing in microorganisms. *Frontiers in Genome Engineering*, *11*, 567.

Guan, Q., Wu, J., Yue, X., Zhang, Y., & Zhu, J. (2013). A nuclear calcium-sensing pathway is critical for gene regulation and salt stress tolerance in *Arabidopsis*. *PLoS Genetics*, *9*(8), e1003755.

Jaganathan, D., Ramasamy, K., Sellamuthu, G., Jayabalan, S., & Venkataraman, G. (2018). CRISPR for crop improvement: an update review. *Frontiers in Plant Science*, *9*, 985.

Khan, M. Z., Zaidi, S. S., Amin, I., & Mansoor, S. (2019). A CRISPR way for fast-forward crop domestication. *Trends in Plant Science*, *24*(5), 293–296.

Komor, A. C., Kim, Y. B., Packer, M. S., Zuris, J. A., & Liu, D. R. (2016). Programmable editing of a target base in genomic DNA without double-stranded DNA cleavage. *Nature*, *533*(7603), 420–424.

Kushwaha, H., Chakraborty, S., & Gupta, S. (2021). Advances in CRISPR-Cas technology for crop improvement. *Critical Reviews in Plant Sciences*, *40*(3), 173–195.

Li, P., Zhang, H., Wang, Y., & Liu, Q. (2024). CRISPR/Cas12b system for multiplexed genome editing and temperature regulation in agricultural crops. *Nature Biotechnology*, *42*(1), 45–56.

Lin, Q., Zong, Y., Xue, C., Wang, S., Jin, S., Zhu, Z., et al. (2020). Prime genome editing in rice and wheat. *Nature Biotechnology, 38*(5), 582–585.

Liu, L., Li, Y., Zheng, H., Zhang, M., & Zhang, W. (2020). CRISPR/Cas9-mediated editing of cold-responsive genes enhances cold tolerance in rice. *Molecular Breeding, 40*(12), 1–12.

Liu, F., Zhang, Y., Li, X., & Wang, J. (2021a). Dual improvement of cold and heat stress tolerance in barley through CRISPR/Cas9-mediated editing of miR397. *Plant Cell Reports, 40*(9), 1575–1586.

Liu, Q., Yang, F., Zhang, J., Liu, H., Rahman, S., Islam, S., Ma, W., & She, M. (2021b). Application of CRISPR/Cas9 in crop quality improvement. *International Journal of Molecular Sciences, 22*(8), 4206.

Liu, Y., Zhang, H., Li, J., & Wang, X. (2022). CRISPR/Cas9-mediated knockdown of lncRNA LAIR enhances cold tolerance in rice by upregulating CBF pathway genes. *Plant Biotechnology Journal, 20*(4), 567–579.

Liu, Y., Zhang, H., & Chen, G. (2023). miRNAs and their roles in plant stress responses: a CRISPR/Cas9 approach. *Journal of Experimental Botany, 74*(3), 801–812.

Makarova, K. S., et al. (2020). Evolutionary classification of CRISPR–Cas systems: a maximum-likelihood approach. *Microbiology and Molecular Biology Reviews, 85*(2), e00032–e00019.

Mittler, R., Finka, A., & Goloubinoff, P. (2018). The heat shock response: a tool for plant stress tolerance. *Annual Review of Plant Biology, 69*, 159–184.

Mittler, R., Finka, A., & Goloubinoff, P. (2022). Heat stress: causes, impacts, and responses in plants. *Annual Review of Plant Biology, 73*, 579–607.

Nguyen, T., Kim, J., & Lee, S. (2024). CRISPR/CasX: an emerging tool for genome-wide screening and diagnostics. *Genomics Advances, 13*(3), 87–98.

Qi, L. S., Larson, M. H., Gilbert, L. A., Doudna, J. A., Weissman, J. S., Arkin, A. P., & Lim, W. A. (2013). Repurposing CRISPR as an RNA-guided platform for sequence-specific control of gene expression. *Cell, 152*(5), 1173–1183.

Rodriguez-Moreno, L., González-Melendi, P., Testillano, P. S., Risueno, M. C., & Medina, F. J. (2020). Synthetic biology for future agriculture. *Plant Biotechnology Journal, 18*(12), 2366–2373.

Smith, J., Li, X., Zhang, P., & Wang, T. (2024). Exploring the potential of CRISPR/Cas12 and Cas13 for precise RNA targeting in plants. *Molecular Plant Biology, 22*(2), 149–161.

Sunkar, R., Huang, P. J., & Zhu, J. K. (2007). Temperature-responsive microRNAs in Arabidopsis. *Plant Cell, 19*(9), 2535–2547.

Sunkar, R., Guo, Y., & Liu, Y. (2024). Emerging roles of non-coding RNAs in plant temperature stress adaptation. *Plant Molecular Biology, 108*, 367–385.

Tang, Y., Zhang, Y., & Yang, J. (2020). CRISPR/Cas9-mediated editing of non-coding RNAs enhances resilience to drought stress in rice. *Frontiers in Plant Science, 11*, 915.

Tang, Y., Chen, Q., & Zhang, Y. (2024). The future of crop breeding: combining CRISPR/Cas systems with ncRNA regulation. *Molecular Plant Biotechnology, 22*(1), 75–90.

Van Vu, T., Das, S., Hensel, G., & Kim, J. Y. (2022). Genome editing and beyond: what does it mean for the future of plant breeding? *Planta, 255*, 130.

Wallace, J. G., Rodgers-Melnick, E., & Buckler, E. S. (2018). On the road to Breeding 4.0: unraveling the good, the bad, and the boring of crop quantitative genomics. *Annual Review of Genetics, 52*, 421–444.

Wang, Y., Zhang, H., Chen, X., & Liu, X. (2019). Long non-coding RNAs in plant immunity: an overview. *Frontiers in Plant Science, 10*, 1213.

Wang, Z., Yang, Y., Zhang, J., & Chen, Y. (2020). CRISPR/Cas9-mediated miR156 editing confers enhanced heat tolerance in maize. *Plant Physiology, 182*(2), 1358–1375.

Wang, T., Zhang, Y., & Chen, J. (2023a). Improving cold tolerance in wheat through CRISPR/Cas9-mediated gene editing of lncRNAs. *Frontiers in Plant Science, 14*, 1025634.

Wang, Z., Li, J., Liu, Q., & Chen, H. (2023b). CRISPR/Cas9-mediated editing of miR156 enhances heat tolerance in tomato by upregulating heat shock protein genes. *Frontiers in Plant Science, 14*, 1267.

Xie, K., Zhang, J., Yang, Y., & Chen, Y. (2020). Multiplexed CRISPR/Cas9 for targeted ncRNA editing in crop improvement. *Current Opinion in Plant Biology, 58*, 74–82.

Yu, J., Hu, S., Wang, J., Wong, G. K. S., Li, S., Liu, B., et al. (2021). A draft sequence of the rice genome (*Oryza sativa* L. ssp. *japonica*). *Nature, 409*(6822), 937–943.

Zetsche, B., Gootenberg, J. S., Abudayyeh, O. O., Slaymaker, I. M., Makarova, K. S., Essletzbichler, P., et al. (2015). Cpf1 is a single RNA-guided endonuclease of a class 2 CRISPR-Cas system. *Cell, 163*(3), 759–771.

Zhang, Z., Zheng, Y., & Ma, Y. (2019). Heat stress responsive ncRNAs and CRISPR approaches for climate-smart crop development. *Journal of Experimental Botany, 70*(16), 4081–4095.

Zhang, J., Khan, Z., Zheng, Y., & Ma, Y. (2020). Heat stress responsive miRNAs in wheat. *BMC Genomics, 21*(1), 905.

Zhang, H., Liu, Y., & Chen, G. (2022). Roles of non-coding RNAs in temperature stress response in plants. *Plant Science, 317*, 111225.

Zhao, H., Wang, J., Liu, Y., & Wei, X. (2017). The role of miR172 in regulating AP2 transcription factors for cold tolerance in barley. *Plant Science, 256*, 61–71.

Zhao, Q., Zhang, H., Li, J., & Wang, L. (2020). CRISPR/Cas9-mediated knockout of lncRNA COLDAIR enhances cold tolerance in wheat by activating CBF pathway genes. *Journal of Experimental Botany, 71*(12), 3466–3479.

Zhao, C., Wang, Y., Wu, Y., & Liu, X. (2021). Impact of global temperature rise on crop production: a review. *Agricultural Systems, 187*, 103016.

Zhou, X., Li, M., & Chen, Y. (2024). Cas13-based RNA editing systems: advancements and potential for therapeutic applications. *Annual Review of Plant Genetics, 75*, 211–223.

13

Technical Advancements in Functional Mutagenesis of Plant Long Noncoding RNAs Using CRISPR/Cas Technology

Tushar K. Dutta

Division of Nematology, ICAR-Indian Agricultural Research Institute, New Delhi, Delhi, India

13.1 Introduction

With major advancements made in high-throughput genome and transcriptome sequencing technologies coupled with stringent bioinformatics pipelines over the last few years, noncoding RNAs (ncRNAs) have emerged as one of the major contributors in regulating plant gene expression that ultimately led to wider genetic and phenotypic diversity across the plant kingdom (Wierzbicki et al., 2021; Domínguez-Rosas et al., 2023). Additionally, ncRNAs play a key role in regulating plant immunity to various abiotic and biotic stresses at transcriptional and posttranscriptional levels (Waititu et al., 2020; Huang et al., 2023). Plant epigenetic regulation is also modulated by ncRNAs that induce DNA methylation and histone acetylation (Wierzbicki et al., 2021; Yang et al., 2023). Intriguingly, approximately 90% of eukaryote genomes are transcribed into RNA, while only 2% of transcribed RNAs code for proteins (Pauli et al., 2011; Rai et al., 2019). Therefore, a major portion of the plant transcriptome belongs to ncRNA types, which were earlier assumed as transcriptional noise because of their inability to code proteins, poor sequence conservation, and low expression levels (Basak & Nithin, 2015; Budak et al., 2020; Urquiaga et al., 2021). According to the origin, biosynthesis, and mode of action, ncRNAs are classified into housekeeping and regulatory ncRNAs. Housekeeping ones perform cellular and ribosomal functions that include ribosomal RNA (rRNA), transfer RNA (tRNA), small nuclear RNA (snRNA), and small nucleolar RNA (snoRNA). While rRNAs can be of variable sizes, tRNAs, snRNAs, and snoRNAs are each of 50–200 nucleotides long. Regulatory ncRNAs comprise microRNA (miRNA – 20–24 nucleotides long), small interfering RNA (siRNA – 20–24 nucleotides long), piwi-interacting RNA (piRNA – 24–32 nucleotides long), and long noncoding RNA (lncRNA – more than 200 nucleotides long). lncRNAs are ubiquitous in their presence because they were detected in prokaryotes, viruses, fungi, animals, and plants (Huang et al., 2023). Using deep sequencing techniques, a repertoire of lncRNA candidates was identified from different plant species such as *Arabidopsis thaliana*, rice, tomato, maize, soybean, cotton, potato, and *Medicago truncatula* (Huang et al., 2023; Bravo-Vázquez et al., 2024). The present chapter provides an overview of lncRNA classes and their functions and discusses the applicability of the clustered regularly interspaced short palindromic repeats/CRISPR-associated (CRISPRs-Cas) genome-editing technique for lncRNA functional validation.

Genome and Epigenome Editing for Stress-Tolerant Crops, First Edition. Edited by Jen-Tsung Chen.

13.2 The Complex Architecture of Plant lncRNAs

Among ncRNA classes, lncRNAs are the most abundant and perform key regulatory roles in different pathways of plant gene expression. As per their relative location and direction to the nearby protein-coding gene/genes on the chromosome, plant lncRNAs are categorized into the following classes (Figure 13.1): (i) Sense lncRNA – positioned on the identical strand of its associated protein-coding gene with partial or complete overlap with the coding region. Examples include *COLDWRAP*, *LDMAR*. (ii) Antisense lncRNA (alternatively known as natural antisense transcripts or NATs) – positioned on the opposite strand of its associated protein-coding gene with partial or complete overlap with the coding region. Examples include *COOLAIR*, *ENOD40*, *MAS*, *nalncFL7*, *SABC1*, *SEAIRa*, *Sho*, *SVALKA*, *TWISTED LEAF*, and *asHSFB2α*. (iii) Intronic lncRNA – positioned within an intron of its associated protein-coding gene. Examples include *COLDAIR*. (iv) Bidirectional lncRNA – positioned on the opposite strand of its associated protein-coding gene at <1 kb distance from the promoter region of the protein-coding gene. (v) Long intergenic ncRNA (lincRNA) – positioned in the intergenic region between two protein-coding genes, Examples include *APOLO*, *At5NC056820*, *DAN1*, *ELENA1*, and *IPS1* (Wierzbicki et al., 2021; Huang et al., 2023; Yang et al., 2023). Apart from these, another class named enhancer lncRNA (eRNA) has also been reported These lncRNAs are present in close proximity to the transcription start site (TSS) or transcription termination site (TTS) of the protein-coding gene (Domínguez-Rosas et al., 2023). To date, lincRNAs followed by antisense lncRNAs constitute the major share of characterized plant lncRNAs (Domínguez-Rosas et al., 2023; Huang et al., 2023).

lncRNA biosynthesis is identical to that of messenger RNAs (mRNAs). RNA polymerase II primarily transcribes lncRNA from intergenic, exonic, or distal protein-coding regions of the genome

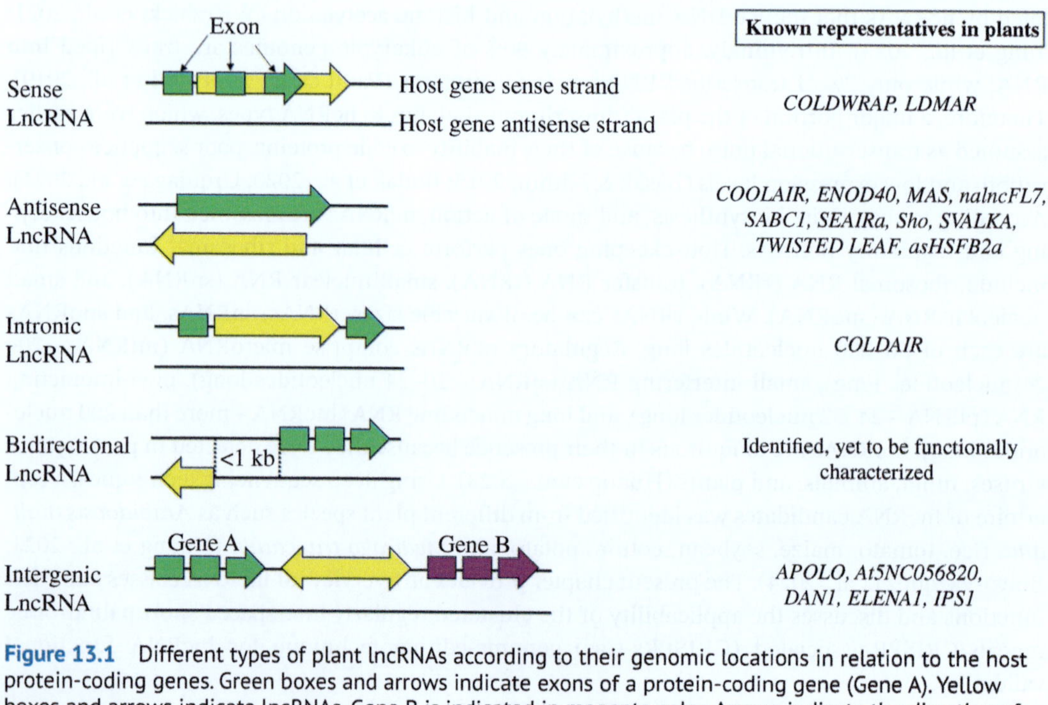

Figure 13.1 Different types of plant lncRNAs according to their genomic locations in relation to the host protein-coding genes. Green boxes and arrows indicate exons of a protein-coding gene (Gene A). Yellow boxes and arrows indicate lncRNAs. Gene B is indicated in magenta color. Arrows indicate the direction of transcription.

(Wierzbicki et al., 2021; Mattick et al., 2023). Post transcription, lncRNA undergoes capping at the 5′ end, polyadenylation at the 3′ end, and occasionally alternative splicing (Quinn & Chang, 2016; Wierzbicki et al., 2021). Apart from RNA polymerase II, RNA polymerases III, IV, and V are capable of transcribing plant lncRNAs (Wierzbicki et al., 2008; Ariel et al., 2015). In *A. thaliana*, RNA polymerase III usually transcribes *AtR8* and *AtR18* lncRNAs. Notably, *AtR8* sequence conservation was documented in *Brassica* spp. and was responsive to hypoxic stresses (Wu et al., 2012). RNA polymerases IV and V transcribe those lncRNAs, which lack polyadenylated tails (Wierzbicki et al., 2008; Zhou & Law, 2015). lncRNAs transcribed by RNA polymerases IV and V are involved in the silencing of transposable elements (TEs). RNA polymerase IV-transcribed lncRNA acts as the template for RNA-dependent RNA polymerase (RDR2) to synthesize 24 nucleotide siRNA precursors that establish and maintain DNA methylation in plants via RNA-directed DNA methylation (RdDM) pathway (Haag et al., 2012; Singh et al., 2019). RNA polymerase V-transcribed lncRNAs act as scaffolds for recruiting 24 nucleotide siRNAs to their corresponding target loci in the genome (Haag et al., 2012; Singh et al., 2019). Due to their lower expression, stability, and sequence conservation, RNA polymerase IV- and V-transcribed lncRNAs are poorly understood compared to the RNA polymerase II-transcribed lncRNAs.

13.3 The Diverse and Dynamic Functions of Plant lncRNAs

lncRNAs regulate the target gene expression in *cis* as well as in *trans*. *cis*-Acting lncRNA regulates gene transcription in close genomic proximity via recruitment or displacement of transcription factors (TFs) at the promoter region of nearby genes (Wierzbicki et al., 2021). In addition, few *cis*-acting lncRNAs can modulate the three-dimensional organization of the genome. These lncRNAs bind to the chromatin-remodeling complex to alter chromatin loops of the corresponding gene and during this process, they can regulate histone modification and epigenetic marks (Huang et al., 2023; Yang et al., 2023). *trans*-Acting lncRNAs regulate gene transcription far from the transcription primary locus of the target gene. They act as scaffolds of protein complexes that deploy transcription- or chromatin-modifying factors. The assembly of protein complexes is also modulated by *trans*-acting lncRNAs (Bardou et al., 2014; Lucero et al., 2021). In addition, lncRNAs bind with proteins and mediate the activity, stability, and subcellular localization of interacting proteins (Wierzbicki et al., 2021; Huang et al., 2023). Further, lncRNAs function as precursors or sponges of some small RNAs (sRNAs) for modulating the mature sRNA expression or act as sRNA decoy for interference in RNA silencing that eventually regulates gene transcription (Wierzbicki et al., 2021; Yang et al., 2023). These diverse functions of plant lncRNAs are schematically represented in Figure 13.2 and described in greater detail in the following paragraphs.

13.3.1 lncRNAs Modulate the Activity of Transcriptional Regulators

TF binds specific DNA sequences via their DNA-binding domain and interacts with proteins of the transcription complex to initiate gene expression (Franco-Zorrilla et al., 2014). lncRNA transcription is regulated by various TFs. Conversely, lncRNAs interact with TFs to regulate the expression of different protein-coding genes (Yang et al., 2023). lncRNAs bind with transcriptional activators or repressors to activate or repress the target gene transcription (Figure 13.2). *TaVRN1* (*VERNALIZATION1*) gene is regulated by APETALA1-like MADS-box TF and transcribed weakly during the vegetative phase of the winter wheat *Triticum aestivum*. However, during the early period of vernalization in *T. aestivum*, *TaVRN1*-derived lncRNA *VAS* (VRN1 alternative splicing) is

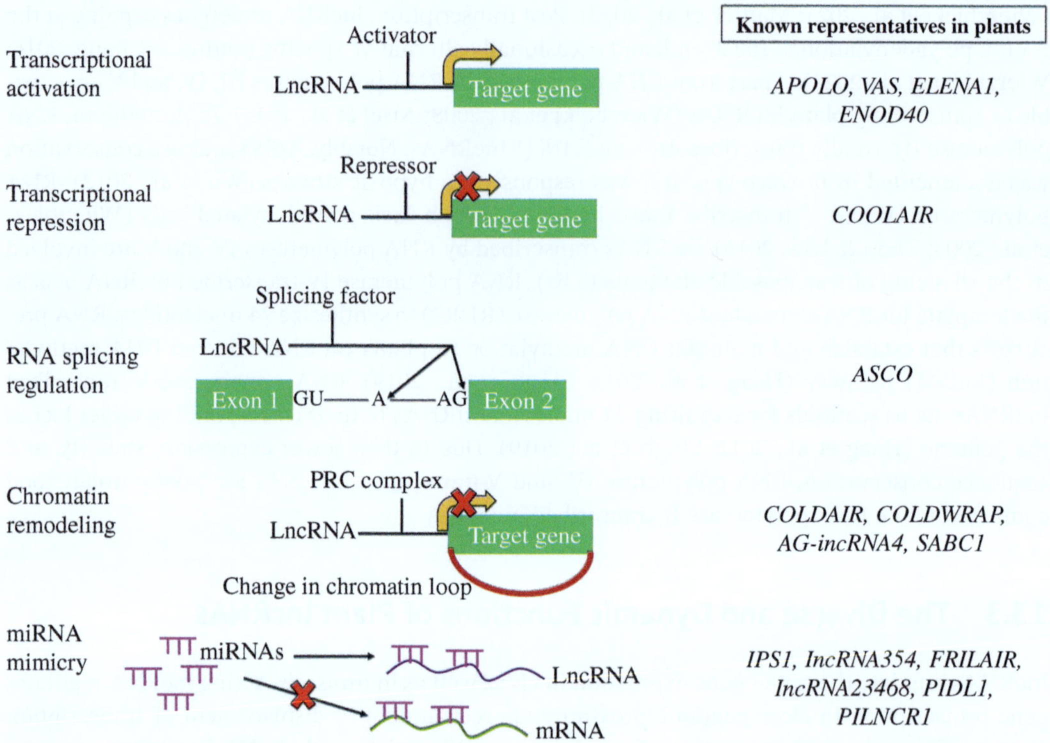

Figure 13.2 lncRNAs regulate plant gene expression via diverse mechanisms. They can interact with transcriptional activators or repressors to activate or repress gene transcription. They control mRNA splicing by binding with the splicing factor or by modulating the precursor mRNA splice junctions. They recruit chromatin-remodeling complexes such as PRC2 for regulating gene expression in the promoter region. They mimic miRNAs and occupy their target sites on the mRNA.

activated. *VAS* interacts with a bZIP TF TaRF2b and aids it in binding and activating *TaVRN1* together with TaRF2a eventually leading to accelerated flowering (Xu et al., 2021a). In *A. thaliana*, lncRNA *APOLO* (auxin-regulated promoter loop) modulates plant response to cold by binding with a TF WRKY42. *APOLO*-WRKY42 ribonucleoprotein (RNP) complex transcriptionally reprograms the root hair development by mediating the expression of *EXT3* (*EXTENSIN3*), *RHD6* (*ROOT HAIR DEFECTIVE 6*), *RSL2* (*ROOT HAIR DEFECTIVE-LIKE*), and *RSL4* genes (Moison et al., 2021; Pacheco et al., 2021). In *A. thaliana*, immunogenic peptide (flg22 and elf18)-induced lncRNA *ELENA1* (*ELF18-INDUCED LONG NON-CODING RNA 1*) interacts with a mediator (a general interactor with TF activation domain) MED19a that affects the saturation of *PR1* (*PATHOGENESIS-RELATED GENE 1*) gene promoter and modulates *PR1* expression for positive regulation of host resistance to *Pseudomonas syringae* pv. *tomato* strain DC3000 (Seo et al., 2017). Thus, lncRNA forms RNP-like macromolecular complexes by binding with TFs and associated proteins for coordinating transcription regulation networks *in planta*.

A number of proteins including importin α, importin β, and ring finger protein modulate nucleo-cytoplasmic partitioning of regulatory proteins in plants (Yang et al., 2023). Similarly, lncRNA *ENOD40* (expressed during the early phase of root nodule development) mediates protein relocalization in leguminous plants. In *M. truncatula*, *ENOD40* lncRNA interacts with MtRBP1 (RNA-binding protein) and relocalizes MtRBP1 from the nucleus to the cytoplasm (Campalans et al., 2004).

13.3.2 lncRNAs Guide Histone Modification

Posttranslational modifications such as histone methylation, acetylation, phosphorylation, ubiquitination, and sumoylation (various types of chemical groups are added to the individual amino acid residues of histone) can affect the chromatin steady state that ultimately alters the binding of different proteins to chromatin and modulates gene transcription status (Hsieh & Fischer, 2005). lncRNAs modulate histone modifications for regulating the target gene transcription (Fonouni-Farde et al., 2021). In *A. thaliana*, an intronic lncRNA *COLDAIR*, transcribes in sense orientation from the first intron of *FLC* (*FLOWERING LOCUS C*) gene, is the first plant lncRNA to be characterized that modulates histone modification (Heo & Sung, 2011). *COLDAIR* interacts with PRC2 (Polycomb Repressive Complex) component CLF (CURLY LEAF) and guides PRC2 to *FLC* chromatin, trimethylates H3K27 (27th lysine residue of H3 histone), and induces silencing of *FLC* while *A. thaliana* plants experience an extended cold environment (Heo & Sung, 2011; Kim et al., 2017). Likewise, another intronic lncRNA *AG-lncRNA4* transcribes from the second intron of *AGAMOUS* (*AG*) gene, recruits PRC2 (via association with CLF) to *AG* chromatin, deposits H3K27me3, and inhibits *AG* expression in *A. thaliana* leaves (Wu et al., 2018). lncRNA *SABC1* (salicylic acid biogenesis controller) fine-tunes SA biosynthesis via recruiting PRC2 to its corresponding gene *NAC3* (encodes a NAC TF) to repress *NAC3* expression by depositing H3K27me3 in the chromatin for *A. thaliana* growth and defense response balancing (Liu et al., 2022). Alternatively, histone marks are also activated by plant lncRNAs. A NAT lncRNA *MAS*, produced from the *MAF4* (*MADS AFFECTING FLOWERING4*) locus, is expressed upon *A. thaliana* exposure to low temperature. *MAS* binds to a COMPASS-LIKE protein complex core component WDR5a (an RNA-binding protein) to tri-methylate H3K4 in *MAF4* chromatin and activates *MAF4* transcription that inhibits *A. thaliana* premature flowering (Zhao et al., 2018). Likewise, a rice lncRNA *LAIR* (*LRK Antisense Intergenic RNA*), transcribes from the antisense strand of *LRK1* (leucine-rich repeat receptor kinase) gene, binds to OsMOF and OsWDR5 to induce H3K4me3 and H4K16ac (acetylation in 16th lysine residue of H4 histone) in LRK1 chromatin to activate *LRK1* expression for improved plant growth and grain yield (Wang et al., 2018).

13.3.3 lncRNAs Modulate Chromatin Loop Formation

The dynamic chromatin loops harbor the protein-coding genes and their associated regulatory elements, which activate or inhibit the target gene transcription. LncRNAs can recruit chromatin-modifying complexes to modulate the three-dimensional chromatin architecture and change gene transcription levels (Yang et al., 2023; Figure 13.2). In *A. thaliana*, upon auxin induction, hypomethylation occurs at lincRNA *APOLO* that opens up the chromatin loop in the promoter region of *APOLO*'s nearby gene *PID* (*PINOID*, a crucial regulator of polar auxin transport), which triggers RNA polymerase II-driven transcription of both *PID* and *APOLO*. Next, RNA polymerase V is recruited to the *APOLO* locus, triggering RdDM. A PRC1 component LHP1 (LIKE HETEROCHROMATIC PROTEIN 1) interacts with *APOLO* Polymerase II transcript to restore chromatin loop formation aided by Polymerase IV and V-mediated DNA methylation (Ariel et al., 2014). Additionally, apart from regulating the chromatin loop of its neighboring *PID* gene, APOLO can also modify the chromatin structure of a repertoire of genes located distantly based on sequence complementarity and R-loop formation (Ariel et al., 2020). Notably, the R-loop is a three-stranded structure containing a DNA-RNA duplex and a displaced single-stranded DNA that has a critical function in organizing genomes and regulating genes (Xu et al., 2017). During vernalization in *A. thaliana*, sense lncRNA *COLDWRAP* together with an intronic lncRNA *COLDAIR* (transcribes from the first intron of floral repressor gene *FLC*) binds to chromatin-modifying complexes PRC2 and CLF. *COLDWRAP* and *COLDAIR* assist in the establishment of an intragenic chromatin loop between *FLC* promoter and the 3′ end of *FLC* first intron ultimately leading to *FLC* silencing (Kim & Sung, 2017). Besides, the R-loop structure formed by an antisense lncRNA *COOLAIR* with the 3′ end of *FLC* gene (stabilized by RNA-binding proteins FCA and FY/WDR33) is also disrupted (via polyadenylation of nascent *COOLAIR* transcript) during vernalization so that chromatin modifiers could be recruited for *FLC* silencing (Xu et al., 2021b). In *A. thaliana*, upon abscisic acid treatment, lncRNA *MARS* (*MARneral Silencing*) activates *MRN1* (*MARNERAL SYNTHASE 1*) transcription by altering its chromatin loop via modulating LHP1 activity (Roulé et al., 2022).

13.3.4 lncRNAs Regulate Alternative Splicing

During posttranscriptional changes in protein-coding genes, precursor mRNAs undergo constitutive splicing (intron is removed and exon is ligated in the sequential order of their location in the gene) and alternative splicing (multiple transcripts are produced from a single gene by exon skipping or intron retention or alternative use of splice sites) to become mature mRNAs. Alternative splicing diversifies the transcriptome and proteome status of a genome (Tognacca et al., 2023). Alternative splicing plays a crucial role in plant organogenesis and response to external stimuli (Staiger & Brown, 2013). lncRNAs play important roles in regulating alternative splicing status in plants (Romero-Barrios et al., 2018; Figure 13.2). In *A. thaliana*, lncRNA *Alternative Splicing Competitor* (*ASCO*) modulates alternative splicing via binding with two nuclear speckle RNA (NSR)-binding proteins, AtNSRa and AtNSRb (Bardou et al., 2014). *ASCO* regulates alternative splicing by inhibiting the binding affinity of different target mRNAs to the NSRs. *ASCO* also binds to two core components of the spliceosomal complex PRP8a and SmD1b, to regulate alternative splicing during *A. thaliana* exposure to flg22 immunopeptide (Rigo et al., 2020).

13.3.5 lncRNAs Regulate Gene Translation

Intriguingly, plant lncRNAs contain evolutionarily conserved small open reading frames (smORFs) of less than 100 codons. Ribosome footprint profiling showed that *A. thaliana* lncRNAs produce detectable peptides, suggesting that lncRNAs are the stockpile of conserved and differentially regulated small peptide-coding genes (Bazin et al., 2017; Fesenko et al., 2021). lncRNAs positively or negatively regulate the gene translation by modulating the polysome (aggregation of a number of ribosomes that actively translate mRNA into a growing polypeptide chain)-associated transcripts. In rice, *OsPHO1.2* (*PHOSPHATE1.2*) in its steady state (an inhibitory structure is formed from the GC-rich region of *OsPHO1.2*) inhibits the binding of 60S subunit of rRNA to its 40S subunit, thereby repressing its translation. During phosphate starvation, antisense lncRNA *cis*-$\text{NAT}_{\text{PHO1.2}}$ alters the inhibitory structure to enhance *OsPHO1.2* translation (Reis et al., 2021).

13.3.6 lncRNAs Act as miRNA Mimics

During miRNA-mediated gene silencing, miRNAs recruit RNA-induced silencing complexes (RISCs) to cleave target mRNA or cause translational inhibition of target mRNA via perfect or imperfect pairing with miRNA recognition element in the target gene (Liu et al., 2017). Plant lncRNAs harbor identical miRNA recognition elements and function as miRNA target mimics (alternatively known as miRNA sponges) that suppress miRNA activity via competitive binding to miRNA and block their interaction with the corresponding target mRNA (Yang et al., 2023; Figure 13.2). In *A. thaliana*, a phosphate deficiency-induced lncRNA *IPS1* (*INDUCED BY PHOSPHATE STARVATION 1*) has a 23 nucleotide motif complementary to miR399 and acts as a decoy to sequester away miR399 from its authentic target gene *AtPHO2*, thus activating *AtPHO2* transcription and modulating inorganic phosphate content in the shoot tissues (Franco-Zorrilla et al., 2007). *LncRNA354* of *Gossypium hirsutum* acts as a noncanonical target mimic of miR160b and suppresses miR160b-regulated degradation of *GhARF17/18* mRNAs to modulate cotton resilience to salt stress (Zhang et al., 2021). Strawberry lncRNA *FRILAIR* harbors the miR397 binding site and acts as an endogenous decoy of miR397 to enhance *LAC11a* mRNA expression for modulating the fruit ripening process (Tang et al., 2021). *LncRNA23468*-miR482b module in *Solanum lycopersicum*, *PIDL1*-miR399 module in *M. truncatula*, and *PILNCR1*-miR399 module in *Zea mays* are the other notable lncRNA-modulated target mimicry systems reported in various plant species (Wang et al., 2017; Du et al., 2018; Jiang et al., 2019).

13.4 Utility and Challenges of CRISPR/Cas System for Deciphering lncRNA Functions

CRISPR/Cas system was earlier identified as a means of bacterial adaptive immunity. The system constitutes tandemly arrayed DNA repeats, which are interspersed with sequences acquired (spacer) from invading bacteriophages (Horvath & Barrangou, 2010; Bhaya et al., 2011). The type II CRISPR/Cas9 system of *Streptococcus pyogenes* bacterium has been extensively investigated and optimized to be used as a classical system for plant genome editing (Cardi et al., 2023). In this system, Cas9 endonuclease guided via a single-guide RNA (sgRNA) duplex creates a double-stranded break (DSB) in the genomic DNA. sgRNA duplex constitutes mature CRISPR RNA (crRNA) and *trans*-acting crRNA (tracrRNA) joined via a linker loop. The 20 nucleotide spacer sequence of crRNA guides sgRNA-Cas9 assembly to the target genomic location via sequence complementarity. Cas9 recognizes and binds to the protospacer adjacent motif (PAM) sequence (5′NGG3′ – immediately preceded by the 20 bp spacer sequence) in the genomic DNA and cleaves both strands (3 bp upstream of PAM) using its HNH and RuvC domains (Figure 13.3a). The sgRNA-Cas9 assembly possesses the unprecedented ability to target and cleave genomic DNA in a sequence-specific manner (Jinek et al., 2012; Mali et al., 2013). Most commonly, induced DSBs are repaired via the error-prone nonhomologous end joining (NHEJ) method that randomly incorporates nucleotide bases via the intrinsic cellular repair machinery. Multiple insertion–deletion (indel) of nucleotides can occasionally cause mutations in the targeted genomic locus due to frameshift in the codons (Uranga & Daròs, 2022; Dutta et al., 2023). On the contrary, the frequency of homology-directed repair (HDR) of DSBs is quite low and in this case, cellular repair machinery must utilize homologous DNA transcripts to perform the precise editing. Homologous transcripts must be exogenously supplied to repair induced DSBs in plant genome editing (Uranga & Daròs, 2022; Dutta et al., 2023).

Similar to mRNA, lncRNAs are mostly transcribed via RNA polymerase II, and their transcription is regulated by TFs that bind to the promoter region to recruit the transcription initiation complex (Lyu et al., 2023). To inactivate the lncRNA expression, the promoter region (approximately 1000–2000 bp upstream sequence from the TSS) must be knocked out using a dual sgRNA-based CRISPR/Cas9 system (Awwad, 2019). Alternatively, the transcription unit of lncRNA (either partial or complete gene sequence) can also be knocked out using a dual sgRNA-based CRISPR/Cas9 system (Figure 13.3b). Notably, longer DNA fragment deletion of plant lncRNAs provided a

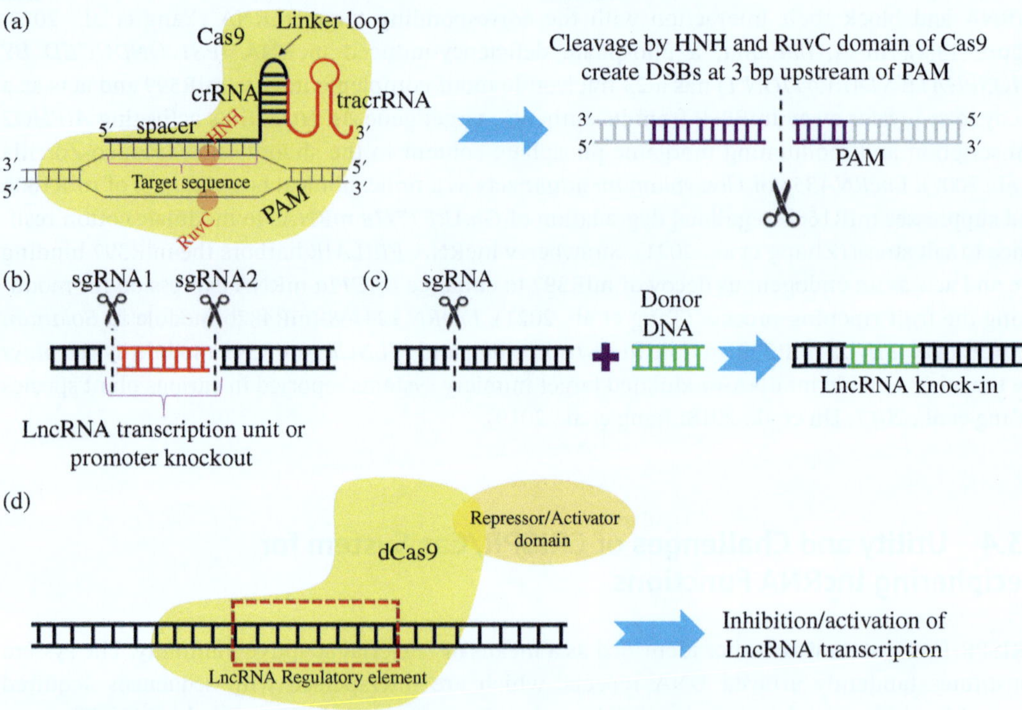

Figure 13.3 lncRNA functional knockout strategies using the CRISPR/Cas system. (a) The canonical CRISPR/Cas9 system that creates double-stranded breaks (DSBs). The single-guide RNA (sgRNA) constitutes CRISPR RNA (crRNA) and *trans*-acting crRNA (tracrRNA) hybrid (joined via linker loop) and binds to the Cas9 nuclease. 20 nucleotide spacer of crRNA guides the sgRNA-Cas9 complex to the target genomic sequence via sequence complementarity. Cas9 recognizes and binds to the PAM sequence in the genomic DNA and cleaves both strands (3 bp upstream of PAM) using its HNH and RuvC domains. (b) Knockout of large DNA fragments from either promoter region or transcription unit of lncRNA using multiple sgRNAs. (c) Introduction of transcription inhibition signal (PolyA sequence) via donor DNA fragment to induce termination of lncRNA expression. (d) Deactivated Cas9 (dCas9) fused with repressor/activator domain can be recruited to the lncRNA promoter region. CRISPR interference (CRISPRi) via dCas9-repressor fusion assembly (interferes with binding of transcription activation factors to the promoter region) confers lncRNA transcriptional inhibition. Conversely, CRISPR activation (CRISPRa) via dCas9-activator fusion assembly induces lncRNA transcription.

greater knockout effect in different plant systems (Bravo-Vázquez et al., 2024). The candidate sgRNA sequences must be designed and optimized via efficient analysis of on-target and off-target editing probabilities. A number of computational algorithms can be utilized for this purpose (Pramanik et al., 2021). The whole sgRNA sequence (20 bp spacer followed by the 76 bp scaffold) can be chemically synthesized and assembled into a Cas9-expressing binary vector. Otherwise, spacer sequences can be assembled into a CRISPR-ready (already contains multiple scaffold sequences in separate expression cassettes) Cas9-expressing binary vector (Dutta et al., 2023). The binary vector can then be transformed into the recipient plant via the *Agrobacterium*-mediated coculture method or particle bombardment/biolistics method. Apart from that, RNP (transformed into protoplasts and plantlets regenerated via tissue culture)-based or viral vector-mediated delivery of CRISPR components has also gained considerable attention (Uranga & Daròs, 2022).

The dual sgRNA-based longer lncRNA fragment deletion (either in promoter or in transcription unit) tactic is quite advantageous. The reasons are as follows: (i) Simplicity in the method, only two sgRNAs are to be assembled, which can be delivered to the plant cell at a single go. (ii) Knockout efficiency is usually high in NHEJ-induced repair mechanisms in plant cells. (iii) Target lncRNA is extremely unlikely to be expressed because its entire promoter region or a large chunk of transcription unit is removed. (iv) Because the editing may generate a ~1 kb gap in the genomic DNA sequence, mutations can be conveniently detected (apart from sequencing or PCR/RFLP or T7E1 assays) using standard PCR amplification followed by the agarose gel electrophoresis. Nevertheless, the major bottleneck associated with this tactic is putative disturbances to the genomic architecture due to induced lncRNA knockout. Since numerous lncRNA loci are either closely overlapped with or included within the genomic loci of the protein-coding genes, CRISPR/Cas-induced removal of the promoter site or the lncRNA transcript may also affect the protein-coding sequences of the lncRNA neighboring genes (Awwad, 2019; Lyu et al., 2023). This can confer change in phenotypes leading to the masking of natural biological functions of candidate lncRNAs. Moreover, lncRNA promoters that are internal or bidirectional would be difficult to knock out. In this case, a partial deletion of selective lncRNA exons that do not intersect with neighboring genes and/or that are distal from their promoter sites can be targeted.

Another CRISPR/Cas strategy for inactivating lncRNA function includes the introduction of a polyadenylation signal sequence downstream of the lncRNA TSS that would cause the dissociation of RNA polymerase II from the lncRNA promoter region ultimately leading to the disassembly of lncRNA transcriptional machinery (Lavalou et al., 2019; Powell et al., 2021). This knock-in strategy encompasses an HDR-based repair mechanism in which a polyadenylation sequence-containing donor template is integrated into the Cas9-induced DSB by homologous recombination (Figure 13.3c). Intriguingly, this tactic curtails the induced disturbances to genomic integrity because only a handful of nucleotides are inserted (Awwad, 2019; Lyu et al., 2023). Nevertheless, a few limitations are also associated with this tactic. For example: (i) Editing efficiency is quite low in HDR-mediated repair mechanism compared to that of NHEJ. (ii) The editing method is somewhat sophisticated because apart from delivering sgRNA and Cas9 editing assemblies, a homolog transcript needs to be delivered separately. (iii) Similar to mRNA, lncRNA undergoes alternative splicing to give rise to multiple splice variants. If a lncRNA transcript harbors an alternative TSS and the transcription termination signal is incorporated either upstream of the second alternative TSS or too far from the TSS, a truncated lncRNA transcript would still be produced and induced knockout might not succeed in entirety (Lyu et al., 2023). Consequently, resultant phenotypes would be falsely attributed to the lncRNA function. In the plant genome-editing landscape, a novel tactic such as prime editing has emerged. Using prime editing, a precise sequence insertion or substitution of DNA sequence can be achieved in any targeted genomic loci. Prime editing uses the assembled prime editing guide RNA (pegRNA) and Cas9 nickase (nCas9), which creates (instead of a DSB) single-stranded nick in the genome DNA. Next, reverse transcriptase (RT, fused with nCas9) along with its RT template (containing the edit) fills the nick via 5′ flap (nascent strand) excision and 3′ flap (edited strand) insertion (Vu et al., 2024).

Another version of Cas9, i.e. dead or deactivated Cas9 (dCas9) can be used to investigate plant lncRNA functions (Figure 13.3d). Inactivation of HNH and RuvC domains of Cas9 resulted in dCas9, which lacks the nuclease activity. Intriguingly, dCas9 maintains its intrinsic property, i.e. RNA-dependent recognition of and binding to the target DNA molecule. Therefore, dCas9 can be customized to be used as TFs by fusing dCas9 with different functional domains such as transcriptional activators or repressors (Perez-Pinera et al., 2013; Qi et al., 2013; Chavez et al., 2015). In a eukaryote cell, dCas9 fused with an inhibitory domain can bind to the promoter region of a target gene that sterically hinders the binding of functional transcription machinery to the promoter region, thereby causing transcriptional repression of the target gene. This mechanism is known as

CRISPR interference or CRISPRi (Awwad, 2019). Alternatively, a fusion of dCas9 with the activator domain can activate the target gene expression in *cis*. This mechanism is known as CRISPR activation or CRISPRa (Awwad, 2019). Both CRISPRa and CRISPRi have negligible off-target effects because these tactics cause minimal alteration in the genomic sequences (Lowder et al., 2015; Piatek et al., 2015). Additionally, epigenetic silencing of plant lncRNAs by using dCas9 appears to be an attractive alternative. Upon fusion with epigenetic effector proteins including DNA methyltransferase or histone modifier, dCas9 can be guided to any lncRNA locus (Pulecio et al., 2017).

13.5 Functional Analysis of Different Plant lncRNAs Using CRISPR/Cas System

Table 13.1 demonstrates a cumulative list of lncRNAs targeted for CRISPR/Cas-based knockout in different plant species. More details are provided in the following paragraphs.

13.5.1 Role of Tomato lncRNAs

In tomatoes, lncRNAs regulate ethylene signaling that eventually mediate fruit ripening (Ma et al., 2020). Approximately 3679 lncRNAs were differentially expressed between wild-type and mutant (for ripening trait) tomato fruits; 490 lncRNAs were upregulated and 187 lncRNAs were downregulated in the mutant fruits indicating their putative role in the ripening process (Zhu et al., 2015). Notably, upon silencing of two lincRNAs such as lncRNA1459 and lncRNA1840 tomato fruit ripening could be deferred (Zhu et al., 2015). Additionally, lncRNAs play a critical role in tomato innate immune response during virus and fungus-induced biotic stresses (Sharma et al., 2022). CRISPR/Cas9 system was utilized to generate loss-of-function mutants of lncRNA1459 in tomato cv. Alisa Craig. Compared to the wild type, in CRISPR-edited lines, the fruit ripening process was significantly delayed. Lycopene biogenesis-related genes such as *PDS* (phytoene desaturase), *PSY* (phytoene synthase), and *ZDS* (zeta-carotene desaturase) were downregulated in the edited lines, while *LCY-b* (lycopene β-cyclase) was upregulated (Li et al., 2018). Likewise, ethylene accumulation was also attenuated in the edited lines as indicated by downregulation of *ACS4* (1-aminocyclopropane-1-carboxylic synthase), *ACS2*, *ACO1* (1-aminocyclopropane-1-carboxylic acid oxidase), and *ERF6* (ethylene response factor TF) genes. *SlDML2* (*DEMETER-LIKE*), a crucial fruit ripening-related gene (associated with DNA hypomethylation), was also repressed in the edited plant. Most importantly, the expression of fruit ripening-related lncRNAs such as lncRNA26, lncRNA28, and lncRNA48 was markedly altered in the edited lines (Li et al., 2018). All these findings suggested that lncRNA1459 is one of the major modulators of the tomato fruit ripening process.

Six tomato lncRNAs such as lncRNA443, lncRNA609, lncRNA1780, lncRNA2155, lncRNA2250, and lncRNA3197 were validated (by qPCR, chromatin-based immunoprecipitation or ChIP-sequencing, and electrophoretic mobility shift assay) as targets of MADS-box TF RIPENING INHIBITOR (RIN). These lncRNAs bound with RIN directly in their promoter regions in vitro and in vivo (Yu et al., 2019). Further, the role of lncRNA2155 in the fruit ripening process was confirmed via CRISPR/Cas9. Using three sgRNAs, 58–65 bp deletion was obtained in the *lncRNA2155* locus resulting in the delayed fruit ripening of tomatoes. Reduced accumulation of ethylene and lycopene was also evident in edited fruits because qPCR assays demonstrated the downregulation of *ACS4*, *ACS2*, *PSY1*, and *ZDS* transcripts. Apart from that, transcripts corresponding to different

Table 13.1 Knockout of different plant lncRNAs using the CRISPR/Cas system.

Edited lncRNA/genomic locus	Host plant	Major finding	Application	References
Chr6:156117287-156117915, Chr6:156118571-156117710	Maize (*Zea mays*)	Advanced designing of sgRNA spacer sequences followed by multiplex editing caused precise chromosomal fragment deletion at targeted lncRNA loci and abolished lncRNA function	lncRNA functional characterization	Qi et al. (2016)
LncRNA1459	Tomato (*Solanum lycopersicum*)	Knockout of target lncRNA suppressed fruit ripening in terms of downregulation in ethylene production, lycopene accumulation, and carotenoid biogenesis	lncRNA functional characterization, crop improvement	Li et al. (2018)
LncRNA2155	Tomato (*S. lycopersicum*)	Multiplex editing using three sgRNAs caused 58–65 bp homozygous deletions in target lncRNA leading to delayed fruit ripening	lncRNA functional characterization, crop improvement	Yu et al. (2019)
LncRNA57811	Rice (*Oryza sativa*)	No obvious alteration in plant reproductive morphology was documented upon target lncRNA knockout (with large deletions), indicating its functional redundancy despite being noncoding in nature	lncRNA functional characterization	Li et al. (2020)
StFLORE	Potato (*S. tuberosum*)	CRISPR-induced deletion 770–952 bp in the promoter region of *StFLORE* antisense lncRNA conferred early tuberization with a greater number of tubers. StFLORE interacts with transcription factor StCDF1, a nonredundant regulator of tuberization	lncRNA functional characterization, crop improvement	Ramirez Gonzales et al. (2021)
LncRNA77580	Soybean (*Glycine max*)	Large DNA fragment deletion (using dual sgRNA) in target lncRNA altered the expression of several nearby protein-coding genes implicated in salt stress response	lncRNA functional characterization, crop improvement	Niu et al. (2021)
LncA07, LncD09	Cotton (*Gossypium hirsutum*)	LncA07 and LncD09 are hub genes of a coexpression network, which regulate cotton defense against sucking pests. Targeted knockout of *lncA07* and *lncD09* attenuated jasmonic acid (JA) production that in turn increased host susceptibility to insects	lncRNA functional characterization, crop improvement	Zhang et al. (2022)

repining regulators including RIN, COLORLESS RIPENING (CNR), and NONRIPENING (NOR) were significantly repressed in the mutant fruits (Yu et al., 2019). Thus, lncRNA2155 can be considered another regulator of tomato fruit ripening.

13.5.2 Role of Potato lncRNAs

In potatoes, CYCLING DOF FACTOR 1 (StCDF1), a TF nonredundantly regulates tuberization and modulates drought tolerance. A NAT lncRNA gene *StFLORE* was demonstrated to be associated with the *StCDF1* gene (Shaikh et al., 2024). Using multiplex CRISPR/Cas9 editing, 770–952 bp long fragment deletion was obtained in the lncRNA locus. Edited plants exhibited early tuberization and a greater number of tubers compared to the wild-type plants. However, edited plants showed slightly higher water loss (with a similar tuber yield) upon drought stress indicating their greater drought susceptibility compared to wild-type plants. Moreover, edited plants were insensitive (as measured by stomatal behavior during drought stress) to abscisic acid treatment (Shaikh et al., 2024).

13.5.3 Role of Soybean lncRNAs

Soybean lncRNAs are involved in various physiological processes including floral morphogenesis, seed production, nodulation, and defense response to salinity stress and biotic stress (Bravo-Vázquez et al., 2024). From in-house RNAseq datasets of 332 samples belonging to *Glycine max* and *Glycine sojae* tissues, 69,000 lncRNAs were identified using a novel integrated reference-based de novo transcript assembly method (Lin et al., 2020). Using dual sgRNAs, large fragment deletion in the *lncRNA77580* locus was obtained that resulted in the transcriptional alteration of nearby protein-coding genes when CRISPR-edited soybean was exposed to salt stress. Moreover, the size of lncRNA77580 fragment deletion was proportional to the dynamics of transcription alteration in neighboring protein-coding genes and target lncRNA suppression (Niu et al., 2021).

13.5.4 Role of Cotton lncRNAs

Cotton lncRNAs are fundamentally important in plant biology because they maintain cell homeostasis. Using de novo transcriptome sequencing, 30,500 lincRNAs and 4718 NAT lncRNAs were identified from cotton. Interestingly, these lncRNAs serve as the precursor of miR397, which regulates the fiber development in *Gossypium* spp. (Wang et al., 2015). By combining strand-specific RNA sequencing (ssRNA-seq), cap analysis gene expression sequencing (CAGE-seq), and polyadenylation site sequencing (PolyA-seq), 9240 lncRNAs were identified from 21 tissue samples of *Gossypium arboreum*. A number of those lncRNAs co-expressed with ovule and fiber development-associated protein-coding genes (Zheng et al., 2021). In another study, 606 differentially expressed lncRNAs were identified from white fly (*Bemisia tabaci*) infected *Gossypium hirsutum* leaves that corresponded to 6651 lincRNA transcripts. Two candidate lncRNAs (lncA07 and lncD09) were identified by using a coexpression network analysis that potentially regulates the cotton defense against white fly and aphid infection (Zhang et al., 2022). Using dual sgRNAs, loss-of-function mutants were obtained for both *lncA07* and *lncD09*. Intriguingly, compared to wild-type plants, CRISPR-edited cotton lines displayed increased susceptibility to the sucking insects. Incidentally, CRISPR-edited lines also displayed reduced accumulation of jasmonic acid (JA), which is a regulator of plant defense response. Thus, it was speculated that lncA07 and lncD09 may positively regulate cotton defense against sap-sucking insects. Notably, lncD09-associated protein-coding genes

Ghir_A01G022270, *Ghir_D04G014430*, and *Ghir_A01G022270* are involved in the JA-regulated signaling pathways (Zhang et al., 2022).

13.5.5 Role of Rice lncRNAs

From autotetraploid rice, 444 differentially expressed lncRNAs (mostly lincRNAs followed by NAT and sense lncRNAs) were identified that are involved in anther and ovary meiosis; 328 lncRNAs were linked with the TEs, which show lower expression during meiosis (Li et al., 2020). Using CRISPR/Cas9, loss-of-function mutants were generated for lncRNA57811, lncRNA130461, and lncRNA55980 in the background of Taichung 65. However, compared to the wild type, edited plants did not show any phenotypic difference in terms of seed setting (Li et al., 2020). This suggests the redundant role of lncRNA57811, lncRNA130461, and lncRNA55980 in rice reproduction and warrants further investigations to explore specific lncRNAs.

13.5.6 Role of Maize lncRNAs

Two overlapping maize lncRNA loci (Chr6:156117287-156 117 915 and Chr6:156118571-156 117 710) were knocked out using a polycistronic tRNA–gRNA-based CRISPR/Cas9 editing system (Qi et al., 2016). These lncRNAs are regulated by the Opaque2 TF that regulates maize endosperm development vis-à-vis the nutritional value of maize seeds (Li et al., 2015). Large chromosomal fragment deletion at intended sites abolished the putative function of targeted lncRNAs as edited plants exhibited considerable developmental delay compared to wild-type plants (Qi et al., 2016).

13.6 Conclusion and Future Directions

lncRNAs are the key regulatory molecules in plant cells. However, the exact biological roles of the majority of the identified lncRNA molecules are not yet unraveled. First, the biological functions of lncRNAs were discerned by analyzing their interactome with proteins. However, lncRNA binds with proteins promiscuously, RNA–protein association is modulated by noncovalent bonds, and RNA tertiary structure is an important determinant of binding stability. Notably, RNA tertiary structure is quite unstable compared to that of protein molecules. The stability of RNA tertiary structure is affected by various factors such as transcriptome, interacting protein partner, and ionic strength. It is quite unlikely to determine the exact tertiary structure of a lncRNA molecule using existing techniques (Lyu et al., 2023). Second, using T-DNA insertional mutagenesis-based forward genetics screening, a number of *Arabidopsis* lncRNAs were functionally validated. However, this technique could not be extended to other plant species. Third, RNA interference (RNAi)-based reverse genetics screening was adopted to silence lncRNA expression posttranscriptionally. Nevertheless, RNAi-induced lncRNA knockdown carries few limitations. In contrast to the cytoplasmic location of mRNAs, lncRNAs are mostly localized to the cell nucleus (Wierzbicki et al., 2021). Even though the RNAi mechanism works in the nucleus, siRNAs targeted to silence nuclear lncRNAs exhibited limited efficacy probably because of no RISC distribution in the nucleus (Lennox & Behlke, 2016). Nuclear lncRNAs can be degraded in a sequence-specific manner by short single-stranded antisense oligonucleotides (ASOs) that recruit RNase H (Bennett, 2019). RNAi-based downregulation does not lead to complete inhibition of the gene function (Dutta et al., 2020), a feedback loop may restore the function of redundant lncRNAs. The transient nature of RNAi treatment has led to ambiguous delineation of lncRNA function. In view of this, CRISPR/Cas-based gene knockout has emerged as the ultimate tool to investigate the lncRNA function.

Interestingly, nuclear lncRNAs are primarily associated with epigenome and transcriptional regulation, while cytosolic lncRNAs are predominantly associated with posttranscriptional regulation and protein–protein interactions (Lyu et al., 2023). Using ChIP sequencing, other immunoprecipitation techniques, and interactome profiling, a number of RNA-binding proteins were identified that aid in lncRNA-mediated posttranscriptional regulation (Domínguez-Rosas et al., 2023). CRISPRi and CRISPRa tools can further be extrapolated to identify specific lncRNAs that modulate epigenetic regulations. Although dual sgRNA-based CRISPR/Cas9-induced deletion of longer lncRNA fragments (including promoter and gene body regions) has emerged as an effective tool to investigate plant lncRNA function, one has to be cautious regarding unintended effects on neighboring protein-coding genes. A careful design of the sgRNA molecule and extensive off-targeting analysis would be imperative. Targeted knock-in or gene replacement (for both gain-of-function and loss-of-function studies) can also be adopted using the highly precise prime editing tactic. Lastly, the sgRNA library-based functional screening would provide a clearer picture of the functional network of various lncRNA molecules. Unraveling the lncRNA function via CRISPR/Cas strategy not only enriches the existing literature on the regulatory role of specific lncRNAs in plant development and stress responses, but also opens up the application part related to the deployment of specific lncRNAs for improving plant traits including yield, nutrient content, reduced concentration of toxins, postharvest shelf life, and resilience to biotic/abiotic stressors.

References

Ariel, F., Jegu, T., Latrasse, D., Romero-Barrios, N., Christ, A., Benhamed, M., & Crespi, M. (2014). Noncoding transcription by alternative RNA polymerases dynamically regulates an auxin-driven chromatin loop. *Molecular Cell*, *55*, 383–396.

Ariel, F., Romero-Barrios, N., Jegu, T., Benhamed, M., & Crespi, M. (2015). Battles and hijacks: noncoding transcription in plants. *Trends in Plant Science*, *20*, 362–371.

Ariel, F., Lucero, L., Christ, A., et al. (2020). R-loop mediated *trans* action of the *APOLO* long noncoding RNA. *Molecular Cell*, *77*, 1055–1065.e4.

Awwad, D. A. (2019). Beyond classic editing: innovative CRISPR approaches for functional studies of long non-coding RNA. *Biology Methods & Protocols*, *4*, bpz017.

Bardou, F., Ariel, F., Simpson, C. G., Romero-Barrios, N., Laporte, P., Balzergue, S., et al. (2014). Long noncoding RNA modulates alternative splicing regulators in Arabidopsis. *Developmental Cell*, *30*, 166–176.

Basak, J., & Nithin, C. (2015). Targeting non-coding RNAs in plants with the CRISPR-Cas technology is a challenge yet worth accepting. *Frontiers in Plant Science*, *6*, 1001.

Bazin, J., Baerenfaller, K., Gosai, S. J., Gregory, B. D., Crespi, M., & Bailey-Serres, J. (2017). Global analysis of ribosome-associated noncoding RNAs unveils new modes of translational regulation. *Proceedings of the National Academy of Sciences of the United States of America*, *114*, E10018–E10027.

Bennett, C. F. (2019). Therapeutic antisense oligonucleotides are coming of age. *Annual Review of Medicine*, *70*, 307–321.

Bhaya, D., Davison, M., & Barrangou, R. (2011). CRISPR-Cas systems in Bacteria and Archaea: versatile small RNAs for adaptive defense and regulation. *Annual Review of Genetics*, *45*, 273–297.

Bravo-Vázquez, L. A., Méndez-García, A., Chamu-García, V., Rodríguez, A. L., Bandyopadhyay, A., & Paul, S. (2024). The applications of CRISPR/Cas-mediated microRNA and lncRNA editing in plant biology: shaping the future of plant non-coding RNA research. *Planta*, *259*, 32.

Budak, H., Kaya, S. B., & Cagirici, H. B. (2020). Long non-coding RNA in plants in the era of reference sequences. *Frontiers in Plant Science*, *11*, 276.

Campalans, A., Kondorosi, A., & Crespi, M. (2004). Enod40, a short open reading frame-containing mRNA, induces cytoplasmic localization of a nuclear RNA binding protein in *Medicago truncatula*. *The Plant Cell*, *16*, 1047–1059.

Cardi, T., Murovec, J., Bakhsh, A., Boniecka, J., Bruegmann, T., Bull, S. E., et al. (2023). CRISPR/Cas-mediated plant genome editing: outstanding challenges a decade after implementation. *Trends in Plant Science*, *28*, P1144–P1165.

Chavez, A., Scheiman, J., Vora, S., et al. (2015). Highly efficient Cas9-mediated transcriptional programming. *Nature Methods*, *12*, 326–328.

Domínguez-Rosas, E., Hernández-Oñate, M. Á., Fernandez-Valverde, S. L., & Tiznado-Hernández, M. E. (2023). Plant long non-coding RNAs: identification and analysis to unveil their physiological functions. *Frontiers in Plant Science*, *14*, 1275399.

Du, Q., Wang, K., Zou, C., Xu, C., & Li, W. X. (2018). The *PILNCR1*-miR399 regulatory module is important for low phosphate tolerance in maize. *Plant Physiology*, *177*, 1743–1753.

Dutta, T. K., Papolu, P. K., Singh, D., Sreevathsa, R., & Rao, U. (2020). Expression interference of a number of *Heterodera avenae* conserved genes perturbs nematode parasitic success in *Triticum aestivum*. *Plant Science*, *301*, 110670.

Dutta, T. K., Ray, S., & Phani, V. (2023). The status of the CRISPR/Cas9 research in plant-nematode interactions. *Planta*, *258*, 103.

Fesenko, I., Shabalina, S. A., Mamaeva, A., et al. (2021). A vast pool of lineage-specific microproteins encoded by long non-coding RNAs in plants. *Nucleic Acids Research*, *49*, 10328–10346.

Fonouni-Farde, C., Ariel, F., & Crespi, M. (2021). Plant long noncoding RNAs: new players in the field of post-transcriptional regulations. *Noncoding RNA*, *7*, 12.

Franco-Zorrilla, J. M., Valli, A., Todesco, M., Mateos, I., Puga, M. I., Rubio-Somoza, I., Leyva, A., Weigel, D., García, J. A., & Paz-Ares, J. (2007). Target mimicry provides a new mechanism for regulation of microRNA activity. *Nature Genetics*, *39*, 1033–1037.

Franco-Zorrilla, J. M., López-Vidriero, I., Carrasco, J. L., Godoy, M., Vera, P., & Solano, R. (2014). DNA-binding specificities of plant transcription factors and their potential to define target genes. *Proceedings of the National Academy of Sciences of the United States of America*, *111*, 2367–2372.

Haag, J. R., Ream, T. S., Marasco, M., Nicora, C. D., Norbeck, A. D., et al. (2012). In vitro transcription activities of Pol IV, Pol V, and RDR2 reveal coupling of Pol IV and RDR2 for dsRNA synthesis in plant RNA silencing. *Molecular Cell*, *48*, 811–818.

Heo, J. B., & Sung, S. (2011). Vernalization-mediated epigenetic silencing by a long intronic noncoding RNA. *Science*, *331*, 76–79.

Horvath, P., & Barrangou, R. (2010). CRISPR/Cas, the immune system of Bacteria and Archaea. *Science*, *327*, 167–170.

Hsieh, T.-F., & Fischer, R. L. (2005). Biology of chromatin dynamics. *Annual Review of Plant Biology*, *56*, 327–351.

Huang, J., Zhou, W., Zhang, X., & Li, Y. (2023). Roles of long non-coding RNAs in plant immunity. *PLoS Pathogens*, *19*, e1011340.

Jiang, N., Cui, J., Shi, Y., Yang, G., Zhou, X., Hou, X., Meng, J., & Luan, Y. (2019). Tomato lncRNA23468 functions as a competing endogenous RNA to modulate *NBS-LRR* genes by decoying miR482b in the tomato-*Phytophthora infestans* interaction. *Horticulture Research*, *6*, 28.

Jinek, M., Chylinski, K., Fonfara, I., et al. (2012). A programmable dual-RNA-guided DNA endonuclease in adaptive bacterial immunity. *Science, 337*, 816–821.

Kim, D. H., & Sung, S. (2017). Vernalization-triggered intragenic chromatin loop formation by long noncoding RNAs. *Developmental Cell, 40*, 302–312.e4.

Kim, D. H., Xi, Y., & Sung, S. (2017). Modular function of long noncoding RNA, COLDAIR, in the vernalization response. *PLoS Genetics, 13*, e1006939.

Lavalou, P., Eckert, H., Damy, L., Constanty, F., Majello, S., Bitetti, A., et al. (2019). Strategies for genetic inactivation of long non-coding RNAs in zebrafish. *RNA, 25*, 897–904.

Lennox, K. A., & Behlke, M. A. (2016). Cellular localization of long non-coding RNAs affects silencing by RNAi more than by antisense oligonucleotides. *Nucleic Acids Research, 44*, 863–877.

Li, C., Qiao, Z., Qi, W., Wang, Q., Yuan, Y., et al. (2015). Genome-wide characterization of cis-acting DNA targets reveals the transcriptional regulatory framework of Opaque2 in maize. *Plant Cell, 27*, 532–545.

Li, R., Fu, D., Zhu, B., et al. (2018). CRISPR/Cas9-mediated mutagenesis of *lncRNA1459* alters tomato fruit ripening. *The Plant Journal, 94*, 513–524.

Li, X., Shahid, M. Q., Wen, M., et al. (2020). Global identification and analysis revealed differentially expressed lncRNAs associated with meiosis and low fertility in autotetraploid rice. *BMC Plant Biology, 20*, 82.

Lin, X., Lin, W., Ku, Y. S., et al. (2020). Analysis of soybean long non-coding RNAs reveals a subset of small peptide-coding transcripts. *Plant Physiology, 182*, 1359–1374.

Liu, S.-R., Zhou, J.-J., Hu, C.-G., Wei, C.-L., & Zhang, J.-Z. (2017). MicroRNA-mediated gene silencing in plant defense and viral counter-defense. *Frontiers in Microbiology, 8*, 1801.

Liu, N., Xu, Y., Li, Q., Cao, Y., Yang, D., Liu, S., et al. (2022). A lncRNA fine-tunes salicylic acid biosynthesis to balance plant immunity and growth. *Cell Host & Microbe, 30*, 1124–1138.e8.

Lowder, L. G., Zhang, D., Baltes, N. J., et al. (2015). A CRISPR/Cas9 toolbox for multiplexed plant genome editing and transcriptional regulation. *Plant Physiology, 169*, 971–985.

Lucero, L., Ferrero, L., Fonouni-Farde, C., & Ariel, F. (2021). Functional classification of plant long noncoding RNAs: a transcript is known by the company it keeps. *The New Phytologist, 229*, 1251–1260.

Lyu, Q. R., Zhang, S., Zhang, Z., & Tang, Z. (2023). Functional knockout of long noncoding RNAs with genome editing. *Frontiers in Genetics, 14*, 1242129.

Ma, L., Mu, J., Grierson, D., et al. (2020). Noncoding RNAs: functional regulatory factors in tomato fruit ripening. *Theoretical and Applied Genetics, 133*, 1753–1762.

Mali, P., Yang, L., Esvelt, K. M., et al. (2013). RNA-guided human genome engineering via Cas9. *Science, 339*, 823–826.

Mattick, J. S., Amaral, P. P., Carninci, P., Carpenter, S., Chang, H. Y., Chen, L. L., et al. (2023). Long non-coding RNAs: definitions, functions, challenges and recommendations. *Nature Reviews. Molecular Cell Biology, 24*, 430–447.

Moison, M., Pacheco, J. M., Lucero, L., et al. (2021). The lncRNA *APOLO* interacts with the transcription factor WRKY42 to trigger root hair cell expansion in response to cold. *Molecular Plant, 14*, 937–948.

Niu, F., Jiang, Q., Sun, X., et al. (2021). Large DNA fragment deletion in *lncRNA77580* regulates neighboring gene expression in soybean (*Glycine max*). *Functional Plant Biology, 48*, 1139–1147.

Pacheco, J. M., Mansilla, N., Moison, M., Lucero, L., Gabarain, V. B., Ariel, F., & Estevez, J. M. (2021). The lncRNA *APOLO* and the transcription factor WRKY42 target common cell wall EXTENSIN encoding genes to trigger root hair cell elongation. *Plant Signaling & Behavior, 16*, 1920191.

Pauli, A., Rinn, J. L., & Schier, A. F. (2011). Non-coding RNAs as regulators of embryogenesis. *Nature Reviews. Genetics*, *12*, 136–149.

Perez-Pinera, P., Kocak, D. D., Vockley, C. M., et al. (2013). RNA-guided gene activation by CRISPR-Cas9-based transcription factors. *Nature Methods*, *10*, 973–976.

Piatek, A., Ali, Z., Baazim, H., et al. (2015). RNA-guided transcriptional regulation *in planta* via synthetic dCas9-based transcription factors. *Plant Biotechnology Journal*, *13*, 578–589.

Powell, G., Pavlovic Djuranovic, S., & Djuranovic, S. (2021). Gene dosage effects of poly(A) track-engineered hypomorphs. *Molecular Therapy - Nucleic Acids*, *26*, 865–878.

Pramanik, D., Shelake, R. M., Kim, M. J., & Kim, J. Y. (2021). CRISPR-mediated engineering across the central dogma in plant biology for basic research and crop improvement. *Molecular Plant*, *14*, 127–150.

Pulecio, J., Verma, N., Mejía-Ramírez, E., Huangfu, D., & Raya, A. (2017). CRISPR/Cas9-based engineering of the epigenome. *Cell Stem Cell*, *21*(4), 431–447.

Qi, L. S., Larson, M. H., Gilbert, L. A., et al. (2013). Repurposing CRISPR as an RNA-guided platform for sequence-specific control of gene expression. *Cell*, *152*, 1173–1183.

Qi, W., Zhu, T., Tian, Z., et al. (2016). High-efficiency CRISPR/Cas9 multiplex gene editing using the glycine tRNA-processing system based strategy in maize. *BMC Biotechnology*, *16*, 58.

Quinn, J. J., & Chang, H. Y. (2016). Unique features of long non-coding RNA biogenesis and function. *Nature Reviews. Genetics*, *17*, 47–62.

Rai, M. I., Alam, M., Lightfoot, D. A., Gurha, P., & Afzal, A. J. (2019). Classification and experimental identification of plant long non-coding RNAs. *Genomics*, *111*, 997–1005.

Ramirez Gonzales, L., Shi, L., Bergonzi, S. B., et al. (2021). Potato CYCLING DOF FACTOR 1 and its lncRNA counterpart *StFLORE* link tuber development and drought response. *The Plant Journal*, *105*, 855–869.

Reis, R. S., Deforges, J., Schmidt, R. R., Schippers, J. H. M., & Poirier, Y. (2021). An antisense noncoding RNA enhances translation via localized structural rearrangements of its cognate mRNA. *The Plant Cell*, *33*, 1381–1397.

Rigo, R., Bazin, J., Romero-Barrios, N., et al. (2020). The *Arabidopsis* lncRNA *ASCO* modulates the transcriptome through interaction with splicing factors. *EMBO Reports*, *21*, e48977.

Romero-Barrios, N., Legascue, M. F., Benhamed, M., Ariel, F., & Crespi, M. (2018). Splicing regulation by long noncoding RNAs. *Nucleic Acids Research*, *46*, 2169–2184.

Roulé, T., Christ, A., Hussain, N., Huang, Y., Hartmann, C., Benhamed, M., Gutierrez-Marcos, J., Ariel, F., Crespi, M., & Blein, T. (2022). The lncRNA *MARS* modulates the epigenetic reprogramming of the marneral cluster in response to ABA. *Molecular Plant*, *15*, 840–856.

Seo, J. S., Sun, H. X., Park, B. S., Huang, C. H., Yeh, S. D., Jung, C., & Chua, N. H. (2017). ELF18-INDUCED LONG-NONCODING RNA associates with mediator to enhance expression of innate immune response genes in Arabidopsis. *The Plant Cell*, *29*, 1024–1038.

Sharma Y, Sharma A, Madhu, et al. (2022). Long non-coding RNAs as emerging regulators of pathogen response in plants. *Non-Coding RNA* 8:4.

Shaikh, M. A., Ramírez-Gonzales, L., Franco-Zorrilla, J. M., Steiner, E., Oortwijn, M., Bachem, C. W., & Prat, S. (2024). StCDF1: A 'jack of all trades' clock output with a central role in regulating potato nitrate reduction activity. *New Phytologist*. https://doi.org/10.1111/nph.20186

Singh, J., Mishra, V., Wang, F., Huang, H.-Y., & Pikaard, C. S. (2019). Reaction mechanisms of Pol IV, RDR2, and DCL3 drive RNA channeling in the siRNA-directed DNA methylation pathway. *Molecular Cell*, *75*(3), 576–589.e5.

Staiger, D., & Brown, J. W. (2013). Alternative splicing at the intersection of biological timing, development, and stress responses. *The Plant Cell*, *25*, 3640–3656.

Tang, Y., Qu, Z., Lei, J., He, R., Adelson, D. L., Zhu, Y., Yang, Z., & Wang, D. (2021). The long noncoding RNA *FRILAIR* regulates strawberry fruit ripening by functioning as a noncanonical target mimic. *PLoS Genetics, 17*, e1009461.

Tognacca, R. S., Rodríguez, F. S., Aballay, F. E., Cartagena, C. M., Servi, L., & Petrillo, E. (2023). Alternative splicing in plants: current knowledge and future directions for assessing the biological relevance of splice variants. *Journal of Experimental Botany, 74*, 2251–2272.

Uranga, M., & Daròs, J. A. (2022). Tools and targets: the dual role of plant viruses in CRISPR–Cas genome editing. *The Plant Genome*, e20220.

Urquiaga, M. C. D. O., Thiebaut, F., Hemerly, A. S., & Ferreira, P. C. G. (2021). From trash to luxury: the potential role of plant lncRNA in DNA methylation during abiotic stress. *Frontiers in Plant Science, 11*, 603246.

Vu, T. V., Nguyen, N. T., Kim, J., Hong, J. C., & Kim, J. Y. (2024). Prime editing: mechanism insight and recent applications in plants. *Plant Biotechnology Journal, 22*, 19–36.

Waititu, J. K., Zhang, C., Liu, J., & Wang, H. (2020). Plant non-coding RNAs: origin, biogenesis, mode of action and their roles in abiotic stress. *International Journal of Molecular Sciences, 21*, 8401.

Wang, M., Yuan, D., Tu, L., et al. (2015). Long noncoding RNAs and their proposed functions in fibre development of cotton (*Gossypium* spp.). *The New Phytologist, 207*, 1181–1197.

Wang, T., Zhao, M., Zhang, X., Liu, M., Yang, C., Chen, Y., Chen, R., Wen, J., Mysore, K. S., & Zhang, W. H. (2017). Novel phosphate deficiency-responsive long non-coding RNAs in the legume model plant *Medicago truncatula*. *Journal of Experimental Botany, 68*, 5937–5948.

Wang, Y., Luo, X., Sun, F., Hu, J., Zha, X., Su, W., & Yang, J. (2018). Overexpressing lncRNA *LAIR* increases grain yield and regulates neighbouring gene cluster expression in rice. *Nature Communications, 9*, 3516.

Wierzbicki, A. T., Haag, J. R., & Pikaard, C. S. (2008). Noncoding transcription by RNA polymerase Pol IVb/Pol V mediates transcriptional silencing of overlapping and adjacent genes. *Cell, 135*, 635–648.

Wierzbicki, A. T., Blevins, T., & Swiezewski, S. (2021). Long noncoding RNAs in plants. *Annual Review of Plant Biology, 72*, 245–271.

Wu, J., Okada, T., Fukushima, T., Tsudzuki, T., Sugiura, M., & Yukawa, Y. (2012). A novel hypoxic stress-responsive long non-coding RNA transcribed by RNA polymerase III in Arabidopsis. *RNA Biology, 9*, 302–313.

Wu, H. W., Deng, S., Xu, H., Mao, H. Z., Liu, J., Niu, Q. W., Wang, H., & Chua, N. H. (2018). A noncoding RNA transcribed from the *AGAMOUS* (*AG*) second intron binds to CURLY LEAF and represses *AG* expression in leaves. *New Phytologist, 219*, 1480–1491.

Xu, W., Xu, H., Li, K., Fan, Y., Liu, Y., Yang, X., & Sun, Q. (2017). The R-loop is a common chromatin feature of the *Arabidopsis* genome. *Nature Plants, 3*, 704–714.

Xu, S., Dong, Q., Deng, M., et al. (2021a). The vernalization-induced long noncoding RNA *VAS* functions with the transcription factor TaRF2b to promote *TaVRN1* expression for flowering in hexaploid wheat. *Molecular Plant, 14*, 1525–1538.

Xu, C., Wu, Z., Duan, H. C., Fang, X., Jia, G., & Dean, C. (2021b). R-loop resolution promotes co-transcriptional chromatin silencing. *Nature Communications, 12*, 1790.

Yang, J., Ariel, F., & Wang, D. (2023). Plant long non-coding RNAs: biologically relevant and mechanistically intriguing. *Journal of Experimental Botany, 74*, 2364–2373.

Yu, T., Tzeng, D. T. W., Li, R., et al. (2019). Genome-wide identification of long non-coding RNA targets of the tomato MADS box transcription factor RIN and function analysis. *Annals of Botany, 123*, 469–482.

Zhang, X., Shen, J., Xu, Q., Dong, J., Song, L., Wang, W., & Shen, F. (2021). Long noncoding RNA lncRNA354 functions as a competing endogenous RNA of miR160b to regulate *ARF* genes in response to salt stress in upland cotton. *Plant, Cell & Environment*, *44*, 3302–3321.

Zhang, J., Li, J., Saeed, S., et al. (2022). Identification and functional analysis of lncRNA by CRISPR/Cas9 during the cotton response to sap-sucking insect infestation. *Frontiers in Plant Science*, *13*, 784511.

Zhao, X., Li, J., Lian, B., Gu, H., Li, Y., & Qi, Y. (2018). Global identification of *Arabidopsis* lncRNAs reveals the regulation of *MAF4* by a natural antisense RNA. *Nature Communications*, *9*, 5056.

Zheng, X., Chen, Y., Zhou, Y., et al. (2021). Full-length annotation with multistrategy RNA-seq uncovers transcriptional regulation of lncRNAs in cotton. *Plant Physiology*, *185*, 179–195.

Zhou, M., & Law, J. A. (2015). RNA Pol IV and V in gene silencing: rebel polymerases evolving away from Pol II's rules. *Current Opinion in Plant Biology*, *27*, 154–164.

Zhu, B., Yang, Y., Li, R., et al. (2015). RNA sequencing and functional analysis implicate the regulatory role of long non-coding RNAs in tomato fruit ripening. *Journal of Experimental Botany*, *66*, 4483–4495.

14

Plant Gene Silencing Modified by CRISPR/Cas-Reprogrammed Small Interfering RNAs for Regulating Abiotic Stress Responses

Noru Rajasekhar Reddy[1], *Janapareddy Rajesh*[2], *Kadiyala Kavya*[3], *Lellapalli Rithesh*[4], *Abhishek Kumar*[5], *and Pooja*[6]

[1] *Department of Genetics and Plant Breeding, College of Agriculture, Kerala Agricultural University, Thiruvananthapuram, Kerala, India*
[2] *Department of Seed Science and Technology, Hemvati Nandan Bahuguna Garhwal University, Srinagar, Uttarakhand, India*
[3] *Department of Genetics and Plant Breeding, College of Post Graduate Studies in Agricultural Sciences, Central Agricultural University (Imphal), Umiam, Meghalaya, India*
[4] *Department of Plant Pathology, Kerala Agricultural University, Thiruvananthapuram, Kerala, India*
[5] *Faculty of Agriculture, Maharishi Markandeshwar University, Mullana, Ambala, Haryana, India*
[6] *Department of Plant Pathology, Chaudhary Charan Singh (CCS) Haryana Agricultural University, Hisar, Haryana, India*

14.1 Introduction

Gene silencing is essential for regulating gene expression and stability in eukaryotes. In plants, gene silencing plays a vital role in development, stress responses, and pathogen defense. The two primary strategies for gene silencing in plants are transcriptional gene silencing (TGS) and post-transcriptional gene silencing (PTGS). Both processes require dense and multimolecular interplay, which helps fine-tune gene expression and protect the genome. Eukaryotic organisms use short RNAs (20–30 nucleotides) to suppress the expression of homologous genes during transcription or posttranscriptional processes. Plants rely on RNA-silencing pathways to regulate development and address both biotic and abiotic stresses (Ashfaq et al., 2020; Pandey et al., 2022). This chapter delves into the methods and strategies for reprogramming small interference RNA (siRNAs) using clustered regularly interspaced short palindromic repeats and CRISPR-associated (CRISPRs/Cas) systems and highlights the mechanisms, applications, and future directions of this innovative field.

14.1.1 Transcriptional Gene Silencing (TGS)

TGS involves silencing gene expression at the transcriptional level, typically through chromatin structure changes or DNA methylation, and is often associated with heterochromatic formation. TGS occurs within the nucleus and establishes and maintains the heterochromatic state in some regions of the genome.

14.1.1.1 DNA Methylation

DNA methylation involves adding methyl groups to cytosine residues, usually within CG dinucleotides (CpG). These epigenetic modifications can inhibit gene expression by blocking transcription factor binding or through methyl-CpG binding domain (MBD) proteins, which restrict

Genome and Epigenome Editing for Stress-Tolerant Crops, First Edition. Edited by Jen-Tsung Chen.

transcript activity and play a key role in plant development and stress responses (Kumar & Mohapatra, 2021).

- **CG Methylation**: This is maintained during DNA replication and is generally associated with gene repression.
- **CHG Methylation**: This type is maintained by specific methyltransferases and is essential for repressing transposable elements (TEs).
- **CHH Methylation**: This form of methylation is typically de novo and is often associated with repetitive sequences and transposons.

Plant DNA methylation is established and maintained by several key enzymes, including DNA methyltransferases such as MET1 (for CG methylation), CMT3 (for CHG methylation), and DRM2 (for CHH methylation). Various factors, including environmental conditions and developmental cues, influence the DNA methylation dynamics.

14.1.1.2 Histone Modifications

Histone modifications are posttranslational alterations in histone proteins that influence chromatin shape and gene expression. Common modifications include acetylation, methylation, phosphorylation, and ubiquitination (Ramazi et al., 2020; Liu et al., 2022a,b,c).

- **Histone Acetylation**: Typically associated with gene activation, histone acetylation is mediated by histone acetyltransferases (HATs) and is removed by histone deacetylases (HDACs).
- **Histone Methylation**: Histone methylation can either activate or repress gene expression depending on the modified residues. Histone methyltransferases and demethylases catalyze this modification.

14.1.2 Posttranscriptional Gene Silencing (PTGS)

PTGS is the process by which mRNA transcripts are degraded. Short RNAs predominantly mediate this process, but may also involve RNA-binding proteins. The PTGS pathway is embedded in the cellular regulatory networks and plays a crucial role in growth and development (Zhong et al., 2023).

14.1.2.1 Antisense RNA Technique

Antisense RNA methods are based on the principle that an antisense RNA strand that includes a specific mRNA sequence binds to a target mRNA. This interaction inhibits the ribosome from translating mRNA into protein. The process may involve the degradation of mRNA by RNase H, an enzyme that specifically breaks down RNA in RNA–DNA hybrids.

14.1.2.2 RNA Interference (RNAi)

RNAi is utilized in functional genomics to silence specific genes and improve pest and disease resistance during crop development. RNAi is a well-established PTGS mechanism that disrupts mRNA expression through sequence-specific interactions. It starts with double-stranded RNA (dsRNA), which is processed into siRNA by the Dicer-like 1 (DCL1) enzyme. The siRNAs are then incorporated into the RNA-induced silencing complex (RISC), where they guide the complex to target and degrade complementary mRNAs. In plants, RNAi can be derived from transgenes or viral vectors and is essential for preserving genomic integrity and protecting against infection. RNAi efficiency is affected by the degree of siRNA accumulation and the presence of RNA-dependent RNA polymerases (RDRs) that amplify the silencing signal (Ray et al., 2022).

MicroRNAs (miRNAs): Primary miRNA transcripts (pri-miRNAs) are generated through processing processes involving Dicer-like proteins and the RNAse III enzyme Drosha miRNAs often attach to complementary regions in mRNA, causing translational suppression or damage. miRNAs are crucial for plant development, stress responses, and pathogen defense, playing key roles in gene regulation and maintaining cellular balance. For instance, miR156 regulates the transition from juvenile to adult stages in plants, while miR399 is important for phosphate homeostasis (Wang et al., 2023a,b).

siRNAs: Dicer-like proteins cleave long dsRNA molecules into 21–24 nucleotide fragments called siRNAs. These siRNAs direct the RISC to target and degrade complementary mRNA sequences (Vaucheret & Voinnet, 2024).

14.1.3 Abiotic Stress and Its Impact on Plant Growth and Productivity

Abiotic causes induce abiotic stress, which can significantly reduce plant development and agricultural yield. These stressors can originate in the natural environment or from human actions and frequently interact to magnify their effects. Abiotic stress has various physiological and biological effects that influence plant health and agricultural productivity. Abiotic stresses such as drought (water stress), extreme temperatures (cold and heat), salinity, and mineral toxicity adversely affect the growth, development, yield, and seed quality of crops and other plants (Kumar, 2020).

Plant metabolism is crucial for signaling, physiological regulation, and defense under challenging environmental conditions. Abiotic stress, in turn, impacts the synthesis, concentration, transport, and storage of both primary and secondary metabolites. Stress-induced metabolic changes often involve specific adjustments in amino acid, carbohydrate, and amine metabolism. Effective activation of early metabolic responses is essential for restoring chemical and energy balance, which is vital for plant adaptation and survival (Carrera et al., 2021).

14.1.4 Types of Abiotic Stress

14.1.4.1 Drought Stress

Drought stress occurs when there is insufficient water for plants. Stress has many physiological effects, including photosynthesis, nutrient absorption, and cell metabolism. Drought decreases cell damage and photosynthetic activity but elevates the levels of reactive oxygen species (ROS), which can lead to oxidative damage (Wahab et al., 2022).

The fundamental physiological responses to drought include the following:

- **Reduced Photosynthesis**: Drought stress decreases stomatal conductance and photosynthetic rates, thereby limiting carbon fixation.
- **Increased ROS**: Increased ROS production from oxidative stress can harm cellular components and impede growth.
- **Altered Hormone Levels**: Hormones such as abscisic acid (ABA) increase during drought, signaling the plant to initiate stress responses such as stomatal closure and expression of stress-related genes.

14.1.4.2 Salinity Stress

Salinity stress arises from the buildup of salts in the soil, leading to osmotic stress and ion toxicity. Elevated salt levels reduce water absorption, hindering plant growth and yield. Toxic ions like sodium chloride accumulate, disrupting enzyme function and metabolic processes (Balasubramaniam et al., 2023). Key impacts include the following:

- **Osmotic Stress**: High salinity reduces water availability, leading to osmotic stress and reduced cell expansion.
- **Ion Toxicity**: Accumulation of toxic ions such as sodium (Na^+) and chloride (Cl^-) can disrupt cellular functions and enzyme activity.
- **Nutrient Imbalance**: Salinity can interfere with the uptake of essential nutrients, leading to deficiencies and imbalances.

14.1.4.3 Extreme Temperature Stress

Extreme temperatures, both high and low, can negatively impact plant growth. Heat stress can cause protein degradation, tissue instability, and disruption of photosynthesis, whereas cold stress affects enzyme function and tissue hydration. Both types of high heat can disrupt flowering, fruit production, and the setting of seeds (Seneviratne et al., 2021).

Heat Stress: Heat stress occurs when temperatures exceed optimal levels for plant growth, leading to cellular damage and impaired physiological processes (Hassan et al., 2021).

- **Reduced Photosynthesis**: High temperatures can denature photosynthetic enzymes and reduce photosynthetic efficiency.
- **Protein Denaturation**: Heat stress can lead to protein denaturation and impair cellular function.
- **Heat Shock Proteins (HSPs)**: Plants produce HSPs to protect and repair damaged proteins in response to heat stress.

Cold Stress: Cold stress arises from exposure to low temperatures, leading to frost damage and altered physiological processes (Atayee & Noori, 2020).

- **Cryoprotection**: Plants may produce cryoprotective compounds such as antifreeze proteins to prevent ice crystal formation inside cells.
- **Membrane Stability**: Cold stress affects membrane fluidity, impairing cellular transport and signaling.

14.1.4.4 Heavy Metal Toxicity

Toxicity induced by heavy metal deposition, including cadmium (Cd), copper (Cu), and arsenic (As), can affect plant growth and development. Heavy metals disrupt critical metabolic processes by reducing nutrient intake, enzyme activity, and cell cohesion. They can also cause oxidative stress and harm the cell structure (Ungureanu et al., 2024).

14.2 CRISPR/Cas System: Principles and Applications

CRISPR/Cas systems have revolutionized genetics, molecular biology, and biotechnology. Originally discovered as a versatile immune mechanism, CRISPR/Cas has been adapted for precise genome editing across various species, including humans. The new CRISPR-Cas system offers greater efficiency and faster results than zinc-finger nucleases (ZFNs) and transcription activator-like effector nucleases (TALENs). This system uses the RNA-guided nuclease Cas9 to create double-strand breaks (DSBs), which are then repaired by cellular DNA repair processes, leading to targeted gene or genome modifications (Afzal et al., 2020; Verma et al., 2023a,b).

CRISPR is an extensive class of short palindromic repeat sequences found in many prokaryotes, including bacteria and archaea. In prokaryotes, these short repetitive sequences complement some foreign DNA sequences, such as viral DNA, which infiltrates bacteria or archaea. When viruses infect bacteria, the bacteria create this sort of DNA and bind to the viral DNAs; the Cas enzyme then cuts the invading DNA into pieces. Thus, CRISPR/Cas is a form of acquired immune defense system that prokaryotes use against viruses. It is also a naturally occurring genome-editing technique (Zhang et al., 2021).

14.2.1 Background and Discovery

The CRISPR/Cas systems' discovery and development are regarded as breakthroughs in molecular biology. Initially found in bacteria as part of their adaptive immune systems (Munawar & Ahmad, 2021), these technologies use small viral DNA fragments to identify and cleave foreign DNA. While various DNA sequences related to CRISPR/Cas were identified in the 1980s, a functional understanding of CRISPR/Cas as an adaptive immune system began to emerge in the early 2000s (Arnesen et al., 2021). Francisco Mojica and colleagues conducted key research and demonstrated that CRISPR sequences integrate "spacers" from viral DNA, allowing bacteria to "remember" and use them to target previously attacked viruses (Nidhi et al., 2021). Subsequent research has enhanced the CRISPR toolset by investigating Cas proteins and enhancing their precise modifications. Recent advancements include base- and prime-editing technologies that provide even greater control of genetic variation, limit off-target effects, and increase the range of potential applications (Nidhi et al., 2021).

14.2.2 Components of the CRISPR/Cas System

CRISPR Arrays and CRISPR RNAs (crRNAs): CRISPR arrays are constructed from repeating DNA sequences, including unique portions known as spacers. These spacers are the molecular remains of viral or plasmid infections that have previously occurred. CRISPR-mediated immune responses transcribe these sequences into CrRNA precursors (pre-crRNAs), which are then processed into tiny crRNAs designated by each crRNA. The crRNA guides Cas protein to its specific DNA target (Liao & Beisel, 2021). The detailed view of the CRISPR structure and its components is depicted in Figure 14.1.

CRISPR-Associated (Cas) Proteins: Cas proteins are molecular scissors that cleave target DNA using crRNA. Cas9, a member of the type II CRISPR system, is the most extensively researched Cas protein. Other notable Cas proteins include Cpf1 (now known as Cas12) and C2c2 (Cas13), which exhibit various features and functions (Kaul et al., 2020). The types of Cas proteins and their respective functions are outlined in Table 14.1.

Guide RNA (gRNA): gRNA is a synthetic RNA molecule that integrates crRNA and *trans*-activating CRISPR RNA (tracrRNA) into one continuous RNA strand. This reduced gRNA structure enables the Cas9 protein to target certain DNA regions effectively. The gRNA has a 20-nucleotide sequence that complements the target DNA and a scaffolding region that interacts with Cas9 (Asmamaw & Zawdie, 2021).

Protospacer Adjacent Motif (PAM): The PAM sequence is a short, conserved DNA motif necessary for Cas9 binding and function. Cas9 identifies the PAM sequence to initiate interaction with the target DNA. The specificity of the CRISPR/Cas system relies partly on PAM sequences, which are crucial for accurate binding and cleavage of the target DNA.

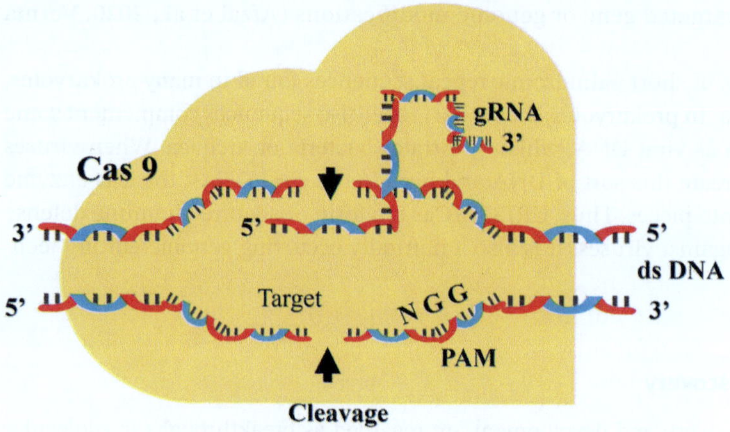

Figure 14.1 Detailed view of the CRISPR structure and its components.

Table 14.1 Types of Cas Proteins and their functions.

Class	Type	Subtype	Signature Cas protein (s)	Function	References
Class 1	Type I	I-A to I-F	Cas3	Multiprotein cascade complex Cas3 has helicase and nuclease activity for degrading foreign DNA	He et al. (2020)
Class 1	Type III	III-A to III-D	Cas10	Csm/Cmr complex targets both DNA and RNA; Cas10 has nuclease and polymerase activity	Marraffini (2022)
Class 1	Type IV	IV	Cas1, CsxCas proteins	Less characterized; typically lacks the Cascade complex, involved in interference and adaptation	Chaudhary et al. (2024)
Class 2	Type II	II-A to II-C	Cas9	Cleaves double-stranded DNA requires tracrRNA to form a complex with crRNA for DNA targeting	Liao & Beisel (2021)
Class 2	Type V	V-A (Cpf1/Cas12a), V-B (C2c1/Cas12b), V-C (C2c3)	Cas12	Cleaves DNA at sites distinct from Cas9; Cas12 can process crRNA without tracrRNA	Bhatia & Yadav (2023)
Class 2	Type VI	VI-A (C2c2/Cas13a), VI-B (C2c4/Cas13b), VI-C (Cas13c), VI-D (Cas13d)	Cas13	Targets and degrades single-stranded RNA; useful for RNA interference applications	Kavuri et al. (2022)

14.2.3 Overview of CRISPR-Mediated Gene Editing in Plants

CRISPR-based genome editing, inspired by the immune system of prokaryotes, is becoming a popular method for making targeted changes in the genome. CRISPR-based technologies are used for various purposes, including gene expression investigation, genome mutation detection, epigenetic regulation, and chromatin imaging. Compared with typical genetic engineering methods, this method is straightforward, cost-effective, and specific for causing genetic changes. Although popular, this approach has disadvantages, such as off-target effects, limited mutagenesis efficiency, and reliance on in vitro regeneration protocols to recover stable plant lines (Ahmad et al., 2020).

14.2.4 Steps in CRISPR-Mediated Gene Editing in Plants

Design of gRNA: The first step in gene engineering using CRISPR is the synthesis of gRNA specific to the target gene. This method requires the identification of a suitable target sequence in the gene of interest, typically a sequence of 20 nucleotides in length, adjacent to the PAM (Liu et al., 2022a,b,c). A detailed view of CRISPR-mediated gene editing is shown in Figure 14.2.

Construction of CRISPR/Cas Expression Constructs: Once gRNA is synthesized, it must be incorporated into a vector containing the Cas9 gene. This expression pattern was introduced into the plant cells.

Steps:

- **Vector Selection**: Select the appropriate vector based on the plant species and rotation method. Typical vectors include binary vectors for *Agrobacterium*-mediated transformation and plasmids for particle bombing (Ali et al., 2024).
- **Cloning**: Clone gRNA and Cas9 coding sequences into vectors using restriction enzymes or Gibson assembly (Deka, 2024).

Delivery of CRISPR/Cas Components: The next step was the delivery of CRISPR/Cas constructs into plant cells. This can be achieved in various ways, depending on the plant species and the desired efficiency.

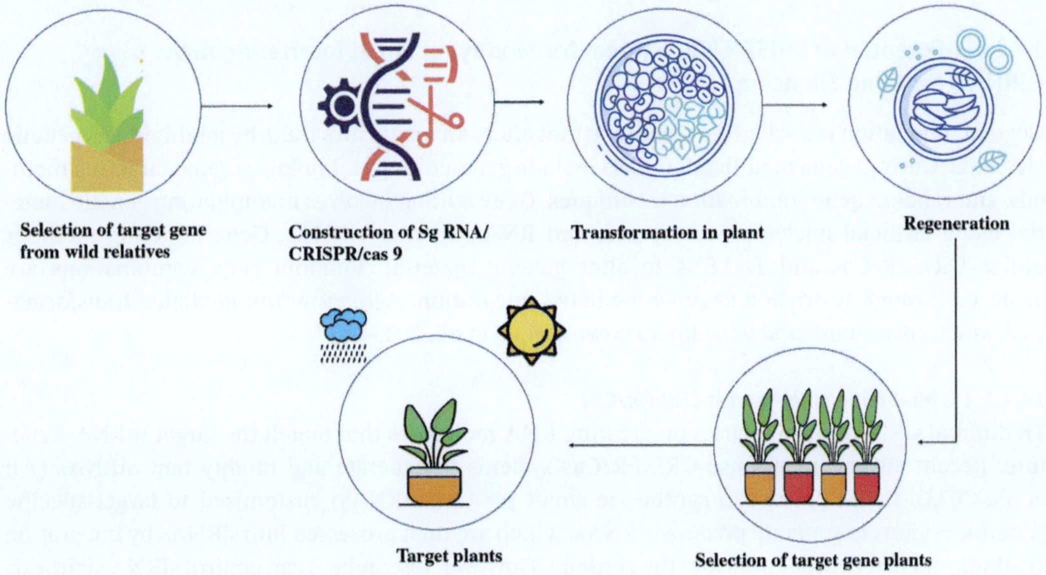

Figure 14.2 Detailed view of CRISPR-mediated gene editing in plants.

Common Methods:

- ***Agrobacterium*-Mediated Transformation**: CRISPR/Cas constructs were integrated into the plant genome using *Agrobacterium tumefaciens*, a technique commonly applied to diploid and monophyletic plants (Aliu et al., 2022).
- **Particle Bombardment (Biolistic)**: DNA-coated gold or tungsten particles are injected into plant cells. This method works well for unicorns and docile species (Begum & Jayawardana, 2021).
- **Viral Vectors**: Plant viruses deliver CRISPR/Cas components into plant cells. This method allows for a shorter exposure time and higher efficiency (Chuang et al., 2021).

Selection and Screening: Once CRISPR/Cas components are administered, modified plant cells or tissues must be selected and tested to ensure successful modification.

Selection Methods:

- **Selectable Markers**: Include antibiotic or herbicide resistance genes in the vector to facilitate the selection of successfully transformed cells (Dormatey et al., 2021).
- **Screening Techniques**: PCR, sequencing, and reporter genes are used to verify the presence of the desired gene mutation. T7E1 assays and Sanger sequencing are commonly employed for this confirmation (Lee et al., 2021).

Regeneration and Verification: The next step is to regenerate the whole plant from modified cells or tissues after successfully identifying a plant. This step is crucial for ensuring that genetic variation is tightly integrated into the plant genome (Zhu et al., 2020).

Regeneration Methods:

- **Tissue Culture**: Tissue culture techniques are used to regenerate whole plants from transformed tissues. This process may involve callus induction, shoot regeneration, and rooting.
- **Verification**: Genetically modified plants have the desired genetic variation and do not have off-target effects by molecular analysis.

14.2.5 Potential of CRISPR/Cas System for Modifying Small Interfering RNAs (siRNAs) for Gene Silencing

Gene manipulation is a scientific technique that alters an organism's traits by modifying its genetic elements. Current gene modification tools include gene-editing technologies, gene targeting methods, and random gene combination techniques. Gene editing involves manipulating genetic material using artificial nucleases, ribozymes, and RNAi, such as siRNAs. Gene targeting methods utilize CRISPR/Cas and TALENs to alter genetic material. Random gene combinations are achieved through restriction enzyme-mediated integration, *Agrobacterium*-mediated transformation, and transposon-based gene knockdown (Wang et al., 2023a,b).

14.2.5.1 Modifying siRNAs with CRISPR/Cas

Traditional siRNA design focuses on creating RNA molecules that match the target mRNA structure. Recent advancements use CRISPR/Cas systems to generate and modify new siRNAs (Ali et al., 2023). CRISPR/Cas can synthesize small RNAs (smRNAs) customized to target specific genomic regions to generate precursor RNAs, which are then processed into siRNAs by integrating synthetic or custom sequences into the genome. However, researchers can control siRNA synthesis precisely (Gerashchenkov et al., 2020). The CRISPR/Cas system can be programmed to regulate the expression of endogenous or synthetic siRNAs in response to specific stimuli. This provides fine control over gene silencing (Lennox & Behlke, 2020).

14.3 Role of Small Interfering RNAs in Gene Silencing

smRNAs control gene expression in eukaryotes throughout different developmental stages and respond to various conditions, including environmental stresses, that disrupt cellular or organismal equilibrium. To overcome these challenges, plants respond to stress by modifying gene expression based on environmental signals. These alterations in gene expression lead to broad and stress-specific molecular responses, allowing the plant to adapt to environmental fluctuations (Qiao et al., 2021; Sihag et al., 2024).

Eukaryotes have three main types of smRNAs: miRNAs, siRNAs, and piwi-interacting RNAs (piRNAs). While miRNAs and siRNAs are common in both plants and animals, piRNAs are found exclusively in animals (Chen, 2009; Wang & Chekanova, 2016). In plants, most smRNAs range from 20 to 24 nucleotides in length and are formed through complex processes. Typically, miRNAs are derived from MIR genes transcribed by RNA polymerase II (Pol II) and emerge from specific dsRNA precursors that can form hairpin loops. Conversely, siRNAs are generated from dsRNA precursors through various pathways, especially in plants, often involving RNA-directed RNA polymerases. Dicer proteins (Dicer-like [DCL] proteins in plants) play a key role in converting these dsRNA precursors into miRNAs and siRNAs. Additionally, in both plants and some animal siRNAs and piRNAs, smRNAs are stabilized by 2′-O-methylation on the 3′ terminal ribose, a process facilitated by the methyltransferase HEN1 (Fukudome & Fukuhara, 2017; Kryovrysanaki et al., 2021).

siRNAs play crucial roles in the complex coordination of gene regulation within cells, mainly via the RNAi pathway. This pathway enables precise PTGS that affects various cellular processes. Extensive research has shed light on their involvement in modulating gene expression, particularly in response to multiple stressors. The detailed process of siRNA synthesis is depicted in Figure 14.3.

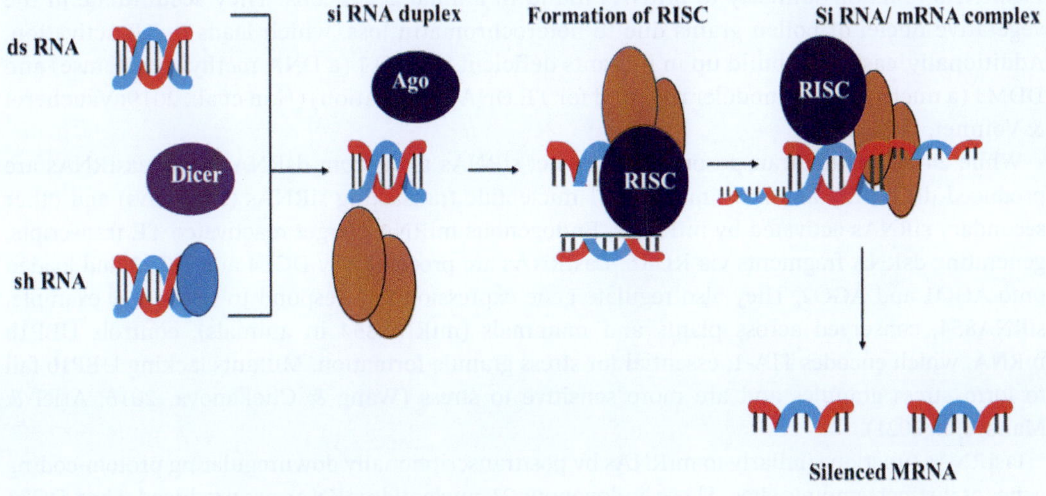

Figure 14.3 A detailed process of siRNA synthesis.

In plants, siRNAs serve a dual function in controlling gene expression, much like miRNAs, by affecting both the translation and cleavage of target RNAs. However, their primary role is to induce TGS by directing repressive epigenetic changes, such as DNA cytosine methylation and histone methylation, to corresponding genomic regions. This group of heterochromatic plant siRNAs acts similarly to nuclear siRNAs that mediate cotranscriptional gene silencing in *Schizosaccharomyces pombe*. Additionally, siRNAs are involved in the repair of DNA double-stranded breaks in both plants and humans (Kasai & Kanazawa, 2013; Jogam et al., 2022).

The main role of plant siRNAs is to trigger TGS by guiding repressive epigenetic modifications to genome regions with sequence homology. This includes the action of heterochromatic siRNAs (het-siRNAs) in RNA-directed DNA methylation (RdDM) and epigenetically activated small interfering RNAs (easiRNAs) in preserving genomic stability by suppressing TEs in germ cells. Moreover, examples illustrate how these smRNAs help plants cope with various stressors. Plant siRNAs are highly diverse and include several subclasses, differentiated by their origin and biogenesis (Wang & Chekanova, 2016; Zavallo et al., 2024).

Het-siRNAs primarily originate from silenced repetitive sequences and TEs. They target the corresponding chromatin to initiate DNA methylation and histone modifications, resulting in transcriptional repression. In plants, this process, known as RdDM, is similar to cotranscriptional gene silencing in *S. pombe*. RdDM requires two plant-specific RNA polymerases, Pol IV and Pol V, with RNA Pol II also contributing to the establishment of DNA methylation and histone modifications. In *S. pombe*, RNA Pol II performs functions comparable to Pol IV and Pol V, leading to H3K9 methylation (Yadav et al., 2021). Most 24-nucleotide het-siRNAs are derived from long noncoding RNAs (lncRNAs) synthesized by Pol IV, which are converted into dsRNAs by RDR2 and processed by DCL3. These het-siRNAs, loaded onto AGO4, recognize lncRNA scaffolds produced by Pol V via sequence complementarity, recruiting DNA methyltransferases and other factors to initiate de novo cytosine methylation, ultimately silencing transcription (Rymen et al., 2020; Rzeszutek & Betlej, 2020).

During germline reprogramming in plants, developmental stress arises because of the erasure of epigenetic marks. In *Arabidopsis*, this stress is linked to the activity of 21–22-nucleotide easiRNAs, which target TE transcripts to prevent their activation during reprogramming. EasiRNAs function similarly to piRNAs found in animal germ cells. They accumulate in the vegetative nuclei of pollen grains due to heterochromatin loss, which leads to TE activation. Additionally, easiRNAs build up in mutants deficient in MET1 (a DNA methyltransferase) and DDM1 (a nucleosome remodeler required for TE DNA methylation) (Han et al., 2019; Vaucheret & Voinnet, 2024).

While 24-nucleotide transposon-silencing het-siRNAs arise from dsRNA, some easiRNAs are produced differently and are similar to 21-nucleotide transacting siRNAs (tasiRNAs) and other secondary siRNAs activated by miRNAs. Endogenous miRNAs target reactivated TE transcripts, generating dsRNA fragments via RDR6. EasiRNAs are processed by DCL4 and DCL2 and loaded onto AGO1 and AGO2. They also regulate gene expression and respond to stress. For example, siRNA854, conserved across plants and mammals (miRNA854 in animals), controls UBP1b mRNA, which encodes TIA-1, essential for stress granule formation. Mutants lacking UBP1b fail to form stress granules and are more sensitive to stress (Wang & Chekanova, 2016; Ariel & Manavella, 2021).

tasiRNAs function similarly to miRNAs by posttranscriptionally downregulating protein-coding genes at distinct genomic sites. These endogenous 21-nucleotide siRNAs are produced when DCL1 cleaves the tasiRNA primary transcript, establishing its 5′ and 3′ ends. RDR6 and SGS3 then convert the transcript into dsRNA, allowing DCL4 to process it into 21-nucleotide tasiRNAs in *Arabidopsis* (Yu et al., 2020; Vigh et al., 2024). Both DCL1 and DCL4, along with RDR6, are essential for tasiRNA production. The process also requires AGO1, HEN1, HYL1, and SGS3. Unlike the het-siRNA pathway, which involves DCL3 and RDR2, the tasiRNA pathway is more akin to the miRNA pathway, relying on AGO1, DCL1, HEN1, and HYL1. In plants, tasiRNA precursors are targeted by specific miRNAs like miR173 and miR390. To date, tasiRNAs have been identified only in plants (Jin et al., 2022).

Genome-wide studies have identified a large number of natural *cis*-antisense gene pairs (NATs) in eukaryotic genomes. *Arabidopsis* contains over 1000 NAT pairs, while rice has around 600, suggesting potential siRNA production from these pairs. Research shows that the overlapping *Arabidopsis* genes P5CDH and SRO5 generate siRNAs in response to salt stress, enhancing salt tolerance (Tiwari, 2021). Additionally, natsiRNA-ATGB2 is induced by *Pseudomonas syringae* carrying the avrRpt2 effector and plays a key role in immunity via the resistance gene RPS2. The 22-nucleotide Pst (avrRpt2)-induced natsiRNA requires DCL1 and RDR6, while the 24-nucleotide SRO5-P5CDH nat-siRNA depends on DCL2 and RDR6 (Wang et al., 2023a,b).

Recognizing the role of siRNAs in regulating gene expression during stress provides valuable insights into cellular resilience and disease development. Exploring siRNA-mediated gene regulation can reveal potential therapeutic targets for stress-related conditions. Understanding siRNAs' diverse functions in stress responses offers potential for advancements in gene regulation research and therapeutic applications.

14.4 CRISPR/Cas-Mediated Reprogramming of siRNAs for Gene Silencing

siRNAs play a key role in gene silencing and regulation through the RNAi pathway. Reprogramming siRNAs holds important potential for therapeutic applications, genetic research, and crop enhancement (Kong et al., 2022). While CRISPR/Cas technology was initially recognized for genome editing, it has recently been investigated as a tool for modulating RNA pathways, including the reprogramming of siRNAs (Loriato et al., 2020).

The CRISPR/Cas system, originating from bacterial adaptive immunity, has transformed genetic engineering. The widely used CRISPR/Cas9, guided by smRNA (gRNA), introduces DSBs at specific DNA sites for targeted modifications. Beyond DNA editing, CRISPR/Cas can also target RNA molecules, providing a means for siRNA reprogramming (Liu et al., 2022a,b,c). This involves redirecting CRISPR machinery to interact with RNA rather than DNA. Key methods include the following:

14.4.1 Using CRISPR/Cas13 for Direct RNA Targeting

The CRISPR/Cas13 system is particularly well suited for reprogramming siRNAs because it naturally targets RNA rather than DNA. crRNA binds specifically to the RNA sequence of interest. The crRNA directs Cas13 to the RNA molecules. Once the Cas13-crRNA complex attaches to the target RNA, Cas13 cuts the RNA, resulting in its degradation. By targeting the precursor or mature form of siRNAs, researchers can effectively control the levels of specific siRNAs within the cell and reprogram their activities).

14.4.2 Using Catalytically Inactive Cas9 (dCas9) to Modulate siRNA Activity

Another approach involves the use of a modified version of Cas9 known as catalytically inactive Cas9 or dCas9. Unlike traditional Cas9, dCas9 does not cut DNA; instead, it binds to specific sequences directed by a gRNA, allowing it to interfere with gene expression or RNA processing without causing DSBs. By fusing dCas9 with RNA-binding proteins, researchers can guide dCas9 to specific RNA sequences, including those involved in siRNA biogenesis or function. This enables the modulation of siRNA processing and stability. dCas9 can be directed to the promoter regions of siRNA-encoding genes or the siRNAs themselves, allowing for either inhibition or enhancement of their functions. This approach is useful for precisely adjusting gene expression without making permanent changes to the genome (Shakirova et al., 2020).

14.4.3 Synthetic gRNAs for siRNA Reprogramming

Synthetic gRNAs can target specific RNA sequences, including those involved in the siRNA pathways. These gRNAs are delivered into the cell along with CRISPR/Cas components to modify siRNA activity. Synthetic gRNAs are designed to bind precisely to the RNA sequences of interest. These gRNAs can either promote cleavage of target RNAs (using Cas13) or interfere with RNA processing (using dCas9). By designing gRNAs that target precursor siRNAs, researchers can control the maturation process of siRNAs, thereby regulating downstream gene-silencing effects. This strategy allows for selective activation or inhibition of specific siRNAs in a controlled manner (Ghavami & Pandi, 2021).

14.5 Regulation of Abiotic Stress Responses Through Gene Silencing

Climate change is a common crisis challenging every corner of life, wherein plants cannot escape but stand against all odds. Frequent droughts in plains, flooding on coasts, and ongoing issues with soil acidity and salinity are common effects of climate change that are placing plants under stress (Eckardt et al., 2023). Besides managing these stress factors, plants also need to meet the demands of a growing global population. Given these challenges, efforts to breed abiotic stress-tolerant crops must be strengthened not only to exploit (Wani et al., 2018; Rajpal et al., 2019) the native adaptation abilities accumulated over the years but also to reprogram novel strategies through genomic engineering tools (Hossain et al., 2021; Yadav et al., 2023).

The consequences of stress-inducing factors can generally be delineated as drought, submergence, light, heat, cold, salinity, acidity, metal toxicity, and nutrient deficiency constraints (Baweja & Kumar, 2020). Despite these challenges, plants gradually accumulate survival tricks through plasticity in their cellular metabolism. Understanding the critical genetic determinants underlying these developmental and behavioral plasticities can help accelerate breeding efforts for engineering solutions. With regard to breeding efforts to improve adaptation strategies of crops, rice has been focused on more than any other crop, being the major food crop in India, and hence, it is the crop that has been most exposed to stress in both scale and diversity (Pathak et al., 2021). Other major food crops, including rice, are also prone to the evolutionary forces of light, temperature, water, and nutrients. Several genes are involved in the acquisition of resistance. Therefore, a list of various gene classes that are known to be differentially regulated in response to different stress factors in plants is provided.

Consistently upgraded omics studies at hierarchically different biological layers of the genome (Naveed et al., 2018), transcriptome (Wei et al., 2023), proteome (Kosova et al., 2018), and metabolome (Tinte et al., 2021) have revealed a sufficiently high number of stress-responsive genes. Genes based on the line of action that receives the signal transduce the signal in cascades and give a final response to the signal, forming a complex regulatory circuit (Trono & Pecchioni, 2022). Transcription factors, detoxification enzymes, protein protectants, osmolyte biosynthesis enzymes, water channel transporters, protein kinases, ABA signaling factors, and proteins involved in phospholipid metabolism are key gene classes linked to stress-signaling networks (Shinozaki & Yamaguchi-Shinozaki, 2022). Stress-responsive transcription factors include bZIP (Liu et al., 2023a,b), bHLH (Guo et al., 2021), NAC (Shao et al., 2015), AP2/ERF (Ma et al., 2024), MYB (Wang et al., 2021), and WRKY (Li et al., 2024). Enzymatic antioxidants like ascorbate peroxidase (APX) (Pandey et al., 2017), superoxide dismutase (SOD) (Szőllősi, 2014), catalase (CAT) (Verma et al., 2023a,b), and glutathione reductase (GR) (Harshavardhan et al., 2017) help detoxify ROS and maintain homeostasis (Mishra et al., 2023).

The response to abiotic stress involves cellular signaling pathways largely regulated by protein kinases such as SnRKs (Sucrose non-fermenting-1-related protein kinase) (Luo et al., 2022), MAPK (Mitogen-activated protein kinase) (Majeed et al., 2023), CDPKs (Calcium-dependent protein kinase)(Dekomah et al., 2022), and RLKs(Receptor-like kinase) (Soltabayeva et al., 2022). LEA proteins (Chen et al., 2019) and HSPs (ul Haq et al., 2019) help protect stress-sensitive metabolic proteins. Key osmolyte biosynthesis genes like P5CS(Δ1-Pyrroline-5-carboxylate synthetase) (Feng et al., 2022), G6PDH(Glucose-6-phosphate dehydrogenase) (Jiang et al., 2022), TPS(Trehalose-6-phosphate synthase)(Kerbler et al., 2023), and BADH(Betaine aldehyde dehydrogenase) (Fitzgerald et al., 2009) mitigate oxidative stress damage. Stress also impacts water and nutrient uptake, with transporters such as NRAMP(Natural resistance-associated macrophage protein) (Bozzi & Gaudet, 2021), ZIP(Zinc/Iron-regulated transporter-like protein) (Ajeesh Krishna et al., 2020), NRT(Nitrate transporter) (Wang et al., 2018), PHT(Phosphate transporter) (Wang et al., 2017a,b), PIP(Plasma membrane intrinsic protein) (Israel et al., 2022), and CAX(Cation/H+ exchanger) (Yang et al., 2022) maintaining ionic and osmotic balance. ABA interacts with both upstream and downstream stress-signaling genes (Vishwakarma et al., 2017), with ZEP (Zeaxanthin epoxidase) (Park et al., 2008), NCED(9-cis-epoxycarotenoid dioxygenase) (Basso et al., 2023), and ABF (ABRE-binding factor)(Liu et al., 2023a,b) playing roles in the ABA response. Understanding these molecular mechanisms and identifying relevant mutants are crucial for developing improved plant cultivars.

14.5.1 Genes of Target for CRISPR/Cas-Mediated Plant Gene Silencing to Engineer Abiotic Stress Tolerance

Despite pressing concerns, gene editing is the current cutting-edge technology in the toolbox for crop improvement and is increasingly being adopted to design better cultivars. CRISPR is widely applied as the best alternative to other gene-editing tools for crop improvement, mainly because of its programmable nuclease (Shen et al., 2017), Successful utilization of siRNAs for targeted gene silencing is described in Table 14.1. However, determining which genes to target for editing to gain the desired effect is the most challenging task, given the complexity of the molecular basis underlying the impact and response to abiotic stress, various genes of target and methods for silencing genes are outlined in Tables 14.2 and 14.3. Gene expression can be regulated in different ways, namely, knocking out, knocking in, and knocking down, and the comparatively lower risk of knocking down makes gene silencing a preferable approach (El-Sappah et al., 2021).

Table 14.2 Successful modification of siRNAs for targeted gene silencing.

S. no.	Target gene	Plant species	Abiotic stress	CRISPR/Cas system used	Modification strategy	Outcome	References
1	*DREB2A* (dehydration-responsive element-binding protein 2A)	*Arabidopsis thaliana*	Drought	CRISPR/Cas13	Cas13-mediated siRNA reprogramming targeting DREB2A	Enhanced drought tolerance; improved survival under water deficit	Farhat et al. (2019)
2	*NAC1* (NAC domain-containing protein)	Rice (*Oryza sativa*)	Salt stress	CRISPR/dCas9	Cas9-siRNA complex targeting NAC1-related siRNA	Increased salt tolerance; better growth in saline environments	Kurowska & Daszkowska-Golec (2023)
3	ABA8ox2 (abscisic acid 8′-hydroxylase)	Tomato (*Solanum lycopersicum*)	Drought	CRISPR/Cas13	Cas13-directed siRNA modulation targeting ABA8ox2	Improved drought resistance; enhanced ABA accumulation and stomatal closure	Wang et al. (2022)
4	CBF1 (C-repeat binding factor)	Wheat (*Triticum aestivum*)	Cold stress	CRISPR/Cas9	Synthetic gRNA-guided siRNA targeting CBF1 pathway	Increased cold tolerance; reduced damage under low-temperature conditions	Hwarari et al. (2022)
5	SOS1 (salt overly sensitive 1)	Soybean (*Glycine max*)	Salt stress	CRISPR/dCas9	dCas9-mediated siRNA reprogramming targeting SOS1-related siRNA	Enhanced salt tolerance; improved ion homeostasis under saline conditions	Erdoğan et al. (2023)
6	MYB4 (myeloblastosis transcription factor 4)	Maize (*Zea mays*)	Cold stress	CRISPR/Cas13	Cas13-mediated siRNA reprogramming targeting MYB4	Enhanced cold tolerance; improved seedling survival under cold stress	Chakraborty et al. (2023)
7	HSP70 (heat shock protein 70)	Rice (*Oryza sativa*)	Heat stress	CRISPR/dCas9	dCas9-siRNA complex targeting HSP70	Enhanced heat tolerance; improved grain filling under high temperature	Urhan (2024)
8	CAT1 (catalase 1)	Soybean (*Glycine max*)	Oxidative stress	CRISPR/Cas13	Cas13-mediated siRNA reprogramming targeting CAT1	Reduced oxidative damage; increased tolerance to abiotic stress	Zhong et al. (2022)
9	TIP1 (tonoplast intrinsic protein 1)	Wheat (*Triticum aestivum*)	Drought	CRISPR/Cas9	Cas9-guided siRNA modulation targeting TIP1	Improved drought tolerance; enhanced water use efficiency	Cevher-Keskin et al. (2019)
10	P5CS (Δ^1-pyrroline-5-carboxylate synthetase)	Tobacco (*Nicotiana tabacum*)	Drought	CRISPR/Cas9	Cas9-mediated siRNA reprogramming targeting P5CS	Improved drought tolerance; increased proline accumulation and stress adaptation	Nasir et al. (2023)

Table 14.3 Different genes of target and methods for silencing genes are outlined.

Crop	Target gene	Technique	Stress	References
Alfalfa	*MsLEA-D34*	RNAi	Osmotic and salt stress	Lv et al. (2021)
Maize	*ZmAMT3;1*	RNAi	N uptake efficiency	Hui et al. (2022)
Rice	*OsABA8ox1*	RNAi	Saline-alkaline stress	Liu et al. (2022a,b,c)
Rice	*OsBBTI5*	RNAi	Salt stress	Lin et al. (2024)
Rice	*NRAMP5*	RNAi	Nutrient toxicity stress	Ishimaru et al. (2012)
Soybean	*GmTLP8*	RNAi	Drought and salt stress	Xu et al. (2022)
Soybean	*GmMIPS1*	RNAi	Mineral bioavailability	Kumar et al. (2019)
Tomato	*SlGRAS10*	RNAi	Salinity, drought, and ABA response	Habib et al. (2021)
Tomato	*HyPRP1*	RNAi	Drought and salt stress	Chen et al. (2022)
Wheat	*IPK1*	RNAi	Micronutrient bioavailability	Aggarwal et al. (2018)
Rice	*miR166*	STTM	Drought	Zhang et al. (2018)
Potato	*miR160a/b*	STTM	Root architecture	Yang et al. (2021)
Rice	*miR5488*	STTM	Auxin response	Guo et al. (2023)
Rice	*miR5144-3p*	STTM	Heat stress	Xia et al. (2018)
Rice	*miR535*	STTM	Salt stress	Yue et al. (2020)
Cotton	*lncRNA973*	VIGS	Salt stress	Zhang et al. (2019a,b)
Birch	*LncY1*	RNAi	Salt stress	Jia et al. (2023)
Wheat	*lncRNAs (L0760, L6247, L0208, L2098, and L3065)*	VIGS	Alkalinity stress	Wei et al. (2022)

STTM, short tandem target mimics; RNAi, RNA interference.

The development of RNAi significantly advanced gene knockdown techniques (Fire et al., 1998). miRNAs and siRNAs, essential to RNAi, play key roles in abiotic stress signaling (Sunkar et al., 2007; Khraiwesh et al., 2012; Zhang et al., 2022). These small noncoding RNAs respond to stress depending on the tissue, age, stress type, and genotype, making their manipulation a promising approach to enhance abiotic stress tolerance (Zhang, 2015). However, RNAi-induced gene silencing methods often encounter off-target effects that lead to unintended phenotypes (Jackson & Linsley, 2010). CRISPR is gaining importance over RNAi, as its toolbox is flexible in mitigating these off-target effects (Naeem et al., 2020). The phenomenon of transient gene knockdown through CRISPR is termed CRISPRi, which is short for CRISPR interference (Unniyampurath et al., 2016).

Repression of gene expression for a transient period is considered safe compared to the lethal effects of gene knockout-mediated gene editing methods (Du, 2020). Among the gene silencing methods, miRNA- and siRNA-mediated RNAi have dominated since their discovery (De Felippes & Waterhouse, 2020). The CRISPR toolbox's efficiency and low error rate make CRISPRi a compelling choice (Stojic et al., 2018; Feder et al., 2020). Previous studies on CRISPR-induced gene silencing (CIGS) focused on its impact on viral accumulation (Aman et al., 2018; Mahas et al., 2019;

Table 14.4 Challenges and perspectives of CRISPRi.

Challenges	Future directions
ncRNAs are found to have diverse functions and yet the techniques to study their functions are limited	Complementing the prediction models with new representative machine learning models (Tan et al., 2022)
Genome-wide distribution of ncRNAs and the complexity of genomic architecture around them reduce the specificity to manipulate both bidirectional and independent promoters	Studying the phenotypes established in *cis* and *trans* effects by checking for reproducible results with RNAi or RNaseH-mediated approaches (Goyal et al., 2017)
Tissue-specific expression of ncRNAs limits the applicability of gene-editing tools	Using CRISPR-tissue-specific knockout (TSKO) with cell-specific promoters helps investigate gene functions (Ali et al., 2020)
Phenotyping analysis of mutants with engineered ncRNAs revealed impaired biogenesis, including DROSHA processing machinery	Posttranscriptional RNA silencing targeting either 5′ or 3′ arms through CRISPRi (Bassett et al., 2014)
The lack of an open reading frame does not necessarily result in a complete functional loss of noncoding genes	Integration of markers into the genome using an HR-mediated targeting pathway (Ho et al., 2015)
Lack of efficient delivery system accompanied by potential regeneration ability	Delivery strategy using nonviral nanomaterials (Sahel et al., 2023)
Short target windows and structural complexity limit the targeting specificity toward ncRNAs	Targeting multiple sites like promoters and hairpin motifs through multiguided CRISPR constructs (Xie et al., 2015)

Zhang et al., 2019a,b). A similar study involving multiguide crRNAs revealed that CIGS uses endogenous RNAi machinery to reduce mRNA availability (Sharma et al., 2022). In mammalian cells, there is rapid momentum through CRISPRi, in contrast to that in plants.

The ENCODE project analysis highlights that approximately 80% of the genome is involved in biochemical activities, with noncoding regions playing a significant role in protein-coding regions (Bernstein et al., 2012). ncRNAs open a new frontier in regulatory biology by acting as essential nodes in molecular interactions through the information residing in their primary and secondary structures (Fu, 2014). The selection of ncRNAs as targets for gene-editing tools is frequent in mammalian cells (Shakirova et al., 2020).

14.5.2 Challenges and Future Perspectives

Despite the tunability and reversibility of CRISPRi, its application in crop improvement is limited; the challenges and future directions are listed in Table 14.4.

14.6 Conclusion

The integration of CRISPR/Cas technology with the modulation of siRNAs represents a transformative approach for enhancing plant resilience to abiotic stress. By precisely targeting and silencing specific genes associated with stress responses, this strategy not only enables plants to better withstand adverse environmental conditions but also offers a powerful tool for the sustainable improvement of crop performance. As research progresses, the potential to fine-tune plant responses at the genetic level opens new avenues for developing crops that are adaptable to the challenges posed by climate change and resource limitations. The advancements highlighted in this chapter emphasize the need for ongoing research, with the goal of a future where CRISPR/Cas-reprogrammed siRNAs are pivotal in enhancing global food security through cutting-edge plant biotechnology.

References

Afzal, S., Sirohi, P., & Singh, N. K. (2020). A review of CRISPR associated genome engineering: application, advances and future prospects of genome targeting tool for crop improvement. *Biotechnology Letters*, *42*, 1611–1632.

Aliu, E., Lee, K., & Wang, K. (2022). CRISPR RNA-guided integrase enables high-efficiency targeted genome engineering in Agrobacterium tumefaciens. *Plant Biotechnology Journal*, *20*(10), 1916–1927.

Aggarwal, S., Kumar, A., Bhati, K. K., Kaur, G., Shukla, V., Tiwari, S., & Pandey, A. K. (2018). RNAi-mediated downregulation of inositol pentakisphosphate kinase (*IPK1*) in wheat grains decreases phytic acid levels and increases Fe and Zn accumulation. *Frontiers in Plant Science*, *9*, 259.

Ahmad, N., Rahman, M. U., Mukhtar, Z., Zafar, Y., & Zhang, B. (2020). A critical look on CRISPR-based genome editing in plants. *Journal of Cellular Physiology*, *235*(2), 666–682.

Ajeesh Krishna, T. P., Maharajan, T., Victor Roch, G., Ignacimuthu, S., & Antony Ceasar, S. (2020). Structure, function, regulation and phylogenetic relationship of ZIP family transporters of plants. *Frontiers in Plant Science*, *11*, 662.

Ali, Z., Mahfouz, M. M., & Mansoor, S. (2020). CRISPR-TSKO: a tool for tissue-specific genome editing in plants. *Trends in Plant Science*, *25*(2), 123–126.

Ali, A., Zafar, M. M., Farooq, Z., Ahmed, S. R., Ijaz, A., Anwar, Z., Abbas, H., Tariq, M. S., Tariq, H., Mustafa, M., & Bajwa, M. H. (2023). Breakthrough in CRISPR/Cas system: current and future directions and challenges. *Biotechnology Journal*, *18*(8), 2200642.

Ali, A., Amir, R., Gul, A., Munir, F., Ahmad, K., & Akram, A. (2024). Genomic engineering in peanut. In: *Targeted Genome Engineering Via CRISPR/Cas9 in Plants*, Academic Press, 159–175.

Aman, R., Ali, Z., Butt, H., Mahas, A., Aljedaani, F., Khan, M. Z., Ding, S., & Mahfouz, M. (2018). RNA virus interference via CRISPR/Cas13a system in plants. *Genome Biology*, *19*, 1–9.

Ariel, F. D., & Manavella, P. A. (2021). When junk DNA turns functional: transposon-derived non-coding RNAs in plants. *Journal of Experimental Botany*, *72*(11), 4132–4143.

Arnesen, J. A., Hoof, J. B., Kildegaard, H. F., & Borodina, I. (2021). Genome editing of Eukarya. In: *Metabolic Engineering: Concepts and Applications, Vol. Vol. 13*, 301–337.

Ashfaq, M. A., Dinesh Kumar, V., Soma Sekhar Reddy, P., Anil Kumar, C., Sai Kumar, K., Narasimha Rao, N., Tarakeswari, M., & Sujatha, M. (2020). Post-transcriptional gene silencing: basic concepts and applications. *Journal of Biosciences*, *45*(1), 128.

Asmamaw, M., & Zawdie, B. (2021). Mechanism and applications of CRISPR/Cas-9-mediated genome editing. *Biologics: Targets and Therapy*, *15*, 353–361.

Atayee, A. R., & Noori, M. S. (2020). Alleviation of cold stress in vegetable crops. *Journal of Scientific Agriculture*, *4*, 38–44.

Balasubramaniam, T., Shen, G., Esmaeili, N., & Zhang, H. (2023). Plants' response mechanisms to salinity stress. *Plants*, *12*(12), 2253.

Bassett, A. R., Azzam, G., Wheatley, L., Tibbit, C., Rajakumar, T., McGowan, S., Stanger, N., Ewels, P. A., Taylor, S., Ponting, C. P., & Liu, J. L. (2014). Understanding functional miRNA–target interactions in vivo by site-specific genome engineering. *Nature Communications*, *5*(1), 4640.

Basso, M. F., Contaldi, F., Celso, F. L., Karalija, E., Paz-Carrasco, L. C., Barone, G., Ferrante, A., & Martinelli, F. (2023). Expression profile of the *NCED/CCD* genes in chickpea and lentil during abiotic stress reveals a positive correlation with increased plant tolerance. *Plant Science*, *336*, 111817.

Baweja, P., & Kumar, G. (2020). Abiotic stress in plants: an overview. In: *Plant Stress Biology: Strategies and Trends*, Springer, Singapore. https://doi.org/10.1007/978-981-15-9380-2_1, 1–15.

Begum, R., & Jayawardana, N. U. (2021). A review of nanotechnology as a novel method of gene transfer in plants. *Journal of Agricultural Sciences – Sri Lanka*, *16*(2), 300–316.

Bernstein, B. E., Birney, E., Dunham, I., Green, E. D., Gunter, C., Snyder, M., & Project Consortium ENCODE (2012). An integrated encyclopedia of DNA elements in the human genome. *Nature*, *489*, 57–74.

Bhatia, S., & Yadav, S. K. (2023). CRISPR-Cas for genome editing: classification, mechanism, designing and applications. *International Journal of Biological Macromolecules*, *238*, 124054.

Bozzi, A. T., & Gaudet, R. (2021). Molecular mechanism of Nramp-family transition metal transport. *Journal of Molecular Biology*, *433*(16), 166991.

Carrera, F. P., Noceda, C., Maridueña-Zavala, M. G., & Cevallos-Cevallos, J. M. (2021). Metabolomics, a powerful tool for understanding plant abiotic stress. *Agronomy*, *11*(5), 824.

Cevher-Keskin, B., Yıldızhan, Y., Yüksel, B., Dalyan, E., & Memon, A. R. (2019). Characterization of differentially expressed genes to Cu stress in Brassica nigra by Arabidopsis genome arrays. *Environmental Science and Pollution Research*, *26*, 299–311.

Chakraborty, A., Choudhury, S., Kar, S. R., Deb, P., & Wylie, S. J. (2023). Gene editing for tolerance to temperature stress in plants: a review. *Plant Gene*, *37*, 100439.

Chaudhary, E., Chaudhary, A., Sharma, S., Tiwari, V., & Garg, M. (2024). Different classes of CRISPR-Cas systems. In: *Gene Editing in Plants: CRISPR-Cas and Its Applications*, Springer Nature Singapore, Singapore, 73–94.

Chen, X. (2009). Small RNAs and their roles in plant development. In: *Annual Review of Cell and Developmental Biology*, *Vol. 25*, 21–44. https://doi.org/10.1146/annurev.cellbio.042308.113417

Chen, Y., Li, C., Zhang, B., Yi, J., Yang, Y., Kong, C., Lei, C., & Gong, M. (2019). The role of the late embryogenesis-abundant (LEA) protein family in development and the abiotic stress response: a comprehensive expression analysis of potato (*Solanum tuberosum*). *Genes*, *10*(2), 148.

Chen, X., Wang, L., Liang, Y., Hu, X., Pan, Q., Ding, Y., & Li, J. (2022). *HyPRP1*, a tomato multipotent regulator, negatively regulates tomato resistance to sulfur dioxide toxicity and can also reduce abiotic stress tolerance of Escherichia coli and tobacco. *Horticulturae*, *8*(12), 1118.

Chuang, Y. F., Phipps, A. J., Lin, F. L., Hecht, V., Hewitt, A. W., Wang, P. Y., & Liu, G. S. (2021). Approach for in vivo delivery of CRISPR/Cas system: a recent update and future prospect. *Cellular and Molecular Life Sciences*, *78*, 2683–2708.

De Felippes, F. F., & Waterhouse, P. M. (2020). The whys and wherefores of transitivity in plants. *Frontiers in Plant Science*, *11*, 579376.

Deka, P. C. (2024). Gene cloning. In: *Research Anthology on Bioinformatics, Genomics, and Computational Biology*, IGI Global, 650–689.

Dekomah, S. D., Bi, Z., Dormatey, R., Wang, Y., Haider, F. U., Sun, C., Yao, P., & Bai, J. (2022). The role of CDPKs in plant development, nutrient and stress signaling. *Frontiers in Genetics*, *13*, 996203.

Dormatey, R., Sun, C., Ali, K., Fiaz, S., Xu, D., Calderón-Urrea, A., Bi, Z., Zhang, J., & Bai, J. (2021). ptxD/Phi as alternative selectable marker system for genetic transformation for bio-safety concerns: a review. *PeerJ*, *9*, e11809.

Du, L. L. (2020). Resurrection from lethal knockouts: bypass of gene essentiality. *Biochemical and Biophysical Research Communications*, *528*(3), 405–412.

Eckardt, N. A., Cutler, S., Juenger, T. E., Marshall-Colon, A., Udvardi, M., & Verslues, P. E. (2023). Focus on climate change and plant abiotic stress biology. *The Plant Cell*, *35*(1), 1–3.

El-Sappah, A. H., Yan, K., Huang, Q., Islam, M. M., Li, Q., Wang, Y., Khan, M. S., Zhao, X., Mir, R. R., Li, J., & El-Tarabily, K. A. (2021). Comprehensive mechanism of gene silencing and its role in plant growth and development. *Frontiers in Plant Science*, *12*, 705249.

Erdoğan, İ., Cevher-Keskin, B., Bilir, Ö., Hong, Y., & Tör, M. (2023). Recent developments in CRISPR/Cas9 genome-editing technology related to plant disease resistance and abiotic stress tolerance. *Biology*, *12*(7), 1037.

Farhat, S., Jain, N., Singh, N., Sreevathsa, R., Dash, P. K., Rai, R., Yadav, S., Kumar, P., Sarkar, A. K., Jain, A., Singh, N. K., & Rai, V. (2019). CRISPR-Cas9 directed genome engineering for enhancing salt stress tolerance in rice. *Seminars in Cell & Developmental Biology*, *96*, 91–99. Academic Press

Feder, A., Jensen, S., Wang, A., Courtney, L., Middleton, L., Van Eck, J., Liu, Y., & Giovannoni, J. J. (2020). Tomato fruit as a model for tissue-specific gene silencing in crop plants. *Horticulture Research*, *7*, 142.

Feng, X., Hu, Y., Zhang, W., Xie, R., Guan, H., Xiong, H., Jia, L., Zhang, X., Zhou, H., Zheng, D., & Wen, Y. (2022). Revisiting the role of delta-1-pyrroline-5-carboxylate synthetase in drought-tolerant crop breeding. *The Crop Journal*, *10*(4), 1213–1218.

Fire, A., Xu, S., Montgomery, M. K., Kostas, S. A., Driver, S. E., & Mello, C. C. (1998). Potent and specific genetic interference by double-stranded RNA in Caenorhabditis elegans. *Nature*, *391*(6669), 806–811.

Fitzgerald, T. L., Waters, D. L., & Henry, R. J. (2009). Betaine aldehyde dehydrogenase in plants. *Plant Biology*, *11*(2), 119–130.

Fu, X. D. (2014). Non-coding RNA: a new frontier in regulatory biology. *National Science Review*, *1*(2), 190–204.

Fukudome, A., & Fukuhara, T. (2017). Plant dicer-like proteins: double-stranded RNA-cleaving enzymes for small RNA biogenesis. *Journal of Plant Research*, *130*, 33–44.

Gerashchenkov, G. A., Rozhnova, N. A., Kuluev, B. R., Kiryanova, O. Y., Gumerova, G. R., Knyazev, A. V., Vershinina, Z. R., Mikhailova, E. V., Chemeris, D. A., Matniyazov, R. T., & Baimiev, A. K. (2020). Design of guide RNA for CRISPR/Cas plant genome editing. *Molecular Biology*, *54*, 24–42.

Ghavami, S., & Pandi, A. (2021). CRISPR interference and its applications. *Progress in Molecular Biology and Translational Science*, *180*, 123–140.

Goyal, A., Myacheva, K., Groß, M., Klingenberg, M., Duran Arqué, B., & Diederichs, S. (2017). Challenges of CRISPR/Cas9 applications for long non-coding RNA genes. *Nucleic Acids Research*, *45*(3), e12.

Guo, J., Sun, B., He, H., Zhang, Y., Tian, H., & Wang, B. (2021). Current understanding of bHLH transcription factors in plant abiotic stress tolerance. *International Journal of Molecular Sciences*, *22*(9), 4921.

Guo, S., Zheng, C., Wang, Y., Xu, Y., Wu, J., Wang, L., Liu, X., & Chen, Z. (2023). *OsmiRNA5488* regulates the development of embryo sacs and targets *OsARF25* in rice (Oryza sativa L.). *International Journal of Molecular Sciences*, *24*(22), 16240.

Habib, S., Lwin, Y. Y., & Li, N. (2021). Down-regulation of *SlGRAS10* in tomato confers abiotic stress tolerance. *Genes*, *12*(5), 623.

Han, Q., Bartels, A., Cheng, X., Meyer, A., An, Y. Q. C., Hsieh, T. F., & Xiao, W. (2019). Epigenetics regulates reproductive development in plants. *Plants*, *8*(12), 564.

Harshavardhan, V. T., Wu, T. M., & Hong, C. Y. (2017). Glutathione reductase and abiotic stress tolerance in plants. In: *Glutathione in Plant Growth, Development, and Stress Tolerance*, Springer, Cham, 265–286.

Hassan, M. U., Chattha, M. U., Khan, I., Chattha, M. B., Barbanti, L., Aamer, M., Iqbal, M. M., Nawaz, M., Mahmood, A., Ali, A., & Aslam, M. T. (2021). Heat stress in cultivated plants: nature, impact, mechanisms, and mitigation strategies—a review. *Plant Biosystems - An International Journal Dealing with all Aspects of Plant Biology*, *155*(2), 211–234.

He, L., St. John James, M., Radovcic, M., Ivancic-Bace, I., & Bolt, E. L. (2020). Cas3 protein—a review of a multi-tasking machine. *Genes*, *11*(2), 208.

Ho, T. T., Zhou, N., Huang, J., Koirala, P., Xu, M., Fung, R., Wu, F., & Mo, Y. Y. (2015). Targeting non-coding RNAs with the CRISPR/Cas9 system in human cell lines. *Nucleic Acids Research*, *43*(3), e17.

Hossain, A., Ahmed, S., Ahmad, Z., Alam, M. J., Moulick, D., Saha, B., Samanta, S., Maitra, S., Bhadra, P., Bukhari, M. A., & Aftab, T. (2021). Next-generation genetic engineering tools for abiotic stress tolerance in plants. In: *Frontiers in Plant-Soil Interaction*, Academic Press, 153–197.

Hui, J., An, X., Li, Z., Neuhäuser, B., Ludewig, U., Wu, X., Schulze, W. X., Chen, F., Feng, G., Lambers, H., & Zhang, F. (2022). The mycorrhiza-specific ammonium transporter *ZmAMT3*; 1 mediates mycorrhiza-dependent nitrogen uptake in maize roots. *The Plant Cell*, *34*(10), 4066–4087.

Hwarari, D., Guan, Y., Ahmad, B., Movahedi, A., Min, T., Hao, Z., Lu, Y., Chen, J., & Yang, L. (2022). ICE-CBF-COR signaling cascade and its regulation in plants responding to cold stress. *International Journal of Molecular Sciences*, *23*(3), 1549.

Ishimaru, Y., Takahashi, R., Bashir, K., Shimo, H., Senoura, T., Sugimoto, K., Ono, K., Yano, M., Ishikawa, S., Arao, T., & Nakanishi, H. (2012). Characterizing the role of rice *NRAMP5* in manganese, iron and cadmium transport. *Scientific Reports*, *2*(1), 286.

Israel, D., Lee, S. H., Robson, T. M., & Zwiazek, J. J. (2022). Plasma membrane aquaporins of the PIP1 and PIP2 subfamilies facilitate hydrogen peroxide diffusion into plant roots. *BMC Plant Biology*, *22*(1), 566.

Jackson, A. L., & Linsley, P. S. (2010). Recognizing and avoiding siRNA off-target effects for target identification and therapeutic application. *Nature Reviews Drug Discovery*, *9*(1), 57–67.

Jia, Y., Zhao, H., Niu, Y., & Wang, Y. (2023). Identification of birch lncRNAs and mRNAs responding to salt stress and characterization of functions of lncRNA. *Horticulture Research*, *10*(2), uhac277.

Jiang, Z., Wang, M., Nicolas, M., Ogé, L., Pérez-Garcia, M. D., Crespel, L., Li, G., Ding, Y., Le Gourrierec, J., Grappin, P., & Sakr, S. (2022). Glucose-6-phosphate dehydrogenases: the hidden players of plant physiology. *International Journal of Molecular Sciences*, *23*(24), 16128.

Jin, L., Chen, M., Xiang, M., & Guo, Z. (2022). RNAi-based antiviral innate immunity in plants. *Viruses*, *14*(2), 432.

Jogam, P., Sandhya, D., Alok, A., Peddaboina, V., Allini, V. R., & Zhang, B. (2022). A review on CRISPR/Cas-based epigenetic regulation in plants. *International Journal of Biological Macromolecules*, *219*, 1261–1271.

Kasai, M., & Kanazawa, A. (2013). Induction of RNA-directed DNA methylation and heritable transcriptional gene silencing as a tool to engineer novel traits in plants. *Plant Biotechnology*, *30*(3), 233–241.

Kaul, T., Sony, S. K., Raman, N. M., Eswaran, M., Verma, R., Prakash, A. T., Bharti, J., Motelb, K. F. A., & Kaul, R. (2020). How crisp is CRISPR? CRISPR-Cas-mediated crop improvement with special focus on nutritional traits. In: *Advancement in Crop Improvement Techniques*, Woodhead Publishing, 159–197.

Kavuri, N. R., Ramasamy, M., Qi, Y., & Mandadi, K. (2022). Applications of CRISPR/Cas13-based RNA editing in plants. *Cells*, *11*(17), 2665.

Kerbler, S. M. L., Armijos-Jaramillo, V., Lunn, J. E., & Vicente, R. (2023). The trehalose 6-phosphate phosphatase family in plants. *Physiologia Plantarum*, *175*(6), e14096.

Khraiwesh, B., Zhu, J. K., & Zhu, J. (2012). Role of miRNAs and siRNAs in biotic and abiotic stress responses of plants. *Biochimica et Biophysica Acta (BBA) - Gene Regulatory Mechanisms, 1819*(2), 137–148.

Kong, X., Yang, M., Le, B. H., He, W., & Hou, Y. (2022). The master role of siRNAs in plant immunity. *Molecular Plant Pathology, 23*(10), 1565–1574.

Kosova, K., Vitamvas, P., Urban, M. O., Prasil, I. T., & Renaut, J. (2018). Plant abiotic stress proteomics: the major factors determining alterations in cellular proteome. *Frontiers in Plant Science, 9*, 122.

Kryovrysanaki, N., James, A., Tselika, M., Bardani, E., & Kalantidis, K. (2021). RNA silencing pathways in plant development and defense. *The International Journal of Developmental Biology, 66*(1-2-3), 163–175.

Kumar, S. (2020). Abiotic stresses and their effects on plant growth, yield and nutritional quality of agricultural produce. *International Journal of Food Science and Agriculture, 4*(4), 367–378.

Kumar, S., & Mohapatra, T. (2021). Dynamics of DNA methylation and its functions in plant growth and development. *Frontiers in Plant Science, 12*, 596236.

Kumar, A., Kumar, V., Krishnan, V., Hada, A., Marathe, A., Jolly, M., & Sachdev, A. (2019). Seed targeted RNAi-mediated silencing of *GmMIPS1* limits phytate accumulation and improves mineral bioavailability in soybean. *Scientific Reports, 9*(1), 7744.

Kurowska, M., & Daszkowska-Golec, A. (2023). Molecular mechanisms of SNAC1 (stress-responsive NAC1) in conferring the abiotic stress tolerance. *Plant Science, 111894*.

Lennox, K. A., & Behlke, M. A. (2020). Chemical modifications in RNA interference and CRISPR/Cas genome editing reagents. In: *RNA Interference and CRISPR Technologies: Technical Advances and New Therapeutic Opportunities*, Springer, 23–55.

Lee, N. E., Ha, D. I., Ko, J. H., & Kim, Y. S. (2021). One-step genotyping method in CRISPR based on short inner primer-assisted, tetra primer-paired amplifications. *Molecular and Cellular Probes, 55*, 101675.

Liao, C., & Beisel, C. L. (2021). The tracrRNA in CRISPR biology and technologies. *Annual Review of Genetics, 55*(1), 161–181.

Lin, Z., Yi, X., Ali, M. M., Zhang, L., Wang, S., Tian, S., & Chen, F. (2024). RNAi-mediated suppression of *OsBBTI5* promotes salt stress tolerance in rice. *International Journal of Molecular Sciences, 25*(2), 1284.

Liu, X., Xie, X., Zheng, C., Wei, L., Li, X., Jin, Y., Zhang, G., Jiang, C. J., & Liang, Z. (2022a). RNAi-mediated suppression of the abscisic acid catabolism gene *OsABA8ox1* increases abscisic acid content and tolerance to saline–alkaline stress in rice (*Oryza sativa* L.). *The Crop Journal, 10*(2), 354–367.

Liu, X., Xiong, W., Qi, Q., Zhang, Y., Ji, H., Cui, S., An, J., Sun, X., Yin, H., Tian, T., & Zhou, X. (2022b). Rational guide RNA engineering for small-molecule control of CRISPR/Cas9 and gene editing. *Nucleic Acids Research, 50*(8), 4769–4783.

Liu, Y., Wang, J., Liu, B., & Xu, Z. Y. (2022c). Dynamic regulation of DNA methylation and histone modifications in response to abiotic stresses in plants. *Journal of Integrative Plant Biology, 64*(12), 2252–2274.

Liu, H., Tang, X., Zhang, N., Li, S., & Si, H. (2023a). Role of bZIP transcription factors in plant salt stress. *International Journal of Molecular Sciences, 24*(9), 7893.

Liu, X., Cheng, R., Chen, Y., Wang, S., Qin, F., Wang, D., Liu, Y., Hu, L., & Meng, S. (2023b). Identification and characterization of the *AREB/ABF/ABI5* gene family in sandalwood (*Santalum album* L.) and its potential role in drought stress and ABA treatment. *Forests, 14*(8), 1691.

Loriato, V. A., Martins, L. G., Euclydes, N. C., Reis, P. A., Duarte, C. E., & Fontes, E. P. (2020). Engineering resistance against geminiviruses: a review of suppressed natural defenses and the use of RNAi and the CRISPR/Cas system. *Plant Science, 292*, 110410.

Luo, Y., Niu, Y., Gao, R., Wang, C., & Liao, W. (2022). Genome-wide identification and expression analysis of SnRK gene family under abiotic stress in cucumber (*Cucumis sativus* L.). *Agronomy, 12*(7), 1550.

Lv, A., Su, L., Wen, W., Fan, N., Zhou, P., & An, Y. (2021). Analysis of the function of the alfalfa MsLEA-D34 gene in abiotic stress responses and flowering time. *Plant and Cell Physiology, 62*(1), 28–42.

Ma, Z., Hu, L., & Jiang, W. (2024). Understanding AP2/ERF transcription factor responses and tolerance to various abiotic stresses in plants: a comprehensive review. *International Journal of Molecular Sciences, 25*(2), 893.

Mahas, A., Aman, R., & Mahfouz, M. (2019). CRISPR-Cas13d mediates robust RNA virus interference in plants. *Genome Biology, 20*, 1–16.

Majeed, Y., Zhu, X., Zhang, N., Ul-Ain, N., Raza, A., Haider, F. U., & Si, H. (2023). Harnessing the role of mitogen-activated protein kinases against abiotic stresses in plants. *Frontiers in Plant Science, 14*, 932923.

Marraffini, L.A., 2022. Mechanism of type III CRISPR-Cas immunity. Crispr: Biology and Applications, Wiley pp. 71–84.

Mishra, N., Jiang, C., Chen, L., Paul, A., Chatterjee, A., & Shen, G. (2023). Achieving abiotic stress tolerance in plants through antioxidative defense mechanisms. *Frontiers in Plant Science, 14*, 1110622.

Munawar, N., & Ahmad, A. (2021). CRISPR/Cas system: an introduction. In: *CRISPR Crops: The Future of Food Security*, Springer, 1–35.

Naeem, M., Majeed, S., Hoque, M. Z., & Ahmad, I. (2020). Latest developed strategies to minimize the off-target effects in CRISPR-Cas-mediated genome editing. *Cells, 9*(7), 1608.

Nasir, A., Shahzadi, I., & Nawaz, I. (2023). Molecular mechanisms and strategies contributing toward abiotic stress tolerance in plants. In: *Abiotic Stress in Plants-Adaptations to Climate Change*, IntechOpen.

Naveed, S. A., Zhang, F., Zhang, J., Zheng, T. Q., Meng, L. J., Pang, Y. L., Xu, J. L., & Li, Z. K. (2018). Identification of QTN and candidate genes for salinity tolerance at the germination and seedling stages in rice by genome-wide association analyses. *Scientific Reports, 8*(1), 6505.

Nidhi, S., Anand, U., Oleksak, P., Tripathi, P., Lal, J. A., Thomas, G., Kuca, K., & Tripathi, V. (2021). Novel CRISPR–Cas systems: an updated review of the current achievements, applications, and future research perspectives. *International Journal of Molecular Sciences, 22*(7), 3327.

Pandey, S., Fartyal, D., Agarwal, A., Shukla, T., James, D., Kaul, T., Negi, Y. K., Arora, S., & Reddy, M. K. (2017). Abiotic stress tolerance in plants: myriad roles of ascorbate peroxidase. *Frontiers in Plant Science, 8*, 581.

Pandey, P., Mysore, K. S., & Senthil-Kumar, M. (2022). Recent advances in plant gene silencing methods. In: *Plant Gene Silencing: Methods and Protocols*, Humana, New York, NY., 1–22.

Park, H. Y., Seok, H. Y., Park, B. K., Kim, S. H., Goh, C. H., Lee, B. H., Lee, C. H., & Moon, Y. H. (2008). Overexpression of Arabidopsis ZEP enhances tolerance to osmotic stress. *Biochemical and Biophysical Research Communications, 375*(1), 80–85.

Pathak, H., Kumar, M., Molla, K. A., & Chakraborty, K. (2021). Abiotic stresses in rice production: impacts and management. *Oryza, 58*, 103–125.

Qiao, Y., Xia, R., Zhai, J., Hou, Y., Feng, L., Zhai, Y., & Ma, W. (2021). Small RNAs in plant immunity and virulence of filamentous pathogens. *Annual Review of Phytopathology, 59*, 265–288.

Rajpal, V. R., Sehgal, D., Kumar, A., & Raina, S. N. eds. (2019). *Genomics Assisted Breeding of Crops for Abiotic Stress Tolerance, Vol. Vol. II*, Springer International Publishing.

Ramazi, S., Allahverdi, A., & Zahiri, J. (2020). Evaluation of post-translational modifications in histone proteins: a review on histone modification defects in developmental and neurological disorders. *Journal of Biosciences*, *45*(1), 135.

Ray, P., Sahu, D., Aminedi, R., & Chandran, D. (2022). Concepts and considerations for enhancing RNAi efficiency in phytopathogenic fungi for RNAi-based crop protection using nanocarrier-mediated dsRNA delivery systems. *Frontiers in Fungal Biology*, *3*, 977502.

Rymen, B., Ferrafiat, L., & Blevins, T. (2020). Non-coding RNA polymerases that silence transposable elements and reprogram gene expression in plants. *Transcription*, *11*(3–4), 172–191.

Rzeszutek, I., & Betlej, G. (2020). The role of small noncoding RNA in DNA double-strand break repair. *International Journal of Molecular Sciences*, *21*(21), 8039.

Sahel, D. K., Vora, L. K., Saraswat, A., Sharma, S., Monpara, J., D'Souza, A. A., Mishra, D., Tryphena, K. P., Kawakita, S., Khan, S., & Azhar, M. (2023). CRISPR/Cas9 genome editing for tissue-specific in vivo targeting: nanomaterials and translational perspective. *Advanced Science*, *10*(19), 2207512.

Seneviratne, S. I., Zhang, X., Adnan, M., Badi, W., Dereczynski, C., Luca, A. D., Ghosh, S., Iskandar, I., Kossin, J., Lewis, S., & Otto, F. (2021). *Weather and Climate Extreme Events in a Changing Climate*, *Vol. 15*, Cambridge University Press, Cambridge, UK, 1513–1766.

Shakirova, K. M., Ovchinnikova, V. Y., & Dashinimaev, E. B. (2020). Cell reprogramming with CRISPR/Cas9 based transcriptional regulation systems. *Frontiers in Bioengineering and Biotechnology*, *8*, 882.

Shao, H., Wang, H., & Tang, X. (2015). NAC transcription factors in plant multiple abiotic stress responses: progress and prospects. *Frontiers in Plant Science*, *6*, 902.

Sharma, V. K., Marla, S., Zheng, W., Mishra, D., Huang, J., Zhang, W., Morris, G. P., & Cook, D. E. (2022). CRISPR guides induce gene silencing in plants in the absence of Cas. *Genome Biology*, *23*, 1–24.

Shen, S., Loh, T. J., Shen, H., Zheng, X., & Shen, H. (2017). CRISPR as a strong gene editing tool. *BMB Reports*, *50*(1), 20.

Shinozaki, K., & Yamaguchi-Shinozaki, K. (2022). Functional genomics in plant abiotic stress responses and tolerance: from gene discovery to complex regulatory networks and their application in breeding. *Proceedings of the Japan Academy, Series B, Physical and Biological Sciences*, *98*(8), 470–492.

Sihag, N., Singh, T., Sheoran, S., Singh, O., Malik, R., Kumar, L., & Singh, J. (2024). Role of RNA interference in drought stress management: physiological, biochemical and molecular approach. *Cisro, Crop & Pasture Science*, *75*(1), CP23189.

Soltabayeva, A., Dauletova, N., Serik, S., Sandybek, M., Omondi, J. O., Kurmanbayeva, A., & Srivastava, S. (2022). Receptor-like kinases (LRR-RLKs) in response of plants to biotic and abiotic stresses. *Plants*, *11*(19), 2660.

Stojic, L., Lun, A. T. L., Mangei, J., Mascalchi, P., Quarantotti, V., Barr, A. R., Bakal, C., Marioni, J. C., Gergely, F., & Odom, D. T. (2018). Specificity of RNAi, LNA and CRISPRi as loss-of-function methods in transcriptional analysis. *Nucleic Acids Research*, *46*(12), 5950–5966.

Sunkar, R., Chinnusamy, V., Zhu, J., & Zhu, J. K. (2007). Small RNAs as big players in plant abiotic stress responses and nutrient deprivation. *Trends in Plant Science*, *12*(7), 301–309.

Szőllősi, R. (2014). Superoxide dismutase (SOD) and abiotic stress tolerance in plants: an overview. In: *Oxidative Damage to Plants*, 89–129.

Tan, J., Li, X., Zhang, L., & Du, Z. (2022). Recent advances in machine learning methods for predicting LncRNA and disease associations. *Frontiers in Cellular and Infection Microbiology*, *12*, 1071972.

Tinte, M. M., Chele, K. H., van Der Hooft, J. J., & Tugizimana, F. (2021). Metabolomics-guided elucidation of plant abiotic stress responses in the 4IR era: an overview. *Metabolites*, *11*(7), 445.

Tiwari, B. (2021). *Identification of Small RNAs During Abiotic Stresses in Arabidopsis thaliana* (Doctoral Dissertation), LMU.

Trono, D., & Pecchioni, N. (2022). Candidate genes associated with abiotic stress response in plants as tools to engineer tolerance to drought, salinity and extreme temperatures in wheat: an overview. *Plants, 11*(23), 3358.

ul Haq, S., Khan, A., Ali, M., Khattak, A. M., Gai, W. X., Zhang, H. X., Wei, A. M., & Gong, Z. H. (2019). Heat shock proteins: dynamic biomolecules to counter plant biotic and abiotic stresses. *International Journal of Molecular Sciences, 20*(21), 5321.

Unniyampurath, U., Pilankatta, R., & Krishnan, M. N. (2016). RNA interference in the age of CRISPR: will CRISPR interfere with RNAi? *International Journal of Molecular Sciences, 17*(3), 291.

Urhan, E. K. (2024). Abiotic stress tolerance in plants by genome editing applications. In: *Applications of Genome Engineering in Plants*, Wiley, ,185–,221.

Vaucheret, H., & Voinnet, O. (2024). The plant siRNA landscape. *The Plant Cell, 36*(2), 246–275.

Verma, D., Kaushal, N., Balhara, R., & Singh, K. (2023a). Genome-wide analysis of catalase gene family reveal insights into abiotic stress response mechanism in *Brassica juncea* and *B. rapa*. *Plant Science, 330*, 111620.

Verma, V., Kumar, A., Partap, M., Thakur, M., & Bhargava, B. (2023b). CRISPR-Cas: a robust technology for enhancing consumer-preferred commercial traits in crops. *Frontiers in Plant Science, 14*, 1122940.

Vigh, M.L., Thieffry, A., Arribas-Hernandez, L. and Brodersen, P., 2024. Arabidopsis terminal nucleotidyl transferases govern secondary siRNA production at distinct steps bioRxiv, pp. 2024-05.

Vishwakarma, K., Upadhyay, N., Kumar, N., Yadav, G., Singh, J., Mishra, R. K., Kumar, V., Verma, R., Upadhyay, R. G., Pandey, M., & Sharma, S. (2017). Abscisic acid signaling and abiotic stress tolerance in plants: a review on current knowledge and future prospects. *Frontiers in Plant Science, 8*, 161.

Ungureanu, L. B., Ghiciuc, C. M., Amalinei, C., Ungureanu, C., Petrovici, C. G., & Stănescu, R. Ș. (2024). Antioxidants as protection against reactive oxygen stress induced by formaldehyde (FA) exposure: a systematic review. *Biomedicines, 12*(8), 1820.

Wahab, A., Abdi, G., Saleem, M. H., Ali, B., Ullah, S., Shah, W., Mumtaz, S., Yasin, G., Muresan, C. C., & Marc, R. A. (2022). Plants' physio-biochemical and phyto-hormonal responses to alleviate the adverse effects of drought stress: a comprehensive review. *Plants, 11*(13), 1620.

Wang, H. L. V., & Chekanova, J. A. (2016). Small RNAs: essential regulators of gene expression and defenses against environmental stresses in plants. *Wiley Interdisciplinary Reviews: RNA, 7*(3), 356–381.

Wang, D., Lv, S., Jiang, P., & Li, Y. (2017a). Roles, regulation, and agricultural application of plant phosphate transporters. *Frontiers in Plant Science, 8*, 817.

Wang, S., Chen, H., Tang, X., Zhang, H., Chen, W., & Chen, Y. Q. (2017b). Molecular tools for gene manipulation in filamentous fungi. *Applied Microbiology and Biotechnology, 101*, 8063–8075.

Wang, Y. Y., Cheng, Y. H., Chen, K. E., & Tsay, Y. F. (2018). Nitrate transport, signaling, and use efficiency. *Annual Review of Plant Biology, 69*(1), 85–122.

Wang, X., Niu, Y., & Zheng, Y. (2021). Multiple functions of MYB transcription factors in abiotic stress responses. *International Journal of Molecular Sciences, 22*(11), 6125.

Wang, Y., Liao, Y., Quan, C., Li, Y., Yang, S., Ma, C., Mo, Y., Zheng, S., Wang, W., Xu, Z., Li, L., Huang, Z., Zhu, J., Ji, X., Ye, X., Yang, Z., Sun, Y., Liu, H., & Chen, R. (2022). C2H2-type zinc finger OsZFP15 accelerates seed germination and confers salinity and drought tolerance of rice seedling through ABA catabolism. *Environmental and Experimental Botany, 199*, 104873.

Wang, X., Yan, F., Ma, G., Li, A., & Liu, L. (2023a). The diverse functions of *Pseudomonas syringae* syringae van Hall effectors in regulating the plant immune response. *Phytopathology Research*, *5*(1), 63.

Wang, Y., Luo, Z., Zhao, X., Cao, H., Wang, L., Liu, S., Wang, C., Liu, M., Wang, L., & Liu, Z. (2023b). Superstar microRNA, miR156, involved in plant biological processes and stress response: a review. *Scientia Horticulturae*, *316*, 112010.

Wani, S. H., Choudhary, M., Kumar, P., Akram, N. A., Surekha, C., Ahmad, P., & Gosal, S. S. (2018). Marker-assisted breeding for abiotic stress tolerance in crop plants. In: *Biotechnologies of Crop Improvement, Volume 3: Genomic Approaches*, Springer, 1–23.

Wei, L., Zhang, R., Zhang, M., Xia, G., & Liu, S. (2022). Functional analysis of long non-coding RNAs involved in alkaline stress responses in wheat. *Journal of Experimental Botany*, *73*(16), 5698–5714.

Wei, J., Luo, B., Kong, S., Liu, W., Zhang, C., Wei, Z., & Min, X. (2023). Screening and identification of multiple abiotic stress responsive candidate genes based on hybrid-sequencing in Vicia sativa. *Heliyon*, *9*(2), 1–9.

Xia, K., Zeng, X., Jiao, Z., Li, M., Xu, W., Nong, Q., Mo, H., Cheng, T., & Zhang, M. (2018). Formation of protein disulfide bonds catalyzed by OsPDIL1; 1 is mediated by *microRNA5144-3p* in rice. *Plant and Cell Physiology*, *59*(2), 331–342.

Xie, K., Minkenberg, B., & Yang, Y. (2015). Boosting CRISPR/Cas9 multiplex editing capability with the endogenous tRNA-processing system. *Proceedings of the National Academy of Sciences*, *112*(11), 3570–3575.

Xu, H. R., Liu, Y., Yu, T. F., Hou, Z. H., Zheng, J. C., Chen, J., Zhou, Y. B., Chen, M., Fu, J. D., Ma, Y. Z., & Wei, W. L. (2022). Comprehensive profiling of tubby-like proteins in soybean and roles of the *GmTLP8* gene in abiotic stress responses. *Frontiers in Plant Science*, *13*, 844545.

Yadav, R. K., Matsuda, A., Lowe, B. R., Hiraoka, Y., & Partridge, J. F. (2021). Subtelomeric chromatin in the fission yeast S. pombe. *Microorganisms*, *9*(9), 1977.

Yadav, R. K., Tripathi, M. K., Tiwari, S., Tripathi, N., Asati, R., Chauhan, S., Tiwari, P. N., & Payasi, D. K. (2023). Genome editing and improvement of abiotic stress tolerance in crop plants. *Life*, *13*(7), 1456.

Yang, J., Zhang, N., Zhang, J., Jin, X., Zhu, X., Ma, R., Li, S., Lui, S., Yue, Y., & Si, H. (2021). Knockdown of *MicroRNA160a/b* by STTM leads to root architecture changes via auxin signaling in Solanum tuberosum. *Plant Physiology and Biochemistry*, *166*, 939–949.

Yang, J., Mathew, I. E., Rhein, H., Barker, R., Guo, Q., Brunello, L., Loreti, E., Barkla, B. J., Gilroy, S., Perata, P., & Hirschi, K. D. (2022). The vacuolar H+/Ca transporter *CAX1* participates in submergence and anoxia stress responses. *Plant Physiology*, *190*(4), 2617–2636.

Yu, Y., Mo, X., & Mo, B. (2020). Introduction to plant small RNAs. In: *Plant Small RNA*, Academic Press, 3–35.

Yue, E., Cao, H., & Liu, B. (2020). OsmiR535, a potential genetic editing target for drought and salinity stress tolerance in *Oryza sativa*. *Plants*, *9*(10), 1337.

Zavallo, D., Cara, N., Leone, M., Crescente, J. M., Marfil, C., Masuelli, R., & Asurmendi, S. (2024). Assessing small RNA profiles in potato diploid hybrid and its resynthesized allopolyploid reveals conserved abundance with distinct genomic distribution. *Plant Cell Reports*, *43*(4), 85.

Zhang, B. (2015). MicroRNA: a new target for improving plant tolerance to abiotic stress. *Journal of Experimental Botany*, *66*(7), 1749–1761.

Zhang, J., Zhang, H., Srivastava, A. K., Pan, Y., Bai, J., Fang, J., Shi, H., & Zhu, J. K. (2018). Knockdown of rice *microRNA166* confers drought resistance by causing leaf rolling and altering stem xylem development. *Plant Physiology*, *176*(3), 2082–2094.

Zhang, T., Zhao, Y., Ye, J., Cao, X., Xu, C., Chen, B., An, H., Jiao, Y., Zhang, F., Yang, X., & Zhou, G. (2019a). Establishing CRISPR/Cas13a immune system conferring RNA virus resistance in both dicot and monocot plants. *Plant Biotechnology Journal*, *17*(7), 1185.

Zhang, X., Dong, J., Deng, F., Wang, W., Cheng, Y., Song, L., Hu, M., Shen, J., Xu, Q., & Shen, F. (2019b). The long non-coding RNA *lncRNA973* is involved in cotton response to salt stress. *BMC Plant Biology*, *19*, 1–16.

Zhang, D., Zhang, Z., Unver, T., & Zhang, B. (2021). CRISPR/Cas: a powerful tool for gene function study and crop improvement. *Journal of Advanced Research*, *29*, 207–221.

Zhang, F., Yang, J., Zhang, N., Wu, J., & Si, H. (2022). Roles of microRNAs in abiotic stress response and characteristics regulation of plant. *Frontiers in Plant Science*, *13*, 919243.

Zhong, X., Hong, W., Shu, Y., Li, J., Liu, L., Chen, X., Islam, F., Zhou, W., & Tang, G. (2022). CRISPR/Cas9 mediated gene-editing of GmHdz4 transcription factor enhances drought tolerance in soybean (*Glycine max* [L.] Merr.). *Frontiers in Plant Science*, *13*, 988505.

Zhong, V., Archibald, B. N., & Brophy, J. A. (2023). Transcriptional and post-transcriptional controls for tuning gene expression in plants. *Current Opinion in Plant Biology*, *71*, 102315.

Zhu, H., Li, C., & Gao, C. (2020). Applications of CRISPR–Cas in agriculture and plant biotechnology. *Nature Reviews Molecular Cell Biology*, *21*(11), 661–677.

15

Functional Analysis and Modification of Plant MicroRNA by CRISR/Cas System for Stress Tolerance

Rida Zahid, Ifrah Imran, Muhammad Waseem Sajjad, Muhammad Arslan Mahmood[#], Anam Ishtiaq, Saqib Siddique, Rabia Rehman, Imran Amin, and Rubab Zahra Naqvi

Agricultural Biotechnology Division, National Institute for Biotechnology and Genetic Engineering (NIBGE), Constitute College of Pakistan Institute of Engineering and Applied Sciences (PIEAS), Faisalabad, Pakistan

15.1 Introduction

Plants are continuously subjected to many environmental stresses, including abiotic stresses (drought, salt stress, and high temperature [HT]) and biotic stresses (pests and pathogens) (Zhu, 2016; Banerjee et al., 2017). These stressors hinder plant growth, yield, and ultimately food security. 50% of global yield losses are caused alone by abiotic stresses (Xia et al., 2015). Climate change is aggravating these issues because it intensifies the stress conditions. Creating stress-tolerant crops is crucial for global food security. Researchers have turned to genetic engineering and clustered regularly interspaced short palindromic repeats/CRISPR-associated (CRISPR/Cas) genome editing to address these complex stress responses, as conventional breeding approaches did not succeed (Esmaeili et al., 2022a). Plants possess a complex innate immunity system to address the stresses, which includes the detection of external stress signals and the initiation of defensive mechanisms (Macho & Zipfel, 2014). Pattern-triggered immunity (PTI), which is activated when plant receptors identify pathogen-associated molecular patterns (PAMPs), and effector-triggered immunity (ETI), which is elicited in response to certain pathogen effectors. Despite these protective mechanisms, at the molecular level stress mitigation involves complex regulatory networks, especially noncoding RNAs (ncRNAs), that precisely modulate gene expression in reaction to stress (Sun, 2012).

Among ncRNAs, microRNAs (miRNAs) are pivotal in modulating gene expression during stress conditions (Rogers & Chen, 2013). miRNAs are generally short ncRNAs, usually 19–24 nucleotides in length, and they function posttranscriptionally by complementary binding to messenger RNAs (mRNAs) at the target region, which leads to translational repression (Xie et al., 2015). This process is vital for modulating plant responses to biotic and abiotic stimuli, as miRNAs control critical pathways associated with growth, development, and stress adaptation (Sunkar et al., 2012). For instance, miR156 regulates flowering and stress tolerance, whereas miR159 and miR393 are essential for drought response and pathogen definition, respectively

[#] Present Address: Plant Sciences Division, Research School of Biology, The Australian National University, Canberra, ACT 2601, Australia.

Genome and Epigenome Editing for Stress-Tolerant Crops, First Edition. Edited by Jen-Tsung Chen.

(Cui et al., 2017). The biosynthesis of miRNAs comprises many stages; RNA polymerase II translates miRNA genes (MIR genes) into primary miRNAs (pri-miRNAs), which assume stem-loop configurations. Dicer-like (DCL) enzymes process these structures into precursor miRNAs (pre-miRNAs) and mature miRNAs (Voinnet, 2009; Bologna & Voinnet, 2014). The RNA-induced silencing complex (RISC) directs mature miRNAs to complementary target mRNAs and initiates posttranscribing, allowing plants to adapt to fluctuating environmental conditions (Sunkar et al., 2012).

Given the pivotal role of miRNAs in modulating stress responses, they have become appealing targets for genetic modification to improve plant stress tolerance. The emergence of CRISPR/Cas-based genome editing has offered a powerful tool for the precise alteration of miRNAs and their target genes (Jones-Rhoades et al., 2006). CRISPR/Cas technology enables the exact insertion of mutations at miRNA loci, allowing for either the whole elimination of the miRNA or alterations to its function, therefore affecting the expression of stress-related genes (Esmaeili et al., 2022b). The deletion of stress-sensitive miRNAs, such as miR164, has shown enhanced drought tolerance in plants. Moreover, CRISPR/Cas can modify the binding sites of miRNAs on target mRNAs, therefore preventing miRNA-mediated degradation and enabling stress-responsive genes to remain operational under adverse conditions (Biswas et al., 2021).

This chapter examines the functional utilities of miRNAs in plant stress responses and emphasizes the application of CRISPR/Cas technology to edit miRNAs for improved stress tolerance. We begin by studying plant innate immunity and how miRNAs modulate it. A full examination of plant miRNA synthesis and mechanism follows. We end by reviewing CRISPR/Cas-mediated miRNA gene-editing developments and demonstrating crop species' stress-tolerance improvements. CRISPR/Cas precise targeting and miRNA function understanding allow the production of crops that are more resilient to environmental shocks and more adaptable to modern agriculture.

15.2 miRNA Role in Plants

The endogenous, noncoding, and regulatory family of RNA called miRNA has a primary purpose to regulate gene expression after transcription (Guo et al., 2021). miRNAs were first identified as elements of the heterochrony pathway in *Caenorhabditis elegans*, and they have since been discovered in numerous sorts of living organisms, particularly plants (Yang et al., 2007). miRNAs are inherent, ncRNAs that range in size from about 19 to 23 nucleotides and employ matching complementarity to decrease the levels of transcription of their target genes. They are critical for all aspects of the ideal development of plants, improvement, and adaptation to stress, both biotic and abiotic (Zhang et al., 2022).

In general, there are two main approaches by which miRNAs act: via translational repression in the second place and first by degradation of their target mRNA. The extent to which miRNA base pairs complement one other and their binding site inside the target mRNA determines the corresponding mechanism of action. Increased homology between the targeted miRNA and the desired mRNA induces the target to cleave, whereas lower or partial complementarity causes translational repression to be the mechanism by which the miRNAs work (Jg, 2003). These next sections will look at miRNA's many activities in plants, with a focus on their importance as transcript regulators, mediators between growing and stressful situations, and methods for exploiting miRNA targets and variations using various experimental methodologies.

15.2.1 Mechanism

miRNAs bind to complementary areas of the desired mRNAs, regulating gene manifestation after transcription. Typically, this interaction causes translational repression or mRNA cleavage, silencing the target gene. The first stage in miRNA production is RNA polymerase II-mediated transcription of pri-miRNAs. Prior to being processed into mature miRNAs, pri-miRNAs are converted into pre-miRNAs by the Dicer-like 1 (DCL1) enzyme complex. The complex that targets mRNAs is directed to its targets by the mature miRNAs included in the RISC (Figure 15.1; Afonso-Grunz & Müller, 2015).

15.2.2 miRNA as a Modulator of Gene Expression

By influencing the quantity and comprehensive generation of target mRNAs, miRNAs in another way control a range of plant activities. The activity of MIR genes, which have been regulated by other miRNAs, can be induced by various transcription factors (TFs). An efficient biotechnology technique for enhancing biotic or abiotic stress tolerance in commercially significant crops is

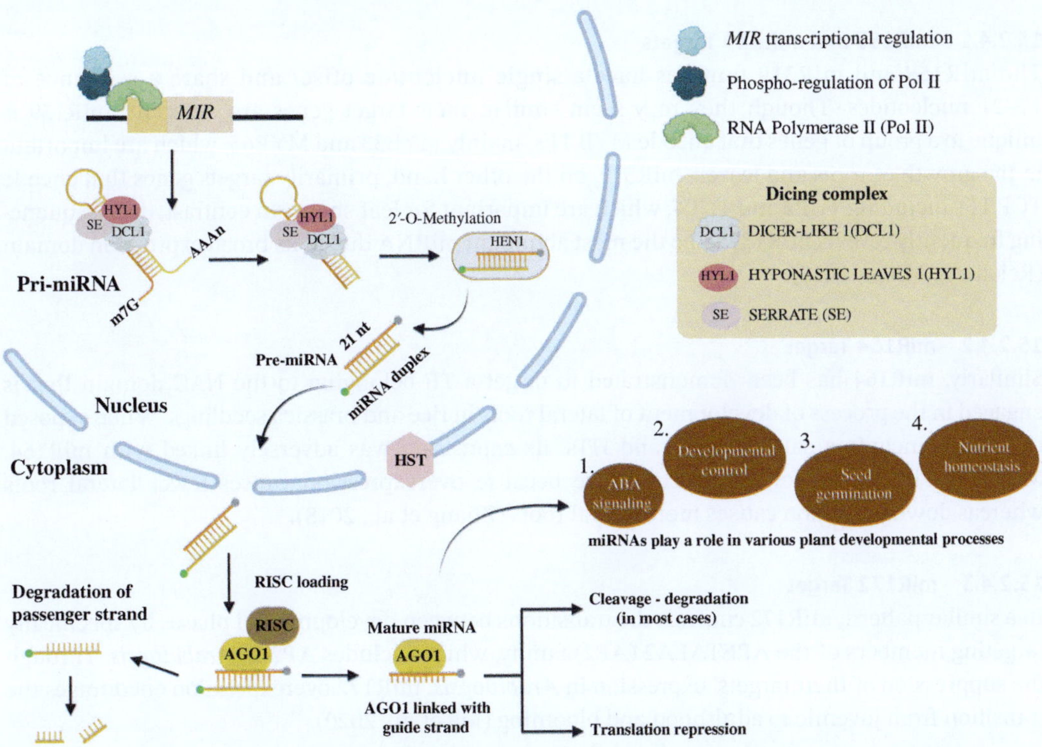

Figure 15.1 miRNA biogenesis and function in plants. The production of plant miRNAs starts in the nucleus, where the miRNA-encoding (MIR) gene is translated to produce a pri-miRNA transcript. The DCL1 complex, together with cofactors HYL1 and SE, processes this precursor by cleaving the pri-miRNA into a pre-miRNA. The pre-miRNA undergoes further processing and methylation by HEN1 to produce an miRNA duplex, which is subsequently transported to the cytoplasm by HASTY (HST). Within the cytoplasm, the miRNA duplex is incorporated into RISC, where AGO interacts with the guide strand. The mature miRNA-RISC modulates gene expression by either facilitating the degradation of the target mRNA or inhibiting its translation. miRNAs in plants are essential for several developmental processes, including abscisic acid signaling, seed germination, nutritional balance, and general developmental regulation.

optimizing MIR genes or miRNAs. To modify the eventual level of these particular miRNAs in the cells, genetic engineering methods are employed in combination with downward or increased expression (Basso et al., 2019).

15.2.3 Regulation of Vital Processes

Numerous genes involved in vital processes like expansion, advancement, and interaction with external stimuli have been demonstrated to be regulated by miRNAs. For instance, protein-like binding to Squamosa Promoter (SPL) are significant modulator of phase transitions of plants, and miR156 targets them. miR156 regulates SPL levels and assists in plants migrating from juvenile to mature stages (Ye et al., 2020).

15.2.4 miRNAs Targeted in Various Crops

As the importance of miRNA is discussed, the following are some examples of miRNA targeted in various crops.

15.2.4.1 miR159 and miR319 Targets

The miR159 and miR319 families have a single nucleotide offset and share a sequence of 17–21 nucleotides. Though they may seem similar, their target genes are different; miR159 is unique to a group of genes that encode MYB TFs, mainly MYB33 and MYB65, which are important in the growth of roots and leaves. miR319, on the other hand, primarily targets genes that encode TCP TFs including TCP2 and TCP4, which are important for leaf shape. In contrast, deep sequencing frequently reveals miR159 to be the most abundant miRNA due to its broad expression domain (Reichel & Millar, 2015).

15.2.4.2 miR164 Target

Similarly, miR164 has been demonstrated to target a TF belonging to the NAC domain that is engaged in the process of development of lateral roots in rice and brassica seedlings. When exposed to stresses including salt, dryness, and HTs, its expression was adversely linked with miR164. MiR164 is important for root architecture because overexpression causes fewer lateral roots whereas downregulation causes more lateral roots (Zhang et al., 2018).

15.2.4.3 miR172 Target

In a similar pattern, miR172 controls the transitions between developmental phases by specifically targeting members of the APETALA2 (AP2) family, which includes AP2 in *Arabidopsis*. Through the suppression of their targets' expression in *Arabidopsis*, miR172 overexpression encourages the transition from juvenile to adulthood and blooming (Ma et al., 2020).

15.2.5 miRNA as a Coordinator Between Development and Stress Response

Numerous abiotic and biotic stresses that plants encounter regularly harm their development and growth. To ensure plant survival and adaptability, miRNAs have been seen playing a very crucial role in managing different equilibriums across growth and stress response. Target genes implicated in stress responses are modulated due to changes in the levels of activity of certain miRNAs under stress (Ding et al., 2013).

15.2.5.1 Coordination by miR398

When there is an abundance of copper (Cu), miR398s (miR398b and miR398c) become abundant and reduce the accumulation of CSD1 and CSD2 transcripts because miR398s mediate their transcript degradation. Conversely, when there is a deficiency of Cu, miR398s experience a decrease in expression, which increases the accumulation of CSD1 and CSD2 transcript and protein (Javed et al., 2019). Similarly, miR393, which targets the auxin receptor TIR1 and adversely inhibits auxin signaling, is increased in response to pathogen infection. This modification aids in strengthening the plant's defenses against infections (Chen et al., 2011a).

15.2.5.2 Regulation by miR160'

The expression of ARF10/16/17, which is reduced by miR160, regulates the levels of heat shock protein (HSP) gene and thermotolerance in *Arabidopsis*. More research is revealing the significance associated with the miR160-ARF pathway in sterility in male plants brought on by prolonged exposure to HTs. More indehiscence was produced by triggering the auxin-mediated signal right after the microsporangium-bearing cell proliferation stage, which increased cotton's susceptibility to heat stress by elevating miR160. When tomatoes are subjected to heat stress, the transition from a cell after meiosis to a mature pollen production is regulated by miR160 and ARF17. This gene pair has been revealed to contribute to maize's response to chilling stress (Hao et al., 2022).

15.2.5.3 Regulation by miR393 Against Bacterial Infection

miR393 protects against bacterial infection in *Arabidopsis*. In its overexpression lines, TIR1, a miR393 target, is substantially less abundant. This decrease in TIR1 levels stabilizes auxin response factors (ARFs) and auxin/indole-3-acetic acid (IAA) proteins, which decrease auxin signaling by heterodimerizing with ARFs and altering the transcription of several essential auxin response genes that ARFs bind to. AXR3/IAA17, BDL/IAA12, and GH3-like are the three primary auxin-responsive genes whose levels were shown to be downregulated in short RNA-mediated defensive and adaptive responses in plants. Furthermore, it was shown by high bacterial titers that the miR393-resistant *Arabidopsis* lines overexpressing AFB1 were susceptible to bacterial infection. The findings demonstrated that ARF inactivation arises from auxin signaling downregulation (Naqvi et al., 2011).

15.2.6 Regulation by miR169 Against Drought Stress

Another situation where miRNAs are essential is during drought, for example, during drought, miR169 is downregulated, which causes its target gene to express more, NF-YA5, a TF that improves drought tolerance. Reduced drought tolerance due to miR169 overexpression emphasizes the protein's function in stress response. In a similar way, miR393 is elevated in response to drought stress, where it modifies auxin signaling to minimize water loss via inhibiting leaf development and encouraging root growth, thus improving water absorption (Zanetti et al., 2017).

15.2.7 Regulation by miR399 Against Phosphorus Deficiency

A different miRNA called miR399 contributes to potassium deficit signaling as a conserved ribonuclease. Three possible target genes of miR399 have been identified in *Arabidopsis*; these genes contain an E2 enzyme that conjugates ubiquitin, a DEAD box helicase, and the Pi transporter (PHT1; 7). E2, a protein having enzymatic activity, is encoded by UBC24 and has been verified by experimentation.

In *Arabidopsis thaliana*, miR399 inhibits enzyme E2, which gives the plant its capacity to adjust to variations in Pi availability in the environment. *A. thaliana* overexpressing miR399 inhibits the build-up of E2 transcripts. Transgenic plants with overexpression of miR399 showed toxic symptoms that resembled the appearance of an E2 enzyme loss-of-function mutant. Upregulated phosphate absorption, translocation of phosphate from roots to shoots, and accumulation of phosphate in shoots all contributed to the development of this toxicity (Sinha et al., n.d.).

15.2.8 Exploitation of Variation in miRNA and Their Target

Utilizing these variations for agricultural enhancement and stress resilience (both biotic and abiotic) is made possible by the inherent variety in miRNA transcripts and their target locations (Hausser & Zavolan, 2014). Several investigations have been carried out to comprehend and utilize this variance, resulting in the creation of plants with desired characteristics (Table 15.1).

Table 15.1 Key experiments exploring the variation in miRNA and their targets.

Plant species	miRNA	Target gene(s)	Phenotype	References
A. thaliana	miR156	SPL	Enriched biomass, delay in flowering	Wang & Wang (2015)
	miR398	CSD1, CSD2	Increased resistance to oxidative stress	Sunkar et al. (2006)
	miR399	PHO2	Increased uptake of phosphate	Bari et al. (2006)
	miR827	NLA, PHO2	Enhanced efficiency of nitrogen utilization	Kant et al. (2011)
	miR172	AP2-like TF	Increased reproductive development, promotion of flowering	Zhu & Helliwell (2011)
	miR164	NAC domain TF	Leaf senescence, regulation of lateral root development	Lee et al. (2017)
	miR159	MYB33, MYB65	Regulation of seed germination and anther development	Millar & Gubler (2005)
	miR162	DCL1	Modulation of biogenesis (miRNA)	Li & Yu (2021)
Brassica napus	miR395	APS, SULTR2, SULTR1	Increased assimilation and sulfate utilization	Liang et al. (2010)
Hordeum vulgare	miR444	MADS-box TF	Increased tolerance against salinity, enhancement of grain yield	Delavari et al. (2021)
Nicotiana tabacum	miR398	CSD1, CSD2	Increased potential for tackling oxidative stress	Noman et al. (2019)
O. sativa	miR396	GRF TF	Reduction in size of life, increase in drought tolerance	Liebsch & Palatnik (2020)
	miR160	ARF10, ARF16, ARF17	High drought tolerance, increased root architecture	Bakhshi et al. (2014)
	miR160	ARF10, ARF16, ARF17	Enhanced drought tolerance, increased root development	Bakhshi et al. (2014)
	miR397	LACCASE genes	Enhancement in straw biomass and yield	Parida et al. (2022)

Table 15.1 (Continued)

Plant species	miRNA	Target gene(s)	Phenotype	References
Solanum lycopersicum	miR319	TCP TF	Increased viral resistance, change in leaf morphology	Fang et al. (2021)
	miR482	NBS	Enhanced resistance against fungal infections	Hong et al. (2019)
	miR2118	LRR	Enhanced resistance against fungal infections	Yang et al. (2021)
Zea mays	miR169	NF-YA TF	Increased tolerance against drought	Ni et al. (2013)
	miR156	SPL TF	Enriched biomass, delay in flowering	Wang & Wang (2015)
	miR396	GRF TF	Reduction in size of leaf, enhancement of drought tolerance	Liebsch & Palatnik (2020)

15.3 In Silico Approaches to Identifying miRNA

Owing to their gene regulatory roles, miRNAs are crucial for the optimum functioning of a plant cell. Identification of miRNA and further analysis would help in understanding and modifying developmental mechanisms, stress responses, and agronomic traits for different crops. The initial processes for miRNA identification were based on direct cloning followed by sequencing. However, these traditional approaches experienced setbacks due to the compact nature of miRNAs, their recurring appearance in the genome, and their methylation profile (Prabu & Mandal, 2010). Advancements in cloning and computational methods led to the development of interdisciplinary pipelines involving both computational and biological methods (Akhtar et al., 2015). Dry-lab pipelines are usually preferred for efficient and cost-effective hypothesis testing prior to wet-lab experiments. Following are some in silico workflows for the identification of miRNAs and their destined targets.

15.3.1 Data Retrieval and Transcriptome Assembly

Targeted sequencing of only small RNA (sRNA) transcriptome has proved more effective because of its peculiarity toward miRNA (Zhou et al., 2020a). The competence levels in miRNA studies are increased because of the integrative execution of revolutionized next-generation sequencing (NGS) (Reddy, 2015). NGS and various bioinformatics can examine strand-specific ncRNA transcripts, even those with low abundance. Large miRNA data from plants has been compiled in multiple databases, owing to the increased sensitivity and cost-effectiveness of NGS (Turner et al., 2012; Li et al., 2013; Wu et al., 2014). The initial step in the identification of miRNA and their targets involves the retrieval of nucleotide sequences followed by assembly of the transcriptome. Sequence retrieval typically makes use of public databases, such as NCBI or Ensembl. Following retrieval, transcriptome assembly is performed using bioinformatics tools that align and merge overlapping transcriptomic sequences, reconstructing complete transcripts. Computational tools have been deployed (Table 15.2) for identifying miRNAs from plants including *Arabidopsis*, *Oryza sativa*, *Zea mays*, *Lycopersicon esculantum*, *Glycine max*, *Setaria italica*, *Brassica napus*, *Vitis vinifera*, *Gossypium hirsutum*, and *V. vinifera* (Banerjee et al., 2016).

Table 15.2 Transcriptome sequence retrieval and assembly tools for miRNA identification.

Tool/database	Link	Description	References
Transcriptome Shotgun Assembly (TSA)	https://www.ncbi.nlm.nih.gov/genbank/tsa/	Compiling transcriptome data and annotations	McIver et al. (2011)
TransAtlasDB	https://modupeore.github.io/TransAtlasDB/	Gene-expression and transcriptomic analysis	Adetunji et al. (2018)
STOmicsDB	https://db.cngb.org/stomics/	To accommodate transcriptomic datasets and multi-omics-related publications	Xu et al. (2023)
Gene Expression Omnibus (GEO)	http://www.ncbi.nlm.nih.gov/geo/	NCBI repository offering users access to transcriptome data from various studies	Edgar et al. (2002)
ArrayExpress	www.ebi.ac.uk/arrayexpress	Transcriptomic data from microarray and RNA-Seq experiments	Athar et al. (2018)
Ensembl	https://www.ensembl.org/	Access to transcriptomic information for various species	Yates et al. (2019)

15.3.2 Preprocessing of Raw Sequence Data

Adapter sequences are added to raw sequencing data during the library preparation process before sequencing. These nongenomic sequences need to be removed to eliminate low-quality reads, effective mapping, and subsequent downstream analysis (Zhou et al., 2020a). Semiglobal or end-space-free alignment utilizing Smith–Waterman or Needleman–Wunsch algorithms is being widely used for this purpose (Jiang et al., 2014). Different tools and toolkits like Trimmomatic, cutadapt, and FASTX-Toolkit are available for cleansing primers, adapter sequences, and polyA tails from raw sequence reads to generate pure transcriptome (Zhou et al., 2020a). Efficient preprocessing guarantees that clean high-quality readings are correctly aligned to facilitate accurate miRNA identification (Table 15.3).

15.3.3 Known Databases for Identification of ncRNAs and Repetitive sRNAs

Noncoding DNA constitutes about 90% of the total plant genome size while the average percentage of various repeats is approximately 45–49% (Mehrotra & Goyal, 2014; Argentin et al., 2023). These repeats and noncoding DNA are the root causes of assembly errors and misassembles in de novo genome assembly, perturbing the accuracy and complicating the interpretation of results (Tørresen et al., 2019). The first step in the removal of such nonfunctional elements of the genome is to annotate the genomic sequences. Different genome databases (Table 15.4) like Ensembl, NCBI, and UCSC provide annotations of coding, noncoding, and repeat regions within the genome using different tools like Glimmer, RepeatMasker, TransDecode, and GeneMark (Chen et al., 2011b; Aken et al., 2016).

15.3.4 Sequence Alignment

Sequence alignment seeks to structure sequences to maximize matching and minimize the occurrence of mismatches and gaps. Alignment helps in the identification of single nucleotide

Table 15.3 Trimming tools for preprocessing of raw sequence data.

Tool/database	Link	Description	References
Trimmomatic (v0.38)	https://www.usadellab.org/cms/?page=trimmomatic	Transcriptome sequences from single/paired ended NGS data (FASTQ)	Bolger et al. (2014)
cutadapt (v1.18)	https://cutadapt.readthedocs.io/en/stable	Locate and eliminate unnecessary adapters, polyA tails and primers from high-throughput sequence reads	Martin (2011)
FASTQ/A Trimmer and FASTQ/A Clipper tool	https://hannonlab.cshl.edu/fastx_toolkit/	Removes barcodes and adapters sequences to get clear and high-quality reads	Gordon & Hannon (2011)
SeqPurge	https://github.com/imgag/ngs-bits	Removes very short adapter sequence(s)	Sturm et al. (2016)
Adapter Removal v2	https://github.com/MikkelSchubert/adapterremoval/	Trims residual adapter sequences and low-quality bases from 3′ end	Schubert et al. (2016)
RepeatMasker	https://www.repeatmasker.org/	Gives small, high-throughput and nonrepetitive RNA reads	Open (2015)
REPET	https://urgi.versailles.inra.fr/REPET	Mask repeats in RNA/DNA-seq data to generate clean datasets	Flutre et al. (2011)
CEN-SOR	https://github.com/andrewsj/CEN-SOR	Identify and mask repeats by comparative analysis of query with reference	Kohany et al. (2006)

Table 15.4 Databases for identifying noncoding and repetitive sRNAs.

Tool/database	Link	Description	References
Repbase	https://www.ncbi.nlm.nih.gov/genbank/tsa/	Represents repetitive DNA elements from different eukaryotes	Bao et al. (2015)
Rfam	https://modupeore.github.io/TransAtlasDB/	Collection of RNA families with multiple sequence alignments, covariance models, and secondary structures	Kalvari et al. (2018)
piRNABank	https://db.cngb.org/stomics/	Access to Piwi-interacting RNAs	Sai Lakshmi &Agrawal (2008)
NONCODE	http://www.ncbi.nlm.nih.gov/geo/	Information system of ncRNAs, especially lncRNAs	Liu et al. (2005)
RNAcentral	www.ebi.ac.uk/arrayexpress	Hub to explore and retrieve ncRNAs of several organisms	Bateman et al. (2011)

polymorphisms (SNPs) and conserved regions, thus providing insights into evolutionary relationships. In miRNA analysis, preprocessed transcriptome sequence alignment against miRNA databases allows the identification of novel miRNAs, regions of similarity, and functional conservation. General purpose aligners such as SOAP2 and Bowtie2 based on Needleman–Wunsch and Smith–Waterman algorithms are used for this purpose (Table 15.5).

Table 15.5 Sequence alignment tools that might help in novel miRNA identification.

Tool/database	Link	Description	References
Bowtie 2	https://bowtie-bio.sourceforge.net/bowtie2/index.shtml	Supports local, gapped, and paired-end mapping of sequences against reference	Langmead & Salzberg (2012)
BWA	https://bio-bwa.sourceforge.net	Aligns low-divergent sequences against reference	Li & Durbin (2009)
SOAP2	https://soap.genomics.org.cn	Offers scaled down and enhanced alignment rates	Li & Durbin (2009)
Spliced Transcripts Alignment to a Reference (STAR)	http://code.google.com/p/rna-star/	Employs seed extension search to locate splice junctions	Dobin et al. (2013)
HISAT2	http://daehwankimlab.github.io/hisat2	Aligns DNA/RNA-seq reads to a reference (SAM format)	Dobin et al. (2013)
TopHat	https://ccb.jhu.edu/software/tophat/index.shtml	Handles complex genomic features and for de novo identification of splice sites	Trapnell et al. (2009)

15.3.5 Established miRNA Databases

Several miRNA databases have been established that provide computationally predicted or experimentally validated miRNA features. Data is usually sourced from the published literature that implemented techniques like microarrays, reporter assays, and high-throughput sequencing. Some databases integrate various data sources, e.g. miRTarBase assimilates data from miRBase and GWAS catalog providing a broad view of miRNA sequences and interactions (Luna Buitrago et al., 2023). While some databases utilize both automated and manual curation methods, the review of the literature by experts and text mining via automated systems provide authentic data while keeping databases updated (Table 15.6).

15.3.6 Plant-Specific miRNA Databases

These databases present a spectrum of data types, constituting sequence information, target predictions, expression profiles, and functional annotations of plant miRNAs. They serve as a key

Table 15.6 Authorized miRNA databases having already authenticated and validated miRNAs.

Tool/database	Link	Description	References
miRBase	https://www.mirbase.org/index.shtml	miRNA annotations, sequences, and predicted target genes	Kozomara et al. (2018)
miRNEST2.0	https://rhesus.amu.edu.pl/mirnest/copy/	miRNAs and their predicted targets	Szcześniak & Makałowska (2013)
miRDeep	https://www.australianprostatecentre.org/researchsoftware/mirdeep-star	Identification of known and novel miRNAs, accepts FASTQ/SAM/BAM	An et al. (2012)
miRDeep-P2	https://sourceforge.net/projects/mirdp2/	Updated miRDeep with annotation and scoring systems	Kuang et al. (2018)

Table 15.7 Plant miRNA databases that help in miRNA prediction.

Tool/database	Link	Description	References
WMP	https://wheat.bioinfo.uqam.ca/index.php	Abiotic stress-responsive miRNAs in wheat and gives access to miRNA prediction software	Remita et al. (2016)
SoyKB	https://soykb.org	OMICS data of soybean	Joshi et al. (2017)
TFGD	https://ted.bti.cornell.edu/cgi-bin/TFGD/sRNAmiRNA.cgi	Large-scale datasets of tomato sRNAs (miRNAs)	Joshi et al. (2017)
SorghumFDB	https://structuralbiology.cau.edu.cn/sorghum/miRNA.php	List miRNAs prepared from miRNA entries of other datasets	Tian et al. (2016)
CSRDB	https://sundarlab.ucdavis.edu/smrnas/	Provides collection of all sRNA in cereals including from mitochondria and chloroplast	Johnson et al. (2006)
SoMART	https://bakerlab.berkeley.edu/somart-webserver-mirna-sirna-analysis	Identification and analysis of sRNAs from *Solanaceae* family	Li et al. (2012)
PMRD	https://bioinformatics.cau.edu.cn/PMRD/	miRNA database compiling evidence of miRNA sequences, targets, secondary structures, and expression profiles	Zhang et al. (2010)
PNRD	https://structuralbiology.cau.edu.cn/PNRD	Revised version of PMRD which features analytical tools along with miRNA prediction toolkit	Yi et al. (2015)

source for miRNA profiles, facilitating research by improving predictive power and designing hypotheses (Table 15.7).

15.3.7 Secondary Structure Prediction for miRNAs

Secondary structures adopted by miRNAs modulate their biological functions and precision for target miRNA (Gangemi et al., 2020). Solid secondary conformations of miRNA, i.e. homoduplexes and hairpins, provide protection against degradation by endogenous nucleases. Secondary structures are also crucial for miRNA biogenesis as only unique structural configurations are selected by Dicer and Drosha enzymes (Morrison et al., 2019). The following are some tools that help in the prediction of secondary structures for miRNAs (Table 15.8).

15.3.8 Analysis of miRNA and mRNA Interaction

As miRNAs modulate gene expression by altering translation efficiency and miRNA target's stability in plants, various computational tools, such as miRDB, TargetScan, and DIANA-TarBase, are harnessed to forecast potential miRNA targets determined by binding site accessibility and sequence complementarity. These tools incorporate extensive databases providing experimentally validated interactions to maximize prediction accuracy (Table 15.9; Agarwal et al., 2015).

Table 15.8 Databases/tools that can provide secondary structure of miRNAs for better evaluation.

Tool/database	Link	Description	References
miRFinder	https://www.bioinformatics.org/mirfnder	Prediction and correlation between mutation and secondary structure	Huang et al. (2007)
miRNAFold	https://evryrna.ibisc.univ-evry.fr/miRNAFold	Identification of miRNA precursors by hairpin structure-based analysis	Tav et al. (2016)
mfold	https://unafold.rna.albany.edu/?q=mfold/RNA-Folding-Form	Predict the most stable secondary structure for nucleic acids	Zuker (2003)
UNAFold	https://unafold.rna.albany.edu	Combination of several web servers for better analysis	Markham & Zuker (2008)
ViennaRNA Package	https://www.tbi.univie.ac.at/RNA	Various tools providing secondary structure, interaction analysis	Lorenz et al. (2011)
RNAstructure (v6.1)	https://rna.urmc.rochester.edu/RNAstructure.html	Secondary structures with probabilities of base pairing and thermodynamic models	Mathews (2014)

Table 15.9 miRNA and mRNA interaction analysis tools.

Tool/database	Link	Description	References
ENCORI/starBase	https://starbase.sysu.edu.cn	Provides extensive annotation and discovery of RNA–RNA and protein–RNA interactions	Yang et al. (2010)
DIANA-TarBase (v7.0)	https://diana.imis.athena-innovation.gr/DianaToolsindex.php?r=tarbase/index	Authenticated miRNA-gene interactions with metadata	Vlachos et al. (2014)
DIANA-microT-CDS	http://diana.imis.athena-innovation.gr/DianaTools/index.php?r=microT_CDS	miRNA-target prediction in 3′ UTR and coding regions	Nazarov & Kreis (2021)
ARMOUR	https://armour.icgeb.trieste.it/login	Validated differential expression profiles of miRNAs under different abiotic stress	Sanan-Mishra et al. (2018)
miRmapper	https://mirmapper.ezlab.org	Determines dominant miRNAs in an miRNA and mRNA network	da Silveira et al. (2018)
MMIA	http://bioinformatics.ustc.edu.cn/MMIA/	Accesses miRNA and mRNA expression data and their biological effects	Nam et al. (2009)

15.3.9 Target Prediction for miRNA

Experiments say that many development-related genes and TFs are the targets for miRNAs (Zhou et al., 2020a). Orthologous searchers for conserved regions in target sequence and multiple instances of target sequence occurrence help in filtering miRNA targets (Dai et al., 2010). Numerous in silico tools are designed for target predictions considering various parameters including seed region conservation and complementarity, site accessibility, free energy, physicochemical properties of miRNA and its candidate targets, and the availability of miRNA-target sites (Table 15.10; Zhou et al., 2020a).

Table 15.10 Prediction tools for miRNA targets.

Tool/database	Link	Description	References
psRNATarget	https://plantgrn.noble.org/psRNATarget/	Matches reverse complementarity between miRNAs and their targets	Dai & Zhao (2011)
Target-align	https://www.leonxie.com/targetAlign.php	Target detection on complementarity-based scoring matrix system to mark optimal local alignment	Xie & Zhang (2010)
miRWalk	http://mirwalk.umm.uni-heidelberg.de/	miRNA-target prediction and analysis in whole transcript	Sticht et al. (2018)
miRU	https://bioinfo3.noble.org/miRU.htm	miRNA-target prediction platform capable of conducting simultaneous numerous analysis	Zhang (2005)
TAPIR	https://bioinformatics.psb.ugent.be/webtools/tapir	Enables two different search engines, FASTA and RNA hybrid	Bonnet et al. (2010)
miRGator	http://mirgator.kobic.re.kr	Brings together predictions from different algorithms, also carries out functional enrichment analysis	Cho et al. (2012)
TarHunter	https://github.com/XMaBio	Prediction with additional focus on consideration for miRNA-target mimics and cross speciation	Ma et al. (2017)
C-mii	https://www.biotec.or.th/isl/c-mii	Provides a holistic view of identified targets with info-graphics and grouping information	Numnark et al. (2012)
DeepMirTar tool	https://github.com/Bjoux2/DeepMirTar_SdA	Deep learning approaches for accurate prediction and analysis of miRNA targets	Wen et al. (2018)
sPARTA	https://github.com/atulkakrana/sPARTA/releases	Utilizes high-throughput sequence data to give genome-wide cleavage products	Kakrana et al. (2014)
miRTour	https://bio2server.bioinfo.uni-plovdiv.bg/miRTour	Provides miRNA similarity search, pre-miRNA selection, target prediction, and annotation	Milev et al. (2011)
miRTarBase	https://mirtarbase.mbc.nctu.edu.tw/php/index.php	Validated miRNA-target interactions with downloadable information	Chou et al. (2017)
miRGate	http://mirgate.bioinfo.cnio.es	List of human, mouse, and rat miRNAs: mRNA targets	Andres-Leon et al. (2015)
Cytoscape	https://cytoscape.org/cy3.html	Visual investigation of biological networks	Su et al. (2014)

15.3.10 Gene Ontology (GO) and Pathway Analysis of Identified miRNA Targets

GO analysis involves the annotation of identified target genes based on their biological processes, molecular functions, and cellular components. GO is followed by signaling or metabolic pathways analysis modulated by identified miRNA. Different computational tools incorporating miRNA-target interaction data from various repositories have been developed (Table 15.11). These tools figure out the statistical significance of GO terms and pathways by hypergeometric tests or some similar methods (Roubelakis et al., 2009; Garcia-Moreno & Carmona-Saez, 2020).

Table 15.11 Functional annotation and pathway analysis of identified miRNA targets.

Tool/ database	Link	Description	References
The GO resource	https://geneontology.org	Access point for gene function with structured gene functions in relation to each other	The Gene Ontology Consortium (2014)
Blast2GO	https://www.blast2go.com	Provides ontology-based data extracted from crude	Conesa et al. (2016)
BiNGO	https://www.psb.ugent.be/cbd/papers/BiNGO/Home.html	Java-based ontology tool embodying significant categories in biological yielding results in map form	Maere et al. (2005)
WEGO	https://wego.genomics.org.cn	Data illustration, comparison, and graphical depiction of ontology results	Ye et al. (2018)
agriGO	https://systemsbiology.cau.edu.cn/agriGOv2	Ontology enrichment analysis of crop varieties and cultivars	Tian et al. (2017)
ClueGO	https://www.ici.upmc.fr/cluego/cluegoDownload.shtml	A unified tool for elucidating functionally grouped gene ontology along with pathway annotation networks	Bindea et al. (2009)
PlantGSEA	https://structuralbiology.cau.edu.cn/PlantGSEA	Tool for interpreting the list of genes by computing overlaps with designated genes	Yi et al. (2013)

15.4 CRISPR-Based Editing of miRNA in Plants for Stress Control

The unpredictable nature of climate change and the concerning rise in global population necessitate scientific advancements in crop breeding to maintain agricultural productivity. Numerous projects have been undertaken in recent decades to enhance crops by traditional and molecular breeding methods. The evolving environment, characterized by biotic and abiotic stressors, demands an enhanced understanding of molecular biology by focusing on specific genes that produce desirable plants. Creating high-yield and stress-resistant crops is crucial for ensuring food security for the increasing global population, particularly under harsh conditions. The readily accessible genome sequence data for significant crops enables the application of genome-editing technology to construct trait-specific genes through innovative mating strategies (Premnath et al., 2022).

Studies have revealed that altering miRNA expression or targets can impact plant productivity and responsiveness to several biotic and abiotic stresses (Djami-Tchatchou et al., 2017). miRNAs target TFs in several plant species, allowing for fine-tuning of TF expression through CRISPR/Cas targeting (Tang & Chu, 2017). To alter the abundance of certain miRNAs, techniques such as short tandem target mimics (STTM), antisense inhibitors, sponges, and CRISPR have been implemented (Reichel et al., 2015). CRISPR-based editing has proved to be effective in modifying the expression and function of certain genes (Chang et al., 2016).

15.4.1 Genome-Editing Techniques

The dawn of technologies such as genome editing has revolutionized the field of agriculture by employing targeted mutagenesis in crop plants. These techniques utilize the different types of nucleases including the zinc finger nucleases (ZFNs) and transcription activator-like effector nucleases (TALENs) for the addition and deletion of the sequences in the genomes of the organisms, especially the plants (Qi et al., 2023; Saraswat et al., 2023; Susmitha & Farooqkhan, 2023).

Bacteria have developed genomic modifications to resist viruses. Major defensive systems include CRISPR/Cas. Bacteria insert viral genetic fragments into their genomes upon infection, termed spacers, to help identify viruses in the future (Horvath & Barrangou, 2010). Multiple repeats of 24–47 bp spacer sequences are inserted into the CRISPR locus. The first CRISPR locus was identified in 1987 (Ishino et al., 1987). Then later in 2005, it was found out that CRISPR is engaged in protecting bacteria from viral invaders, which we can control and use for crop improvement (Bolotin et al., 2005).

15.4.2 CRISPR/Cas9 System in Plants

In plants, the CRISPR/Cas system has been employed by the manipulation of gRNAs designed for the targeting of particular genes. The Cas endonuclease then generates a double-stranded break which in turn is repaired by cellular machinery either by homology-directed repair or nonhomologous end joining leading to the silencing or knockout of the targeted gene (Asmamaw & Zawdie, 2021). Different mechanisms can be utilized for the delivery of CRISPR/Cas constructs, which include the biolistic method, PEG-mediated delivery, and *Agrobacterium tumefaciens*-mediated delivery (Veillet et al., 2019). Genome editing of *SlLBD40* (Liu et al., 2020), *SlMAPK3* (Tang et al., 2017a), *SlARF4* (Barrangou, 2015), and *SlNPR1* (Konermann et al., 2015) genes in tomatoes and *OsDIS1* (Fang & Xiong, 2015), *OsSRFP1* (Santosh Kumar et al., 2020), *OsERA1* (Ogata et al., 2020), and *OsSAPK2* (Lou et al., 2017) genes in rice, by the utilization of the CRISPR/Cas system, resulted in enhanced tolerance against drought stress. Compared to knockout, knock-in using CRISPR/Cas system is more complicated. In a recent study, scientists achieved around 19% knock-in efficacy by employing geminiviral vectors to deliver donor DNA along with CRISPR/Cas system in rice (Wang et al., 2017). This efficiency improved to 37% when the ribonucleoprotein (RNP) complex was used in conjugation with the donor DNA (Kim et al., 2020).

15.4.3 Modulation of miRNA via CRISPR/Cas Systems

The understanding of the application of CRISPR/Cas systems in the modification of noncoding miRNAs has been employed to plant species, such as *Arabidopsis*, wheat, maize, soybean, tomato, and rice (Table 15.12). The level of miRNA increased/upregulated under stress can fine-tune their target mRNA concentration, which may negatively regulate the adaptation during stress. On the contrary, some miRNAs repressed/downregulated during stress can increase the level of target transcript(s), contributing positively to stress tolerance. In addition, one miRNA may contribute to one or several genes and some genes may be regulated by numerous miRNAs (Singroha et al., 2021). In this way, miRNA regulates several plant processes, such as growth and development, stress regulation, DNA methylation, various metabolic processes, and even their biogenesis (Song et al., 2019).

There are several sites in MIR genes that can serve as potential targets for miRNA modification via CRISPR/Cas systems. However, the homology among most of the members of MIR gene families makes their fine-tuning a bit redundant (Deng et al., 2022). As miR319 in *A. thaliana* regulates leaf development, knocking out of either miR319a or miR319b gene moderately declines the serrated leaf formation. While multiplex mutant plant inhibits leaf serrations resulting in smooth leaf formation (Gramzow & Theißen, 2019; Guo et al., 2020). Because of high similarity and common target genes, miRNAs can be assembled into superfamilies such as miR156 in *A. thaliana* and *O. sativa*. There are 10 and 13 members of miR156 in both crops, respectively (Guo et al., 2020).

Table 15.12 miRNAs targeted using CRISPR/Cas systems.

Plant species	miRNAs	Function	Mutant phenotype	References
Triticum aestivum	miR156	Regulates developmental timing and stress responses	Enhanced drought and salt tolerance	Gupta et al. (2023)
Solanum lycopersicum	miR482b, miR482c	Involved in defense responses against pathogens	Increased resistance to late blight	Hong et al. (2021)
	miR164a, miR164b	Regulate growth and development	Stunted growth	Gupta et al. (2021)
	miR160	Auxin-driven leaf blade outgrowth	Reduced floral organ development	Hendelman et al. (2012)
Glycine max	miR1514, miR1509	Involved in stress responses and regulation of developmental processes	Improved resistance to biotic stresses and enhanced yield traits	Jacobs et al. (2015)
	miR398c	Negatively regulates drought tolerance	Helpful to scavenge O_2 and contributed to increased drought resistance	Zhou et al. (2020b)
	miR169c	Regulatory role in nodulation	Nodulation is promoted	Xu et al. (2021)
Arabidopsis thaliana	miR169a, miR827a	Regulates flowering time and developmental processes	Increased drought resistance	Zhao et al. (2016)
	miR159b	Role in plant growth and development and environmental stresses	Lower expression of miR159b in transgenic lines	Tang et al. (2017b)
Malus domestica	miR171i	Targeted and degraded SCARECROW-LIKE PROTEINS26.1 (MsSCL26.1) transcripts	Enhances drought stress tolerance	Wang et al. (2020)
Oryza sativa	miR535	Induced under stressed condition	Improved drought and salinity tolerance	Yue et al. (2020)
	miR528, miR408	Positive regulator of salt stress	Susceptibility to salt stress increased	Zhou et al. (2017)
	miR168a	Regulates AGO1 family	Stunted growth and enhanced stress susceptibility	Zhou et al. (2022)
	miR396	Regulate GRF4 and GRF8	Increased grain size and improved brown plant hopper (BPH) resistance	Lin et al. (2021)
	miR171b	Enhanced defense response	Increased blast disease susceptibility	Li et al. (2022)
	miR159a	Positive regulator of immunity against rice blast	Susceptibility to rice blast increased	Chen et al. (2021)
	miR390	Lateral shoot development	Defect in shoot apical meristem (SAM), shootless phenotype	Bi et al. (2019)
Zea mays	miR159	Role in endosperm development	Increased grain size	Wang et al. (2023)
Solanum tuberosum	miR160a/b, miR390a	Involved in potato immune response	Abnormal floral organ development	Lukan et al. (2022)

To infer the variation in expression, the superfamily can be further parted into subgroups, depending on matured miRNAs and their flanking sequences (Miao et al., 2019; Deng et al., 2022).

The single-nucleotide indels, several kinds of mutations like segment deletion, and replacement in MIR genes have been observed in these species that are employed by CRISPR/Cas9 (Chung et al., 2020). The feasibility and effectiveness of using the marvelous strategy of CRISPR/Cas-based editing for enhancing traits in plant species pave the way to a sustainable future for agriculture and a healthy environment. 33 or 71 bp large segment deletion for knocking out of miR168a gene in *O. sativa* creates pleiotropic effects on mutant lines using two single-guide RNAs (sgRNAs) (Zhou et al., 2022). Targeting miR169c enhanced nodulation in soybean by the upregulation of its target gene (Xu et al., 2021). In addition, CRISPR-based edited tomato (*Solanum lycopersicum*) showed improved resistance against late blight disease (*Phytophthora infestans*) via targeting miR482b and miR482c genes (Hong et al., 2021).

The CRISPR/Cas system facilitates the accurate targeting of particular miRNAs in plants, allowing for the suppression of either one or several miRNAs, hence influencing the control of essential genes involved in protein synthesis. This approach enables partial or total loss of function and a decrease in gene expression, hence enabling targeted genetic changes for crop enhancement (Figure 15.2). We can predict potential target sites for miRNA modulation experimentally using RNA immunoprecipitation (RIP), crosslinking immunoprecipitation (CLIP), and crosslinking and

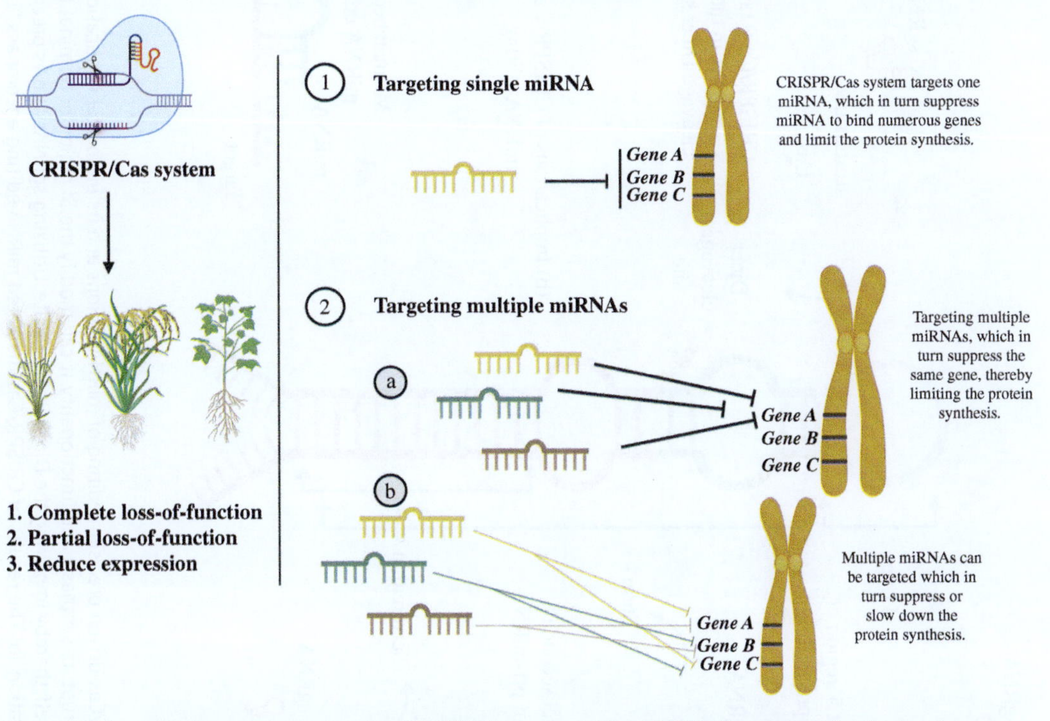

Figure 15.2 miRNA targeting using the CRISPR/Cas system. The CRISPR/Cas system facilitates the targeting and modification of specific miRNAs in plants, thereby influencing gene expression. (1) Targeting an individual miRNA inhibits its capacity to bind to various genes (e.g. Gene A, Gene B, and Gene C), consequently diminishing or abolishing the associated protein synthesis. (2) Targeting multiple miRNAs can be achieved in two manners: (a) by targeting distinct miRNAs that regulate the same genes, further constraining protein synthesis or (b) by targeting miRNAs that govern multiple genes, thereby decelerating or modifying their collective protein synthesis. This system allows complete or partial loss-of-function or decreased gene expression in plants, thereby enabling precise genetic modifications for crop enhancement.

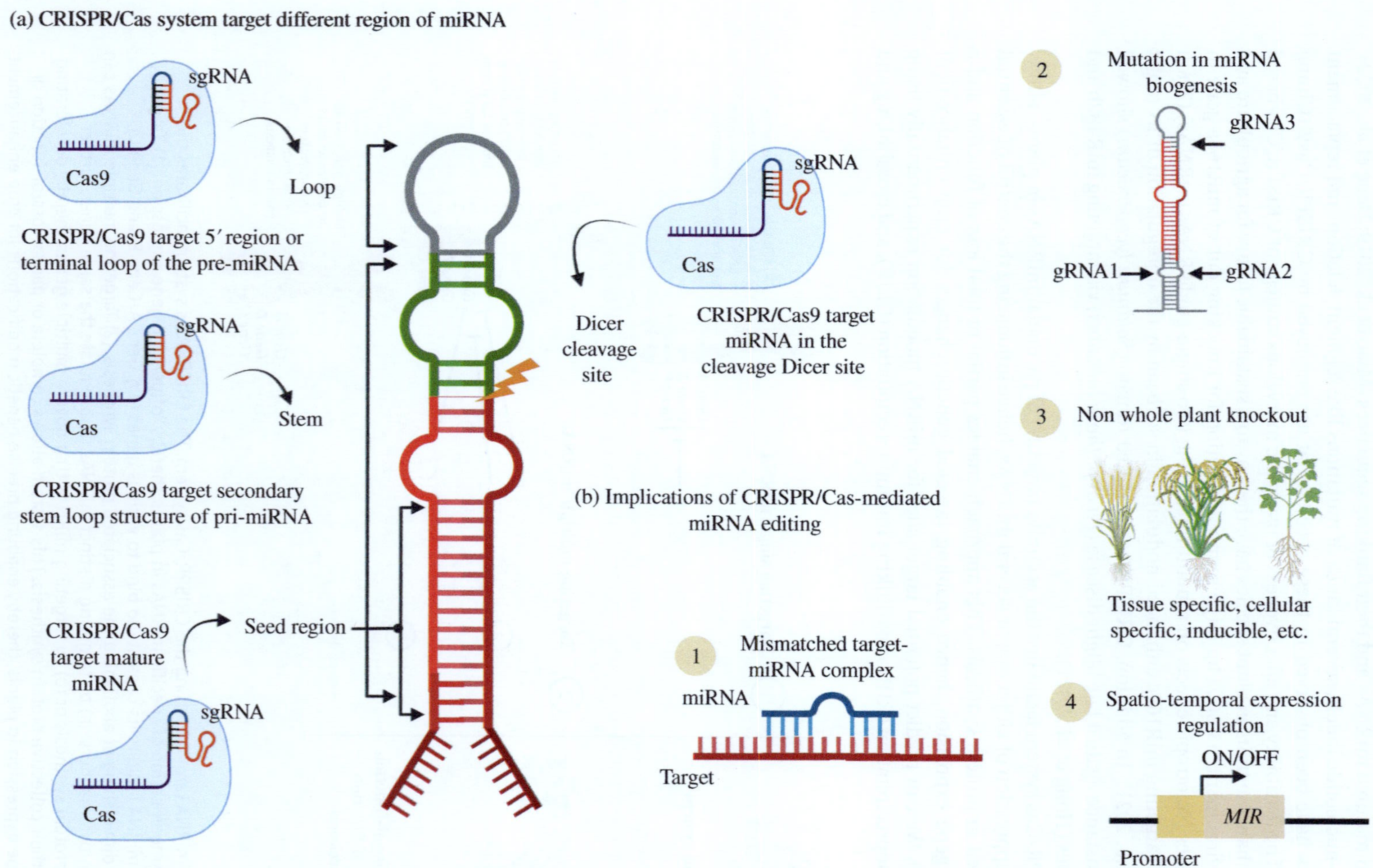

Figure 15.3 CRISPR/Cas-driven precision editing of miRNA regions and its biological implications. (a) The CRISPR/Cas system may specifically target certain areas of miRNA to disrupt its synthesis and functionality. It can modify the 5′ region or terminal loop of the pre-miRNA, the secondary stem loop of pri-miRNA, or the mature miRNA itself, thereby impairing the Dicer cleavage site. Utilizing sgRNAs allows precise targeting of critical regions of the miRNA, influencing its processing and maturation. (b) The results of CRISPR/Cas-mediated miRNA editing encompass (1) mismatched target-miRNA interactions that modify gene silencing, (2) mutations in miRNA biogenesis through alterations to the stem-loop structure, (3) facilitation of specific tissue- or cell-level knockouts rather than whole-plant knockouts, and (4) regulation of spatiotemporal miRNA expression, permitting dynamic on/off expression based on plant requirements.

sequencing of hybrids (CLASHs), which can be used to examine the AGO protein interaction with mRNA (Helwak & Tollervey, 2014; Lu & Leslie, 2016; Lee & Ule, 2018). Currently, computational tools are also available that use several algorithms to identify miRNAs (Zheng et al., 2020). For example, MicroMUMMIE, MIRZA, MirCure, and CRUIS, in combination with CRISPR, are used as a tracker for targeted mRNA and identified by mass spectrometry (Majoros et al., 2013; Pla et al., 2018). Although these tools are limited to a small number of plant species, further progress is being carried out by adopting in-depth learning-based approaches and developing software for accurate prediction of miRNA-target sites interaction and resulting edition in plant species (Pla et al., 2018; Deng et al., 2022).

15.5 Conclusion and Future Prospects

Traditional plant breeding methods, including chemical mutagenesis and intergenic crosses, exhibit errors and frequently lead to the unwanted inclusion of undesired traits due to linkage drag. Conversely, the CRISPR/Cas system facilitates precise, dependable, and rapid genome editing, rendering it a flexible and potent instrument in crop resistance development. This approach provides substantial benefits compared to conventional methods by facilitating accurate targeting of miRNAs, which are essential in gene control. CRISPR/Cas may target several areas of miRNA, including the 5′ region, terminal loops of pre-miRNA, secondary stem loops of pri-miRNA, or mature miRNAs, thereby affecting their processing and maturation. Utilizing sgRNAs enables the introduction of particular mutations in essential miRNA biogenesis areas, hence affecting gene silencing and miRNA production. Moreover, tissue-specific, non-whole-plant knockouts and spatiotemporal modulation of miRNA expression provide enhanced control over gene activity in crops (Figure 15.3). The future potential of CRISPR/Cas technology in plant breeding encompasses the creation of multiplexed systems that can concurrently target numerous miRNA families for genetic research and agricultural enhancement. Advanced techniques, such as prime editing, can be utilized to create specific changes in miRNAs, facilitating tolerance to biotic and abiotic stressors. The application of RNPs will enable the formation of transgene-free mutants, hence broadening the potential for agricultural improvement and sustainability.

References

Adetunji, M. O., Lamont, S. J., & Schmidt, C. J. (2018). TransAtlasDB: an integrated database connecting expression data, metadata and variants. *Database, 2018*, 1–15.

Afonso-Grunz, F., & Müller, S. (2015). Principles of miRNA–mRNA interactions: beyond sequence complementarity. *Cellular and Molecular Life Sciences, 72*, 3127–3141.

Agarwal, V., et al. (2015). Predicting effective microRNA target sites in mammalian mRNAs. *eLife, 4*, 1–38.

Aken, B. L., et al. (2016). The Ensembl gene annotation system. *Database: The Journal of Biological Databases and Curation, 2016*, 1–19.

Akhtar, M. M., et al. (2015). Bioinformatic tools for microRNA dissection. *Nucleic Acids Research, 44*(1), 24–44.

An, J., et al. (2012). miRDeep*: an integrated application tool for miRNA identification from RNA sequencing data. *Nucleic Acids Research, 41*(2), 727–737.

Andres-Leon, E., et al. (2015). miRGate: a curated database of human, mouse and rat miRNA–mRNA targets. *Database, 2015*, bav035.

Argentin, J., et al. (2023). Comparative analysis of repeat content in plant genomes, large and small. *Frontiers in Plant Science*, *14*, 1103035.

Asmamaw, M., & Zawdie, B. (2021). Mechanism and applications of CRISPR/Cas-9-mediated genome editing. *Biologics*, *15*, 353–361.

Athar, A., et al. (2018). ArrayExpress update – from bulk to single-cell expression data. *Nucleic Acids Research*, *47*(D1), D711–D715.

Bakhshi, B., et al. (2014). Characterization of three key microRNAs in rice root architecture under drought stress using in silico analysis and quantitative real-time PCR. *Biosciences, Biotechnology Research Asia*, *11*(2), 555–565.

Banerjee, A., Roychoudhury, A., & Krishnamoorthi, S. (2016). Emerging techniques to decipher microRNAs (miRNAs) and their regulatory role in conferring abiotic stress tolerance of plants. *Plant Biotechnology Reports*, *10*(4), 185–205.

Banerjee, S., et al. (2017). Role of small RNAs in abiotic stress responses in plants. *Plant Gene*, *11*, 1–40.

Bao, W., Kojima, K. K., & Kohany, O. (2015). Repbase update, a database of repetitive elements in eukaryotic genomes. *Mobile DNA*, *6*(1), 11.

Bari, R., et al. (2006). PHO2, microRNA399, and PHR1 define a phosphate-signaling pathway in plants. *Plant Physiology*, *141*(3), 988–999.

Barrangou, R. (2015). Diversity of CRISPR-Cas immune systems and molecular machines. *Genome Biology*, *16*, 1–11.

Basso, M. F., et al. (2019). MicroRNA s and new biotechnological tools for its modulation and improving stress tolerance in plants. *Plant Biotechnology Journal*, *17*(8), 1482–1500.

Bateman, A., et al. (2011). RNAcentral: a vision for an international database of RNA sequences. *RNA*, *17*(11), 1941–1946.

Bi, H., et al. (2019). Disruption of miRNA sequences by TALENs and CRISPR/Cas9 induces varied lengths of miRNA production. *Plant Biotechnology Journal*, *18*, 1–11.

Bindea, G., et al. (2009). ClueGO: a Cytoscape plug-in to decipher functionally grouped gene ontology and pathway annotation networks. *Bioinformatics*, *25*(8), 1091–1093.

Biswas, D., Saha, S. C., & Dey, A. (2021). CRISPR-Cas genome-editing tool in plant abiotic stress-tolerance. *Plant Gene*, *26*, 100286.

Bolger, A. M., Lohse, M., & Usadel, B. (2014). Trimmomatic: a flexible trimmer for Illumina sequence data. *Bioinformatics*, *30*(15), 2114–2120.

Bologna, N. G., & Voinnet, O. (2014). The diversity, biogenesis, and activities of endogenous silencing small RNAs in Arabidopsis. *Annual Review of Plant Biology*, *65*, 473–503.

Bolotin, A., et al. (2005). Clustered regularly interspaced short palindrome repeats (CRISPRs) have spacers of extrachromosomal origin. *Microbiology*, *151*(8), 2551–2561.

Bonnet, E., et al. (2010). TAPIR, a web server for the prediction of plant microRNA targets, including target mimics. *Bioinformatics*, *26*(12), 1566–1568.

Chang, H., et al. (2016). CRISPR/cas9, a novel genomic tool to knock down microRNA in vitro and in vivo. *Scientific Reports*, *6*(1), 22312.

Chen, Z.-H., et al. (2011a). Regulation of auxin response by miR393-targeted transport inhibitor response protein 1 is involved in normal development in Arabidopsis. *Plant Molecular Biology*, *77*, 619–629.

Chen, C. J., et al. (2011b). Plant noncoding RNA gene discovery by "single-genome comparative genomics". *RNA*, *17*(3), 390–400.

Chen, J. F., et al. (2021). Fine-tuning roles of Osa-miR159a in rice immunity against *Magnaporthe oryzae* and development. *Rice (N Y)*, *14*(1), 26.

Cho, S., et al. (2012). MiRGator v3. 0: a microRNA portal for deep sequencing, expression profiling and mRNA targeting. *Nucleic Acids Research, 41*(D1), D252–D257.

Chou, C.-H., et al. (2017). miRTarBase update 2018: a resource for experimentally validated microRNA-target interactions. *Nucleic Acids Research, 46*(D1), D296–D302.

Chung, P. J., et al. (2020). Efficiency of recombinant CRISPR/rCas9-mediated miRNA gene editing in rice. *International Journal of Molecular Sciences, 21*(24), 9606.

Conesa, A., et al. (2016). A survey of best practices for RNA-seq data analysis. *Genome Biology, 17*(1), 13.

Cui, J., You, C., & Chen, X. (2017). The evolution of microRNAs in plants. *Current Opinion in Plant Biology, 35*, 61–67.

da Silveira, W. A., et al. (2018). miRmapper: a tool for interpretation of miRNA-mRNA interaction networks. *Genes (Basel), 9*(9), 1–19.

Dai, X., & Zhao, P. X. (2011). psRNATarget: a plant small RNA target analysis server. *Nucleic Acids Research, 39*(suppl_2), W155–W159.

Dai, X., Zhuang, Z., & Zhao, P. X. (2010). Computational analysis of miRNA targets in plants: current status and challenges. *Briefings in Bioinformatics, 12*(2), 115–121.

Delavari, A., et al. (2021). Integrating expression data and genomic sequences to investigate transcriptional regulation in barley in response to abiotic stress. *Biotechnologia, 102*(1), 21.

Deng, F., et al. (2022). Molecular evolution and functional modification of plant miRNAs with CRISPR. *Trends in Plant Science, 27*(9), 890–907.

Ding, Y., Tao, Y., & Zhu, C. (2013). Emerging roles of microRNAs in the mediation of drought stress response in plants. *Journal of Experimental Botany, 64*(11), 3077–3086.

Djami-Tchatchou, A. T., et al. (2017). Functional roles of microRNAs in agronomically important plants—potential as targets for crop improvement and protection. *Frontiers in Plant Science, 8*, 1–24.

Dobin, A., et al. (2013). STAR: ultrafast universal RNA-seq aligner. *Bioinformatics, 29*(1), 15–21.

Edgar, R., Domrachev, M., & Lash, A. E. (2002). Gene expression omnibus: NCBI gene expression and hybridization array data repository. *Nucleic Acids Research, 30*(1), 207–210.

Esmaeili, N., Shen, G., & Zhang, H. (2022). Genetic manipulation for abiotic stress resistance traits in crops. *Frontiers in Plant Science, 13*, 1–17.

Fang, Y., & Xiong, L. (2015). General mechanisms of drought response and their application in drought resistance improvement in plants. *Cellular and Molecular Life Sciences, 72*, 673–689.

Fang, Y., et al. (2021). Roles of miR319-regulated TCPs in plant development and response to abiotic stress. *The Crop Journal, 9*(1), 17–28.

Flutre, T., et al. (2011). Considering transposable element diversification in de novo annotation approaches. *PLoS One, 6*(1), e16526.

Gangemi, C. M. A., et al. (2020). Endogenous and artificial miRNAs explore a rich variety of conformations: a potential relationship between secondary structure and biological functionality. *Scientific Reports, 10*(1), 453.

Garcia-Moreno, A., & Carmona-Saez, P. (2020). Computational methods and software tools for functional analysis of miRNA data. *Biomolecules, 10*(9), 1–17.

Gordon, A. and G. Hannon, FASTX-Toolkit Computer program distributed by the author. 2011.

Gramzow, L., & Theißen, G. (2019). Plant miRNA conservation and evolution. In: *Plant MicroRNAs: Methods and Protocols*, Springer Nature Link, 41–50.

Guo, Z., et al. (2020). PmiREN: a comprehensive encyclopedia of plant miRNAs. *Nucleic Acids Research, 48*(D1), D1114–D1121.

Guo, Z., et al. (2021). PmiREN2.0: from data annotation to functional exploration of plant microRNAs. *Nucleic Acids Research, 50*(D1), D1475–D1482.

Gupta, S. K., et al. (2021). CRISPR/Cas9 mutants of tomato MICRORNA164 genes uncover their functional specialization in development. *Plant Physiology, 187*(3), 1636–1652.

Gupta, A., et al. (2023). CRISPR-induced miRNA156-recognition element mutations in TaSPL13 improve multiple agronomic traits in wheat. *Plant Biotechnology Journal, 21*(3), 536–548.

Hao, K., et al. (2022). miR160: an indispensable regulator in plant. *Frontiers in Plant Science, 13,* 833322.

Hausser, J., & Zavolan, M. (2014). Identification and consequences of miRNA–target interactions—beyond repression of gene expression. *Nature Reviews Genetics, 15*(9), 599–612.

Helwak, A., & Tollervey, D. (2014). Mapping the miRNA interactome by cross-linking ligation and sequencing of hybrids (CLASH). *Nature Protocols, 9*(3), 711–728.

Hendelman, A., et al. (2012). Inhibition of lamina outgrowth following *Solanum lycopersicum* AUXIN RESPONSE FACTOR 10 (SlARF10) derepression. *Plant Molecular Biology, 78*(6), 561–576.

Hong, Y.-H., et al. (2019). Overexpression of MiR482c in tomato induces enhanced susceptibility to late blight. *Cells, 8*(8), 822.

Hong, Y., et al. (2021). Editing miR482b and miR482c simultaneously by CRISPR/Cas9 enhanced tomato resistance to *Phytophthora infestans. Phytopathology, 111*(6), 1008–1016.

Horvath, P., & Barrangou, R. (2010). CRISPR/Cas, the immune system of bacteria and archaea. *Science, 327*(5962), 167–170.

Huang, T.-H., et al. (2007). MiRFinder: an improved approach and software implementation for genome-wide fast microRNA precursor scans. *BMC Bioinformatics, 8*(1), 341.

Ishino, Y., et al. (1987). Nucleotide sequence of the iap gene, responsible for alkaline phosphatase isozyme conversion in Escherichia coli, and identification of the gene product. *Journal of Bacteriology, 169*(12), 5429–5433.

Jacobs, T. B., et al. (2015). Targeted genome modifications in soybean with CRISPR/Cas9. *BMC Biotechnology, 15*(1), 16.

Javed, M., Sinha, A., & Shukla, L. I. (2019). Evaluation of mature miR398 family, expression analysis and the post-transcriptional regulation evidence in gamma-irradiated and nitrogen-stressed *Medicago sativa* seedlings. *International Journal of Radiation Biology, 95*(5), 585–596.

Jg, D. (2003). siRNAs can function as miRNAs. *Genes & Development, 17,* 438–442.

Jiang, H., et al. (2014). Skewer: a fast and accurate adapter trimmer for next-generation sequencing paired-end reads. *BMC Bioinformatics, 15*(1), 182.

Johnson, C., et al. (2006). CSRDB: a small RNA integrated database and browser resource for cereals. *Nucleic Acids Research, 35*(suppl_1), D829–D833.

Jones-Rhoades, M. W., Bartel, D. P., & Bartel, B. (2006). MicroRNAs and their regulatory roles in plants. *Annual Review of Plant Biology, 57,* 19–53.

Joshi, T., et al. (2017). The evolution of soybean knowledge base (SoyKB). In: *Plant Genomics Databases: Methods and Protocols,* Springer Nature Link, 149–159.

Kakrana, A., et al. (2014). sPARTA: a parallelized pipeline for integrated analysis of plant miRNA and cleaved mRNA data sets, including new miRNA target-identification software. *Nucleic Acids Research, 42*(18), e139–e139.

Kalvari, I., et al. (2018). Non-coding RNA analysis using the Rfam database. *Current Protocols in Bioinformatics, 62*(1), e51.

Kant, S., M. Peng, and S. Rothstein, Genetic Regulation by NLA and MicroRNA827 for Maintaining Nitrate-Dependent Phosphate Homeostasis in Arabidopsis. *PLOS Genetics* 7(3): e1002021. 2011. https://doi.org/10.1371/journal.pgen.1002021.

Kim, J., et al. (2020). Site-specific gene knock-out and on-site heterologous gene overexpression in *Chlamydomonas reinhardtii* via a CRISPR-Cas9-mediated knock-in method. *Frontiers in Plant Science*, *11*, 306.

Kohany, O., et al. (2006). Annotation, submission and screening of repetitive elements in Repbase: RepbaseSubmitter and Censor. *BMC Bioinformatics*, *7*(1), 474.

Konermann, S., et al. (2015). Genome-scale transcriptional activation by an engineered CRISPR-Cas9 complex. *Nature*, *517*(7536), 583–588.

Kozomara, A., Birgaoanu, M., & Griffiths-Jones, S. (2018). miRBase: from microRNA sequences to function. *Nucleic Acids Research*, *47*(D1), D155–D162.

Kuang, Z., et al. (2018). miRDeep-P2: accurate and fast analysis of the microRNA transcriptome in plants. *Bioinformatics*, *35*(14), 2521–2522.

Langmead, B., & Salzberg, S. L. (2012). Fast gapped-read alignment with Bowtie 2. *Nature Methods*, *9*(4), 357–359.

Lee, F. C., & Ule, J. (2018). Advances in CLIP technologies for studies of protein-RNA interactions. *Molecular Cell*, *69*(3), 354–369.

Lee, M. H., et al. (2017). An Arabidopsis NAC transcription factor NAC4 promotes pathogen-induced cell death under negative regulation by microRNA164. *New Phytologist*, *214*(1), 343–360.

Li, H., & Durbin, R. (2009). Fast and accurate short read alignment with Burrows–Wheeler transform. *Bioinformatics*, *25*(14), 1754–1760.

Li, M., & Yu, B. (2021). Recent advances in the regulation of plant miRNA biogenesis. *RNA Biology*, *18*(12), 2087–2096.

Li, F., Orban, R., & Baker, B. (2012). SoMART: a web server for plant miRNA, tasiRNA and target gene analysis. *The Plant Journal*, *70*(5), 891–901.

Li, J.-H., et al. (2013). starBase v2.0: decoding miRNA-ceRNA, miRNA-ncRNA and protein–RNA interaction networks from large-scale CLIP-Seq data. *Nucleic Acids Research*, *42*(D1), D92–D97.

Li, Y., et al. (2022). The rice miR171b–SCL6-IIs module controls blast resistance, grain yield, and flowering. *The Crop Journal*, *10*(1), 117–127.

Liang, G., Yang, F., & Yu, D. (2010). MicroRNA395 mediates regulation of sulfate accumulation and allocation in *Arabidopsis thaliana*. *The Plant Journal*, *62*(6), 1046–1057.

Liebsch, D., & Palatnik, J. F. (2020). MicroRNA miR396, GRF transcription factors and GIF co-regulators: a conserved plant growth regulatory module with potential for breeding and biotechnology. *Current Opinion in Plant Biology*, *53*, 31–42.

Lin, Y., et al. (2021). Derepression of specific miRNA-target genes in rice using CRISPR/Cas9. *Journal of Experimental Botany*, *72*(20), 7067–7077.

Liu, C., et al. (2005). NONCODE: an integrated knowledge database of non-coding RNAs. *Nucleic Acids Research*, *33*(suppl_1), D112–D115.

Liu, L., et al. (2020). CRISPR/Cas9 targeted mutagenesis of SlLBD40, a lateral organ boundaries domain transcription factor, enhances drought tolerance in tomato. *Plant Science*, *301*, 110683.

Lorenz, R., et al. (2011). ViennaRNA package 2.0. *Algorithms for Molecular Biology*, *6*(1), 26.

Lou, D., et al. (2017). OsSAPK2 confers abscisic acid sensitivity and tolerance to drought stress in rice. *Frontiers in Plant Science*, *8*, 993.

Lu, Y., & Leslie, C. S. (2016). Learning to predict miRNA-mRNA interactions from AGO CLIP sequencing and CLASH data. *PLoS Computational Biology*, *12*(7), e1005026.

Lukan, T., et al. (2022). CRISPR/Cas9-mediated fine-tuning of miRNA expression in tetraploid potato. *Horticulture Research*, *9*, uhac147.

Luna Buitrago, D., Lovering, R. C., & Caporali, A. (2023). Insights into online microRNA bioinformatics tools. *Noncoding RNA*, *9*(2), 1–15.

Ma, X., et al. (2017). TarHunter, a tool for predicting conserved microRNA targets and target mimics in plants. *Bioinformatics*, *34*(9), 1574–1576.

Ma, J., et al. (2020). The control of developmental phase transitions by microRNAs and their targets in seed plants. *International Journal of Molecular Sciences*, *21*(6), 1971.

Macho, A. P., & Zipfel, C. (2014). Plant PRRs and the activation of innate immune signaling. *Molecular Cell*, *54*(2), 263–272.

Maere, S., Heymans, K., & Kuiper, M. (2005). BiNGO: a Cytoscape plugin to assess overrepresentation of gene ontology categories in biological networks. *Bioinformatics*, *21*(16), 3448–3449.

Majoros, W. H., et al. (2013). MicroRNA target site identification by integrating sequence and binding information. *Nature Methods*, *10*(7), 630–633.

Markham, N. R., & Zuker, M. (2008). UNAFold. In: J. M. Keith (Ed) *Bioinformatics: Structure, Function and Applications*, Humana Press, Totowa, NJ, 3–31.

Martin, M. (2011). Cutadapt removes adapter sequences from high-throughput sequencing reads. *EMBnet.journal*, *17*(1), 10–12.

Mathews, D. H. (2014). RNA secondary structure analysis using RNA structure. *Current Protocols in Bioinformatics*, *46*(1), 12.6.1–12.6.25.

McIver, L., et al. (2011). Evaluation of microsatellite variation in the 1000 genomes project pilot studies is indicative of the quality and utility of the raw data and alignments. *Genomics*, *97*(4), 193–199.

Mehrotra, S., & Goyal, V. (2014). Repetitive sequences in plant nuclear DNA: types, distribution, evolution and function. *Genomics, Proteomics & Bioinformatics*, *12*(4), 164–171.

Miao, C., et al. (2019). The grain yield modulator miR156 regulates seed dormancy through the gibberellin pathway in rice. *Nature Communications*, *10*(1), 3822.

Milev, I., et al. (2011). miRTour: plant miRNA and target prediction tool. *Bioinformation*, *6*(6), 248.

Millar, A. A., & Gubler, F. (2005). The Arabidopsis GAMYB-like genes, MYB33 and MYB65, are microRNA-regulated genes that redundantly facilitate anther development. *The Plant Cell*, *17*(3), 705–721.

Morrison, K., et al. (2019). The oncogenic Kaposi's sarcoma-associated herpesvirus encodes a mimic of the tumor-suppressive miR-15/16 miRNA family. *Cell Reports*, *29*(10), 2961–2969.e6.

Nam, S., et al. (2009). MicroRNA and mRNA integrated analysis (MMIA): a web tool for examining biological functions of microRNA expression. *Nucleic Acids Research*, *37*(Web Server issue), W356–W362.

Naqvi, A. R., Choudhury, N. R., & Haq, Q. M. R. (2011). Small RNA-mediated defensive and adaptive responses in plants. *Genetics, Biofuels and Local Farming Systems*, 129–160.

Nazarov, P. V., & Kreis, S. (2021). Integrative approaches for analysis of mRNA and microRNA high-throughput data. *Computational and Structural Biotechnology Journal*, *19*, 1154–1162.

Ni, Z., et al. (2013). GmNFYA3, a target gene of miR169, is a positive regulator of plant tolerance to drought stress. *Plant Molecular Biology*, *82*, 113–129.

Noman, A., et al. (2019). Crosstalk between plant miRNA and heavy metal toxicity. In: *Plant Metallomics and Functional Omics: A System-Wide Perspective*, Springer International Publishing, 145–168.

Numnark, S., et al. (2012). C-mii: a tool for plant miRNA and target identification. In: *BMC Genomics*, Springer.

Ogata, T., et al. (2020). CRISPR/Cas9-targeted mutagenesis of OsERA1 confers enhanced responses to abscisic acid and drought stress and increased primary root growth under nonstressed conditions in rice. *PLoS One*, *15*(12), e0243376.

Saha, S., Bridges, S., Magbanua, Z. V., & Peterson, D. G. (2008). Empirical comparison of ab initio repeat finding programs. *Nucleic Acids Research, 36*(7), 2284–2294. https://doi.org/10.1093/nar/gkn064. Epub 2008 Feb 20. PMID: 18287116; PMCID: PMC2367713

Open, R., 4.0 [http://www. repeatmasker. org]. 2015, Accessed.

Parida, A. K., et al. (2022). Effect of panicle morphology on grain filling and rice yield: genetic control and molecular regulation. *Frontiers in Genetics, 13*, 876198.

Pla, A., Zhong, X., & Rayner, S. (2018). miRAW: a deep learning-based approach to predict microRNA targets by analyzing whole microRNA transcripts. *PLoS Computational Biology, 14*(7), e1006185.

Prabu, G. R., & Mandal, A. K. A. (2010). Computational identification of miRNAs and their target genes from expressed sequence tags of tea (*Camellia sinensis*). *Genomics, Proteomics & Bioinformatics, 8*(2), 113–121.

Premnath, A., et al. (2022). Genome editing and miRNA-based approaches in cereals under abiotic stress. In: A. A. H. A. Latef (Ed) *Sustainable Remedies for Abiotic Stress in Cereals*, Springer Nature Singapore, Singapore, 647–673.

Qi, Q., et al. (2023). Advances in plant epigenome editing research and its application in plants. *International Journal of Molecular Sciences, 24*(4), 3442.

Reddy, K. B. (2015). MicroRNA (miRNA) in cancer. *Cancer Cell International, 15*(1), 38.

Reichel, M., & Millar, A. A. (2015). Specificity of plant microRNA target MIMICs: cross-targeting of miR159 and miR319. *Journal of Plant Physiology, 180*, 45–48.

Reichel, M., et al. (2015). Inhibiting plant microRNA activity: molecular SPONGEs, target MIMICs and STTMs all display variable efficacies against target microRNAs. *Plant Biotechnology Journal, 13*(7), 915–926.

Remita, M. A., et al. (2016). A novel comprehensive wheat miRNA database, including related bioinformatics software. *Current Plant Biology, 7*, 31–33.

Rogers, K., & Chen, X. (2013). Biogenesis, turnover, and mode of action of plant microRNAs. *Plant Cell, 25*(7), 2383–2399.

Roubelakis, M. G., et al. (2009). Human microRNA target analysis and gene ontology clustering by GOmir, a novel stand-alone application. In: *BMC Bioinformatics*, Springer.

Sai Lakshmi, S., & Agrawal, S. (2008). piRNABank: a web resource on classified and clustered Piwi-interacting RNAs. *Nucleic Acids Research, 36*(suppl_1), D173–D177.

Sanan-Mishra, N., et al. (2018). ARMOUR – a rice miRNA: mRNA interaction resource. *Frontiers in Plant Science, 9*, 602.

Santosh Kumar, V., et al. (2020). CRISPR-Cas9 mediated genome editing of drought and salt tolerance (OsDST) gene in indica mega rice cultivar MTU1010. *Physiology and Molecular Biology of Plants, 26*, 1099–1110.

Saraswat, P., Chaturvedi, A., & Ranjan, R. (2023). Zinc finger nuclease (ZFNs) and transcription activator-like effector nucleases (TALENs) based genome editing in enhancement of anticancer activity of plants. In: *Plant-Derived Anticancer Drugs in the OMICS Era*, Apple Academic Press, 281–293.

Schubert, M., Lindgreen, S., & Orlando, L. (2016). AdapterRemoval v2: rapid adapter trimming, identification, and read merging. *BMC Research Notes, 9*, 1–7.

Singroha, G., Sharma, P., & Sunkur, R. (2021). Current status of microRNA-mediated regulation of drought stress responses in cereals. *Physiologia Plantarum, 172*(3), 1808–1821.

Paul, S., Datta, S. K., & Datta, K. (2015). miRNA regulation of nutrient homeostasis in plants. *Frontiers in Plant Science, 6*.

Song, X., et al. (2019). MicroRNAs and their regulatory roles in plant–environment interactions. *Annual Review of Plant Biology, 70*, 489–525.

Sticht, C., et al. (2018). miRWalk: an online resource for prediction of microRNA binding sites. *PLoS One, 13*(10), e0206239.

Sturm, M., Schroeder, C., & Bauer, P. (2016). SeqPurge: highly-sensitive adapter trimming for paired-end NGS data. *BMC Bioinformatics, 17*(1), 208.

Su, G., et al. (2014). Biological network exploration with Cytoscape 3. *Current Protocols in Bioinformatics, 47*(1), 8.13.1–8.13.24.

Sun, G. (2012). MicroRNAs and their diverse functions in plants. *Plant Molecular Biology, 80*(1), 17–36.

Sunkar, R., Kapoor, A., & Zhu, J.-K. (2006). Posttranscriptional induction of two Cu/Zn superoxide dismutase genes in Arabidopsis is mediated by downregulation of miR398 and important for oxidative stress tolerance. *The Plant Cell, 18*(8), 2051–2065.

Sunkar, R., Li, Y.-F., & Jagadeeswaran, G. (2012). Functions of microRNAs in plant stress responses. *Trends in Plant Science, 17*(4), 196–203.

Susmitha, J., & Farooqkhan, P. (2023). *Genome Editing Techniques in Plants*, Archives of Agriculture and Allied Sciences.

Szcześniak, M. W., & Makałowska, I. (2013). miRNEST 2.0: a database of plant and animal microRNAs. *Nucleic Acids Research, 42*(D1), D74–D77.

Tang, J., & Chu, C. (2017). MicroRNAs in crop improvement: fine-tuners for complex traits. *Nature Plants, 3*(7), 17077.

Tang, X., et al. (2017). A CRISPR–Cpf1 system for efficient genome editing and transcriptional repression in plants. *Nature Plants, 3*(3), 1–5.

Tav, C., et al. (2016). miRNAFold: a web server for fast miRNA precursor prediction in genomes. *Nucleic Acids Research, 44*(W1), W181–W184.

The Gene Ontology Consortium (2014). Gene ontology consortium: going forward. *Nucleic Acids Research, 43*(D1), D1049–D1056.

Tian, T., et al. (2016). SorghumFDB: sorghum functional genomics database with multidimensional network analysis. *Database, 2016*, 1–16.

Tian, T., et al. (2017). agriGO v2.0: a GO analysis toolkit for the agricultural community, 2017 update. *Nucleic Acids Research, 45*(W1), W122–W129.

Tørresen, O. K., et al. (2019). Tandem repeats lead to sequence assembly errors and impose multi-level challenges for genome and protein databases. *Nucleic Acids Research, 47*(21), 10994–11006.

Trapnell, C., Pachter, L., & Salzberg, S. L. (2009). TopHat: discovering splice junctions with RNA-Seq. *Bioinformatics, 25*(9), 1105–1111.

Turner, M., Yu, O., & Subramanian, S. (2012). Genome organization and characteristics of soybean microRNAs. *BMC Genomics, 13*(1), 169.

Veillet, F., et al. (2019). Transgene-free genome editing in tomato and potato plants using agrobacterium-mediated delivery of a CRISPR/Cas9 cytidine base editor. *International Journal of Molecular Sciences, 20*(2), 402.

Vlachos, I. S., et al. (2014). DIANA-TarBase v7.0: indexing more than half a million experimentally supported miRNA:mRNA interactions. *Nucleic Acids Research, 43*(D1), D153–D159.

Voinnet, O. (2009). Origin, biogenesis, and activity of plant microRNAs. *Cell, 136*(4), 669–687.

Wang, H., & Wang, H. (2015). The miR156/SPL module, a regulatory hub and versatile toolbox, gears up crops for enhanced agronomic traits. *Molecular Plant, 8*(5), 677–688.

Wang, M., et al. (2017). Gene targeting by homology-directed repair in rice using a geminivirus-based CRISPR/Cas9 system. *Molecular Plant, 10*(7), 1007–1010.

Wang, Y., et al. (2020). The apple microR171i-SCARECROW-LIKE PROTEINS26.1 module enhances drought stress tolerance by integrating ascorbic acid metabolism. *Plant Physiology, 184*(1), 194–211.

Wang, Q., et al. (2023). zma-miR159 targets ZmMYB74 and ZmMYB138 transcription factors to regulate grain size and weight in maize. *Plant Physiology*, *193*(4), 2430–2441.

Wen, M., et al. (2018). DeepMirTar: a deep-learning approach for predicting human miRNA targets. *Bioinformatics*, *34*(22), 3781–3787.

Wu, J., et al. (2014). Identification of miRNAs involved in pear fruit development and quality. *BMC Genomics*, *15*(1), 953.

Xia, X.-J., et al. (2015). Interplay between reactive oxygen species and hormones in the control of plant development and stress tolerance. *Journal of Experimental Botany*, *66*(10), 2839–2856.

Xie, F., & Zhang, B. (2010). Target-align: a tool for plant microRNA target identification. *Bioinformatics*, *26*(23), 3002–3003.

Xie, M., Zhang, S., & Yu, B. (2015). microRNA biogenesis, degradation and activity in plants. *Cellular and Molecular Life Sciences*, *72*, 87–99.

Xu, H., et al. (2021). miR169c-NFYA-C-ENOD40 modulates nitrogen inhibitory effects in soybean nodulation. *New Phytologist*, *229*(6), 3377–3392.

Xu, Z., et al. (2023). STOmicsDB: a comprehensive database for spatial transcriptomics data sharing, analysis and visualization. *Nucleic Acids Research*, *52*(D1), D1053–D1061.

Yang, T., Xue, L., & An, L. (2007). Functional diversity of miRNA in plants. *Plant Science*, *172*(3), 423–432.

Yang, J.-H., et al. (2010). starBase: a database for exploring microRNA–mRNA interaction maps from Argonaute CLIP-Seq and Degradome-Seq data. *Nucleic Acids Research*, *39*(suppl_1), D202–D209.

Yang, X., et al. (2021). miRNA mediated regulation and interaction between plants and pathogens. *International Journal of Molecular Sciences*, *22*(6), 2913.

Yates, A. D., et al. (2019). Ensembl 2020. *Nucleic Acids Research*, *48*(D1), D682–D688.

Ye, J., et al. (2018). WEGO 2.0: a web tool for analyzing and plotting GO annotations, 2018 update. *Nucleic Acids Research*, *46*(W1), W71–W75.

Ye, B. B., Zhang, K., & Wang, J. W. (2020). The role of miR156 in rejuvenation in *Arabidopsis thaliana*. *Journal of Integrative Plant Biology*, *62*(5), 550–555.

Yi, X., Du, Z., & Su, Z. (2013). PlantGSEA: a gene set enrichment analysis toolkit for plant community. *Nucleic Acids Research*, *41*(W1), W98–W103.

Yi, X., et al. (2015). PNRD: a plant non-coding RNA database. *Nucleic Acids Research*, *43*(D1), D982–D989.

Yue, E., Cao, H., & Liu, B. (2020). OsmiR535, a potential genetic editing target for drought and salinity stress tolerance in *Oryza sativa*. *Plants*, *9*(10), 1337.

Zanetti, M. E., Rípodas, C., & Niebel, A. (2017). Plant NF-Y transcription factors: key players in plant-microbe interactions, root development and adaptation to stress. *Biochimica et Biophysica Acta (BBA) - Gene Regulatory Mechanisms*, *1860*(5), 645–654.

Zhang, Y. (2005). miRU: an automated plant miRNA target prediction server. *Nucleic Acids Research*, *33*(suppl_2), W701–W704.

Zhang, Z., et al. (2010). PMRD: plant microRNA database. *Nucleic Acids Research*, *38*(suppl_1), D806–D813.

Zhang, L., et al. (2018). Lateral root development in potato is mediated by stu-mi164 regulation of NAC transcription factor. *Frontiers in Plant Science*, *9*, 383.

Zhang, L., et al. (2022). Mechanisms of microRNA biogenesis and stability control in plants. *Frontiers in Plant Science*, *13*, 844149.

Zhao, Y., et al. (2016). An alternative strategy for targeted gene replacement in plants using a dual-sgRNA/Cas9 design. *Scientific Reports*, *6*(1), 23890.

Zheng, X., et al. (2020). Prediction of miRNA targets by learning from interaction sequences. *PLoS One*, *15*(5), e0232578.

Zhou, J., et al. (2017). CRISPR-Cas9 based genome editing reveals new insights into microRNA function and regulation in rice. *Frontiers in Plant Science*, *8*, 1–12.

Zhou, X., Khare, T., & Kumar, V. (2020a). Recent trends and advances in identification and functional characterization of plant miRNAs. *Acta Physiologiae Plantarum*, *42*, 1–21.

Zhou, Y., et al. (2020b). Integration of sRNA, degradome, transcriptome analysis and functional investigation reveals gma-miR398c negatively regulates drought tolerance via GmCSDs and GmCCS in transgenic Arabidopsis and soybean. *BMC Plant Biology*, *20*(1), 190.

Zhou, J., et al. (2022). CRISPR-Cas9 mediated OsMIR168a knockout reveals its pleiotropy in rice. *Plant Biotechnology Journal*, *20*(2), 310–322.

Zhu, J.-K. (2016). Abiotic stress signaling and responses in plants. *Cell*, *167*(2), 313–324.

Zhu, Q.-H., & Helliwell, C. A. (2011). Regulation of flowering time and floral patterning by miR172. *Journal of Experimental Botany*, *62*(2), 487–495.

Zuker, M. (2003). Mfold web server for nucleic acid folding and hybridization prediction. *Nucleic Acids Research*, *31*(13), 3406–3415.

16

Employment of RNA Interference and CRISPR/Cas Machinery for Disease Resistance in Crops

Ifrah Imran, Muhammad Jawad Akbar Awan, Mariam Akhtar, Muhammad Waseem Sajjad, Muhammad Ismail Buzdar, Rida Zahid, Jahan Khan, Imran Amin, and Rubab Zahra Naqvi

Agricultural Biotechnology Division, National Institute for Biotechnology and Genetic Engineering (NIBGE), Constituent College of Pakistan Institute of Engineering and Applied Sciences (PIEAS), Faisalabad, Pakistan

16.1 Introduction

16.1.1 The Importance of Crops

Crops provide necessary nutrients for human consumption, feed for animals, and raw materials for many industries. These are essential for ensuring economies, supporting agricultural livelihoods, and providing food security for growing global populations. As the population of the world is predicted to reach over nine billion people by 2050, the pressure on agricultural systems to boost production is also intensifying (Godfray et al., 2010). The Food and Agriculture Organization (FAO) estimates that taking into account population growth and dietary changes, the production of food needs to be increased by 70% to meet the demand of the future (Alexandratos & Bruinsma, 2012). The requirement to attain such development and to achieve such a big challenge on a restricted amount of arable land is exacerbated by unfavorable climatic circumstances (Ray et al., 2013).

Over the past century, agricultural productivity has increased significantly using conventional breeding methods, but these techniques lack in many ways like in their speed, adaptability, accuracy, and precision (Nerkar et al., 2022). As traditional breeding methods rely on genetic diversity that is already present within crops, it makes it difficult to quickly introduce new traits or features in response to the increasing threats. In this context, sophisticated biotechnological tools like RNA interference (RNAi) and clustered regularly interspaced short palindromic repeats (CRISPRs)-CRISPR-associated (Cas) protein systems usher a new era of crop improvement (Liu et al., 2022). These cutting-edge methods can precisely and directly manipulate plant genomes, giving scientists and farmers unprecedented control over important plant traits like environmental stress tolerance and disease resistance.

Genome and Epigenome Editing for Stress-Tolerant Crops, First Edition. Edited by Jen-Tsung Chen.

16.1.2 Risks to Food Security

The World's agricultural production faces two of the main issues, abiotic and biotic stresses, that threaten food security around the globe. All major crops face significant yield losses due to biotic stressors like pathogens, pests, and weeds. It is estimated that up to 40% of agricultural production is lost due to these biotic factors every year, which has adverse effects on the livelihood of farmers, causes economic losses, and reduces food availability (Savary et al., 2019). Historically, major crops like wheat, rice, and maize have been devastated by fungal diseases, especially rust, bacterial blight, and viral infections, these pathogens are constantly evolving which makes it increasingly difficult to manage them using conventional methods (Fisher et al., 2012).

Extreme temperature, drought, and salinity are the abiotic stressors of crops, which are equally concerning, especially as these events become more severe and often worsen with climate change (Lesk et al., 2016). These environmental factors can affect a crop in many ways such as influencing the growth of plants, lessening its yield, and affecting the quality of production. The unpredictable climate patterns add to the risks, so it is crucial to develop high yielding plants that are resistant to these stresses. There is an urgent need for more advanced, effective, and crop–resilience-boosting techniques that can safeguard global food security. This has led to the exploration of new genetic engineering techniques that can offer accuracy, precision, and flexibility beyond the reach of traditional methods.

16.1.3 Novel Genetic Engineering Approaches

Modern biotechnology enables us to improve our crops, where RNAi and CRISPR-Cas are particularly promising techniques. Crop genomes can be targeted and modified very accurately to enhance traits like stress tolerance, and resilience to disease, and increase crop yield. RNAi occurs spontaneously when small RNA (sRNA) molecules neutralize specific mRNA molecules to stop gene expression or translation. This approach has been improved to benefit crops by silencing pest and disease susceptibility genes. RNAi can provide crops with the ability to withstand bacteria, viruses, and fungi (Koch & Wassenegger, 2021). It offers a sustainable alternative to chemical controls by targeting the genes necessary for the survival or virulence of pests and pathogens. It is made possible by RNAi to silence multiple genes simultaneously for broad-spectrum pest and disease management. Nonetheless, CRISPR-Cas can introduce precise genome modifications, and this system has completely revolutionized the field of genetic engineering. In contrast to RNAi, which suppresses the expression of genes, CRISPR-Cas allows for direct editing or disruption of genes. With the ability to knockout genes that make crops susceptible to pathogens or to introduce desirable traits such as increased nutritional content or tolerance to stress, this technique has become a potent tool in crop breeding (Jaganathan et al., 2018). Multiple genes can also be edited simultaneously to address complex traits that are regulated by multiple genes (Zhang et al., 2020). Later in this chapter, the mechanisms and applications of the RNAi and CRISPR-Cas systems in crops will be discussed in detail.

16.1.4 RNAi and CRISPR-Cas Technology Used in Harmony

RNAi and CRISPR-Cas system-based techniques are powerful tools that have the potential to transform plant research and breeding (Rajput et al., 2021). Targeting gene expression and pest control with RNAi is highly effective; however, CRISPR-Cas technology enables permanent genetic modifications to improve traits and enhance crop resilience. Through the integration of these technologies, scientists can design tailored crops with durable and broad-spectrum protection from stresses (Zotti & Smagghe, 2015). For instance, RNAi can target virus or pest genes for rapid defense, and CRISPR-Cas can edit plant genes to drive immune response for long-term resistance. This combination can lower the likelihood of resistance breakdown and perhaps lower the use of chemical pesticides, offering a sustainable solution compared to single-method approaches.

16.1.5 The Future of Crop Engineering

The development and deployment of crops enhanced by RNAi and CRISPR-Cas technologies are increasingly important as we address challenges posed by population growth, climate change, and the need for sustainable agriculture. These tools open a new door for crop resilience and yield, as they enable precise modification and quick response to the threats faced by crop varieties. This chapter will dive deep into the mechanisms of RNAi and CRISPR-Cas in further sections. The roles, achievements, and possible synergies of these two techniques in crop development will be discussed. The goal is to grasp how these cutting-edge genetic techniques may revolutionize modern agriculture and enhance global food security.

16.2 The Editors of Nature

The global agricultural landscape is shifting toward sustainability, necessitating effective technologies to combat plant diseases. Since the onset of the Green Revolution in the 1960s, various initiatives have been undertaken in the agriculture sector to prevent diseases and mitigate the damage caused by pathogens. Traditionally, farmers relied on chemicals to protect their crops. However, as the risks posed by these chemicals to human health and the ecosystem became evident, agricultural research shifted focus to safer alternatives. Innovative techniques can reduce the inherent risks associated with conventional methods and transgenic crops, which can be cultivated under regulatory guidelines (Sharma et al., 2013).

16.2.1 A Simplified Exploration of the Genetic-Editing Tools Shaping the Future of Agriculture

Employing genome-editing approaches to delete, insert, or replace the DNA of any organism with a high specificity to revamp the phenotype or genotype is termed genome editing (Andergassen & Rinn, 2022). Ingenious crop development strategies are pressingly needed to fulfill the growing demand for food due to the expeditiously increasing population (de Janvry & Sadoulet, 2020). See Figure 16.1a for an overview of the challenges faced by crop varieties. Traditional methods of crop improvement are labor-intensive and cannot counterbalance the growing demand for crops (Charania & Li, 2020). Endless efforts are ongoing to increase crop output, nutrient content, and pest and disease resistance through conventional means of crop improvement. Figure 16.1b and c illustrates the conventional genome engineering methods (Kumar et al., 2020). However, methods such as hybridization for natural mutations, induced mutations via chemical factors like diepoxy butane and sodium azide, physical factors such as fast neurons, and X-rays can induce a wide spectrum of random mutations, including deletions, point mutations, and chromosomal translocations or loss (Zhang et al., 2017). Despite offering a diverse range of autogenous and nonautogenous variations, these breeding methods are not applicable taking into account the

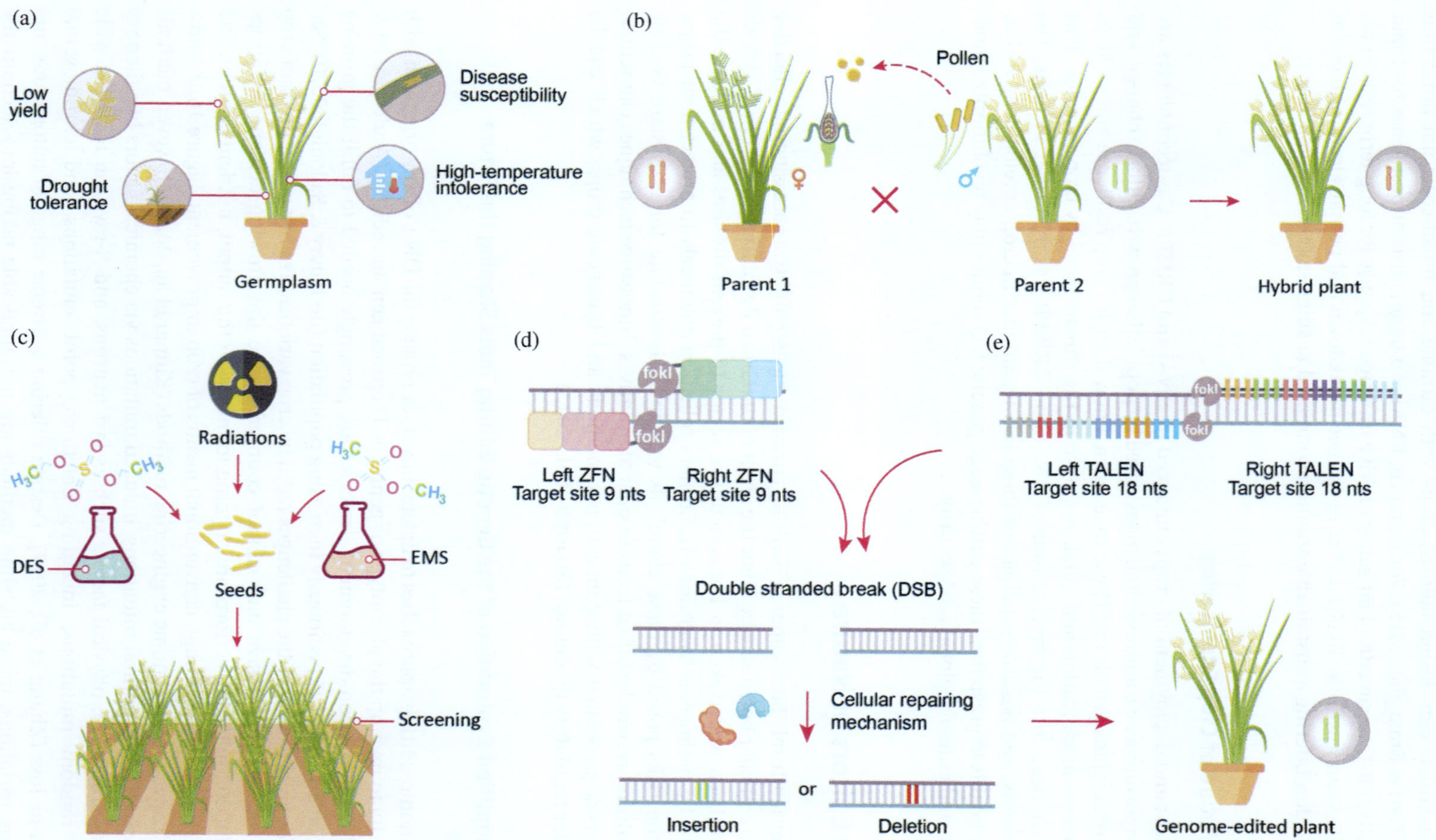

Figure 16.1 Techniques in plant breeding and genome editing for the enhancement of crop traits. (a) Summary of common challenges related to germplasm, encompassing low yield, drought resistance, disease susceptibility, and sensitivity to high temperatures. (b) Conventional plant breeding by cross-pollination of Parent 1 and Parent 2 to generate a hybrid plant having desirable traits. (c) Mutation breeding employs chemical mutagens like diethyl sulfate (DES) and ethyl methane sulfonate (EMS), or radiation, to produce genetic variants, thereafter screening for desirable traits. (d) ZFN genome-editing technology targets specific DNA regions to induce DSBs, which are then repaired by the cell's intrinsic processes, including insertion or deletion. (e) TALEN is a genome-editing technique that targets extended DNA segments (17–18 nucleotides) to facilitate precise alterations, yielding genome-edited plants with better traits.

escalating population's demand and requirements of genome modification, as they are time-consuming, slow, and arduous (Jansing et al., 2019). Additionally, polyethylene glycol for protoplast fusion, Sendai virus induction, and electronic stimulus have also been utilized due to their effectiveness in hybridization (Figure 16.1b and c) (Zhang et al., 2017; Swapnil et al., 2020). The stipulation of efficient and site-directed mutagenesis has led to the emanation of a new boulevard in this domain: the modernization of artificial nucleases (Ilyas et al., 2024).

The newly emerged genome engineering technologies can be widely classified into three systems depending upon their ability to recognize and cleave the target site creating single-stranded breaks (SSBs) or double-stranded breaks (DSBs) promoting random mutations via nonhomologous end joining (NHEJ) or targeted mutations via homologous recombination (HR) through targeted mutation. Figure 16.1d and e shows the mechanism of zinc-finger nucleases (ZFNs) and transcription-activator-like effector nucleases (TALEN) in genome editing (Gu et al., 2021; Carusillo & Mussolino, 2020). First is a DNA cleavage protein-dependent system, such as TALENs and ZFNs; second is the RNA-dependent DNA cleavage system, including CRSIPR-Cas 12, CRISPR-Cas 9, and CRISPR-Cas C2c1; and the third system is the RNA-dependent RNA cleavage method, i.e. RNAi (Punetha et al., 2018; Shamshirgaran et al., 2022).

16.2.2 An Overview of RNAi and CRISPR-Cas Techniques

The currently available RNA-dependent RNA cleavage method is RNAi, which is a naturally occurring gene regulation event in eukaryotes that has been conserved throughout evolution. It has cognate to shield cells from foreign DNA intrusion (Vologodskii, 2022; Zhang et al., 2022). Additionally, it aids in preserving genome stability, epigenetic modifications, regulation of transposon stability, and control of transcription and translational processes (Ugolini & Halic, 2018; Choi & Lee, 2020; Lax et al., 2020). The term RNAi was first coined by Andrew Fire when double-stranded RNA (dsRNA) was introduced into *Caenorhabditis elegans* cells, and it showed impressive gene silencing (Fire et al., 1998). Another unexpected phenomenon of gene silencing was observed while experimenting on *Petunia* flowers when scientists were attempting to alleviate the expression of a gene accountable for pigment production (Napoli et al., 1990). Surprisingly instead of producing the predicted deep purple blooms, this is culminated in flowers with variegated patterns. The researchers ascertained that the introduced transgene suppressed the expression of the homologous native gene, leading to the name of this mechanism as "cosuppression" (Napoli et al., 1990).

RNAi knockdown results in DNA sequence diversification, and knockouts are not permanent or complete (Hentzschel et al., 2020). The most straightforward and common method to comprehend and clench functional genomics and experimental intonation of gene expression at the posttranscriptional level is through loss of function studies, which has made reverse genetics methods achievable (Figure 16.2a) (Shi et al., 2021). RNAi is a dynamic transient system for gene expression, which has incredible applications in studying genotype–phenotype relationships. For this, researchers have employed knockout approaches to modify DNA sequences for analyzing target genes. Many significant agronomical traits have been improved by site-directed mutagenesis in plants (Tyurin et al., 2020; Tsai et al., 2022). The ability to alter gene expression and content has been compelling to understand the role of genes in biological pathways and their association with disease phenotypes, for instance, when rice plants were transformed with an S8-RNAi vector, they showed resistance to rice black-streaked dwarf virus (RBSDV) under infection stress in fields, subsequently improving the overall crop health. A fundamental leverage of RNAi is that this mechanism requires no precedent genetic manipulation of target regions, a simple small interfering RNA (siRNA) transfection can result in a silenced phenotype (Dubrovna et al., 2023).

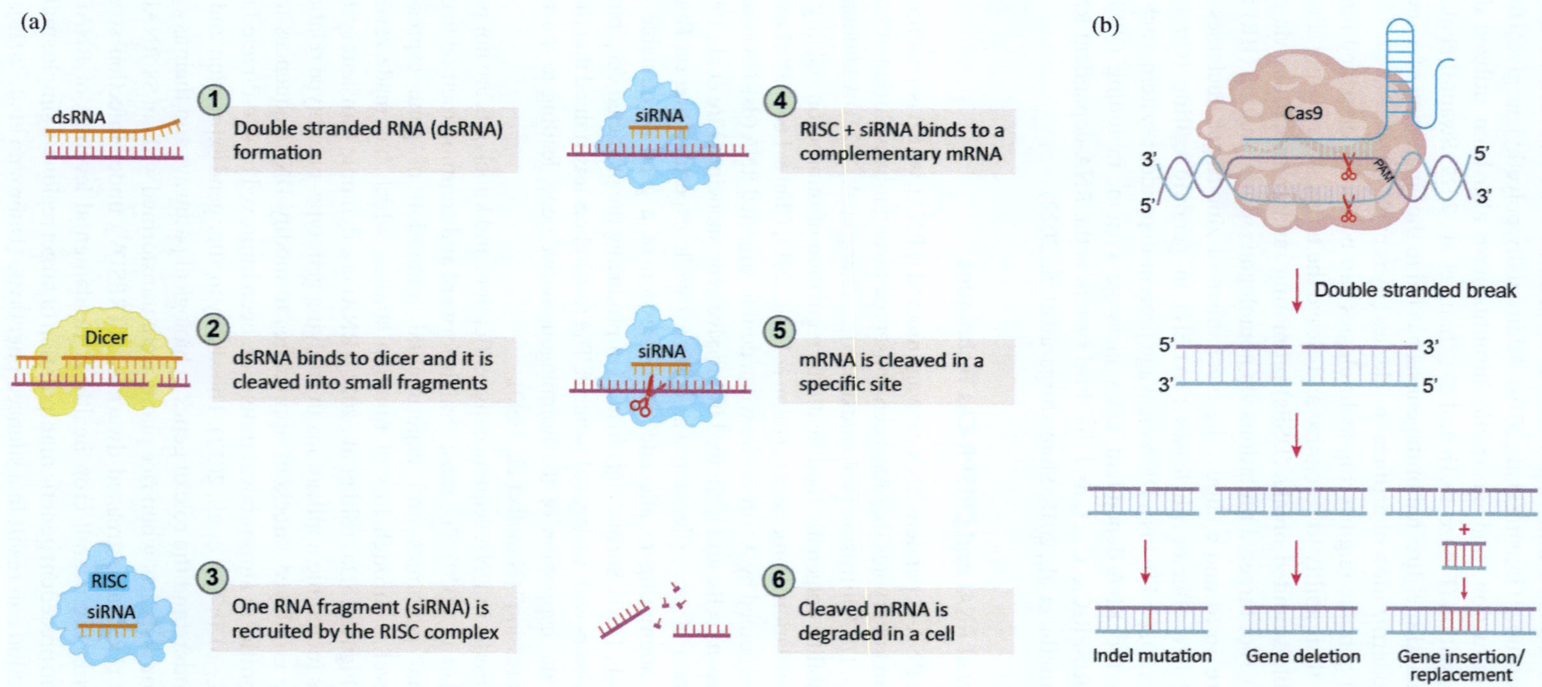

Figure 16.2 RNAi and CRISPR-Cas9 genome-editing mechanisms. (a) RNAi pathway: (1–3) Dicer converts dsRNA into siRNA, (4) which is subsequently integrated into the RISC. The siRNA leads RISC to the corresponding mRNA, (5, 6) cleaving and degrading it. (b) CRISPR-Cas9 mechanism: A gRNA guides Cas9 to DNA sequences, causing double-strand breaks. NHEJ or HDR repairs the break, causing indel mutations, gene loss, or gene insertion/replacement.

CRISPR-Cas genome-editing technology presents new routes for developing disease-resistant crops. The Type II CRISPR-Cas9 system from *Streptococcus pyogenes*, presently the most extensively used system for genome engineering in eukaryotic cells, utilizes a single-guide RNA (sgRNA) consisting of two independently expressed RNAs: a *trans*-activating RNA (tracrRNA) and CRISPR RNA (crRNA), processed by native bacterial machinery to produce a mature gRNA (Burmistrz et al., 2020; Wang & Li, 2021). It was demonstrated that a single chimeric sgRNA, i.e. a combination of tracrRNA and crRNA, could function as a mature sgRNA without the need for further rectification. This chimeric sgRNA mirrors the completely processed endogenous bacterial sgRNA, featuring a constant region for Cas9 interaction and a variable region for the target genome site (Figure 16.2b) (Park et al., 2021; Zhao et al., 2023). Current chimeric sgRNAs used for genome manipulation typically include a 17–20 nucleotide variable region complementary to the target genomic sequence. A refined chimeric sgRNA design was developed to reinforce the interaction with gRNA and expression. This involved eliminating a potential PolIII terminator from the sgRNA's constant region and extending a hairpin by 5 bp to reinstate the complete *S. pyogenes* hairpin sequence identified. An important specificity factor for Cas9 is the protospacer adjacent motif (PAM), a brief sequence immediately following the target sequence in the 3′ direction (Zhao et al., 2023). For *S. pyogenes* Cas9, the PAM sequence is NGG, but it differs for Cas proteins from different species. Other examples of these motifs include NAAR, NGGN, and NNAGAA (Wang et al., 2021).

16.3 The Mechanisms of Silence: Understanding RNA Interference

RNAi appears to be the most straightforward method for creating hypomorphic knockdowns. Unlike CRISPR, RNAi does not require the introduction of additional genes into the cell before knockdown, as most of the cells already possess complete RNAi machinery. This makes RNAi implementation quick and easy. RNAi can target sequences conserved among gene family members and transcript variants. A single siRNA can target all transcripts with conserved sequences, and this feature gives RNAi an advantage over CRISPR. Moreover, RNAi targets RNA transcripts in the cytoplasm rather than genomic DNA in the nucleus. Consequently, its accessibility does not seem to be hindered by chromatin conditions, which is a challenge observed with CRISPR-based approaches (Huang et al., 2020; Loriato et al., 2020; Graca et al., 2021).

16.3.1 Phases of RNAi

RNAi, a gene silencing phenomenon, involves three definite phases. The first phase of RNAi, known as initiation, occurs when dsDNA is identified and cleaved by an enzyme called dicer, similar to RNase III, into 21–23 nucleotide fragments called siRNA. These siRNA fragments require 5′ phosphate ends and 2 nucleotide 3′ overhangs to function appropriately. The second phase involves the integration of siRNA with an RNA-induced silencing complex (RISC), for recognition of target mRNA, and the third is the multiplication of dsRNA using target mRNA as a template (Hung & Slotkin, 2021; Neumeier & Meister, 2021; Paturi & Deshmukh, 2021).

16.3.2 Biogenesis and Mechanism of siRNA-Mediated Gene Silencing

siRNAs toolkit for gene knockdown are 21–23 nucleotides long dsRNA molecules known as siRNA. Upon introduction of siRNAs into the cell, they are loaded onto the RISC, a site for Dicer cleavage activity and initiation for the RNAi pathway (Iwakawa & Tomari, 2022). Dicer pertains to the class III RNAse enzyme, consisting of four domains: a Piwi Argonaute Zwille (PAZ) domain, an N-terminal helicase domain, a dsRNA-binding domain, and dual RNAse III domains (Wang et al., 2021). Argonaute proteins, which direct the siRNA to target mRNA, are among the proteins that make up the RISC (Liu et al., 2022). The target mRNA molecule's corresponding sequence is recognized and bound to by the guide strand of siRNA present in the RISC. The siRNA guide strand partners with the target mRNA to generate a stable RNA–RNA duplex. The target mRNA's cleavage is catalyzed by the bound RISC. The RISC's argonaute protein has endonuclease activity, enabling it to break the target mRNA molecule located inside the siRNA-binding site (Wang et al., 2021). Gene silencing occurs due to this cleavage, which hinders the translation of mRNA into a functional protein. The RISC can recycle and continue to target and silence new RNA molecules after mRNA has been chopped down and degraded. The guide strand of siRNA usually stays connected to the RISC, whereas the passenger strand is freed and ravaged (Iwakawa & Tomari, 2022). The siRNA length is a pivotal aspect for determining the results, as more than 30 nucleotides long siRNAs lead to immune reactions and interferon induction through Toll-like receptors (TLRs) activation (Kaushal, 2023). This causes consequent cell death by global gene silencing. siRNA genesis that is Dicer independent has also been proclaimed in *Arabidopsis*, *Neurospora*, and *C. elegans*, which mostly originate from intergenic elements, transposable elements, and transgenes (Hill et al., 2022).

16.3.3 Biogenesis and Mechanism of miRNA-Mediated Gene Silencing

In 1993, lin-4, the first short RNA was discovered in nematodes via genetic screening (Lee et al., 1993). Later that year, the regulation of lin-14 by lin-4 validated the regulatory role of short RNAs. It is now understood that short lin-4 RNA is the origin of regulatory elements, microRNA (miRNA). miRNAs are ~22 nucleotides long noncoding RNA (lncRNA) molecules processed through the same mechanism as siRNAs, but they come from somewhat distinct origins (McGeary et al., 2022). RNA polymerase II synthesizes primordial miRNA (pri-miRNA) transcripts from miRNA transcribed from DNA (Hynes & Kakumani, 2024). An enzyme known as Drosha and its cofactor DGCR8 recognizes and processes the stem-loop structure seen in pri-miRNA in the nucleus. A pre-miRNA as a hairpin is resulting from this processing stage. The exportin (XPO5) protein transports the pre-miRNA from the nucleus to the cytoplasm. A tiny RNA duplex is created in the cytoplasm when the Dicer enzyme breaks the pre-miRNA (Ergin & Çetinkaya, 2022). This duplex comprises a mature miRNA strand and its complementary strand. The mature mRNA strand is preferentially integrated into the RISC, whereas the miRNA strand is usually destroyed. This process is known as strand selection of miRNA duplex. The RISC is made when argonaute family proteins bind with the mature miRNA strand (Kirstein et al., 2023). RISC facilitates precise targeting of miRNA to its intended mRNA by sequence-specific recognition. The inhibition of translation or degradation of mRNA results in posttranscriptional gene silencing. While miRNAs have unambiguous locus annotations, siRNAs dawdle in this aspect, regardless of these miRNAs represent only a small fragment of the complete sRNA pool, also exhibiting preeminent conservation across species in comparison with siRNA (Dexheimer & Cochella, 2020; Leng et al., 2020). The use of polycistronic miRNA precursors allows for the concurrent silencing of diverse targets. Additionally, research shows that the separation of RNAi transgene can sequel in nongenetic MUTL protein homolog 1 (MSH1) memory reprogramming, which can be transmitted through many generations. MSH1 is a protein specific to plants targeting both plastids and mitochondria, and the suppression of MSH1 might lead to fickleness in phenotypes associated with stress response and developmental pathways (Kundariya et al., 2020).

16.4 Advancing Agriculture: Engineered RNA Interference for Enhanced Crop Performance

Living organisms, exclusively bacteria, viruses, fungi, nematodes, insects, and weeds are key sources of biotic stress in plants. Roughly 40% of the total yield of valuable cash and food crops forfeits because of these organisms' mediated stresses. New avenues for crop conservation and protection against these stressors have been made achievable by RNAi technology (He et al., 2022). RNAi enables the development of ingenious and sustainable methods for plant protection. Crops have been genetically engineered to express dsRNA parade compelling products for executing insect pests (Christiaens et al., 2020). In addition, several nontransformative methods, such as trunk injection, foliar spraying, and irrigation, offer practical alternatives for implementation in the real world (Cagliari et al., 2019). Crop-damaging insects pose a serious threat to food production, primarily due to their ability to reduce quality and yield. These pests cause a potential yield loss of around 18% in major crops (Sharma et al., 2017).

There are mainly three RNAi pathways in insects: miRNA-mediated pathway, siRNA/dsRNA-mediated pathway, and piwi-interacting RNA (piRNA) pathway, which play roles in the regulation of gene expression by miRNA pathway, defense against transposable elements and viruses via siRNA pathway, and suppression of expression of germline transposons by piRNA pathway (Kutter & Svoboda, 2008; Sioud, 2021). The essential machinery involved in the RNAi pathway can fluctuate among insect species. In dipteran and coleopteran pests, dsRNA uptake occurs via clathrin-dependent endocytosis, while siRNA is efficient in other insects (Abbasi et al., 2020). Gene silencing in insects is usually initiated by the introduction of exogenous dsRNA, although some species are unresponsive to RNAi through feeding. This contradicts between plants and fungi, where siRNA is usually used to trigger RNAi. Insecticidal dsRNA must penetrate target cells to be successful in practical use (Flynt, 2021). Direct uptake methods, such as genetically topical applications and modified crops, show potential for throughput use and have offered significant output. However, topical application results in poor dsRNA delivery efficiency and direct uptake methods such as conventional transgenesis combat to produce a sufficient amount of stable dsRNA because of plant RNAi mechanisms (Bennett et al., 2020; Nitnavare et al., 2021). Alternatively, dsRNAs can enter the plant vascular system and get absorbed by pests, a process known as direct uptake, and dsRNAs confined to xylem vessels, limit their distribution within the plant system, hindering the persuasiveness of the indirect uptake method. The preface of transplastomic technology may help overcome these challenges, likely increasing RNAi efficiency and implementation of RNAi-based pest management strategies (He et al., 2022).

16.4.1 Exogenous RNAi: A Novel Epigenetic Tool for Precise Plant Modifications

Another study examines the use of exogenous RNA, a compelling transgene-free tool as an inventive approach for inflating alterations in epigenetically modified plants. These modifications result from exo-RNAi application to selectively trigger the silencing of homologous regions that remain a largely unexplored domain. The aim was to elucidate how different RNAi delivery methods affect the effectiveness and likelihood of off-target effects on plant genomes (Dalakouras & Papadopoulou, 2020). DNA methylation, an imperative epigenetic modification, involves the attachment of a methyl group on the fifth carbon of a ring nucleotide, cytosine. Earlier believed to be triggered by DNA : DNA interactions, a revolutionary study in viroid-infected tobacco plants depicted that RNA : DNA interaction can also provoke DNA methylation, which was later termed

RNA-directed DNA methylation (RdRM) (Das & Sherif, 2020). While the precise mechanisms of RdRM triggered are still obscure, the current study proposes that 24-nt siRNAs determine the DNA region to be methylated by binding wither to the DNA strand or its incipient transcript generated by RNA polymerase V (PolV). The interaction between PolV transcript and 24-nt siRNA attracts other factors, including domain-rearranged methyltransferase 2 (DRM2), to perform de novo methylation of cytosines in the corresponding DNA. RNA : DNA interaction recruits DRM2 to inaugurate a new methylation in both strands of DNA. Finally, PolIV and PolV are recruited to hemimethylated DNA. PolIV gives rise to short transcripts (40 nt), later transcribed by RDR2 into 40 bp dsRNAs. Dicer-like 3 (DCL3) then processes them into 24-nt siRNAs, which are then loaded onto Argonaute RISC Component 4 (AGO4). These siRNAs intermingle with PolV transcript, nominating DRM2 to the hemimethylated locus to enhance methylation (León-Ruiz et al., 2023; Xie et al., 2024).

A transgene-free and effective substitute for genetically modified (GM) crops, exo-RNAi has been applied in plants through many different methods such as petiole uptake, spraying, and root absorption. These methods aim to tweak and modify pant gene expression and also confer resistance against insects, viruses, and fungi. Shortly, exo-RNAi can replace traditional insecticides, herbicides, and fungicides, with so much effort being hoarded by crop industries in this direction (Das & Sherif, 2020; Qiao et al., 2021). For the feasibility of field scale applications, it is important to optimize dsRNA delivery, uptake, and stability and also reduce dsRNA production outlay. Many agro-industrial companies like AgroRNA and RNAgri are now offering large quantities of dsRNA at cheap prices, while various carrier compounds like chitosan nanoparticles, nanosheets, and liposomes have been developed for the improvement of dsRNA uptake, stability, and efficiency. As a result, Monsanto-Bayer's exo-RNAi commercial product "BioDirect," anticipated for control of insects, viruses, and weeds is almost ready to hit the market (Dalakouras & Papadopoulou, 2020; Halder et al., 2022).

When in vitro generated NPTII and GFP dsRNA were applied to transgenic *Arabidopsis* harboring NPTII/GFP cassette, scientists observed that the corresponding mRNA was downregulated, and the DNA of that coding region was methylated seven days later. These results highlighted the need for careful evaluation of potential epigenetic modifications in exo-RNAi applications. The plants treated with exo-RNA may be epigenetically modified, but they are still considered transgene-free (Dubrovina & Kiselev, 2019). Another recent study exhibited that the exogenous application of dsRNA molecules on plants targeting genes linked to fungal growth and virulence can attenuate the disease-causing pathogens, including *Sclerotinia sclerotiorum*, *Botrytis cinerea*, and *Fusarium graminearum*. These findings demonstrate the significant potential of RNAi for the management of RNAi-mediated plant pathogenic fungal diseases.

In conclusion, it is justifiable to believe that dsRNA-induced RNAi can open doors to longer-lasting and environmentally benign methods of gene regulation involved in plant development, disease management, and crop improvement. However, for high-throughput agriculture applications, open field trials are necessary to address the questions raised above (Gebremichael et al., 2021; List et al., 2022).

16.5 RNAi Approaches: Targeted Silence to Combat Crop Diseases

Different RNAi approaches have been used to combat the pathogens targeting the crops. Of these approaches, virus-induced gene silencing (VIGS) is the one that is considered the most established and prominent, whereas other approaches such as host-induced gene silencing (HIGS) and spray-induced gene silencing (SIGS) are also available.

16.5.1 VIGS (The Silent Guardian): Exploring ncRNAs and Their Role in Crop Protection

A mechanism that utilizes the plant's native RNAi-mediated antiviral defensive system is called VIGS. The process attacks the viral DNA exclusively in plants attacked by intact viruses (Lange et al., 2013). On the other hand, the process may also be directed against the equivalent host mRNAs when viral vectors include sequences that are generated from host genes. Employing the microbial pathogen *Agrobacterium tumefaciens* to transfer, via its Ti plasmid, a virus with recombinant characteristics encoding all or part of the gene sequence that is targeted for suppression (see Figure 16.3a), VIGS has now been utilized for highly effective functional genetics in different plants. The intrinsic plant RNAi mechanism handles the remainder of the systemic viral transmission (Liu et al., 2002).

By studying the physiological consequences of reducing gene expression, this approach enables researchers to investigate the functioning of the genes in plants. Apart from its application in the study of genomics, VIGS has proven to be efficacious in conferring resilience against many plant diseases (Bennypaul et al., 2012). VIGS has been used to study and manipulate ncRNAs, especially miRNAs, in plants. A variant of the virus carrying a section identical to the gene or ncRNA of interest is used to "silence" specifically that target gene or ncRNA. This in turn triggers the RNAi pathway of the plant (Brant & Budak, 2018). In plants, VIGS has been used to both study and manipulate ncRNA, especially miRNA. A mutant VIGS virus is used that contains a sequence identical to the ncRNA, or gene, of interest. This is then "silenced" and serves as a template to activate the siRNA mechanism (Nejat & Mantri, 2018). Moreover, the discovery of these regulators and exploring their roles by using VIGS to identify functions in ncRNAs has provided new avenues for crop protection. It was brought into use as a means to increase the tolerance of crops land from diseases that the miRNAs which act on related to defense can be useful targets for VIGS (Table 16.1) (Zaynab et al., 2018). Following are a few major types of ncRNA, implemented through VIGS, that function in crop protection.

16.5.2 miRNAs

miRNAs, having an average length of 17–24 nucleotides, are small strands of RNAs that do not code for any functional proteins. However, they play a very important role in the physiological development of plants, controlling the expression of genes, and facilitating the behavioral adaptation of plants against different biotic and abiotic stresses. Through translation suppression or desired mRNA degrading, miRNA can prevent the expression of genes in different plants (Tang et al., 2010). VIGS involving miRNA is a recently established method for reverse genetic investigations of plants. The indigenous siRNA-mediated gene suppression mechanism of the plant host expresses the big gene fragment introduced during conventional VIGS tests and subsequently converts it into sRNAs. On the other hand, well-designed miRNAs are used by miRNAs engaged via VIGS to cause RNA-mediated silencing of the gene being targeted (Tang et al., 2013).

16.5.3 siRNAs

The sources of siRNAs in plants include transposons, centromeres, repeats in the inverted form involving transgene, invasive viruses as well as some other sequences in repetitive form. One typical entrance element in the siRNA route is the lengthy dsRNA derived through a transgenic construct (Zhang, 2014). By examining the functions of siRNAs in antiviral defense, VIGS has highlighted the processes occurring at the molecular level underpinning the interaction between

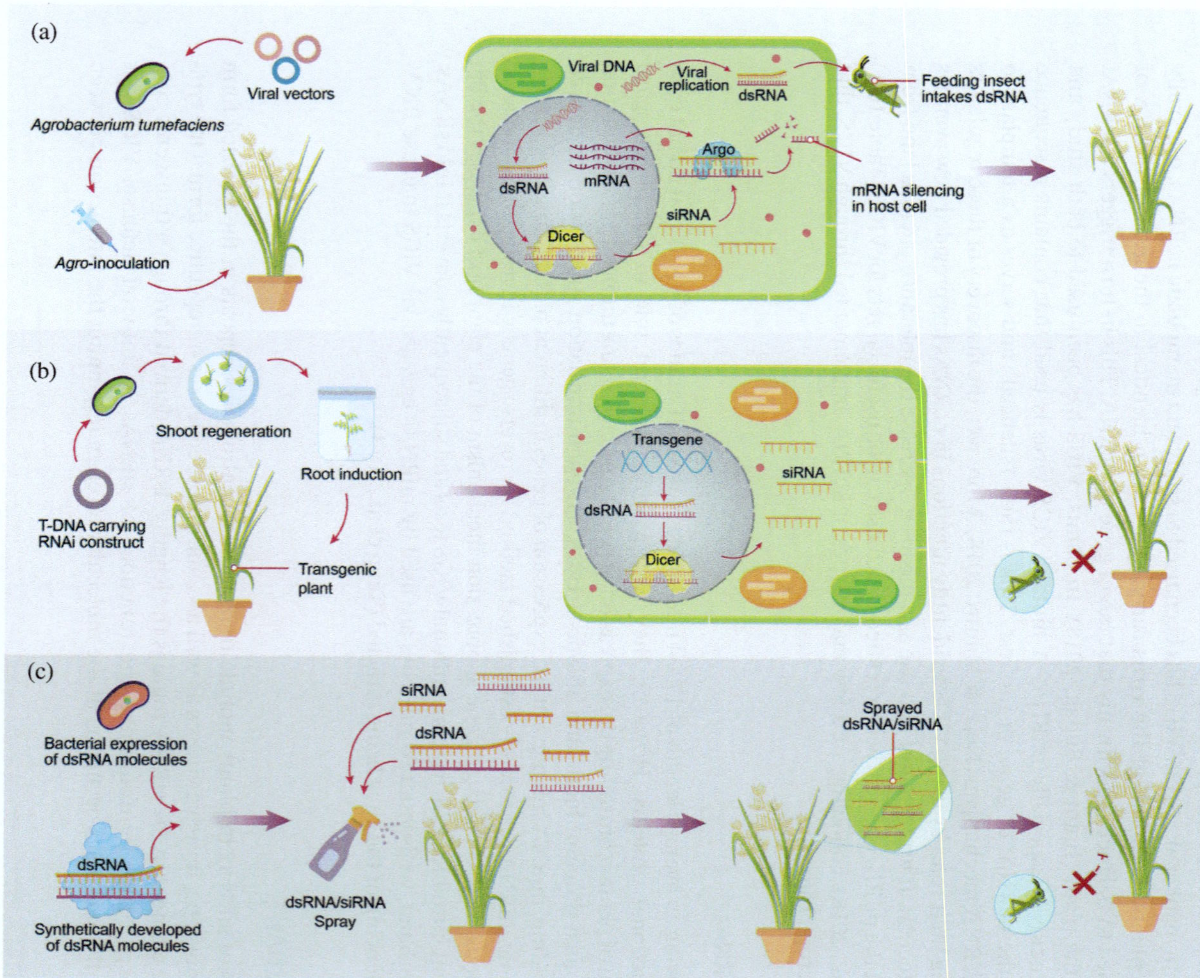

Figure 16.3 Comparative approaches of gene silencing for crop improvement. (a) VIGS: Viral vectors or *A. tumefaciens* are employed to deliver dsRNA into the plant by agro-inoculation. The viral DNA replicates within the host cell, generating dsRNA, which Dicer then processes into siRNA, which induces gene silencing in the host plant, and insects that consume the plant absorb the dsRNA, resulting in the silencing of critical insect genes. (b) HIGS: A transgenic method involving the incorporation of T-DNA constructs containing RNAi sequences into the genome of the host plant. The plant synthesizes dsRNA, which is then processed into siRNA, therefore silencing target genes in both the plant and the pests that consume it. This offers enduring, inheritable defense against pests or diseases. (c) SIGS: This nontransgenic technique involves the external application of dsRNA or siRNA molecules to the plant by spraying. dsRNA is converted into siRNA within the plant or in the pests that ingest it, resulting in targeted gene silencing without genetic alteration.

Table 16.1 Different developments in crop plants using various types of nc-RNAs through VIGS.

Crop	ncRNA type	Phenotypes	References
Arabidopsis	miR398	Improved tolerance against oxidative stress as well as pathogens	Jagadeeswaran et al. (2009)
	miR168	Improved resistance through miRNA-mediated silencing	Kung et al. (2014)
	miR164	Improved resistance to bacterial pathogens	Lee et al. (2017)
	miR398	Improved stress tolerance and pathogen resistance	Zhu et al. (2011)
Cassava	siRNA	Suppressed disease spread and viral infection	Anjanappa et al. (2018)
Rice	miR393	Decreased susceptibility against blast disease in rice	Chandran et al. (2019)
	lncRNA	Induction of immune response and enhanced resistance	Yu et al. (2020)
	miR444	Decreased susceptibility to rice blast through gene modulation	Li et al. (2019)
	miR393	Enhanced resistance through hormone signaling control	Arora et al. (2021)
Soybean	lncRNA	Potential discovery of new resistance pathways	SinghMohan & BhallaPrem (2022)
Tobacco	miR319	Reduced susceptibility to viral attack	Liu et al. (2020)
	miR168	Improved plant defense and decreased viral multiplication	Mengistu & Tenkegna (2021)
Tomato	miR482	Modulation of immune receptor expression leading to improved resistance	Jiang et al. (2018)
	siRNA	Improved health of plants as well as decreased viral load	Zhao et al. (2022)
	miR168	Increased antiviral defense through miRNA regulation	Li et al. (2023)
	siRNA	Reduced crop damage and nematode infection	Iqbal et al. (2020)
Wheat	miR159	Reduction in symptoms of Fusarium head blight	Jin et al. (2020)

plants and viruses. Through VIGS, there is an accumulation of siRNAs that are directed against the virus's DNA. Several crop species have shown success in reducing viral infections with this strategy (Baulcombe, 2015).

16.5.4 lncRNAs

A transcript a little longer than 200 nucleotides that are unable to assemble a complete protein is referred to as a lncRNA. The primary difference between lncRNAs and mRNAs is their apparent lack of coding capability. Like mRNAs, the majority of lncRNAs undergo 5′-end topping, and alternate splicing along with the insertion of 3′ tail in the form of poly A during RNA synthesis by RNA polymerase II. RNA Polymerase IV and RNA Polymerase V are two more polymerases that can transcribe plant lncRNAs. These lncRNAs are not as well defined as Pol II transcripts, and they vary structurally in that they do not have poly-A tails. Recent research indicates that lncRNAs support plant defense mechanisms (Budak et al., 2020). The experimental background behind using the lncRNAs in plants for defense is not properly understood. VIGS may investigate the roles of the lncRNAs in crop protection, opening a door to the further development of new disease-resistant cultivars (Wu et al., 2020).

16.5.5 HIGS

Apart from VIGS, some other RNAi approaches like HIGS have also been used but they have not gained as much fame as VIGS. They have gained some success stories but a lot of work still has to be done to make it as successful as VIGS. Their detailed mechanisms and pathways are poorly understood. From protecting themselves against invasive viruses, plants natively create a defense system that depends on the RNA suppression mechanism. To combat additional plant diseases, HIGS technology, HIGS expresses an RNAi design against certain genes that are indigenous to the infectious agent in its target plant, hence enabling the knockdown of transcripts in plant pathogens. Plants through their RNAi process produce small strands of RNA that through HIGS silence or downregulate the expression of the genes present in attacking insects that injure the plants (Figure 16.3b) (Huang et al., 2006). It has been seen in many transgenic plants that they produce small, tiny RNAs through their dsRNA. In recent research, *Agrobacterium* or different pathogens that multiply through the dsRNA are also able to introduce the dsRNA into the plant cells. This mechanism has been artificially engineered to decrease susceptibility against worms and fungal and bacterial diseases (Ghag, 2017).

16.5.6 Other RNAi Approaches

Other than VIGS and HIGS, there have also been some other RNAi mechanisms that safeguard the plants from different pathogens. The few other mechanisms through which certain pathogenic genes have been targeted are the use of siRNAs, SIGS (Figure 16.3c), and artificial microRNAs (amiRNAs) (Bilir et al., 2022). These different types of RNAi molecules that combat different pathogens and provide a sustainable approach for the survival of plants are applied through different techniques like foliar sprays, genetic transformation, and root soaking (Hoang et al., 2022).

The employment of various RNAi strategies including VIGS, HIGS, and amiRNA for crop improvement in various plant crops to battle various infections and diseases using respective genes is demonstrated in Table 16.2.

16.6 Sculpting Genetic Frontiers: CRISPR and Its Triumphs in Crop Disease Management

CRISPR is one of the latest technologies that revolutionized the concept of genome editing in the field of genetic engineering. It has an unparalleled precision in the genome editing. It has been employed in the management of crop diseases to engineer different plants that are resistant to specific diseases. This technology has carved out a path for sustainable agriculture by enabling the development of plants that are resistant to pathogens and also have a high yield (Rehman et al., 2022). CRISPR-Cas9, a widely used system, is a part of the bacterial adaptive immune system. Bacteria have a habit of storing fragments of DNA of bacteriophages in the form of an array, which allows bacteria to recognize and trigger an immune response against that specific bacteriophage. It

Table 16.2 Deployment of disease resistance in various crops through RNAi approaches.

Crop	Target genes	RNAi approach	Phenotype	References
Banana	*TR4*	HIGS	Decreased susceptibility to Fusarium wilt	Siamak & Zheng (2018)
Barley	*NHR, QR*	HIGS	Decreased susceptibility to powdery mildew	Douchkov et al. (2014)
Citrus	*TFIIAγ*	VIGS	Decreased susceptibility to citrus canker disease	Tang et al. (2021)
Cotton	*GhPGIP1*	HIGS	Enhanced resistance against wilt disease	Song et al. (2020)
	COI	VIGS	Reduced crop damage and Cotton bollworm colonization	Li et al. (2016)
Cucumber	*ZmIspH*	VIGS	Decreased spread and replication of Cucumber mosaic virus	Li et al. (2021)
Grape	*VvMLO*	SIGS	Improved plant physiology and controlled Powdery mildew stress	Amaro et al. (2023)
Grapes	*MrRPV1*	HIGS	Decreased susceptibility to downy mildew	Prajongjai et al. (2014)
Olives	*cox*	HIGS	Decreased susceptibility to wilt disease	Díaz-Rueda et al. (2021)
Pea	*DRR230*	HIGS	Improved crop yield and decreased disease severity by Fusarium	Williamson-Benavides et al. (2021)
Potato	*R1*	HIGS	Decreased susceptibility to late blight disease	Akino et al. (2014)
Rice	*Osa-miR167h-3p*	amiRNA	Increased resistance against bacterial blight	Lu et al. (2021)
	AVR	HIGS	Increase in yield and reduction in blast lesions	Devanna et al. (2022)
	R	HIGS	Reduced sheath blight infection by *Rhizoctonia solani*	Senapati et al. (2022)
Soybean	*mic-3*	amiRNA	Decreased damage and infection from Root-knot nematodes	Pan et al. (2019)
	PI567	HIGS	Reduction in loss of yield and symptoms of sudden death syndrome	Kolander et al. (2012)
Sunflower	*Ss-oah1*	VIGS	Increase in yield and decreased severity of disease by Sclerotinia	McCaghey et al. (2019)
Tomato	*NLR*	VIGS	Reduced pathogenicity and etiology of Tomato Spotted Wilt Virus (TSWV)	Qi et al. (2022)
Wheat	*Fhb1*	HIGS	Reduced pathogenicity and levels of toxins of Fusarium head blight	Dweba et al. (2017)
	Yr27	HIGS	Decreased susceptibility to yellow rust	Khanfri et al. (2018)
	Mi1.2	amiRNA	Reduced crop damage and nematode infection	Jones & Fosu-Nyarko (2014)

has two major components: gRNA and Cas9 protein. gRNA is a short nucleotide sequence that is designed as a specific strand to our target DNA sequence. Cas9 protein is an endonuclease enzyme that cuts at the DNA location guided by the gRNA. Once the cleavage has been done by Cas9, the plant's innate repair mechanism will either insert or delete the nucleotide sequence at the breakage site, which will produce a probability of getting mutations on the cleaved site while repairing (Mir et al., 2018). It is considered a very powerful tool because of its precision in editing the gene but has a limitation of off-target effects. Mainly, it accelerates the process of the development of disease-resistant crops, which is not that fast with traditional methods. Moreover, it can also be used to target or edit multiple genes at a time (Tyagi et al., 2020).

Various plant diseases due to viruses, bacteria, fungi, or other pathogens are primary contributors to substantial declines in agricultural yields. CRISPR provides a method for inserting resistant genes into plants or silencing vulnerable genes to cultivate crops that can endure diseases and infections. For instance, CRISPR has been used to target the *OsMPK5* gene in rice, conferring resistance to bacterial leaf blight (BLB) (Zhang et al., 2022). The *Pi21* gene has been modified to provide resistance to rice blast, a destructive fungal disease (Angeles-Shim et al., 2020). CRISPR-mediated modification of the *ZmFER1* gene in maize has conferred resistance to a fungal disease, ear rot (Liu et al., 2022). Targeting the *ZmJAZ15* gene has offered protection against Gibberella stalk rot (Ma et al., 2021). Similarly, wheat has also benefited from the targeting of the *TaPOD70* gene to improve resistance to powdery mildew, a prevalent fungal disease that significantly affects yields (Li et al., 2023). These examples demonstrate the significance of CRISPR technology in achieving disease-resistant crops that decrease loss of yield and lessen the need for chemical treatments, therefore mitigating the environmental effect of agriculture. Table 16.3 presents an extensive compilation of the crops, target genes, and corresponding phenotypes for disease resistance, emphasizing advancements in CRISPR-Cas technology to address specific plant diseases, which will reduce the use and environmental impact of chemical methods (Zaynab et al., 2020).

Table 16.3 Target genes and corresponding phenotypes for disease resistance in diverse crops through CRISPR-Cas genome-editing system.

Crop	Target genes	Phenotypes	References
Rice	*OsMPK5*	Resistance against bacterial leaf blight (BLB)	Zhang et al. (2022)
	Pi21	Resistance against rice blast	Angeles-Shim et al. (2020)
	OsSWEET11	Resistance against BLB	Roviqowati et al. (2024)
	osERF922	Resistance against blast	Mengchao et al. (2023)
	eIF4G	Resistance against Tungro virus	Zafirov et al. (2023)
	ALB1, RSY1	Resistance rice blast	Zhu et al. (2021)
	OsHPPO4	Resistance against root-knot nematode (RKN)	Huang et al. (2023)
Maize	*bW2, bE1*	Resistance against corn smut	Shi et al. (2017)
	ZmFER1	Resistance against ear rot	Liu et al. (2022)
	ZmLOX3	Resistance against RKN	Gao et al. (2008)
	ZmJAZ15	Resistance against Gibberella stalk rot (GSR)	Ma et al. (2021)

Table 16.3 (Continued)

Crop	Target genes	Phenotypes	References
Wheat	*TaPOD70*	Resistance against powdery mildew (PM)	Li et al. (2023)
	TaEDR1	Resistance against PM	Zhang et al. (2017)
	TaMlo	Resistance against PM	Demirer (2022)
	MLO-A1, *B1*, and *D1*	Resistance against PM	Ingvardsen et al. (2023)
	TaWRKY19	Resistance against yellow rust (YR)	Wang et al. (2022)
	TaPsIPK1	Resistance against YR	Liu et al. (2024)
	TaHRC	Resistance against Fusarium head blight (FHB)	Chen et al. (2022)
Barley	*MORC1*	Resistance against PM	Galli et al. (2022)
	MORC6a	Resistance against PM	
	PDIL5-1	Resistance against virus	Hoffie et al. (2023)
Grapevine	*VvMLO3*	Resistance against PM	Wan et al. (2020)
Papaya	*alEPIC8*	Resistance against root rot	Wang et al. (2019)
Citrus	*CsLOB1*	Resistance against citrus canker	Zou et al. (2021)
	CsWRKY22	Resistance against citrus canker	Long et al. (2021)
Banana	*eBSV*	Resistance against banana streak virus	Cote et al. (2010)
Apple	*DIPM-1*, *DIPM-2DIPM-4*	Resistance against fire blight	Malnoy et al. (2016)
	MdMLO19	Resistance against PM	Pessina et al. (2016)
Tomato	*SlJAZ7*	Resistance against late blight	Chenyu et al. (2024)
	SlTGA7	Resistance against late blight	
	SlJAZ2	Resistance against bacterial speck	Ortigosa et al. (2019)
Water melon	*Clpsk1*	Resistance against Fusarium	Zhang et al. (2020)
Soybean	*Avr4/6*	Disease resistance seed decay/damping off	Zhang et al. (2020)
	NPR3	Resistance against seed decay/damping off	Zhang et al. (2020)
	GmMLO	Resistance against PM	Bui et al. (2023)
	GmTAP1	Resistance against seed decay/damping off	Liu et al. (2023)
	α-SNAP	Resistance against SCN	Usovsky et al. (2023)
Potato	*StDMR6-1*	Resistance against late blight	Karlsson et al. (2024)
Rapeseed	*BnQCR8*	Broad spectrum resistance	Zhang et al. (2021)

Recent advances in the delivery of CRISPR-Cas system through viral vectors and grafting open new avenues to acquire heritable mutagenesis in plants that are tissue culture recalcitrant (Awan et al., 2023a, b; Liu et al., 2023; Yang et al., 2023). The augmentation of mobile elements with transcripts of sgRNA facilitates their trafficking toward flowering tissues, eventually allowing the development of seeds with heritable mutations. The use of viral vectors, such as adenoviruses, lentiviruses, and adeno-associated viruses (AAVs), as effective vehicles for the introduction of genome-editing tools like CRISPR-Cas9, ZFNs, and TALENs into cells. Adenoviruses, known for their capacity to transport substantial genetic payloads, provide transitory production of editing tools but can cause robust immune responses. Lentiviruses incorporate into the host genome, facilitating stable, long-term expression, but with the danger of insertional mutagenesis. AAVs provide a safer option owing to their minimal immunogenicity and nonintegrating characteristics, although they are constrained by their reduced carrying capacity. Although viral vectors are potent instruments for genome editing, several hurdles must be resolved for secure clinical use. This encompasses possible immunological responses, particularly with adenoviruses, as well as the possibility of off-target consequences. The use of vectors such as lentiviruses raises concerns over inadvertent alterations in the DNA. Notwithstanding these obstacles, viral vectors are being engineered for therapeutic use in conditions such as muscular dystrophy and cancer. Current research seeks to enhance their accuracy, efficacy, and safety, positioning them as viable options for the advancement of gene therapy and precision medicine (Awan et al., 2023a).

Grafting is a traditional method that entails joining two plants to develop as one plant. In this context, the researchers employ it as a technique to transmit certain genetic features from one region of the plant (the donor) to another (the receptor) without permanently modifying the recipient's genome by direct genetic alteration. The research emphasizes the capability of this technology to generate genome-edited plants devoid of the regulatory and public apprehensions linked to transgenic species. This approach uses targeted gene-editing technologies like CRISPR to induce modifications in the donor plant's tissues, thereby affecting the receptor at the graft junction. The proposed technology may significantly impact agriculture by enabling the cultivation of plants with advantageous characteristics such as enhanced disease resistance, drought resilience, or increased yield, circumventing the ethical and legal complexities associated with genetically modified organisms (GMOs) (Awan et al., 2023b). The absence of foreign DNA integration into the recipient plant's genome allows the resultant plants to potentially circumvent GMO restrictions in several countries worldwide. This discovery facilitates novel approaches to crop enhancement while alleviating public concerns around GMOs, possibly revolutionizing agricultural biotechnology. These genotype-independent systems confer reproducible alternatives to laborious tissue culture techniques (Awan et al., 2023b; Biswas et al., 2022).

16.7 Conclusion

For RNAi, predesigned reagents are promptly accessible and they are very effective for deciphering the impact of fundamental genes on phenotypes and are notably essential in experiments that require loss of function studies. Disadvantages include the potential obligation for a narrow window assay because of temporary gene disruption, off-target effects, the possibility that insufficient silencing may not produce a strong signal, meticulous analysis for confirmation of silencing, and silencing of multiple targets. Contrastingly, CRISPR endeavors precise genome editing with minimum off-target effects, lower hazard of immune response, and permanent gene disruption or knockout. Limitations of CRISPR include challenges in studying functionally essential genes, mosaicism and incomplete editing, potential immunogenic responses, and the need for laborious verification of indels (Figure 16.4).

As CRISPR technology continues to evolve, its capacity to control pathogens affecting crops is expanding. Researchers are now utilizing various Cas enzymes, such as Cas12 and Cas13, which offer enhanced editing capabilities and the ability to target RNA. Targeting RNA can significantly improve control over viral diseases in different crops. Additionally, CRISPR is being integrated with technologies like high-throughput phenotyping to accelerate the identification of gene edits.

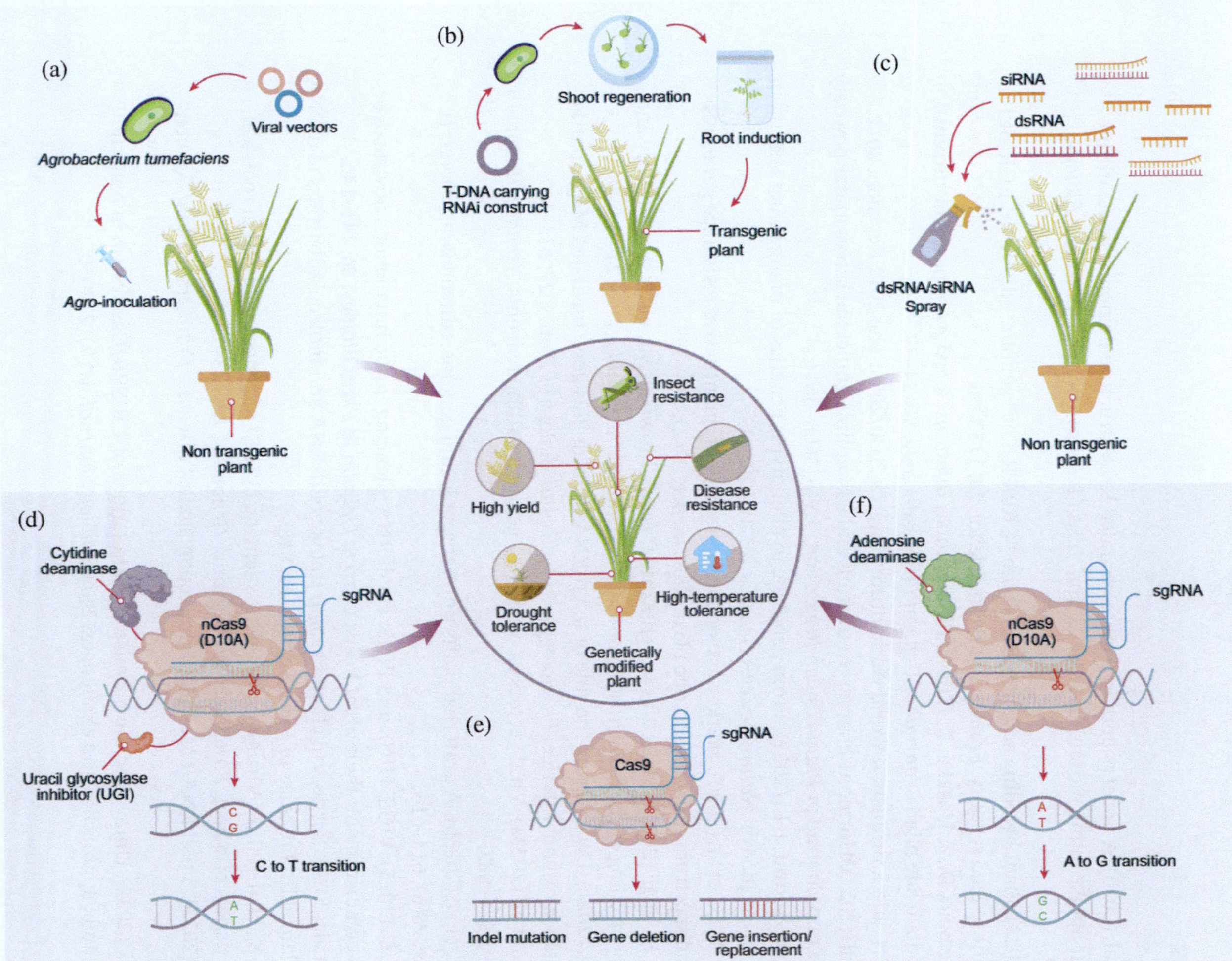

Figure 16.4 Techniques for gene silencing and gene editing in crop improvement. (a) VIGS: Agro-inoculation or viral vectors facilitate gene silencing, conferring resistance to pests and diseases. (b) HIGS: A transgenic method utilizing T-DNA constructs including RNAi sequences that imparts resistance to pests and diseases, resulting in improved resilience. (c) SIGS: Nontransgenic plants are subjected to externally administered dsRNA/siRNA molecules, conferring resistance without genetic alteration. (d) Cytidine base editing (CBE): Cytidine deaminase is combined with an engineered nCas9 (D10A) to transform a C•G base pair into T•A at a specific locus, enabling accurate genetic modifications, including improved agricultural characteristics. (e) Traditional CRISPR-Cas9 genome editing: Cas9 creates DSBs in plant DNA, which are repaired by either NHEJ or HR, resulting in gene changes. (f) Adenine base editing (ABE): Adenosine deaminase is conjugated to nCas9 (D10A) to transform an A•T base pair into G•C, allowing accurate base editing in crops to enhance characteristics such as yield, disease resistance, and abiotic stress tolerance.

Overall, CRISPR has the potential to transform agriculture by ensuring food security in a world facing a growing population and changing climate conditions (Yan et al., 2019; Ahmad, 2023). While CRISPR is an emerging technique that offers enormous promise, there are still challenges that need to be overcome to enhance its efficiency. These challenges include off-target effects, regulatory hurdles, and public perception. A primary problem resides in the regulatory frameworks that regulate transgenic and genome-edited crops, which differ across different regions. Some nations impose flexible laws on genome-edited plants, while others classify them as GMOs, hence enforcing stricter regulations. The lack of worldwide regulatory uniformity, along with public acceptability, is a substantial obstacle to the extensive use of CRISPR-edited crops (Jones, 2015). However, various scientists claim that they will address these issues in the coming years by developing practical solutions (Biswas et al., 2022).

References

Abbasi, R., et al. (2020). A novel paperclip double-stranded RNA structure demonstrates clathrin-independent uptake in the mosquito *Aedes aegypti*. *Insect Biochemistry and Molecular Biology*, *127*, 103492.

Ahmad, M. (2023). Plant breeding advancements with "CRISPR-Cas" genome editing technologies will assist future food security. *Frontiers in Plant Science*, *14*, 1133036.

Akino, S., Takemoto, D., & Hosaka, K. (2014). *Phytophthora infestans*: a review of past and current studies on potato late blight. *Journal of General Plant Pathology*, *80*, 24–37.

Alexandratos, N. and J. Bruinsma, World Agriculture Towards 2030/2050: The 2012 Revision. 2012.

Amaro, R., et al. (2023). Hormone changes in tolerant and susceptible grapevine leaves under powdery mildew infection. *Journal of Plant Growth Regulation*, *42*(6), 3606–3614.

Andergassen, D., & Rinn, J. L. (2022). From genotype to phenotype: genetics of mammalian long non-coding RNAs in vivo. *Nature Reviews Genetics*, *23*(4), 229–243.

Angeles-Shim, R. B., et al. (2020). Marker-assisted introgression of quantitative resistance gene pi21 confers broad spectrum resistance to rice blast. *Rice Science*, *27*(2), 113–123.

Anjanappa, R. B., et al. (2018). Molecular insights into Cassava brown streak virus susceptibility and resistance by profiling of the early host response. *Molecular Plant Pathology*, *19*(2), 476–489.

Arora, K., et al. (2021). Deciphering the role of microRNAs during Pi54 gene mediated Magnaporthe oryzae resistance response in rice. *Physiology and Molecular Biology of Plants*, *27*, 633–647.

Awan, M. J. A., et al. (2023a). Viral vectors as carriers of genome-editing reagents. *Trends in Plant Science*, *28*(9), 981–983.

Awan, M. J. A., et al. (2023b). A graft that crafts nontransgenic and genome-edited plants. *Trends in Plant Science*, *28*(6), 614–616.

Baulcombe, D. C. (2015). VIGS, HIGS and FIGS: small RNA silencing in the interactions of viruses or filamentous organisms with their plant hosts. *Current Opinion in Plant Biology*, *26*, 141–146.

Bennett, M., et al. (2020). Barriers to efficient foliar uptake of dsRNA and molecular barriers to dsRNA activity in plant cells. *Frontiers in Plant Science*, *11*, 816.

Bennypaul, H. S., et al. (2012). Virus-induced gene silencing (VIGS) of genes expressed in root, leaf, and meiotic tissues of wheat. *Functional & Integrative Genomics*, *12*, 143–156.

Bilir, Ö., et al. (2022). Small RNA-based plant protection against diseases. *Frontiers in Plant Science*, *13*, 951097.

Biswas, P., et al. (2022). Unraveling the promise and limitations of CRISPR/Cas system in natural product research: approaches and challenges. *Biotechnology Journal*, *17*(7), 2100507.

Brant, E. J., & Budak, H. (2018). Plant small non-coding RNAs and their roles in biotic stresses. *Frontiers in Plant Science, 9*, 1038.

Budak, H., Kaya, S. B., & Cagirici, H. B. (2020). Long non-coding RNA in plants in the era of reference sequences. *Frontiers in Plant Science, 11*, 276.

Bui, T. P., et al. (2023). Enhancing powdery mildew resistance in soybean by targeted mutation of MLO genes using the CRISPR/Cas9 system. *BMC Plant Biology, 23*(1), 533.

Burmistrz, M., Krakowski, K., & Krawczyk-Balska, A. (2020). RNA-targeting CRISPR-Cas systems and their applications. *International Journal of Molecular Sciences, 21*(3), 1122.

Cagliari, D., et al. (2019). Management of pest insects and plant diseases by non-transformative RNAi. *Frontiers in Plant Science, 10*, 1319.

Carusillo, A., & Mussolino, C. (2020). DNA damage: from threat to treatment. *Cells, 9*(7), 1665.

Chandran, V., et al. (2019). miR396-OsGRFs module balances growth and rice blast disease-resistance. *Frontiers in Plant Science, 9*, 1999.

Charania, I., & Li, X. (2020). Smart farming: agriculture's shift from a labor intensive to technology native industry. *Internet of Things, 9*, 100142.

Chen, H., et al. (2022). TaHRC suppresses the calcium-mediated immune response and triggers wheat Fusarium head blight susceptibility. *Plant Physiology, 190*(3), 1566–1569.

Chenyu, L., et al. (2024). Disease resistance of SlJAZ7 and its interaction with SlTGA7 in tomato. *Journal of South China Agricultural University*, 525–534.

Choi, J. Y., & Lee, Y. C. G. (2020). Double-edged sword: the evolutionary consequences of the epigenetic silencing of transposable elements. *PLoS Genetics, 16*(7), e1008872.

Christiaens, O., Niu, J., & Taning, C. N. T. (2020). RNAi in insects: a revolution in fundamental research and pest control applications. *Insects*, 415.

Cote, F. X., et al. (2010). Micropropagation by tissue culture triggers differential expression of infectious endogenous Banana streak virus sequences (eBSV) present in the B genome of natural and synthetic interspecific banana plantains. *Molecular Plant Pathology, 11*(1), 137–144.

Dalakouras, A., & Papadopoulou, K. K. (2020). Epigenetic modifications: an unexplored facet of exogenous RNA application in plants. *Plants, 9*(6), 673.

Das, P. R., & Sherif, S. M. (2020). Application of exogenous dsRNAs-induced RNAi in agriculture: challenges and triumphs. *Frontiers in Plant Science, 11*, 946.

de Janvry, A., & Sadoulet, E. (2020). Using agriculture for development: supply- and demand-side approaches. *World Development, 133*, 105003.

Demirer, G. S. (2022). Wonder wheat: a disease-resistant crop without growth tradeoff. *GEN Biotechnology, 1*(2), 133–134.

Devanna, B. N., et al. (2022). Understanding the dynamics of blast resistance in rice-Magnaporthe oryzae interactions. *Journal of Fungi, 8*(6), 584.

Dexheimer, P. J., & Cochella, L. (2020). MicroRNAs: from mechanism to organism. *Frontiers in Cell and Developmental Biology, 8*, 409.

Díaz-Rueda, P., et al. (2021). Wild olive genotypes as a valuable source of resistance to defoliating *Verticillium dahliae*. *Frontiers in Plant Science, 12*, 662060.

Douchkov, D., et al. (2014). Discovery of genes affecting resistance of barley to adapted and non-adapted powdery mildew fungi. *Genome Biology, 15*, 1–18.

Dubrovina, A. S., & Kiselev, K. V. (2019). Exogenous RNAs for gene regulation and plant resistance. *International Journal of Molecular Sciences, 20*(9), 2282.

Dubrovna, O., Mykhalska, S., & Komisarenko, A. (2023). Use of RNA interference technology for improving economically valuable traits of cereal crops. *Cytology and Genetics, 57*(6), 587–610.

Dweba, C., et al. (2017). Fusarium head blight of wheat: pathogenesis and control strategies. *Crop Protection*, *91*, 114–122.

Ergin, K., & Çetinkaya, R. (2022). Regulation of microRNAs. miRNomics: MicroRNA Biology and Computational. *Analysis*, 1–32.

Fire, A., et al. (1998). Potent and specific genetic interference by double-stranded RNA in *Caenorhabditis elegans*. *Nature*, *391*(6669), 806–811.

Fisher, M. C., et al. (2012). Emerging fungal threats to animal, plant and ecosystem health. *Nature*, *484*(7393), 186–194.

Flynt, A. S. (2021). Insecticidal RNA interference, thinking beyond long dsRNA. *Pest Management Science*, *77*(5), 2179–2187.

Galli, M., et al. (2022). CRISPR/Sp Cas9-mediated double knockout of barley Microrchidia MORC1 and MORC6a reveals their strong involvement in plant immunity, transcriptional gene silencing and plant growth. *Plant Biotechnology Journal*, *20*(1), 89–102.

Gao, X., et al. (2008). Maize 9-lipoxygenase ZmLOX3 controls development, root-specific expression of defense genes, and resistance to root-knot nematodes. *Molecular Plant-Microbe Interactions*, *21*(1), 98–109.

Gebremichael, D. E., et al. (2021). RNA interference strategies for future management of plant pathogenic fungi: prospects and challenges. *Plants*, *10*(4), 650.

Ghag, S. B. (2017). Host induced gene silencing, an emerging science to engineer crop resistance against harmful plant pathogens. *Physiological and Molecular Plant Pathology*, *100*, 242–254.

Godfray, H. C., et al. (2010). Food security: the challenge of feeding 9 billion people. *Science*, *327*(5967), 812–818.

Graca, F. A., et al. (2021). A large-scale transgenic RNAi screen identifies transcription factors that modulate myofiber size in Drosophila. *PLoS Genetics*, *17*(11), e1009926.

Gu, S., et al. (2021). Base editors: expanding the types of DNA damage products harnessed for genome editing. *Gene and Genome Editing*, *1*, 100005.

Halder, K., et al. (2022). RNA interference for improving disease resistance in plants and its relevance in this clustered regularly interspaced short palindromic repeats-dominated era in terms of dsRNA-based biopesticides. *Frontiers in Plant Science*, *13*, 885128.

He, L., Huang, Y., & Tang, X. (2022). RNAi-based pest control: production, application and the fate of dsRNA. *Frontiers in Bioengineering and Biotechnology*, *10*, 1080576.

Hentzschel, F., et al. (2020). Gene knockdown in malaria parasites via non-canonical RNAi. *Nucleic Acids Research*, *48*(1), e2.

Hill, A. C., et al. (2022). Chemically synthesized, self-assembling small interfering RNA-prohead RNA molecules trigger Dicer-independent gene silencing. *Chemistry A European Journal*, *28*(8), e202103995.

Hoang, B. T. L., et al. (2022). RNAi as a foliar spray: efficiency and challenges to field applications. *International Journal of Molecular Sciences*, *23*(12), 6639.

Hoffie, R. E., et al. (2023). Novel resistance to the Bymovirus BaMMV established by targeted mutagenesis of the PDIL5-1 susceptibility gene in barley. *Plant Biotechnology Journal*, *21*(2), 331–341.

Huang, G., et al. (2006). Engineering broad root-knot resistance in transgenic plants by RNAi silencing of a conserved and essential root-knot nematode parasitism gene. *Proceedings of the National Academy of Sciences*, *103*(39), 14302–14306.

Huang, X., Chen, Z., & Liu, Y. (2020). RNAi-mediated control of CRISPR functions. *Theranostics*, *10*(15), 6661.

Huang, Q., et al. (2023). CRISPR/Cas9-mediated mutagenesis of the susceptibility gene OsHPP04 in rice confers enhanced resistance to rice root-knot nematode. *Frontiers in Plant Science*, *14*, 1134653.

Hung, Y.-H., & Slotkin, R. K. (2021). The initiation of RNA interference (RNAi) in plants. *Current Opinion in Plant Biology*, *61*, 102014.

Hynes, C., & Kakumani, P. K. (2024). Regulatory role of RNA-binding proteins in microRNA biogenesis. *Frontiers in Molecular Biosciences*, *11*, 1374843.

Ilyas, H., et al. (2024). Commands of synthetic biology to modernize and re-design the biological systems. *Advancements in Life Sciences*, *11*(3), 549–557.

Ingvardsen, C. R., et al. (2023). Highly effective MLO-based powdery mildew resistance in hexaploid wheat without pleiotropic effects. *Plant Science*, *335*, 111785.

Iqbal, S., Fosu-Nyarko, J., & Jones, M. G. (2020). Attempt to silence genes of the RNAi pathways of the root-knot nematode, *Meloidogyne incognita* results in diverse responses including increase and no change in expression of some genes. *Frontiers in Plant Science*, *11*, 328.

Iwakawa, H.-O., & Tomari, Y. (2022). Life of RISC: formation, action, and degradation of RNA-induced silencing complex. *Molecular Cell*, *82*(1), 30–43.

Jagadeeswaran, G., Saini, A., & Sunkar, R. (2009). Biotic and abiotic stress down-regulate miR398 expression in Arabidopsis. *Planta*, *229*, 1009–1014.

Jaganathan, D., et al. (2018). CRISPR for crop improvement: an update review. *Frontiers in Plant Science*, *9*, 985.

Jansing, J., et al. (2019). Genome editing in agriculture: technical and practical considerations. *International Journal of Molecular Sciences*, *20*(12), 2888.

Jiang, N., et al. (2018). Function identification of miR482b, a negative regulator during tomato resistance to *Phytophthora infestans*. *Horticulture Research*, 5.

Jin, X., et al. (2020). Identification of *Fusarium graminearum*-responsive miRNAs and their targets in wheat by sRNA sequencing and degradome analysis. *Functional & Integrative Genomics*, *20*, 51–61.

Jones, H. D. (2015). Regulatory uncertainty over genome editing. *Nature Plants*, *1*(1), 14011.

Jones, M., & Fosu-Nyarko, J. (2014). Molecular biology of root lesion nematodes (*Pratylenchus* spp.) and their interaction with host plants. *Annals of Applied Biology*, *164*(2), 163–181.

Karlsson, M., et al. (2024). CRISPR/Cas9 genome editing of potato StDMR6-1 results in plants less affected by different stress conditions. *Horticulture Research*, *11*(7), 1–10.

Kaushal, A. (2023). Innate immune regulations and various siRNA modalities. *Drug Delivery and Translational Research*, *13*(11), 2704–2718.

Khanfri, S., Boulif, M., & Lahlali, R. (2018). Yellow rust (*Puccinia striiformis*): a serious threat to wheat production worldwide. *Notulae Scientia Biologicae*, *10*(3), 410–423.

Kirstein, N., et al. (2023). The integrator complex regulates microRNA abundance through RISC loading. *Science Advances*, *9*(6), eadf0597.

Koch, A., & Wassenegger, M. (2021). Host-induced gene silencing – mechanisms and applications. *The New Phytologist*, *231*(1), 54–59.

Kolander, T., et al. (2012). Symptomatic and asymptomatic host range of Fusarium virguliforme, the causal agent of soybean sudden death syndrome. *Plant Disease*, *96*(8), 1148–1153.

Kumar, S., et al. (2020). Nanovehicles for plant modifications towards pest-and disease-resistance traits. *Trends in Plant Science*, *25*(2), 198–212.

Kundariya, H., et al. (2020). MSH1-induced heritable enhanced growth vigor through grafting is associated with the RdDM pathway in plants. *Nature Communications*, *11*(1), 5343.

Kung, Y.-J., et al. (2014). Genetic analyses of the FRNK motif function of Turnip mosaic virus uncover multiple and potentially interactive pathways of cross-protection. *Molecular Plant-Microbe Interactions*, *27*(9), 944–955.

Kutter, C., & Svoboda, P. (2008). *miRNA, siRNA, piRNA: Knowns of the Unknown*, Taylor & Francis.

Lange, M., et al. (2013). Virus-induced gene silencing (VIGS) in plants: an overview of target species and the virus-derived vector systems. *Methods in Molecular Biology*, 1–14.

Lax, C., et al. (2020). The evolutionary significance of RNAi in the fungal kingdom. *International Journal of Molecular Sciences*, *21*(24), 9348.

Lee, R. C., Feinbaum, R. L., & Ambros, V. (1993). The *C. elegans* heterochronic gene lin-4 encodes small RNAs with antisense complementarity to lin-14. *Cell*, *75*(5), 843–854.

Lee, M. H., et al. (2017). An Arabidopsis NAC transcription factor NAC4 promotes pathogen-induced cell death under negative regulation by microRNA164. *New Phytologist*, *214*(1), 343–360.

Leng, Q., Chen, L., & Lv, Y. (2020). RNA-based scaffolds for bone regeneration: application and mechanisms of mRNA, miRNA and siRNA. *Theranostics*, *10*(7), 3190.

León-Ruiz, J., et al. (2023). RETINOBLASTOMA-RELATED interactions with key factors of the RNA-directed DNA methylation (RdDM) pathway and its influence on root development. *Planta*, *257*(6), 105.

Lesk, C., Rowhani, P., & Ramankutty, N. (2016). Influence of extreme weather disasters on global crop production. *Nature*, *529*(7584), 84–87.

Li, J., et al. (2016). Transcriptome analysis reveals a comprehensive insect resistance response mechanism in cotton to infestation by the phloem feeding insect *Bemisia tabaci* (whitefly). *Plant Biotechnology Journal*, *14*(10), 1956–1975.

Li, Y., et al. (2019). The roles of rice microRNAs in rice-Magnaporthe oryzae interaction. *Phytopathology Research*, *1*(1), 1–12.

Li, H., et al. (2021). Efficient and high-throughput pseudorecombinant-chimeric Cucumber mosaic virus-based VIGS in maize. *Plant Physiology*, *187*(4), 2865–2876.

Li, Q., et al. (2023). miRNAs and lncRNAs in tomato: roles in biotic and abiotic stress responses. *Frontiers in Plant Science*, *13*, 1094459.

Li, R., et al. (2023). Wheat Class III Peroxidase TaPOD70 is a potential susceptibility factor negatively regulating wheat resistance to *Blumeria graminis* f. sp. *tritici*. *Phytopathology®*, *113*(5), 873–883.

List, F., et al. (2022). RNA meets toxicology: efficacy indicators from the experimental design of RNAi studies for insect pest management. *Pest Management Science*, *78*(8), 3215–3225.

Liu, Y., Schiff, M., & Dinesh-Kumar, S. P. (2002). Virus-induced gene silencing in tomato. *The Plant Journal*, *31*(6), 777–786.

Liu, J., et al. (2020). Genome-wide microRNA profiling using oligonucleotide microarray reveals regulatory networks of microRNAs in *Nicotiana benthamiana* during Beet necrotic yellow vein virus infection. *Viruses*, *12*(3), 310.

Liu, H., et al. (2022). CRISPR/Cas9 technology and its utility for crop improvement. *International Journal of Molecular Sciences*, *23*(18), 1–21.

Liu, W., et al. (2022). The mechanisms of siRNA selection by plant Argonaute proteins triggering DNA methylation. *Nucleic Acids Research*, *50*(22), 12997–13010.

Liu, C., et al. (2022). Engineering null mutants in ZmFER1 confers resistance to ear rot caused by *Fusarium verticillioides* in maize. *Plant Biotechnology Journal*, *20*(11), 2045.

Liu, T., et al. (2023). CRISPR/Cas9-mediated editing of GmTAP1 confers enhanced resistance to *Phytophthora sojae* in soybean. *Journal of Integrative Plant Biology*, *65*(7), 1609–1612.

Liu, Q., et al. (2023). Engineered biocontainable RNA virus vectors for non-transgenic genome editing across crop species and genotypes. *Molecular Plant*, *16*(3), 616–631.

Liu, S., et al. (2024). CRISPR-targeted mutagenesis of mitogen-activated protein kinase phosphatase 1 improves both immunity and yield in wheat. *Plant Biotechnology Journal*, 1929–1941.

Long, Q., et al. (2021). Transcription factor WRKY22 regulates canker susceptibility in sweet orange (*Citrus sinensis* Osbeck) by enhancing cell enlargement and CsLOB1 expression. *Horticulture Research*, 8.

Loriato, V. A., et al. (2020). Engineering resistance against geminiviruses: a review of suppressed natural defenses and the use of RNAi and the CRISPR/Cas system. *Plant Science*, *292*, 110410.

Lu, J., et al. (2021). Comparative microRNA profiling reveals microRNAs involved in rice resistant response to bacterial blight. *The Crop Journal*, *9*(4), 834–842.

Ma, L., et al. (2021). Genome-wide characterization of jasmonates signaling components reveals the essential role of ZmCOI1a-ZmJAZ15 action module in regulating maize immunity to gibberella stalk rot. *International Journal of Molecular Sciences*, *22*(2), 870.

Malnoy, M., et al. (2016). DNA-free genetically edited grapevine and apple protoplast using CRISPR/Cas9 ribonucleoproteins. *Frontiers in Plant Science*, *7*, 1904.

McCaghey, M., et al. (2019). The complexity of the *Sclerotinia sclerotiorum* pathosystem in soybean: virulence factors, resistance mechanisms, and their exploitation to control Sclerotinia stem rot. *Tropical Plant Pathology*, *44*, 12–22.

McGeary, S. E., et al. (2022). MicroRNA 3′-compensatory pairing occurs through two binding modes, with affinity shaped by nucleotide identity and position. *eLife*, *11*, e69803.

Mengchao, Q., et al. (2023). Transcriptome analysis of oserf922 mutants reveals new insights into rice blast resistance. *Rice Science*, *30*(5), 374.

Mengistu, A. A., & Tenkegna, T. A. (2021). The role of miRNA in plant-virus interaction: a review. *Molecular Biology Reports*, *48*(3), 2853–2861.

Mir, A., et al. (2018). Type II-C CRISPR-Cas9 biology, mechanism, and application. *ACS Chemical Biology*, *13*(2), 357–365.

Napoli, C., Lemieux, C., & Jorgensen, R. (1990). Introduction of a chimeric chalcone synthase gene into petunia results in reversible co-suppression of homologous genes in trans. *The Plant Cell*, *2*(4), 279–289.

Nejat, N., & Mantri, N. (2018). Emerging roles of long non-coding RNAs in plant response to biotic and abiotic stresses. *Critical Reviews in Biotechnology*, *38*(1), 93–105.

Nerkar, G., et al. (2022). Advances in crop breeding through precision genome editing. *Frontiers in Genetics*, *13*, 880195.

Neumeier, J., & Meister, G. (2021). siRNA specificity: RNAi mechanisms and strategies to reduce off-target effects. *Frontiers in Plant Science*, *11*, 526455.

Nitnavare, R. B., et al. (2021). Next generation dsRNA-based insect control: success so far and challenges. *Frontiers in Plant Science*, *12*, 673576.

Ortigosa, A., et al. (2019). Design of a bacterial speck resistant tomato by CRISPR/Cas9-mediated editing of Sl JAZ 2. *Plant Biotechnology Journal*, *17*(3), 665–673.

Pan, X., et al. (2019). MicroRNA-target gene responses to root knot nematode (*Meloidogyne incognita*) infection in cotton (Gossypium hirsutum L.). *Genomics*, *111*(3), 383–390.

Park, J., et al. (2021). Enhanced genome editing efficiency of CRISPR PLUS: Cas9 chimeric fusion proteins. *Scientific Reports*, *11*(1), 16199.

Paturi, S., & Deshmukh, M. V. (2021). A glimpse of "Dicer Biology" through the structural and functional perspective. *Frontiers in Molecular Biosciences*, *8*, 643657.

Pessina, S., et al. (2016). The knock-down of the expression of MdMLO19 reduces susceptibility to powdery mildew (*Podosphaera leucotricha*) in apple (*Malus domestica*). *Plant Biotechnology Journal*, *14*(10), 2033–2044.

Prajongjai, T., et al. (2014). Evaluation of grapevines for resistance to downy mildew (*Plasmopara viticola*) under laboratory and field conditions. *South African Journal of Enology and Viticulture*, *35*(1), 43–50.

Punetha, A., et al., Cutting It Right: Plasticity and Strategy of CRISPR RNA Specific Nucleases. 2018.

Qi, S., et al. (2022). A new NLR gene for resistance to tomato spotted wilt virus in tomato (*Solanum lycopersicum*). *Theoretical and Applied Genetics*, *135*(5), 1493–1509.

Qiao, L., et al. (2021). Spray-induced gene silencing for disease control is dependent on the efficiency of pathogen RNA uptake. *Plant Biotechnology Journal, 19*(9), 1756–1768.

Rajput, M., et al. (2021). RNA interference and CRISPR/Cas gene editing for crop improvement: paradigm shift towards sustainable agriculture. *Plants (Basel), 10*(9), 1–31.

Ray, D. K., et al. (2013). Yield trends are insufficient to double global crop production by 2050. *PLoS One, 8*(6), e66428.

Rehman, R. S., et al. (2022). CRISPR-Cas mediated genome editing: a paradigm shift towards sustainable agriculture and biotechnology. *Asian Plant Research Journal, 9*(1), 27–49.

Roviqowati, F., et al. (2024). Introduction of CRISPR/Cas9 with the target genes to improve agronomic traits and leaf blight resistance in rice. *SABRAO Journal of Breeding and Genetics, 56*(1), 29–44.

Savary, S., et al. (2019). The global burden of pathogens and pests on major food crops. *Nature Ecology & Evolution, 3*(3), 430–439.

Senapati, M., et al. (2022). *Rhizoctonia solani* Kühn pathophysiology: status and prospects of sheath blight disease management in rice. *Frontiers in Plant Science, 13*, 881116.

Shamshirgaran, Y., et al. (2022). Tools for efficient genome editing; ZFN, TALEN, and CRISPR. *Applications of Genome Modulation and Editing*, 29–46.

Sharma, V., et al. (2013). RNA interference: a novel tool for plant disease management. *African Journal of Biotechnology, 12*(18), 1–20.

Sharma, S., Kooner, R., & Arora, R. (2017). Insect pests and crop losses. In: *Breeding Insect Resistant Crops for Sustainable Agriculture*, 45–66.

Shi, J., et al. (2017). ARGOS 8 variants generated by CRISPR-Cas9 improve maize grain yield under field drought stress conditions. *Plant Biotechnology Journal, 15*(2), 207–216.

Shi, G., et al. (2021). A methodological advance of tobacco rattle virus-induced gene silencing for functional genomics in plants. *Frontiers in Plant Science, 12*, 671091.

Siamak, S. B., & Zheng, S. (2018). Banana Fusarium wilt (*Fusarium oxysporum* f. sp. *cubense*) control and resistance, in the context of developing wilt-resistant bananas within sustainable production systems. *Horticultural Plant Journal, 4*(5), 208–218.

SinghMohan, B., & BhallaPrem, L. (2022). Biological parts for engineering abiotic stress tolerance in plants. *Biodesign Research.*

Sioud, M. (2021). RNA interference: story and mechanisms. In: *Design and Delivery of SiRNA Therapeutics*, 1–15.

Song, R., et al. (2020). An overview of the molecular genetics of plant resistance to the Verticillium wilt pathogen *Verticillium dahliae. International Journal of Molecular Sciences, 21*(3), 1120.

Swapnil, Singh, D., Ekka, J. P., & Kumari, P. (2020). Somatic hybridization: an effective tool for crop improvement. *Food and Scientific Reports, 1*(7), 21–23.

Tang, Y., et al. (2010). Virus-based MicroRNA expression for gene functional analysis in plants. *Plant Physiology, 153*(2), 632–641.

Tang, Y., Lai, Y., & Liu, Y. (2013). Virus-induced gene silencing using artificial miRNAs in *Nicotiana benthamiana*. *Methods in Molecular Biology*, 99–107.

Tang, X., et al. (2021). Natural variations of TFIIAγ gene and LOB1 promoter contribute to citrus canker disease resistance in *Atalantia buxifolia*. *PLoS Genetics, 17*(1), e1009316.

Tsai, H.-C., et al. (2022). Current strategies employed in the manipulation of gene expression for clinical purposes. *Journal of Translational Medicine, 20*(1), 535.

Tyagi, S., et al. (2020). CRISPR-Cas9 system: a genome-editing tool with endless possibilities. *Journal of Biotechnology, 319*, 36–53.

Tyurin, A. A., et al. (2020). Transient gene expression is an effective experimental tool for the research into the fine mechanisms of plant gene function: advantages, limitations, and solutions. *Plants, 9*(9), 1187.

Ugolini, I., & Halic, M. (2018). Fidelity in RNA-based recognition of transposable elements. *Philosophical Transactions of the Royal Society, B: Biological Sciences, 373*(1762), 20180168.

Usovsky, M., et al. (2023). Loss-of-function of an α-SNAP gene confers resistance to soybean cyst nematode. *Nature Communications, 14*(1), 7629.

Vologodskii, A. (2022). Biological defense I: protection from foreign nucleic acids. In: *The Basics of Molecular Biology*, Springer, 131–136.

Wan, D.-Y., et al. (2020). CRISPR/Cas9-mediated mutagenesis of VvMLO3 results in enhanced resistance to powdery mildew in grapevine (*Vitis vinifera*). *Horticulture Research*, 7.

Wang, G., & Li, J. (2021). Review, analysis, and optimization of the CRISPR *Streptococcus pyogenes* Cas9 system. *Medicine in Drug Discovery, 9*, 100080.

Wang, T., Zhang, H., & Zhu, H. (2019). CRISPR technology is revolutionizing the improvement of tomato and other fruit crops. *Horticulture Research*, 6.

Wang, J., et al. (2021). Engineering a PAM-flexible SpdCas9 variant as a universal gene repressor. *Nature Communications, 12*(1), 6916.

Wang, Q., et al. (2021). Mechanism of siRNA production by a plant Dicer-RNA complex in dicing-competent conformation. *Science, 374*(6571), 1152–1157.

Wang, N., et al. (2022). Inactivation of a wheat protein kinase gene confers broad-spectrum resistance to rust fungi. *Cell, 185*(16). 2961–2974.e19

Williamson-Benavides, B. A., et al. (2021). Identification of root rot resistance QTLs in pea using *Fusarium solani* f. sp. pisi-responsive differentially expressed genes. *Frontiers in Genetics, 12*, 629267.

Wu, L., et al. (2020). Research progress on plant long non-coding RNA. *Plants, 9*(4), 408.

Xie, G., et al. (2024). Molecular mechanisms of the RNA polymerases in plant RNA-directed DNA methylation. *Trends in Biochemical Sciences.*

Yan, F., Wang, W., & Zhang, J. (2019). *CRISPR-Cas12 and Cas13: The Lesser Known Siblings of CRISPR-Cas9*, Springer, 489–492.

Yang, L., et al. (2023). Heritable transgene-free genome editing in plants by grafting of wild-type shoots to transgenic donor rootstocks. *Nature Biotechnology, 41*(7), 958–967.

Yu, Y., et al. (2020). Transcriptional landscape of pathogen-responsive lncRNA s in rice unveils the role of ALEX 1 in jasmonate pathway and disease resistance. *Plant Biotechnology Journal, 18*(3), 679–690.

Zafirov, D., et al. (2023). Molecular crosstalk between plant translation initiation complexes influences the outcome of virus infection. *bioRxiv*. 2023.05.12.540471

Zaynab, M., et al. (2018). Long non-coding RNAs as molecular players in plant defense against pathogens. *Microbial Pathogenesis, 121*, 277–282.

Zaynab, M., et al. (2020). CRISPR/Cas9 to generate plant immunity against pathogen. *Microbial Pathogenesis, 141*, 103996.

Zhang, Z. J. (2014). Artificial trans-acting small interfering RNA: a tool for plant biology study and crop improvements. *Planta, 239*(6), 1139–1146.

Zhang, K., et al. (2017). Progress in genome editing technology and its application in plants. *Frontiers in Plant Science, 8*, 177.

Zhang, Y., et al. (2017). Simultaneous modification of three homoeologs of Ta EDR 1 by genome editing enhances powdery mildew resistance in wheat. *The Plant Journal, 91*(4), 714–724.

Zhang, Y., et al. (2020). A CRISPR way for accelerating improvement of food crops. *Nature Food, 1*(4), 200–205.

Zhang, M., et al. (2020). CRISPR/Cas9-mediated mutagenesis of Clpsk1 in watermelon to confer resistance to *Fusarium oxysporum* f. sp. *niveum*. *Plant Cell Reports, 39*, 589–595.

Zhang, X., et al. (2021). Editing homologous copies of an essential gene affords crop resistance against two cosmopolitan necrotrophic pathogens. *Plant Biotechnology Journal, 19*(11), 2349–2361.

Zhang, H., et al. (2022). RNA-dependent RNA polymerase could extend the lasting validity period of exogenous dsRNA. *Pest Management Science, 78*(11), 4569–4578.

Zhang, G., et al. (2022). Phosphorylation of OsABA2 at Ser197 by OsMPK1 regulates abscisic acid biosynthesis in rice. *Biochemical and Biophysical Research Communications, 586*, 68–73.

Zhao, S., et al. (2022). The novel C5 protein from tomato yellow leaf curl virus is a virulence factor and suppressor of gene silencing. *Stress Biology, 2*(1), 19.

Zhao, L., et al. (2023). PAM-flexible genome editing with an engineered chimeric Cas9. *Nature Communications, 14*(1), 6175.

Zhu, C., Ding, Y., & Liu, H. (2011). MiR398 and plant stress responses. *Physiologia Plantarum, 143*(1), 1–9.

Zhu, S., et al. (2021). Role refinement of melanin synthesis genes by gene knockout reveals their functional diversity in *Pyricularia oryzae* strains. *Microbiological Research, 242*, 126620.

Zotti, M. J., & Smagghe, G. (2015). RNAi technology for insect management and protection of beneficial insects from diseases: lessons, challenges and risk assessments. *Neotropical Entomology, 44*(3), 197–213.

Zou, X., et al. (2021). CsLOB1 regulates susceptibility to citrus canker through promoting cell proliferation in citrus. *The Plant Journal, 106*(4), 1039–1057.

17

Antiabiotic Stress Capacities of Crops Gained by CRISPR/Cas-Edited Small RNAs

Arfa Saifullah[1], Asad Azeem[2], Ghulam Mustafa[1], Muhammad Sarwar Khan[1], Sehrish Shehzadi[1], Aqsa Parvaiz[3], Maria Rehman[1,4], and Faiz Ahmad Joyia[1]

[1] *Centre of Agricultural Biochemistry and Biotechnology (CABB), University of Agriculture, Faisalabad, Pakistan*
[2] *Department of Plant Breeding and Genetics, University of Layyah, Layyah, Pakistan*
[3] *Department of Biochemistry and Biotechnology, The Women University Multan, Multan, Pakistan*
[4] *Department of Biotechnology, University of Sargodha, Sargodha, Pakistan*

17.1 Introduction

The notion that atmospheric changes and global climate have the power to drastically change biological diversity and crop yield has received more scientific validation in the last few years. Environmental stress refers to any disruption or change in a plant's metabolic homeostasis caused by changes in its environmental factors (Šamec et al., 2021). Both abiotic and biotic factors have an important impact on the development and growth of plants (Gamit & Amaresan, 2023). Around the world, abiotic stresses such as excessive salinity, drought, and cold are frequent unfavorable environmental circumstances that have a huge impact on plant growth and productivity. Significant crop losses and decreased productivity are the results of growth restrictions and stress brought on by these environmental changes (Lata & Prasad, 2011). Globally, abiotic stresses affecting nearly 90% of cropland that result in losses of yield occur of up to 70% when one or more of the abiotic stresses listed above occur (Waqas et al., 2019; Liu et al., 2020; Gui et al., 2021). The estimated causes of these losses are water stress (9%), heat stress (20%), cold stress (7%), and other stresses (4%). These stresses cost the country hundreds of millions of dollars every year. Abiotic stresses account for about 90% of the arable land that is highly vulnerable. In the main food crops grown worldwide, abiotic stressors have the potential to reduce crop yields by approximately 51–82%. Therefore, it is essential to have a mechanistic explanation of abiotic stress in global agriculture in order to better develop climate-resilient crops and to adopt policies to deal with changing climate conditions (Nehra et al., 2024). Stress-resistant crops are essential under stress to guarantee stable yields and to reduce the environmental or ecological impacts of crop production (Zhang et al., 2018).

In response to abiotic stresses, plants can undergo a range of adoptive modifications involving molecular, cellular, and physiological changes as well as genetic manipulation or reprogramming of gene expression. A physiological adaptation involves modifications to proteins at both the transcriptional and posttranslational levels, which results in a specific phenotype that largely avoids stressful environments. One such method involves altering the expression of genes and activating small RNAs (sRNAs), which activate protective mechanisms by upregulating genes

Genome and Epigenome Editing for Stress-Tolerant Crops, First Edition. Edited by Jen-Tsung Chen.

involved in defense systems and downregulating genes involved in toxicity (Patel et al., 2020). For plants' defense mechanisms, improving against abiotic stresses is essential for environmental sustainability and maintaining agricultural as well as for crop yields (Maqsood et al., 2023).

MicroRNAs (miRNAs) are 20–24 nucleotide endogenous sRNAs a class that do not code and role in plant stress tolerance. Gene expression is regulated by miRNAs, which are tiny molecules that target particular genes. In plant growth, these targeted genes play essential roles in responding to environmental cues and development. MiRNAs assist plants in adjusting biotic stresses like pests and infections and abiotic stresses like heat, cold, salt, and drought (Zhang et al., 2022; Samynathan et al., 2023). Intense heat is caused by global warming that damages the crops globally. Multiple research finding have demonstrated that sRNAs which are small molecules play important roles like show that against heat stress how plants react. These sRNAs assist the plants to survive and adapt in high-temperature conditions (Zuo et al., 2021). In various plant types, it has been found that harsh environmental conditions (without living factors) influence and change the small genetic regulators which are called miRNAs. In plant growth and development, these regulators play an important role. Modifying the genes of crop plants is the ultimate goal so that they can handle these challenging conditions better (Shriram et al., 2016).

17.2 CRISPR/Cas Technology: Mechanism and Applications

Clustered regularly interspaced short palindromic repeats and CRISPR-associated (CRISPRs-Cas) systems for archaea and bacteria are like the immune system but they can also be used for gene editing. CRISPR-Cas is used by scientists to find, change, and edit specific RNA or DNA in living cells. It is very versatile and easy to use, which makes it a powerful tool for the development of new treatments for different types of diseases and for studying biology (Pickar-Oliver & Gersbach, 2019). Some prokaryotes have widespread CRISPR arrays and Cas proteins especially archaea which combine to form a potent acquired immunity system that helps protect them from external invaders (Ishino et al., 2018).

17.2.1 Mechanism

CRISPR-Cas systems have three stages elaborated below (Barrangou, 2013). **Adaptation**: From invading entities by acquiring new spacer sequences, the CRISPR-Cas system updates its memory (e.g. viruses). **crRNA Production**: The genetic information from CRISPR regions into RNA molecules is transcribed by the CRISPR-Cas system, which is then diced into smaller crRNAs. These crRNAs guide the system to particular DNA or RNA targets. **Interference**: Cas proteins precisely cleave and target DNA or RNA sequences that match the crRNA pattern using crRNAs as guides. Against harmful elements, this destructive process helps them to protect.

17.2.2 CRISPR/Cas System and Its Applications

Endonuclease is used by the CRISPR genome editing; a protein associated with gene editing. This technique uses Cas proteins to make accurate changes in genes. In various fields, CRISPR has been mostly used like the food industry, bioenergy, infectious disease research, medicine, and agriculture. Internal DNA or RNA is reprogrammed by using this technique based on a specific sequence, enabling accurate genetic modifications (Nidhi et al., 2021). CRISPR-Cas systems were classified in 2020 according to their vast variety: Main classes: Class (one) and Class (two) and have some types: within each class have 6 different types, subtypes: 33 subtypes further categorizing within these types, variants: different variations exist within the subtypes (Makarova et al., 2020). To guide precise gene alteration enabling customizable and efficient genome-editing CRISPR/Cas technology uses short RNA molecules (Figure 17.1). In various organisms, it has been applied successfully like insects, rodents, fish, and plants. Beyond gene mutation in genome manipulation, CRISPR/Cas offers unique applications. Its RNA-guided design enables it to target not only individual genes but also entire genomes (Zhang & Zhou, 2014). The way of CRISPR-Cas defense is as follows: When a foreign DNA (from bacteria or a virus) enters a host as an outsider it is recognized. This foreign DNA has a small part, which is called a (spacer) in the host's DNA gets copied and stored like a record of the infection. (i) If the host encounters the same foreign DNA again, the spacer acts as a trigger. (ii) This triggers the host's defense system to recognize and cut the foreign DNA, preventing an infection. (iii) This process involves proteins (Cas) and the stored spacers, which help the host develop immunity against specific foreign DNA sequences, making it harder for infections to succeed (Bhaya et al., 2011). CRISPR/Cas is a powerful molecular tool that has revolutionized molecular detection because of its ability to precisely cut DNA and RNA and with specific genetic sequences they attach. On these techniques, these detection methods rely on target fluorescence analysis, DNA amplification, and binding (Figure 17.1). For the development of a versatile molecular identification platform with CRISPR/Cas, three strategies should be considered which are as follows: (i) Improve the system's functionality, (ii) the interpretation of the detection signal should be enhanced, and (iii) collaborate with different reaction systems (He et al., 2023).

Two components are involved in CRISPR technology: a guide RNA (gRNA) and an enzyme which is called CRISPR-associated nuclease (Cas protein). The gRNA contains a user-defined spacer sequence that determines the specific DNA target to be modified. This spacer sequence also

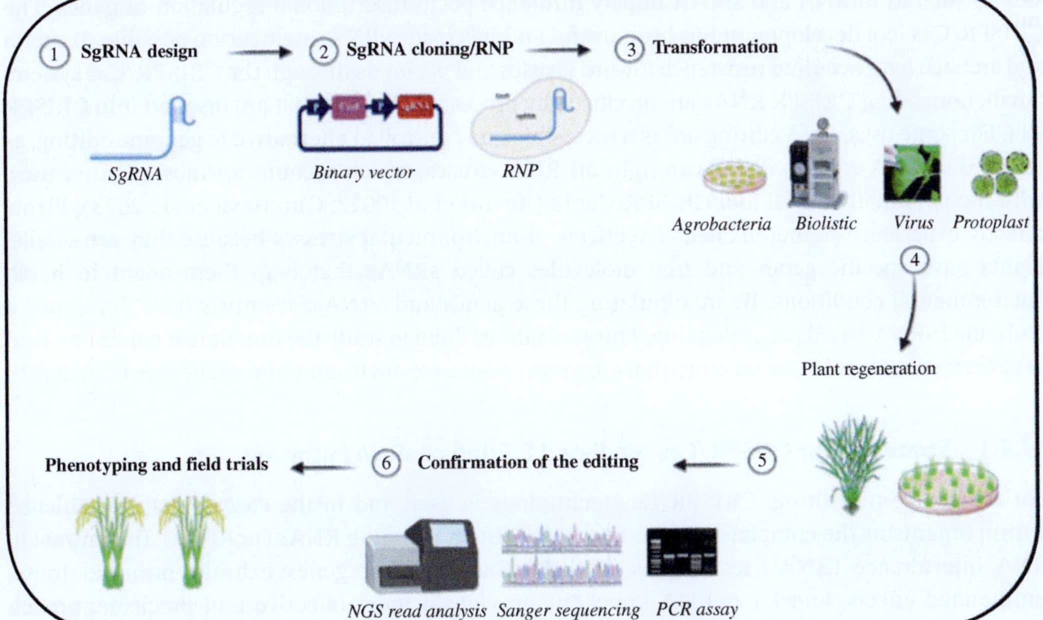

Figure 17.1 Overview of CRISPR technology.

guides the attachment of the Cas protein to the target DNA. In the gRNA, by simply modifying the target sequence researchers can redirect the Cas protein to target different genomic regions. Notably, CRISPR/Cas systems can be utilized to target genes that code for miRNAs or their regulatory sequences, thereby enabling researchers to manipulate miRNA expression levels (Ramesh et al., 2021).

17.2.3 CRISPR/Cas System for Crop Improvement

Stress tolerance is directly enhanced by proteins encoded by structural genes, while other genes involved in various cellular processes are indirectly regulated by regulatory genes. Small noncoding RNAs called miRNAs target specific mRNAs and regulate the posttranscriptional expression of numerous genes. Through complementary base pairing with mRNA transcripts, miRNAs work to silence genes and control various biological processes. It has been reported that certain miRNAs control the abiotic stress response, which targets particular genes that respond to stress (Zafar et al., 2020). Integrating traditional plant breeding techniques with advanced methods like molecular and targeted gene editing will enhance crop production and address global food scarcity. Targeted genome editing has significantly improved crop traits such as size, weight, yield, protein content, plant spread, and grain count. Additionally, it has increased crop resistance to diseases, providing a more sustainable and productive agricultural solution (Nidhi et al., 2021), crop quality improvement, enhancement of crop tolerance to herbicides (Liu et al., 2021), multiplexing abilities (Rukavtsova et al., 2022), and genetic manipulation (Arora & Narula, 2017). These applications are proof that CRISPR applicability in the field of agriculture is very vast and can be extended.

17.3 CRISPR/Cas-Mediated Editing of sRNA for Stress Tolerance

sRNAs such as miRNA and siRNA hugely influence posttranscriptional regulation of genes. The CRISPR/Cas tool development has been useful and has made sRNA engineering possible. Bacteria and archaea have acquired resistance toward viruses and plasmids through the CRISPR/Cas system. Small, noncoding CRISPR RNAs are produced by processing spacers that are inserted into CRISPR loci. For some uses, RNA editing offers a reversible and controlled alternative to genome editing, as opposed to DNA editing. Plants can fight off RNA viruses, which account for most plant viruses, with the potentially useful tool CRISPR/Cas13 (Kavuri et al., 2022; Gunitseva et al., 2023). Plants directly experience numerous negative effects of environmental stresses because they are sessile. Plants have specific genes and tiny molecules called sRNAs that help them adapt to harsh environmental conditions. By manipulating these genes and sRNAs, scientists have developed a tool called sRNA-based gene silencing. This tool allows them to study the function of genes involved in stress response and to create crops that are more resistant to environmental challenges (Cui, 2023).

17.3.1 Strategies for CRISPR/Cas-Mediated Editing of sRNA Pathways

For targeted gene editing, CRISPR/Cas technology is used and in the case of results achieved within organisms the complete elimination of specific noncoding RNAs (ncRNAs). In contrast to RNA interference (RNAi) techniques, CRISPR/Cas on other genes exhibits minimal to no unintended effects, and for ncRNA knockout provides a more effective and precise approach (Bravo-Vázquez et al., 2024). In plants, different genomic sites can be edited together by multiple CRISPR gRNAs (Basak & Nithin, 2015). In a single generation, new varieties of CRISPR/Cas are used to create quick and precise modifications (Ahmad, 2023). CRISPR/Cas technology serves an important role in engineering desired traits in plants by manipulating ncRNAs. Without causing long-term genome changes, CRISPR/Cas introduces RNA manipulations. For precise manipulation of RNA Class 2 type IV CRISPR, the Cas13 system serves a vital role (Jacinto et al., 2020). In diverse organisms for manipulating RNA, Type VI CRISPR/Cas systems offer a versatile toolbox. Advanced RNA-based technologies are enabled by them like identifying RNA molecules (RNA detection), blocking gene expression (RNAi), targeting specific RNAs, and altering RNA sequences (RNA editing) (Wang et al., 2019; Burmistrz et al., 2020; Kordyś et al., 2022). Most plant viruses consist of genetic material which is called RNA. CRISPR/Cas13 is a powerful or versatile technology that allows scientists to modify plant genes to enhance their resistance against RNA viruses (Granados-Riveron & Aquino-Jarquin, 2018; Huynh et al., 2020) Scientists can modify Cas13a, which is called C2c2 which is a type of CRISPR-Cas system that targets the RNA. To reduce the amount of RNA called (RNA knockdown), this modification allows Cas13a to be used in mammalian cells or to bind to specific RNA molecules (Abudayyeh et al., 2017). After transcription, the alteration of RNA at specific locations is called RNA editing. For RNA editing, the programmable replacement of adenosine to inosine (A to I) also called REPAIR is used (Cox et al., 2017). In addition to converting cytosine (C) to uracil (U), a specialized RNA editing is performed and this process is known as RESCUE (RNA editing for specific C to U exchange) (Abudayyeh et al., 2019).

17.3.2 Targeting Key Genes and Regulatory Elements in Stress Response

Increasing crop productivity and plant stress tolerance should be intentionally focused on important genes and regulatory elements for this achievement (Zhang et al., 2005; Vihervaara et al., 2018). For focusing on regulatory elements and relevant genes, understanding the vital regulatory pathways that govern plant stress responses is essential. Identifying the *cis*-regulatory components and genes associated with them means factors involved in transcription (Lindemose et al., 2013). In hormone signaling and stress reactions, the regulatory components of AP2/ERF(APETALA2/Ethylene Response Factor) transcription factors (TFs) play an important role (Xie et al., 2019). For the stress response, *cis* elements like ABRE (Abscisic Acid Responsive Element) and CE (Coupling Element) are important because they regulate gene expression (Zhang et al., 2005).

17.4 Application of CRISPR/Cas-Edited sRNA in Stress Tolerance

sRNAs have become one of the main actors in the modulation of gene expression in plants under stress and during developmental stages. They can affect both when genes are turned on (transcription) and how they are used after being turned on (posttranscription). The main components in the process of sRNA creation are RNA-dependent RNA (RDR) polymerase, argonaute (AGO), and dicer-like (DCL), which are essential for defending plants against harmful factors like diseases and environmental stresses (Garg et al., 2017). In crops, stress-responsive sRNAs are differently identified when a particular gene family (the Specific Recognition Sequence (SRS) gene family) is examined. By the regulation of these genes, plant development and growth as well as the response to ecological stress are regulated (Ma et al., 2022). In plant stress responses, identifying roles of sRNA CRISPR/Cas and single cell sequencings have helped (Lu et al., 2023).

17.4.1 Role of sRNA in Plant Responsiveness to Abiotic Stress

sRNAs (miRNA, siRNA) have significantly important functions in plant responsiveness against environmental stresses. The sRNAs can control the gene implicated in response to these stresses (Table 17.1). These genes can either be silenced or degraded (Lu et al., 2023). For instance, miRNA156: controls plant growth and stress responses by targeting SPL (SQUAMOSA Promoter Binding Protein-Like) factors (Xie et al., 2012). miRNA159: involved in stress response modulation by regulating MYB TF (**Myeloblastosis** Transcription Factor) (Ding et al., 2011). Small interfering RNAs target endogenous genes, viral RNAs, and transposable elements to silence genes by directing the RNA interference pathway (Ding et al., 2011; Matzke & Mosher, 2014). Genes responding to stress and involved in the process of development are controlled by transposon-derived siRNA (Yan et al., 2020). Stress-responsive gene expression is influenced by interaction between miRNA and siRNA. For example, against oxidative stress, miR398-mediated regulation of CSD(Cold Shock Domain) genes may interact with siRNA-mediated defense mechanisms (Mondal et al., 2023).

17.4.2 Role of Small RNAs in Stress-Tolerance Pathways

sRNAs for plants to respond to environmental stresses are essential. RNAi is controlled by them which regulates the expression of genes that engage in stress responses. In particular, ncRNAs have been identified, for example, miRNAs and siRNAs, which play an important role in plant stress tolerance. For enhancing crop resistance to stress, RNAi technology has become a valuable tool by modulating gene expression (Tiwari & Rajam, 2022).

17.4.3 Heat Stress

By heat stress, plant development and growth can be significantly affected. sRNAs play a vital role in plants that adapt to stress by regulating gene expression. They guide RNAi that regulates vital genes in biological processes. Specific sRNAs such as siRNAs and miRNAs have appeared as key

Table 17.1 Abiotic stress and crop response gained by edited small RNAs.

Drought stress			
Abiotic stress	**miRNA/sRNA**	**Target**	**References**
Drought	miR156, miR159, miR160, miR169, miR390, miR393	SPL transcription factors, MYB transcription factors, auxin response factors, NF-YA transcription factor, ARF2, ARF3, ARF4, TIR1	Liu et al. (2008), Eldem et al. (2012) and Xie et al. (2012)
Salt	miR159, miR164, miR166, miR169, miR319	MYB transcription factors, NAC transcription factors, HD-ZIPIII transcription factors, NF-YA transcription factors, OsPCF5 and OsPCF8	Eldem et al. (2012), Khraiwesh et al. (2012), Shuai et al. (2013) and Yang et al. (2022)
Cold	miR171, miR319, miR812	GRAS transcription factors, OsPCF5, OsPCF8, CIPK10	Shriram et al. (2016), Abla et al. (2019) and Yang et al. (2022)
Heat	miR319, miR396, miR408	TCP transcription factors, growth-regulating factors, plantacyanins	Sunkar et al. (2006) and Hackenberg et al. (2013)

regulators for plants that respond to environmental stresses like heat. For example: To control heat shock proteins (HSPs): miRNAs miR160 and miR393 target genes that are essential for heat stress protection of plants. When heat stress is applied, other miRNAs such as miR319 and miR398 help in triggering ROS-scavenging mechanisms to reduce oxidative damage (Zuo et al., 2021).

Scientist studies different plant species like *Arabidopsis*, *Populus trichocarpa*, and rice to understand how their sRNA molecules (microRNAs or miRNAs) change or adjust when exposed to water scarcity (drought stress). Researchers have identified in these plants a collection of miRNAs that react to drought stress, indicating that during or when plants deal with stressful situations these miRNAs might share similar functions or have the same role (Mirlohi & He, 2016). Under drought stress in Zea maize, the proline dehydrogenase (PDH) target miR474 was upregulated (Wei et al., 2009).

17.4.4 Salt Stress

For plants in the soil, high levels of salt pose a major challenge. Plants produce a special type of RNA molecule in response to this "salinity stress," which is called "lncRNA DRIR." This molecule works with a protein together, which is called llamado "AtCOG5." Together, they activate genes that help the plant to survive under salt stress, such as the "RD29" gene. Other lncRNAs, such as "StCDF1-AS," are also involved in regulating how plants behave against salt stress (Jin et al., 2024).

17.4.5 Cold Stress

To survive the chilly weather, certain plants develop a greater tolerance to cold. It is called cold acclimation when this happens. In plants experiencing cold stress, numerous genes regulating the cold have been discovered recently. One of the best-known plant pathways for cold tolerance is the C-repeat binding factor (CBF) cold-responsive pathway (Van Buskirk & Thomashow, 2006). In the plant species *Arabidopsis thaliana*, several miRNAs from different families increase in abundance under cold stress conditions. These miRNAs include miR165/166, miR393, miR396, and miR408. On the other hand, other miRNAs, such as miR156/157, miR159/319, miR164, miR394, and miR398, either show subtle changes or temporary alterations in their levels in return to cold stress treatment (Sunkar & Zhu, 2004). These short, ncRNAs support transcriptional regulation carried out by TFs by offering adaptable, effective, and frequently beneficial control over gene expression.

17.5 Challenges and Future Prospects of CRISPR/Cas-Edited Small RNAs for Abiotic Stress Resistance in Crops

Newly developed RNA-based Cas-effector technologies offer a means of editing sRNAs utilizing the CRISPR/Cas tool, however promising, there remain challenges and opportunities ahead (Sahel et al., 2023). Precision targeting is still a major challenge as off-target effects can end up creating unintended genetic changes, which could interfere with other critical plant functions (Asmamaw Mengstie et al., 2024). Such precision is not achievable with current CRISPR technologies, even enhanced variants such as high-fidelity Cas and prime editing, in part because existing sRNA pathways often work redundantly to other stress responses making them virtually irreplaceable. In addition, the acceptance and deployment of genetically edited crops are constrained by regulatory ethical constraints as well as biosafety concerns (Azeem et al., 2023). However, this relies on transparent and standardized regulatory frameworks as well as robust risk assessments in order to allow for responsible implementation. Given the environmental impacts, especially on biodiversity receptors, safety studies are needed to ensure that activities such as carefully controlled trials. Costs are high and intellectual property bottlenecks the free access, especially in developing regions. The major challenges and future prospects of CRISPR/Cas-edited sRNAs are listed below.

17.5.1 Technical Challenges

The sophisticated webs of genetic factor regulation make it problematic to pinpoint precise sRNAs concerned in abiotic stress reactions. Several genetic factors can be besieged by a single sRNA, and numerous sRNAs have an impression on a solitary gene (Mukherjee et al., 2024). Consequently, far-reaching research using omics and inheritable factor web analysis are essential to recognize sRNAs that significantly improve stress resilience. Involuntarily fluctuating unsolicited localities in the genetic material can upshot in off-target modifications when CRISPR/Cas9 is applied. Unpremeditated deviations made while editing short RNAs may obstruct other monitoring pathways, which could end product in deprived crop excellence as well as growth aberrations. Even though high-fidelity Cas9 disparities are presently being developed by scientists as well as researchers, absolute specificity is still challenging to achieve (Choudry et al., 2024). It is problematic to supply CRISPR/Cas9 machinery to plants, especially for crops that are unaffected by complicated genomes (Li et al., 2024a). Inadequate transgene constancy and unpredictable incorporation are the prime shortcomings of performances approximating particle bombardment (PB) and *Agrobacterium*-mediated transformation (AMT) (Ramarajan, 2020). The constant face of reformed sRNAs is determined by the establishment of reliable as well as operational delivery systems of CRISPR/Cas.

The term outline of sRNAs is particularly unique and governed by environmentally friendly factors, growth phase, and tissue type. An exhaustive regulator-terminated excision process is necessary to ascertain that CRISPR reforms work at its best in bull's eye matters and under demanding conditions. Moreover, the excision process is likely to be challenging because of the information that stress-responsive sRNAs behave differently under different scenarios (Li et al., 2024b).

17.5.2 Regulatory and Ethical Concerns

The positions of the international regulatory agencies differ in addressing genome-edited crops (Lakhotia et al., 2020). The regulatory status for the CRISPR/Cas-edited crop remains a gray area, especially if these edits target noncoding regions such as sRNAs. In some places, edited crops might incur the same stringent regulations meted out to transgenic crops, resulting in postponed approval and further limited commercial release (Hundleby & Harwood, 2022). Currently, public sentiment regarding genetically engineered crops remains diverse, with mixed concerns involving environmental and human health repercussions (Beghin et al., 2024). Acceptance, including in food and feed aspects, will vary with what the public will be increasingly informed about the safety, and benefits, of these GMOs(Genetically Modified Organisms), particularly because it brings more exact changes over traditional genetic manipulation (Goodman, 2024). The CRISPR/Cas9 system is generally considered a highly challenging technology to use commercially and researchwise, primarily because the system has multiple patents involved (Arif et al., 2024). The technology will find its access restricted by third-world countries that would benefit highly from crop stress resilience only through the costs of its licensing and intellectual property rights.

17.5.3 Biological and Ecological Challenges

The tolerance developed by sRNAediting for abiotic stress may not be sustained over generations or under different environmental conditions. The changed sRNAs' performance in the field could also be influenced by factors such as epigenetic effects and environmental impacts on levels of expression of sRNA; hence, a long-term study is required to ascertain stability in traits (Ghosh & Dey, 2022). Apart from the stress response, sRNAs frequently have functions in growth regulation, development, and metabolism. Manipulation of sRNAs to create a tolerance to stresses may lower the yields or change the plant shape. For example, enhancing drought tolerance could lower the photosynthetic efficiency or the total rate of growth of the plant (Mondal et al., 2023).

Unintentional ecological impacts are a concern when releasing CRISPR/Cas-edited crops into the environment. Modified stress reactions could impact the soil nitrogen cycle or interactions between plants and pollinators. Maintaining ecological equilibrium requires tracking the effects of CRISPR-edited crops on the ecosystem (Ahmad et al., 2021).

17.5.4 Future Prospects and Opportunities

Base editors and prime editing, two developments in CRISPR technology, enable even more accurate alterations with less off-target effects. These developments minimize unwanted effects on nontarget genes or pathways by enabling very selective editing of short RNAs (Tan et al., 2019). Bioinformatics techniques and high-throughput sequencing technology can find new stress-responsive sRNAs in various crop species. By combining CRISPR with multi-omics techniques, distinct sRNA targets may be found, allowing for the creation of crops with enhanced stress tolerance suited to certain environmental circumstances (Thriveni et al., 2024). Many stress-responsive genes can be controlled at once using synthetic biology techniques, such as creating synthetic sRNAs. Regulating several genes involved in stress adaptation presents a viable path for improving complex traits like heat and drought resistance (Ebeed & Ceasar, 2024). The development of transgene-free editing techniques, such as CRISPR/Cas9 ribonucleoprotein (RNP) delivery, may allay regulatory worries and increase public acceptance. By avoiding the steady integration of foreign DNA, transgene-free techniques result in plants that are more likely to withstand regulatory scrutiny and be accepted by the market more quickly (Inam et al., 2024).

17.6 Conclusion

sRNAs modified with CRISPR/Cas hold promising chances to improve crops' abiotic stress tolerance. Technical, legal, and environmental barriers are still there, yet most of them are addressed by continuous development in the field of synthetic biology as well as in CRISPR technology. With climate change, stable, targeted, and environmentally safe alterations in sRNAs may result in resilient crop types that could thrive under various environmental stressors, thereby significantly enhancing global food security. The precision targeting of sRNA pathways could therefore ensure crop resilience against drought, salinity, extreme temperatures, and other environmental stresses. Technical challenges concerning precision targeting, regulatory hurdles, and high costs notwithstanding, steady breakthroughs in CRISPR technology, and the incorporation of bioinformatics are slowly overcoming the constraints. The future prospects of these crops look bright in meeting the demands of an increasing population under a changing climate. Together with acceptance and regulatory frameworks, CRISPR innovations could play their role toward sustainable agriculture practices, having farmers around the globe grow crops that maintain yield and quality even in hostile environmental conditions. This final point is that the improvement of food security is a very significant aspect, in addition to the direction for a resilient and sustainable future agricultural sector.

References

Abla, M., Sun, H., Li, Z., Wei, C., Gao, F., Zhou, Y., & Feng, J. (2019). Identification of miRNAs and their response to cold stress in *Astragalus membranaceus. Biomolecules, 9*, 182.

Abudayyeh, O. O., Gootenberg, J. S., Essletzbichler, P., Han, S., Joung, J., Belanto, J. J., Verdine, V., Cox, D. B. T., Kellner, M. J., & Regev, A. (2017). RNA targeting with CRISPR–Cas13. *Nature, 550*, 280–284.

Abudayyeh, O. O., Gootenberg, J. S., Franklin, B., Koob, J., Kellner, M. J., Ladha, A., Joung, J., Kirchgatterer, P., Cox, D. B. T., & Zhang, F. (2019). A cytosine deaminase for programmable single-base RNA editing. *Science, 365*, 382–386.

Ahmad, M. (2023). Plant breeding advancements with "CRISPR-Cas" genome editing technologies will assist future food security. *Frontiers in Plant Science, 14*, 1133036.

Ahmad, A., Ghouri, M. Z., Munawar, N., Ismail, M., Ashraf, S., & Aftab, S. O. (2021). Regulatory, ethical, and social aspects of CRISPR crops. In: *CRISPR Crops: The Future of Food Security*, 261–287.

Arif, A., Munir, A., Noman, M., Munawar, N., Abd-Elsalam, K. A., Qari, S. H., & Ahmad, A. (2024). Global patent landscape in CRISPR-Cas. In: *CRISPRized Horticulture Crops*, Academic Press.

Arora, L., & Narula, A. (2017). Gene editing and crop improvement using CRISPR-Cas9 system. *Frontiers in Plant Science, 8*, 296675.

Asmamaw Mengstie, M., Teshome Azezew, M., Asmamaw Dejenie, T., Teshome, A. A., Tadele Admasu, F., Behaile Teklemariam, A., Tilahun Mulu, A., Mekonnen Agidew, M., Adugna, D. G., Geremew, H., & Abebe, E. C. (2024). Recent advancements in reducing the off-target effect of CRISPR-Cas9 genome editing. *Biologics: Targets and Therapy, 18*, 21–28.

Azeem, A., Ul-Allah, S., Sher, A., Ijaz, M., Sattar, A., Ahmad, W., & Qayyum, A. (2023). Biosafety and biosecurity in genetically modified crops. *Sustainable Agriculture in the Era of the OMICs Revolution*. Springer

Barrangou, R. (2013). CRISPR-Cas systems and RNA-guided interference. *Wiley Interdisciplinary Reviews: RNA, 4*, 267–278.

Basak, J., & Nithin, C. (2015). Targeting non-coding RNAs in plants with the CRISPR-Cas technology is a challenge yet worth accepting. *Frontiers in Plant Science, 6*, 166365.

Beghin, J., This, P., & Jones, K. (2024). *Genome Editing, R&D, and Heterogeneous Regulatory Regimes: Implications for Innovation and Agricultural Trade*, Center for Agricultural and Rural Development.

Bhaya, D., Davison, M., & Barrangou, R. (2011). CRISPR-Cas systems in bacteria and archaea: versatile small RNAs for adaptive defense and regulation. *Annual Review of Genetics, 45*, 273–297.

Bravo-Vázquez, L. A., Méndez-García, A., Chamu-García, V., Rodríguez, A. L., Bandyopadhyay, A., & Paul, S. (2024). The applications of CRISPR/Cas-mediated microRNA and lncRNA editing in plant biology: shaping the future of plant non-coding RNA research. *Planta, 259*, 1–21.

Burmistrz, M., Krakowski, K., & Krawczyk-Balska, A. (2020). RNA-targeting CRISPR–Cas systems and their applications. *International Journal of Molecular Sciences, 21*, 1122.

Choudry, M. W., Riaz, R., Nawaz, P., Ashraf, M., Ijaz, B., & Bakhsh, A. (2024). CRISPR-Cas9 mediated understanding of plants' abiotic stress-responsive genes to combat changing climatic patterns. *Functional & Integrative Genomics, 24*, 132.

Cox, D. B. T., Gootenberg, J. S., Abudayyeh, O. O., Franklin, B., Kellner, M. J., Joung, J., & Zhang, F. (2017). RNA editing with CRISPR-Cas13. *Science, 358*, 1019–1027.

Cui, J. (2023). Small RNA transgenesis for abiotic stress tolerant food crops. In: *Plant Small RNA in Food Crops*, Elsevier, 507–543.

Ding, Y., Chen, Z., & Zhu, C. (2011). Microarray-based analysis of cadmium-responsive microRNAs in rice (Oryza sativa). *Journal of Experimental Botany, 62*, 3563–3573.

Ebeed, H. T., & Ceasar, S. A. (2024). Genomics and bioinformatics applications in crop stress: unraveling molecular mechanisms for sustainable agriculture. *South African Journal of Botany, 170*, 38–47.

Eldem, V., U. Çelikkol Akçay, E. Ozhuner, Y. Bakır, S. Uranbey and T. Unver. (2012). Genome-wide identification of miRNAs responsive to drought in peach (*Prunus persica*) by high-throughput deep sequencing. *PLoS ONE*, 7, e50298. https://doi.org/10.1371/journal.pone.0050298

Gamit, H., & Amaresan, N. (2023). Role of methylotrophic bacteria in managing abiotic stresses for enhancing agricultural production. *Pedosphere, 33*, 49–60.

Garg, V., Agarwal, G., Pazhamala, L. T., Nayak, S. N., Kudapa, H., Khan, A. W., Doddamani, D., Sharma, M., Kavi Kishor, P. B., & Varshney, R. K. (2017). Genome-wide identification, characterization, and expression analysis of small RNA biogenesis purveyors reveal their role in regulation of biotic stress responses in three legume crops. *Frontiers in Plant Science, 8*, 238864.

Ghosh, S., & Dey, G. (2022). Biotic and abiotic stress tolerance through CRISPR-Cas mediated genome editing. *Journal of Plant Biochemistry and Biotechnology, 31*, 227–238.

Goodman, R. E. (2024). Twenty-eight years of GM food and feed without harm: why not accept them? *GM Crops & Food, 15*, 40–50.

Granados-Riveron, J. T., & Aquino-Jarquin, G. (2018). CRISPR–Cas13 precision transcriptome engineering in cancer. *Cancer Research, 78*, 4107–4113.

Gui, Y.-W., Sheteiwy, M. S., Zhu, S.-G., Batool, A., & Xiong, Y.-C. (2021). Differentiate effects of non-hydraulic and hydraulic root signaling on yield and water use efficiency in diploid and tetraploid wheat under drought stress. *Environmental and Experimental Botany, 181*, 104287.

Gunitseva, N., Evteeva, M., Borisova, A., Patrushev, M., & Subach, F. (2023). RNA-dependent RNA targeting by CRISPR-cas systems: characterizations and applications. *International Journal of Molecular Sciences, 24*, 6894.

Hackenberg, M., Shi, B.-J., Gustafson, P., & Langridge, P. (2013). Characterization of phosphorus-regulated miR399 and miR827 and their isomirs in barley under phosphorus-sufficient and phosphorus-deficient conditions. *BMC Plant Biology, 13*, 1–17.

He, Y., Yan, W., Long, L., Dong, L., Ma, Y., Li, C., Xie, Y., Liu, N., Xing, Z., & Xia, W. (2023). The CRISPR/Cas system: a customizable toolbox for molecular detection. *Genes (Basel), 14*, 850.

Hundleby, P., & Harwood, W. (2022). Regulatory constraints and differences of genome-edited crops around the globe. In: *Genome Editing: Current Technology Advances and Applications for Crop Improvement*, Springer.

Huynh, N., Depner, N., Larson, R., & King-Jones, K. (2020). A versatile toolkit for CRISPR-Cas13-based RNA manipulation in Drosophila. *Genome Biology, 21*, 279.

Inam, S., Muhammad, A., Irum, S., Rehman, N., Riaz, A., Uzair, M., & Khan, M. R. (2024). Genome editing for improvement of biotic and abiotic stress tolerance in cereals. *Functional Plant Biology, 51*, FP24092.

Ishino, Y., Krupovic, M., & Forterre, P. (2018). History of CRISPR-Cas from encounter with a mysterious repeated sequence to genome editing technology. *Journal of Bacteriology, 200*, 10–1128.

Jacinto, F. V., Link, W., & Ferreira, B. I. (2020). CRISPR/Cas9-mediated genome editing: from basic research to translational medicine. *Journal of Cellular and Molecular Medicine, 24*, 3766–3778.

Jin, X., Wang, Z., Li, X., Ai, Q., Wong, D. C. J., Zhang, F., Yang, J., Zhang, N., & Si, H. (2024). Current perspectives of lncRNAs in abiotic and biotic stress tolerance in plants. *Frontiers in Plant Science, 14*, 1334620.

Kavuri, N. R., Ramasamy, M., Qi, Y., & Mandadi, K. (2022). Applications of CRISPR/Cas13-based RNA editing in plants. *Cells, 11*, 2665.

Khraiwesh, B., Zhu, J.-K., & Zhu, J. (2012). Role of miRNAs and siRNAs in biotic and abiotic stress responses of plants. *Biochimica et Biophysica Acta-Gene Regulatory Mechanisms, 1819*, 137–148.

Kordyś, M., Sen, R., & Warkocki, Z. (2022). Applications of the versatile CRISPR-Cas13 RNA targeting system. *Wiley Interdisciplinary Reviews: RNA, 13*, e1694.

Lakhotia, S. C., Mallick, B. & Roy, J. 2020. Non-coding RNAs: ever-expanding diversity of types and functions. RNA-Based Regulation in Human Health and Disease, Elsevier, 5–57.

Lata, C., & Prasad, M. (2011). Role of DREBs in regulation of abiotic stress responses in plants. *Journal of Experimental Botany, 62*, 4731–4748.

Li, L., Zhang, D., Zhang, Z., & Zhang, B. (2024a). CRISPR/Cas: a powerful tool for designing and improving oil crops. *Trends in Biotechnology*.

Li, Q., Wang, Y., Sun, Z., Li, H., & Liu, H. (2024b). The biosynthesis process of small RNA and its pivotal roles in plant development. *International Journal of Molecular Sciences, 25*, 7680.

Lindemose, S., O'Shea, C., Jensen, M. K., & Skriver, K. (2013). Structure, function and networks of transcription factors involved in abiotic stress responses. *International Journal of Molecular Sciences, 14*, 5842–5878.

Liu, H.-H., Tian, X., Li, Y.-J., Wu, C.-A., & Zheng, C.-C. (2008). Microarray-based analysis of stress-regulated microRNAs in Arabidopsis thaliana. *RNA, 14*, 836–843.

Liu, Z., Li, G., Zhang, H., Zhang, Y., Zhang, Y., Duan, S., Sheteiwy, M. S. A., Zhang, H., Shao, H., & Guo, X. (2020). TaHsfA2-1, a new gene for thermotolerance in wheat seedlings: characterization and functional roles. *Journal of Plant Physiology, 246*, 153135.

Liu, Q., Yang, F., Zhang, J., Liu, H., Rahman, S., Islam, S., Ma, W., & She, M. (2021). Application of CRISPR/Cas9 in crop quality improvement. *International Journal of Molecular Sciences, 22*, 4206.

Lu, W., Wang, Y., Shi, Y., Liang, Q., Lu, X., Su, D., Xu, X., Pirrello, J., Gao, Y., & Huang, B. (2023). Identification of SRS transcription factor family in Solanum lycopersicum, and functional characterization of their responses to hormones and abiotic stresses. *BMC Plant Biology, 23*, 495.

Ma, B., Nian, L., ul Ain, N., Liu, X., Yang, Y., Zhu, X., Haider, F. U., Lv, Y., Bai, P., & Zhang, X. (2022). Genome-wide identification and expression profiling of the SRS gene family in Melilotus albus reveals functions in various stress conditions. *Plants, 11*, 3101.

Makarova, K. S., Wolf, Y. I., Iranzo, J., Shmakov, S. A., Alkhnbashi, O. S., Brouns, S. J. J., Charpentier, E., Cheng, D., Haft, D. H., & Horvath, P. (2020). Evolutionary classification of CRISPR–Cas systems: a burst of class 2 and derived variants. *Nature Reviews. Microbiology, 18*, 67–83.

Maqsood, M. F., Shahbaz, M., Khalid, F., Rasheed, Y., Asif, K., Naz, N., Zulfiqar, U., Zulfiqar, F., Moosa, A., & Alamer, K. H. (2023). Biogenic nanoparticles application in agriculture for ROS mitigation and abiotic stress tolerance: a review. *Plant Stress, 10*, 100281.

Matzke, M. A., & Mosher, R. A. (2014). RNA-directed DNA methylation: an epigenetic pathway of increasing complexity. *Nature Reviews. Genetics, 15*, 394–408.

Mirlohi, S., & He, Y. (2016). Small RNAs in plant response to abiotic stress. In: *Abiotic and Biotic Stress in Plants - Recent Advances and Future Perspectives*, Intech, Croat, 63–80.

Mondal, R., Das, A., & Bandyopadhyay, A. (2023). Implications of small RNAs in plant development, abiotic stress response and crop improvement in changing climate. *The Nucleus, 66*, 321–339.

Mukherjee, A., Das, S., Mondal, N. H., Choudhury, A., & Banerjee, S. (2024). Transgenesis: in the drug discovery process, including target identification and target validation. In: *Concepts in Pharmaceutical Biotechnology and Drug Development*, Springer.

Nehra, A., Kalwan, G., Gill, R., Nehra, K., Agarwala, N., Jain, P. K., Naeem, M., Tuteja, N., Pudake, R. N., & Gill, S. S. (2024). Status of impact of abiotic stresses on global agriculture. In: *Nanotechnology for Abiotic Stress Tolerance and Management in Crop Plants*, Elsevier, 1–21.

Nidhi, S., Anand, U., Oleksak, P., Tripathi, P., Lal, J. A., Thomas, G., Kuca, K., & Tripathi, V. (2021). Novel CRISPR–Cas systems: an updated review of the current achievements, applications, and future research perspectives. *International Journal of Molecular Sciences, 22*, 3327.

Patel, A., Tiwari, S., Singh, M., & Prasad, S. M. (2020). Role of sRNAs in abiotic stress tolerance. In: *Plant Life Under Changing Environment*, Elsevier, 467–480.

Pickar-Oliver, A., & Gersbach, C. A. (2019). The next generation of CRISPR–Cas technologies and applications. *Nature Reviews. Molecular Cell Biology, 20*, 490–507.

Ramarajan, M. (2020). *Development of a Molecular Toolkit to Genetically Engineer the Microalga Nannochloropsis gaditana CCMP526 for Biotechnology Applications*, University of Technology Sydney (Australia).

Ramesh, S. V., Mangruthia, S. K., & Praveen, S. (2021). Genetic modifications for disease resistance in crops. In: *Genetically Modified Crops in Asia Pacific*, 273.

Rukavtsova, E. B., Zakharchenko, N. S., Lebedev, V. G., & Shestibratov, K. A. (2022). CRISPR-Cas genome editing for horticultural crops improvement: advantages and prospects. *Horticulturae, 9*, 38.

Sahel, D. K., Vora, L. K., Saraswat, A., Sharma, S., Monpara, J., D'Souza, A. A., Mishra, D., Tryphena, K. P., Kawakita, S., & Khan, S. (2023). CRISPR/Cas9 genome editing for tissue-specific in vivo targeting: nanomaterials and translational perspective. *Advanced Science, 10*, 2207512.

Šamec, D., Karalija, E., Šola, I., Vujčić Bok, V., & Salopek-Sondi, B. (2021). The role of polyphenols in abiotic stress response: the influence of molecular structure. *Plants, 10*, 118.

Samynathan, R., Venkidasamy, B., Shanmugam, A., Ramalingam, S., & Thiruvengadam, M. (2023). Functional role of microRNA in the regulation of biotic and abiotic stress in agronomic plants. *Frontiers in Genetics, 14*, 1272446.

Shriram, V., Kumar, V., Devarumath, R. M., Khare, T. S., & Wani, S. H. (2016). MicroRNAs as potential targets for abiotic stress tolerance in plants. *Frontiers in Plant Science, 7*, 198276.

Shuai, P., Liang, D., Zhang, Z., Yin, W., & Xia, X. (2013). Identification of drought-responsive and novel Populus trichocarpa microRNAs by high-throughput sequencing and their targets using degradome analysis. *BMC Genomics, 14*, 1–14.

Sunkar, R., & Zhu, J.-K. (2004). Novel and stress-regulated microRNAs and other small RNAs from Arabidopsis. *Plant Cell, 16*, 2001–2019.

Sunkar, R., Kapoor, A., & Zhu, J.-K. (2006). Posttranscriptional induction of two Cu/Zn superoxide dismutase genes in Arabidopsis is mediated by downregulation of miR398 and important for oxidative stress tolerance. *Plant Cell, 18*, 2051–2065.

Tan, J., Zhang, F., Karcher, D., & Bock, R. (2019). Engineering of high-precision base editors for site-specific single nucleotide replacement. *Nature Communications, 10*, 439.

Thriveni, V., Teotia, J., Hazra, S., Bharti, T., Kumar, M., Lallawmkimi, M. C., & Panwar, D. (2024). A review on integrating bioinformatics tools in modern plant breeding. *Archives of Current Research International, 24*, 293–308.

Tiwari, R., & Rajam, M. V. (2022). RNA- and miRNA-interference to enhance abiotic stress tolerance in plants. *Journal of Plant Biochemistry and Biotechnology, 31*, 689–704.

Van Buskirk, H. A., & Thomashow, M. F. (2006). Arabidopsis transcription factors regulating cold acclimation. *Physiologia Plantarum, 126*, 72–80.

Vihervaara, A., Duarte, F. M., & Lis, J. T. (2018). Molecular mechanisms driving transcriptional stress responses. *Nature Reviews. Genetics*, *19*, 385–397.

Wang, F., Wang, L., Zou, X., Duan, S., Li, Z., Deng, Z., Luo, J., Lee, S. Y., & Chen, S. (2019). Advances in CRISPR-Cas systems for RNA targeting, tracking and editing. *Biotechnology Advances*, *37*, 708–729.

Waqas, M. A., Kaya, C., Riaz, A., Farooq, M., Nawaz, I., Wilkes, A., & Li, Y. (2019). Potential mechanisms of abiotic stress tolerance in crop plants induced by thiourea. *Frontiers in Plant Science*, *10*, 1336.

Wei, L., Zhang, D., Xiang, F., & Zhang, Z. (2009). Differentially expressed miRNAs potentially involved in the regulation of defense mechanism to drought stress in maize seedlings. *International Journal of Plant Sciences*, *170*, 979–989.

Xie, K., Shen, J., Hou, X., Yao, J., Li, X., Xiao, J., & Xiong, L. (2012). Gradual increase of miR156 regulates temporal expression changes of numerous genes during leaf development in rice. *Plant Physiology*, *158*, 1382–1394.

Xie, Z., Nolan, T. M., Jiang, H., & Yin, Y. (2019). AP2/ERF transcription factor regulatory networks in hormone and abiotic stress responses in Arabidopsis. *Frontiers in Plant Science*, *10*, 437723.

Yan, S., Ren, B., Zeng, B., & Shen, J. (2020). Improving RNAi efficiency for pest control in crop species. *Biotechniques*, *68*, 283–290.

Yang, Z., Dong, T., Dai, X., Wei, Y., Fang, Y., Zhang, L., Zhu, M., Nawaz, G., Cao, Q., & Xu, T. (2022). Comparative analysis of salt responsive microRNAs in two sweet potato [Ipomoea batatas (L.) Lam.] cultivars with different salt stress resistance. *Frontiers in Plant Science*, *13*, 879819.

Zafar, S. A., Zaidi, S. S.-A., Gaba, Y., Singla-Pareek, S. L., Dhankher, O. P., Li, X., Mansoor, S., & Pareek, A. (2020). Engineering abiotic stress tolerance via CRISPR/Cas-mediated genome editing. *Journal of Experimental Botany*, *71*, 470–479.

Zhang, L., & Zhou, Q. (2014). CRISPR/Cas technology: a revolutionary approach for genome engineering. *Science China. Life Sciences*, *57*, 639–640.

Zhang, W., Ruan, J., Ho, T. D., You, Y., Yu, T., & Quatrano, R. S. (2005). Cis-regulatory element based targeted gene finding: genome-wide identification of abscisic acid- and abiotic stress-responsive genes in Arabidopsis thaliana. *Bioinformatics*, *21*, 3074–3081.

Zhang, H., Li, Y., & Zhu, J.-K. (2018). Developing naturally stress-resistant crops for a sustainable agriculture. *Nature Plants*, *4*, 989–996.

Zhang, F., Yang, J., Zhang, N., Wu, J., & Si, H. (2022). Roles of microRNAs in abiotic stress response and characteristics regulation of plant. *Frontiers in Plant Science*, *13*, 919243.

Zuo, Z.-F., He, W., Li, J., Mo, B., & Liu, L. (2021). Small RNAs: the essential regulators in plant thermotolerance. *Frontiers in Plant Science*, *12*, 726762.

Index

Genome and Epigenome Editing for Stress-Tolerant Crops, First Edition. Edited by Jen-Tsung Chen.

f

g

h

i

j

l

m

s

t

u

V

W

Z